BIOLOGICAL INHIBITORS

Studies in Medicinal Chemistry
A series of books presenting various aspects of the research on medicinal chemistry and providing comprehensive accounts of recent and important developments in the field.

Edited by Atta-ur-Rahman, H.E.J. Research Institute of Chemistry, University of Karachi, Pakistan

---

Volume 1
Progress in Medicinal Chemistry
edited by M. Iqbal Choudhary

Volume 2
Biological Inhibitors
edited by M. Iqbal Choudhary

Volumes in preparation

Antibodies in Diagnosis and Therapy: Technologies, Mechanisms and Clinical Data
S. Matzku and R.A. Stahel

The Guanidine Entity in Medicinal Chemistry and in Molecular Recognition
I. Agranat and S. Topiol

BIOLOGICAL INHIBITORS

Edited by

M. Iqbal Choudhary

H.E.J. Research Institute of Chemistry
University of Karachi
Pakistan

harwood academic publishers
Australia • Canada • China • France • Germany • India • Japan • Luxembourg
Malaysia • The Netherlands • Russia • Singapore • Switzerland • Thailand
United Kingdom

Emmaplein 5
1075 AW Amsterdam
The Netherlands

British Library Cataloguing in Publication Data

Biological inhibitors. – (Studies in medicinal
   chemistry; v. 2, ISSN 1024–8056)
   1. Clinical inhibitors 2. Pharmaceutical chemistry
   I. Series II. Choudhary, M. Iqbal
   615.1'9

ISBN 3-7186-5879-8

CONTENTS

FOREWORD

For Volume 2 (**Biological Inhibitors**) of the new series **Studies in Medicinal Chemistry**, Dr. Iqbal Choudhary (Volume Editor) and Prof. Atta-ur-Rahman (Series Editor) have assembled an impressive group of academic and industrial scientists as contributors. The seven chapters describe in detail important biological inhibitors that reflect drug discovery concerning the treatment of various human diseases, inclusive of cardiovascular, central nervous system, inflammatory, hormonal, and metabolic processes. Also covered herein are compounds of use in the therapy of fungal diseases and HIV/AIDS. Each chapter is thoroughly written, up-to-date, liberally illustrated, and extensively referenced. Overall, thousands of synthetic and natural product molecules are considered from the viewpoints of mechanism-of-action, metabolism, and structure-activity relationships.

The volume contributors and editors are to be congratulated for their efforts in producing such a high-quality and timely text. The book should be useful in advanced undergraduate and graduate courses in biochemistry, biology, chemistry, medicinal chemistry, pharmacy and pharmacology. Moreover, specific chapters will serve as helpful contemporary reviews for established and beginning researchers in the particular areas covered.

A. Douglas Kinghorn
University of Illinois, Chicago

PREFACE TO THE SERIES

The spectacular advances in the field of medicinal chemistry in the last few decades have been triggered by a greater understanding of cellular processes at the molecular level. These advances have provided a fresh impetus in the area of drug development and design, which is reflected by the rapidly growing number of new medicines introduced clinically every year. The growing knowledge about disease complexities and challenges of discovering cures for such ailments as AIDS, cancer and arthritis have made the field of medicinal chemistry an exciting and vibrant discipline.

During the past few years, medicinal chemistry has undergone a healthy renaissance. Due to its multidisciplinary nature, medicinal chemistry is becoming increasingly multi-faceted involving interactions of chemists, pharmacologists and physicians. The series **Studies in Medicinal Chemistry** will hopefully fulfil an important need of the scientific community by providing comprehensive reviews on different fields of current interest.

Every volume, after the first introductory volume, is devoted to a specific field of medicinal chemistry and will hence provide a focused account of recent development in each field.

I extend my warmest thanks to our volume editors for undertaking the important task of organising each volume and cooperating in preparing the content lists. The staff of Harwood Academic Publishers who have made this project a success also deserve appreciation. I would also like to thank Mr. Mahmood Alam for secretarial help.

Atta-ur-Rahman
Series Editor

INTRODUCTION

The second volume in the series **Studies in Medicinal Chemistry** contains seven articles which review recent developments in the field of biological inhibition.

Biological inhibitors provide an exciting area of research for medicinal chemists and biochemists. The greater understanding of cellular processes at the receptor level has revolutionized the entire field. Seven articles on different biological inhibitors review the recent research in the area in a systematic and comprehensive fashion. The medicinal chemistry of squalene epoxidase inhibitors reviewed by Stütz *et al.* provides a deep insight in different biological processes involved in the biosynthesis of squalene epoxide which is the bioprecursor of cholesterol in the body. Chemistry and biology of dual inhibitors of 5-lipoxygenase and cyclooxygenase is reviewed by Connor and Boschelli. Inhibitors of these enzymes inhibit the biosynthesis of proinflammatory prostaglandins and leukotrienes and thus downregulate an inflammatory response. Reviews on the so-called glamour areas of medicinal chemistry *i.e.* cholesterol biosynthesis and HIV-proteinase inhibitors are contributed by Sliskovic and Roth, and Thomas respectively. A chapter contributed by Lowe *et al.* comprehensively reviews the current research in the area of nonpeptide antagonists at peptide receptors. Craik *et al.* have contributed an interesting article on the conformation of hormones and their binding interactions with various receptors as well as the structures and conformations of a range of molecules which interact with thyroid hormones. The last chapter contributed by Zhang *et al.* describes in detail the various stages involved in the development of inclusive pharmacophores for the benzodiazepine receptor.

It is hoped that the present volume will be received with enthusiasm by the community of medicinal chemists. I wish to thank my student Miss Farzana Akhtar and our librarian Mr. Ejaz Soofi for their assistance in the preparation of the index, Mr. M. Asif for the typing work and Mr. Mahmood Alam for secretarial help.

M. Iqbal Choudhary
Volume Editor

CONTRIBUTORS

Michael S. Allen
Department of Chemistry
University of Wisconsin-Milwaukee
Milwaukee WI 53201
USA

Diane H. Boschelli
Department of Chemistry
Parke-Davis Pharmaceutical Research
  Division
Warner-Lambert Company
2800 Plymouth Road
Ann Arbor MI 48105-2430
USA

Philip A. Carpino
Central Research Division
Pfizer Inc.
Eastern Point Road
Groton CT 06340
USA

David T. Connor
Department of Chemistry
Parke-Davis Pharmaceutical Research
  Division
Warner-Lambert Company
2800 Plymouth Road
Ann Arbor MI 48105-2430
USA

James M. Cook
Department of Chemistry
University of Wisconsin-Milwaukee
Milwaukee WI 53201
USA

David J. Craik
Centre for Drug Design and
  Development

University of Queensland
Brisbane 4072
Queensland
Australia

Hernando Diaz-Arauzo
Department of Chemistry
University of Wisconsin-Milwaukee
Milwaukee WI 53201
USA

Brendan M. Duggan
Victorian College of Pharmacy
School of Pharmaceutical Chemistry
381 Royal Parade
Parkville, Victoria 3052
Australia

Konrad F. Koehler
Istituto di Ricerche di Biologia
    Moleculare P. Angeletti (IRBM)
Via Pontina Km. 30,600
00040 Pomezia (Roma)
Italy

John A. Lowe III
Central Research Division
Pfizer Inc.
Eastern Point Road
Groton CT 06340
USA

Sharon L.A. Munro
Victorian College of Pharmacy
School of Pharmaceutical Chemistry
381 Royal Parade
Parkville, Victoria 3052
Australia

Peter Nussbaumer
Department of Dermatology
Sandoz Forschungsinstitut
Postfach 80
Brunner Strasse 59
Vienna A-1235
Austria

Brian T. O'Neill
Central Research Division
Pfizer Inc.
Eastern Point Road
Groton CT 06340
USA

Bruce D. Roth
Department of Chemistry
Parke-Davis Pharmaceutical Research
  Division
Warner-Lambert Company
2800 Plymouth Road
Ann Arbor MI 48105-2430
USA

Neil S. Ryder
Department of Dermatology
Sandoz Forschungsinstitut
Postfach 80
Brunner Strasse 59
Vienna A-1235
Austria

Drago R. Sliskovic
Department of Chemistry
Parke-Davis Pharmaceutical Research
  Division
Warner-Lambert Company
2800 Plymouth Road
Ann Arbor MI 48105-2430
USA

Anton Stütz
Department of Dermatology
Sandoz Forschungsinstitut
Postfach 80
Brunner Strasse 59
Vienna A-1235
Austria

Gareth J. Thomas
Roche Research Centre
Roche Products Ltd
40 Broadwater Road
Welwyn Garden City
Hertfordshire AL7 3AY
UK

Weijiang Zhang
Department of Chemistry
University of Wisconsin-Milwaukee
Milwaukee WI 53201
USA

1. MEDICINAL CHEMISTRY OF SQUALENE EPOXIDASE INHIBITORS

PETER NUSSBAUMER, NEIL S. RYDER, and ANTON STÜTZ

*Department of Dermatology, Sandoz Research Institute Vienna, Brunnerstraße 59,
A-1235 Vienna, Austria*

1. INTRODUCTION

Squalene epoxidase (SE) is a key enzyme in the sterol biosynthesis pathway of mammals, fungi, and plants.[1] Acetyl coenzyme A is processed in a multi-step sequence leading to ergosterol (in fungi), cholesterol (in animals), or several plant sterols.[2] These sterols differ structurally by the degree of unsaturation in the steroidal B ring and in the aliphatic side chain and by optional substitution at C-24 and may have specific biological functions. Scheme 1 summarises the biosynthesis of ergosterol and cholesterol featuring only those steps for which inhibitors of therapeutic relevance exist at present. Sterol biosynthesis inhibitors, shown in Scheme 1 by compound class, are used clinically to reduce cholesterol levels or to treat fungal infections[3] (Table 1). Azole and morpholine derivatives are also widely used in agriculture against phytopathogenic fungi.[4,5] Additional efforts to produce antihypercholesterolemic or antifungal drugs have focussed on other enzymes of the sterol biosynthesis pathway,[6] including squalene synthetase,[7,8] oxidosqualene cyclase,[8] and sterol 24-methyltransferase.[9] These concepts are, however, still in an exploratory phase and have found no clinical applications up to now.

The medicinal chemistry of SE inhibitors started with naftifine (Figure 1, 1), the first representative of the allylamine antimycotics. The finding that naftifine exerted its potent antifungal activity by selectively blocking squalene epoxidation[10,11] revealed SE for the first time as a therapeutic target. Starting with naftifine as the lead compound, an extensive medicinal chemistry programme resulted in the elucidation of detailed structure-activity relationships (SAR) and substantial improvement of biological properties.[12] Interference with sterol biosynthesis by inhibiting squalene epoxidation is attractive for two reasons: (1) cholesterol lowering agents which inhibit early steps in the biosynthetic pathway, such as the HMG-CoA reductase inhibitors (Scheme 1), could also reduce the synthesis of other key biological substances like dolichol and ubiquinone, whereas SE acts past the key branchpoint for these compounds; (2) despite its epoxidation activity, SE does not belong to the cytochrome P-450 superfamily,[1] members of which are involved in many other biological oxidation processes, so SE inhibitors do not have the selectivity problems associated with inhibitors of cytochrome P-450 enzymes. In the case of antimycotic agents which act as ergosterol biosynthesis inhibitors, potential effects on mammalian cholesterol biosynthesis also need to be considered in order to avoid undesirable side-effects after systemic administration. As squalene epoxidase is involved in both pathways

Table 1   Sterol biosynthesis inhibitors of therapeutic use

Substance Class	Inhibition of	Indication	Topical Drugs	Systemic Drugs
HMG-CoA Red. Inhibitors	HMG-CoA reductase	cholesterol reduction	—	compactin fluvastatin lovastatin pravastatin
Allylamines	squalene epoxidase	antifungal	naftifine terbinafine	
Thiocarbamates	squalene epoxidase	antifungal	tolciclate tolnaftate	—
Azoles	C-14-demethylation	antifungal	bifonazole clotrimazole econazole	fluconazole itraconazole ketoconazole miconazole
Morpholines	Δ14-reduction + Δ8-Δ7-isomerisation	antifungal	amorolfine	—

(Scheme 1), SE inhibitors used as antifungal drugs are required to possess a high degree of species selectivity. Both the allylamines[13,14] and the thiocarbamates[15] (Table 1) are highly selective inhibitors of the fungal SE and affect the corresponding mammalian enzyme only at several orders of magnitude higher concentrations.

More recently, structural modifications at one of the two hypothesised binding regions[16] of allylamines to SE led to the discovery of analogues which selectively inhibit mammalian SE and lack antifungal activity.[17,18] Thus medicinal chemistry provided the basis for a two binding-site model of SE inhibition, with clearcut differences in structural requirements at one binding region for potent and selective inhibition of fungal versus mammalian SE.

2.   SQUALENE EPOXIDASE

The hydrocarbon squalene is the first irreversibly committed intermediate in the sterol biosynthesis pathway. Introduction of the 2,3-epoxide is an essential pre-condition for subsequent cyclisation to the rigid tetracyclic steroid structure. The epoxidation is carried out by a microsomal enzyme complex consisting of a flavo-protein with NAD(P)H cytochrome C reductase activity and a terminal oxidase which, unusually for an epoxidation reaction, is not a cytochrome P-450. Squalene epoxidase was first described by Yamamoto and Bloch,[19] using rat liver microsomes, and subsequently purified by Ono *et al.*[20] Similar enzymes have been characterised in a number of fungi, including *Candida albicans*,[21,22] *C. parapsilosis*[13] and *Saccharomyces cerevisiae*.[23-26] Interestingly, SE is the first enzyme of the sterol pathway to require molecular oxygen and is therefore an important regulatory step in the pathway, particularly in those fungi which are able to grow under anaerobic conditions.

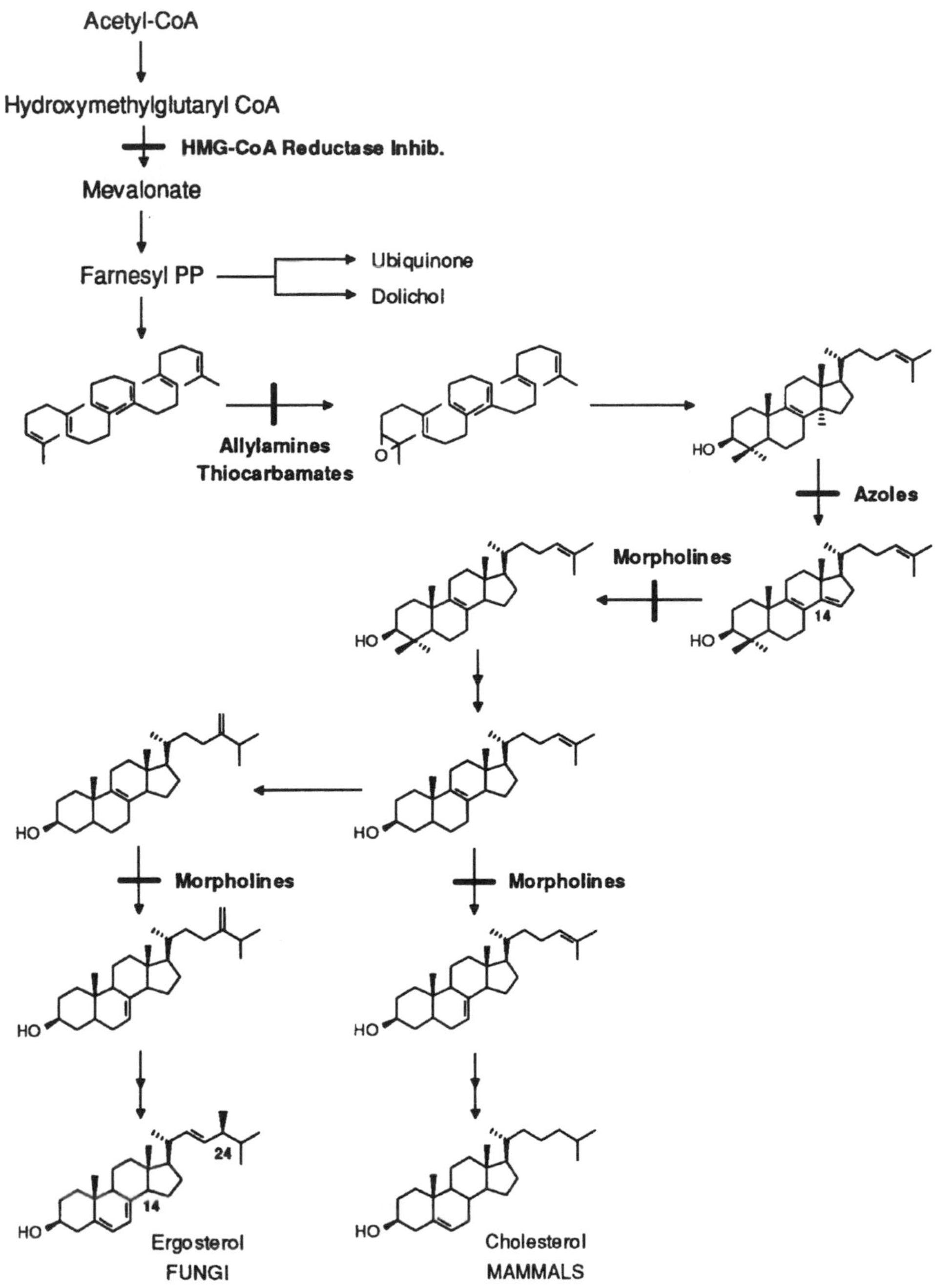

Acetyl-CoA

Hydroxymethylglutaryl CoA

HMG-CoA Reductase Inhib.

Mevalonate

Farnesyl PP ⟶ Ubiquinone
⟶ Dolichol

**Allylamines
Thiocarbamates**

Azoles

Morpholines

14

Morpholines

Morpholines

24

14

Ergosterol
FUNGI

Cholesterol
MAMMALS

Scheme 1 Simplified scheme of ergosterol and cholesterol biosynthesis showing the steps inhibited by clinically used substance classes.

The discovery that the topical antimycotic naftifine (Figure 1, **1**) acts by inhibition of fungal SE[10,11] greatly stimulated research into this area. Subsequent work[28,29] has shown that all derivatives of the allylamine type displaying antifungal activity are SE inhibitors, including the clinically used drug terbinafine (Figure 1, **2**).[30] Naftifine and terbinafine are active *in vitro* against a wide range of pathogenic fungi and exert a primary fungicidal action in most cases.[31-34] Extensive biochemical evidence[1,30,35,36] indicates that inhibition of SE is the sole mechanism responsible for the antifungal activity of these drugs, resulting in ergosterol deficiency and the build-up of high intracellular squalene concentrations. The accumulation of squalene appears to be responsible for the characteristic fungicidal action.[37,38] A second class of antifungals, the thiocarbamates, was subsequently found also to inhibit SE. Examples of this class

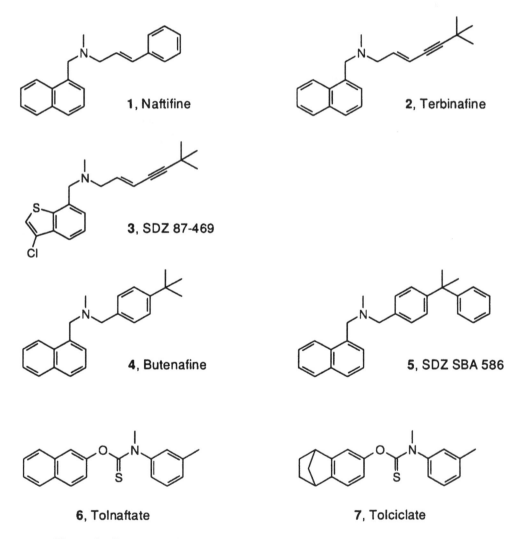

Figure 1   Representative structures of fungal squalene epoxidase inhibitors.

include the topical antimycotics tolnaftate,[15,39,40] tolciclate[15] and piritetrate.[41] The action of these compounds is primarily restricted to dermatophyte fungi, apparently due to poor penetration of the cell envelope in other fungi.[15]

Both naftifine and terbinafine cause a reversible inhibition of fungal SE, which is non-competitive with respect to the substrate squalene and the cofactors NADH and FAD.[13,42] However, the exact mechanism of inhibition is still not known. Both allylamines[13,42] and thiocarbamates[15] are selective in action and display only very low activity against mammalian squalene epoxidases, but the mechanism of this selectivity is also currently unknown. Terbinafine was found to be a competitive inhibitor of rat liver SE at very high concentrations,[13] as was the potent inhibitor NB-598,[17] suggesting a possible difference in the mechanism of inhibition of the fungal and mammalian SE enzymes.

3. PROTOTYPES OF SQUALENE EPOXIDASE INHIBITORS (SEI)

3.1 Inhibitors of Fungal Squalene Epoxidase

Selective inhibitors of fungal SE are of medical importance, because they can be used for the treatment of fungal infections. At present, three substance classes of potent and selective fungal SEI are known (Figure 1): the allylamines (1–3), the benzylamines (4,5), and the thiocarbamates (6,7).

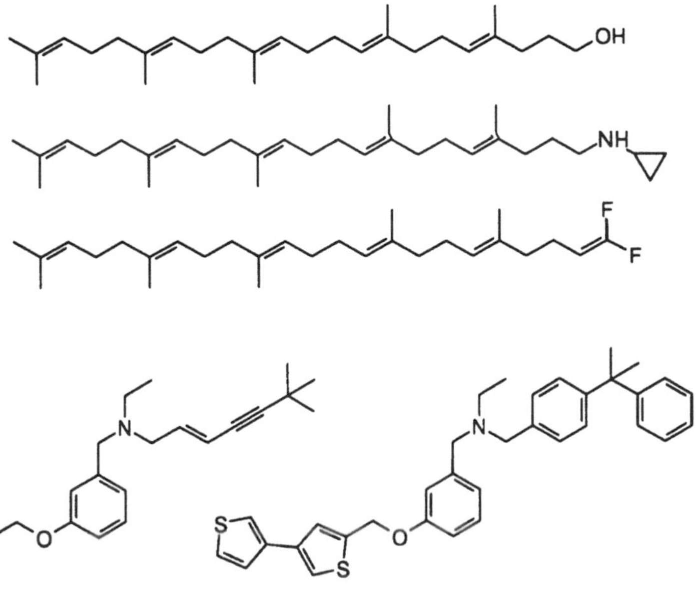

Figure 2 Representative structures of mammalian squalene epoxidase inhibitors.

3.2   Inhibitors of Mammalian Squalene Epoxidase

Inhibition of mammalian SE has been envisaged as a therapeutic principle for the treatment of hypercholesterolemia. The first compounds exhibiting a moderate inhibitory potential against mammalian SE were squalene analogues, represented by structures 8–10 in Figure 2. The recent discovery of highly potent inhibitors, the allylamine (11) and the benzylamine derivatives (12), might lead to drugs for clinical use.

4.   BIOLOGICAL TESTING

Due to the distinct potential applications of squalene epoxidase inhibitors — treatment of fungal infections by selective blocking of ergosterol biosynthesis in fungi, or lowering of cholesterol levels by inhibition of human sterol biosynthesis — differential biological profiles of the drugs and, hence, different test systems were required for the respective indication.

4.1   Testing of Antifungal Squalene Epoxidase Inhibitors

For routine screening and the establishment of SARs the determination of the antifungal activity *in vitro* against a range of human pathogenic fungi was used. Structural prototypes and highly potent compounds were subsequently tested for their intrinsic biochemical activity using microsomal preparations of the fungal enzyme. Comparison of the results revealed that inhibition of SE is required for antifungal activity, but does not guarantee it. Poor penetration of the compounds through the cell envelope seems to account chiefly for divergent results. All derivatives showing reasonable *in vitro* activity were studied in various animal models using both topical and oral administration. The most effective compounds were also tested against mammalian SE to investigate their selectivity for the fungal enzyme. Several additional assays were used to study activity of selected compounds against a broader range of fungi including less common strains, fungicidal versus fungistatic action,[32] specificity of ergosterol biosynthesis inhibition,[30,36] lack of effects on cytochrome P-450 enzymes,[43,44] etc.

4.1.1   Sterol Biosynthesis Inhibition

Three experimental systems were employed to measure the activity of antifungal compounds as ergosterol biosynthesis inhibitors:

i. The whole-cell test, measuring incorporation of radiolabelled acetate into sterols and sterol precursors in *C. albicans* cells.[11,30]
ii. The cell-free test, measuring incorporation of radiolabelled mevalonate into sterols and sterol precursors in cell-free extract of *C. albicans*.[11,30]

iii. The SE enzyme assay, measuring SE activity in microsomal preparations of *C. albicans* by incorporation of radiolabelled squalene into squalene epoxide and sterol.[13,21]

Generally, very good agreement between values from cell-free and SE tests was observed. The possibility of inhibiting steps other than squalene epoxidase in the ergosterol biosynthesis pathway was investigated in a range of cell-free tests using different labelled sterol precursors.[30,36] For species specificity studies, microsomal epoxidase from rat liver[13] and guinea pig liver[42] was used.

4.1.2 Antifungal Activity In vitro

The *in vitro* antifungal activity of the test compounds was investigated against a broad range of human pathogenic fungi, including dermatophytes, molds, biphasic fungi, and yeasts.[32] In early studies (Figure 3, Table 2) *Trichophyton mentagrophytes* (T. ment.), *Epidermophyton floccosum* (Epid. fl.), *Microsporum canis* (M. can.), *Sporothrix schenckii* (Sp. sch.), and *Candida parapsilosis* (C. par.) were used in the screening program. The spectrum of strains was then broadened in routine testing, but for simplification the present discussion of SARs in sections 5 and 7 deals only with the *in vitro* potencies against the following selected test organisms: *Trichophyton mentagrophytes* (T. ment.), *Microsporum canis* (M. can.), *Apergillus fumigatus* (A. fum.), *Sporothrix schenckii* (Sp. sch.), *Candida albicans* Δ124 (C. alb.), and *Candida parapsilosis* Δ39 (C. par.). Serial dilution tests were used to determine the MIC (minimum inhibitory concentration) of the test compound.[32] The MIC was defined as the lowest substance concentration at which fungal growth was macroscopically undetectable. The fungal strains were obtained from the American Type Culture Collection (Rockville, MD), the Centraalbureau voor Schimmelcultures (Baarn, The Netherlands), the Hygiene-Institut Würzburg (FRG), or the II. Universitäts-Hautklinik (Vienna, Austria).

Additional *in vitro* tests used to assess inhibitory potency were the agar diffusion test[32] and the germ tube formation of *Candida albicans* after induction of hyphal growth by either serum or N-acetylglucosamine.[45]

4.1.3 Antifungal Efficacy In vivo

Compounds showing high antifungal *in vitro* activity were evaluated in several animal models using topical and oral application.[33,46,47] Topical efficacy was studied primarily in the guinea pig trichophytosis model (experimental infection with *Trichophyton mentagrophytes*). In secondary test systems, guinea pig dermatophytosis caused by *Microsporum canis*, skin infections of guinea pigs caused by *Candida albicans*, and vaginal candidiasis in rats were treated topically with the test compounds in comparison to clinical standards. Systemic efficacy was investigated in the guinea pig trichophytosis and microsporosis models.

Allylamine antimycotics generally display a primary fungicidal action (i.e. fungicidal activity at the MIC) against dermatophytes, other filamentous fungi, *Sporothrix schenckii,* and *Candida parapsilosis* resulting in killing of the fungal organism. This property is thought to contribute significantly to the high therapeutic efficacy of the allylamine derivatives and to lead to rapid regression of symptoms accompanying skin mycoses, in particular inflammation. Measurement of skin temperature in a guinea pig model of epidermal dermatophytosis permits a quantitative assessment of this latter effect.[33,47]

4.2   Testing of Cholesterol-Lowering Squalene Epoxidase Inhibitors

For compounds inhibiting the mammalian SE, only data obtained in enzymatic (microsomal SE from rat liver or Hep G2 cells) and cellular test sytems (cholesterol synthesis in Hep G2 cells) are discussed. These assays based on incorporation of radiolabelled precursors are very similar to the corresponding SE and whole-cell tests used for the evaluation of antifungal SE inhibitors. The potencies are given as IC_{50} values (drug concentrations at which 50% of the enzyme activity or cholesterol production, respectively, is inhibited) and are consistently taken from the literature, except the data listed in Table 22, which have been produced at the Sandoz Research Institute, East Hanover.

5.   MEDICINAL CHEMISTRY OF THE ALLYLAMINE AND BENZYLAMINE ANTIMYCOTICS

5.1   The Discovery of the Allylamines

The first representative of the allylamine antimycotics, naftifine (Figure 1, 1), was obtained accidentally as the result of an unexpected chemical reaction[48] at Sandoz-Wander, Berne, in 1974. In a routine screening program at the Sandoz Research Institute in Vienna, 1 was found to be highly active against a number of human pathogenic fungi *in vitro.*[31] High antifungal efficacy could be confirmed in animal models after topical application, whereas oral activity was only observed at very high doses of no therapeutic relevance. Subsequent investigations revealed a novel mode of action, namely inhibition of fungal squalene epoxidase.[10] In extended pre-clinical and clinical trials 1 proved to be an efficacious antimycotic for topical treatment of dermatomycoses[49,50] and was first marketed in 1985 (Exoderil®).

The first findings of the antifungal activity of naftifine (1) initiated a medicinal chemistry programme aiming at providing answers to the following questions:

– Is naftifine unique or a representative of a new class of active compounds?
– Which structural elements are essential for antifungal activity, and can structure-activity relationships (SAR) be defined?
– Is it possible to improve antifungal potency and spectrum, in particular oral efficacy, by structural modification?

Chemical derivatisation of 1 soon revealed that the antifungal activity was specifically linked to certain structural elements that could not be related to any

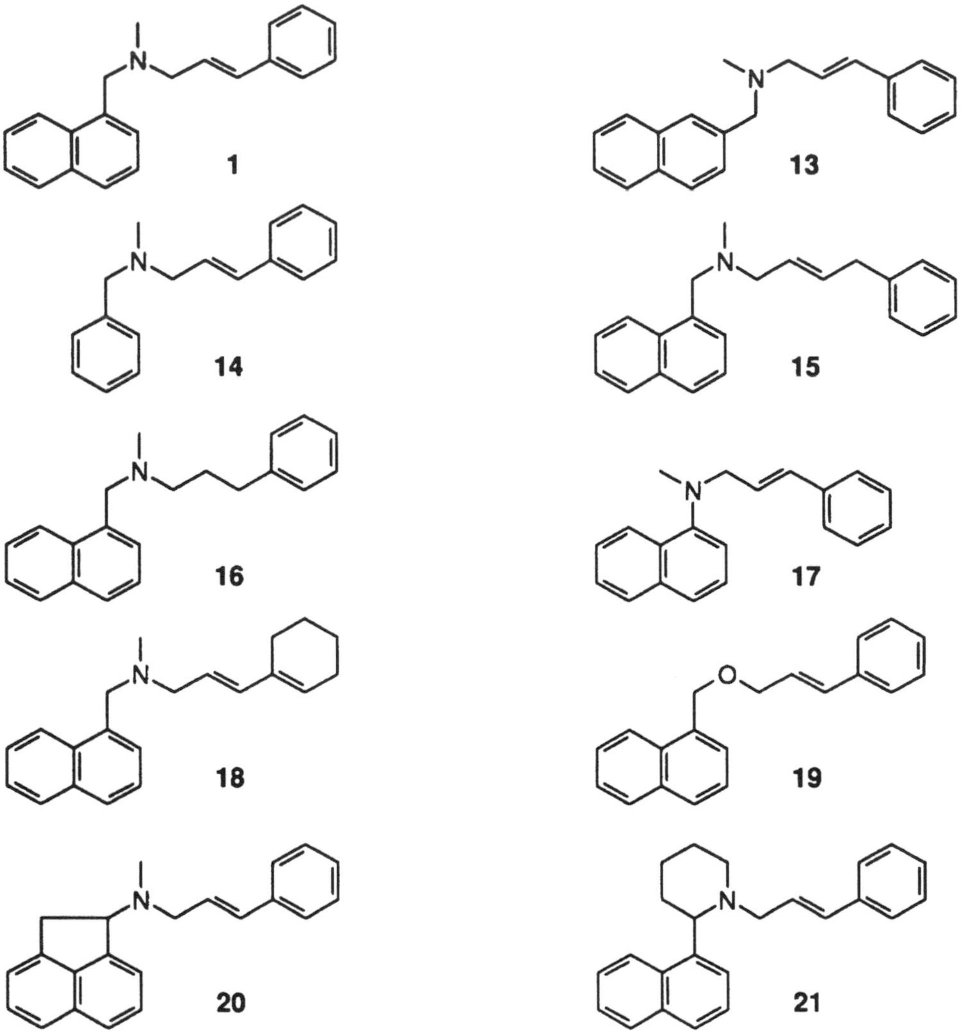

Figure 3 Basic structural modifications of naftifine (1).

known classes of antimycotics.[51] From the analogues 13–21 shown in Figure 3, only compounds 18, 20, and 21 demonstrated antifungal potency comparable to 1, whereas the other derivatives were almost or completely devoid of activity. Thus, it was established that 1 was not a solitary finding but the first representative of a new class of synthetic antimycotics. Since the tertiary allylamine function was common to all active structures, the new substance class was termed "allylamine derivatives".[27] Piperidine compound 21, together with 20 the only active structure out of a series of semirigid naftifine analogues, provided the answer to the third question. Due to its interesting spectrum of activity, the racemic 21 was resolved into its enantiomers.

Animal experiments showed that efficacy was restricted to (R)-**21** (determined by X-ray structure analyses), which was the first derivative to exhibit substantially improved oral activity (guinea pig trichophytosis model).[52] Consequently, the main effort of the continued medicinal chemistry programme was directed to the identification of allylamine derivatives with enhanced systemic activity.

5.2   From Naftifine to Terbinafine, From Topical to Oral Activity

Systematic derivatisation of **1** suggested that modifications of the cinnamyl moiety were most promising for possible enhancement of antifungal efficacy. Therefore, analogues of the cyclohexenyl compound **18**, which was equipotent to **1**, with linear allylamine side chains were synthesised and their activity against various human pathogenic fungi evaluated (Table 2).[52]

Antifungal activity was strongly dependent on the chain length and on the number of conjugated double bonds with a maximum potency for C_9- and C_{10}-2,4-alkadienyl side chains (**24**, **25**). Both shortening (**22**, **23**) and elongation (**26**) of

Table 2   *In vitro* antifungal activity of linear side chain allylamines in comparison with **1** and **18** (for abbreviations see Section 4.1.2.)

	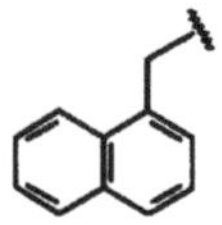	MIC [mg/L]				
		T. ment.	Epid. fl.	M. can.	Sp. sch.	C. par.
1	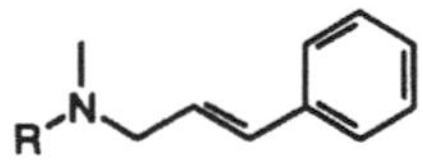	0.05	0.2	0.1	1.6	1.6
18	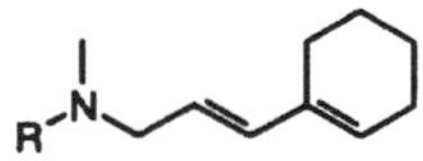	0.05	0.1	0.1	0.8	0.8
22	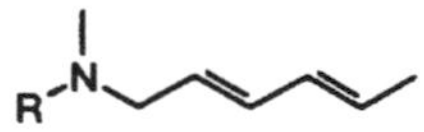	1.6	50	25	100	>100
23	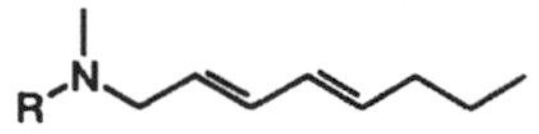	0.1	0.8	0.8	6.2	12.5
24	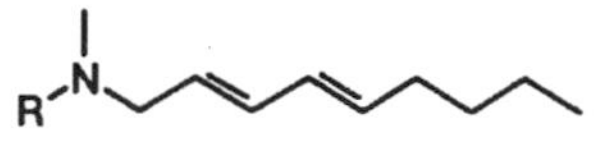	0.05	0.2	0.1	3.1	1.6
25	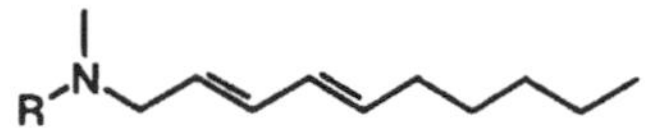	0.05	0.1	0.1	1.6	1.6
26	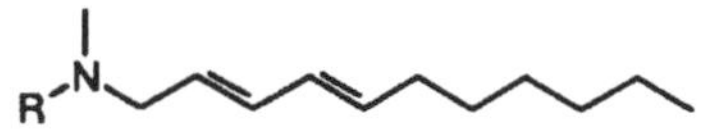	0.2	0.8	0.4	3.1	100
27	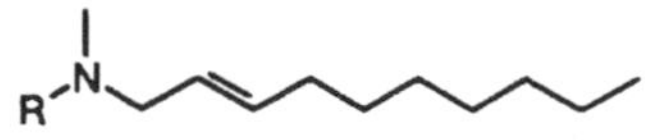	1.6	25	50	>100	>100
28	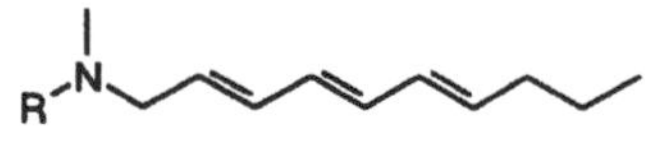	1.6	3.1	3.1	100	>100

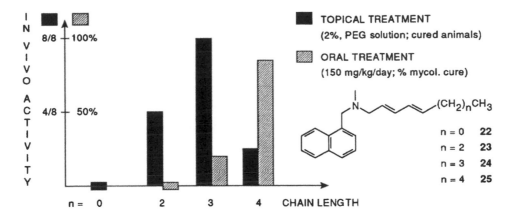

Figure 4   *In vivo* activity of *N*-2,4-alkadienyl derivatives **22–25**.[52]

the 2,4-alkadienyl chain led to considerable decrease in activity. Pronounced reduction in potency was also observed for analogues **27** and **28** bearing side chains of optimum length, but with one double bond more or less relative to **25**, respectively. Evaluation of this series of compounds in animal models revealed that the *in vivo* efficacy was even more affected by variation of the chain length (Figure 4). Among these derivatives, only compound **24** (C_9-2,4-alkadienyl side chain) completely cured the animals using topical treatment with a 2 % solution, whereas homologue **25** was the most effective compound after oral administration.

Substitution of the C4–C5 double bond in **24** by a triple bond (**29**) resulted in substantial increase in antifungal activity both *in vitro* (Table 3) and *in vivo* (Table 4). Further structural modifications revealed that the (*E*)-1,3-enyne element in **29** was responsible for this improvement. For example, the (*Z*)-1,3-enyne compound **32** and the (*E*)-2-alkyn-4-enyl derivative **30** (with inverted enyne group) displayed considerably reduced antifungal properties (Table 3).[53]

The most dramatic enhancement in efficacy of the (*E*)-2-alken-4-ynyl derivative **29** was observed after oral administration. Whereas the oral dose of 75 mg/kg of either **1** or **24** was completely ineffective in the guinea pig trichophytosis model, the same dose of **29** produced 100% mycological cure in all animals (Table 4).

Consequently the (*E*)-1,3-enyne feature, which represented a novel structural element in pharmaceutical agents, was kept constant in the subsequent SAR studies. Increased branching of the alkyl substituent adjacent to the triple bond (*n*-butyl, **29** ⟹ *s*-butyl, **31** ⟹ *t*-butyl, **2**) finally led to the discovery of terbinafine[53] (Table 3, **2**) and the breakthrough for oral efficacy of allylamine antimycotics.[33,46] Comparative studies with terbinafine and analogue **29** proved **2** to be superior by more than one order of magnitude after oral administration (Table 4). The tremendous improvement of the oral activity within the allylamine antimycotics is furthermore clearly reflected in the ED_{50} values (substance dose required for 50% cure of all animals) in the guinea pig trichophytosis model of naftifine (292 mg/kg), terbinafine (2.8 mg/kg), and the clinical standard griseofulvin (40 mg/kg). Extensive clinical

Table 3 Antifungal activity of 1,3-enyne compounds **2, 29–32** (for abbreviations see Section 4.1.2.)

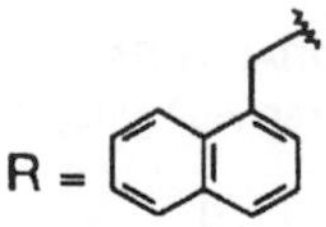 R = (naphthyl-CH₂)	MIC [mg/L]					
	T. ment.	*M. can.*	*A. fum.*	*Sp. sch.*	*C. alb.*	*C. par.*
24	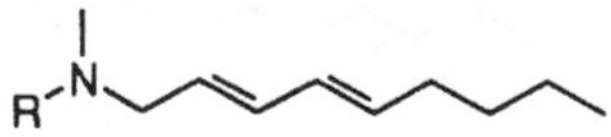 0.05	0.1	12.5	3.1	>100	1.6
29	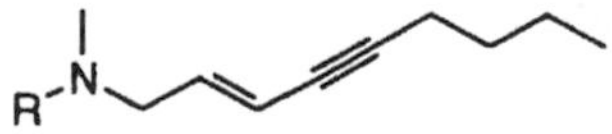 0.01	0.01	100	1.6	100	0.4
30	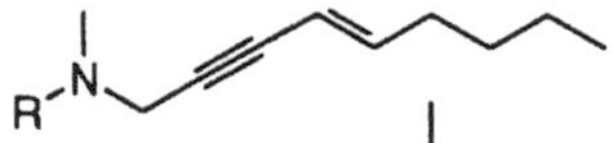 0.2	0.2	>100	>100	>100	>100
31	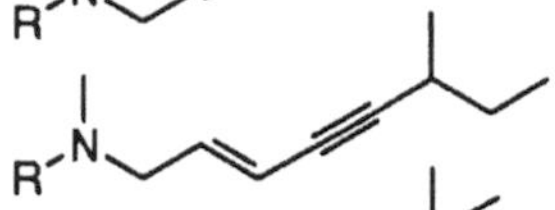 0.01	0.006	100	0.8	100	0.4
2	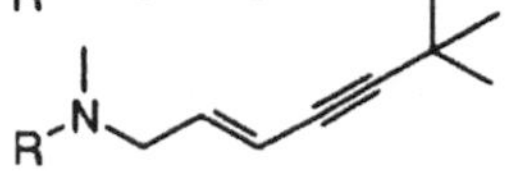 0.003	0.006	0.8	0.4	25	0.4
32	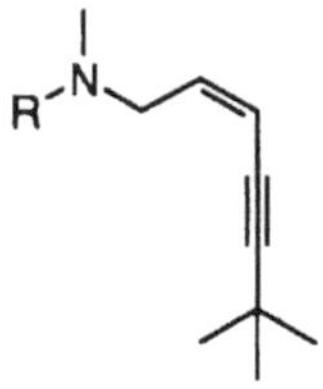 0.1	0.1	>100	3.1	>100	12.5

trials revealed terbinafine (**2**) to be well tolerated in man and to be extremely potent in the topical and oral treatment of mycoses of the skin and its appendages, especially in the therapy of onychomycosis.[54-57] Since 1991 terbinafine is marketed under the name Lamisil®.

Table 4 The improvement of *in vivo* efficacy (guinea pig trichophytosis model, % mycological cure) by (*E*)-2-alken-4-ynyl compounds **29, 31**, and **2** (for structures see Table 3)

	Topical Activity		Oral Activity		
	0.125%	*0.5%*	*5 mg/kg*	*10 mg/kg*	*75 mg/kg*
1	35%	65%	—	—	0%
24	40%	69%	—	—	0%
29	83%	100%	0%	15%	100%
31	—	—	23%	75%	100%
2	100%	—	98%	100%	—

5.3 SAR of Terbinafine Analogues

Analysis of the antifungal spectrum of **2** (Table 3) and the progenitor allylamine derivatives revealed that the *t*-butyl group was also a prerequisite for susceptibility of *Aspergillus fumigatus* and *Candida albicans in vitro*. The results obtained with further (*E*)-2-alken-4-ynyl derivatives (Table 5, **33–42**) featuring various substitution at the triple bond confirmed the importance of a tertiary alkyl group at this position for high antifungal activity.[12,52] Thus, the phenylacetylene analogue **34** was almost

Table 5   *In vitro* activity of differently substituted (*E*)-2-alken-4-ynyl derivatives

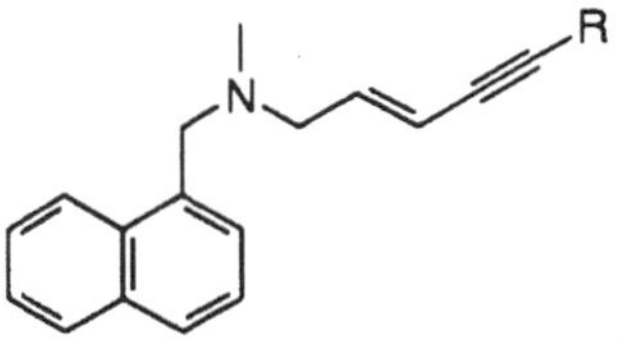

	R	MIC [mg/L]					
		T. ment.	*M. can.*	*A. fum.*	*Sp. sch.*	*C. alb.*	*C. par.*
2		0.003	0.006	0.8	0.4	25	0.4
33	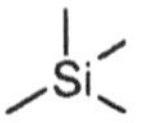	0.05	0.05	>100	3.1	25	1.6
34	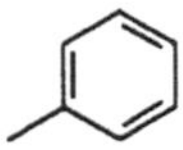	3.1	3.1	>100	>100	>100	>100
35	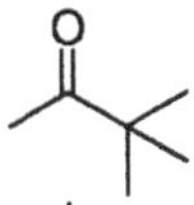	6.2	50	>100	>100	>100	25
36	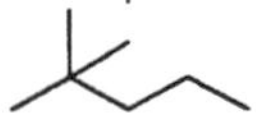	0.003	0.006	0.4	0.2	6.2	0.1
37	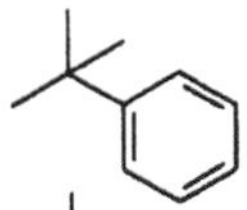	0.003	0.003	12.5	0.2	12.5	0.2
38	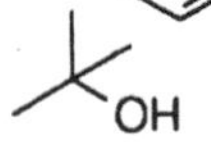	0.8	3.1	>100	50	>100	>100
39	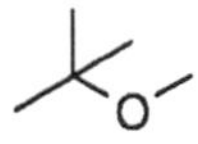	0.01	0.05	3.1	6.2	>100	12.5
40	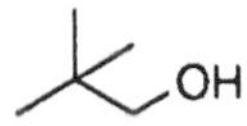	0.4	0.8	>100	25	>100	>100
41	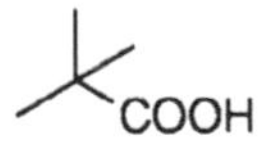	>100	>100	>100	>100	>100	>100
42		0.003	0.006	>100	1.6	>100	0.8

Table 6 Antifungal activity of terbinafine (**2**) analogues modified at the amino function

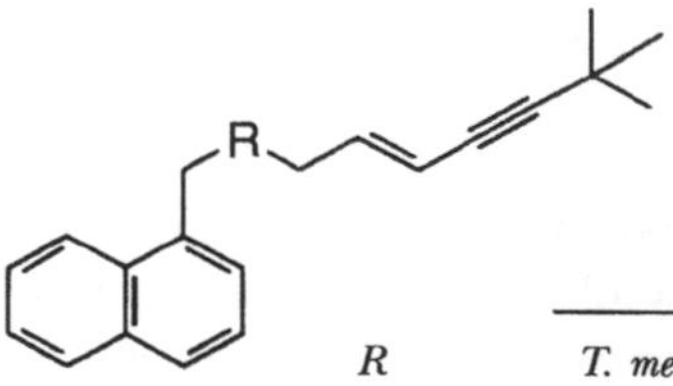

	R	MIC [mg/L]					
		T. ment.	M. can.	A. fum.	Sp. sch.	C. alb.	C. par.
2	N-Me	0.003	0.006	0.8	0.4	25	0.4
43	N-H	0.4	0.8	>100	6.2	>100	>100
44	N-Et	0.01	0.02	1.6	1.6	12.5	0.8
45	N-cPr	3.1	6.2	>100	>100	>100	>100
46	N-CHO	6.2	50	>100	>100	>100	>100
47	O	0.1	0.1	>100	100	>100	>100
48	CH$_2$	0.8	12.5	>100	>100	>100	>100
49	CH-Me	0.4	3.1	>100	>100	>100	>100
50	C-ME$_2$	>100	>100	>100	>100	>100	>100

completely inactive, whereas compound **37** with the phenyl ring incorporated in a tertiary alkyl substituent turned out to be one of the most active allylamine antimycotics. Analogue **33**, obtained by replacement of the *t*-butyl group in **2** by a trimethylsilyl residue, showed only weak activity *in vitro*. The *in vivo* efficacy of **33**, however, was comparable to that of **2**.[12,52]

Additional valuable SAR could be extracted from the results listed in Table 5. Increase of the bulkiness in the *t*-butyl region of **2** as in analogues **36** and **37** resulted in maintained or even enhanced *in vitro* potency. In contrast, introduction of very polar substituents, such as hydroxy (**38**, **40**)[12,58] or carboxy groups (**41**),[58] caused drastic reduction or even suppression of antifungal activity. This implied that the terbinafine metabolites **40** and **41** do not contribute to the therapeutic effect of terbinafine. Methylation of the free hydroxy functions to afford the corresponding ether (**39**) or ester derivative (**42**) substantially reduced the polarity of the substituents and restored high *in vitro* potency. These findings demonstrated clearly that the antifungal activity of the allylamine derivatives is strongly dependent on their lipophilicity.

The *in vitro* results obtained for terbinafine analogues modified at the polar centre of the molecule (Table 6, **43–50**) confirmed our initial observations with naftifine derivatives (Figure 3), that optimal activity was achieved when R was the N-methyl group.[51] The N-demethyl product **43**, which was identified as one of the main metabolites of **2**, displayed only moderate activity against dermatophytes (*T. ment. M. canis*) and *Sporothrix schenckii*.[12] Increase in bulkiness of the N-alkyl substituent (**44**, **45**) resulted in reduced potency, as did replacement by the formyl group (**46**). Selective growth inhibition of dermatophytes only was observed for the corresponding ether derivative **47** and the carba analogues **48** and **49**.[58]

Table 7 Inhibition of *C. albicans* sterol biosynthesis by carba analogues of **2**

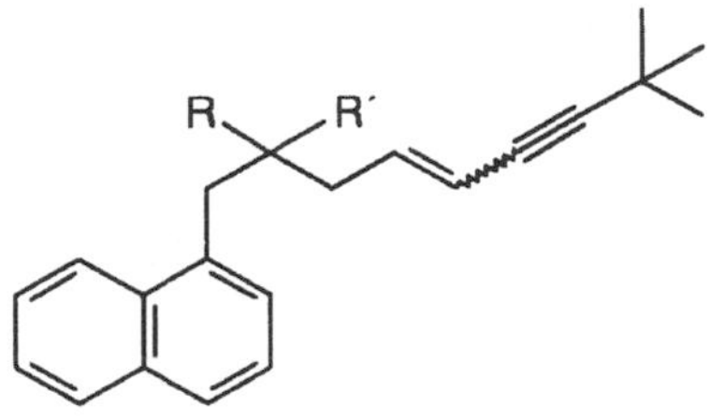

				IC_{50} [μM]	
	E/Z	R	R'	Cells	Cell-free
48	E	H	H	>30.0	16.3
49	E	Me	H	2.7	0.2
51	Z	Me	H	>30.0	2.5
50	E	Me	Me	—	>30.0
52	Z	Me	Me	—	>30.0
2	E	N-Me		0.03	0.03
43	E	N-H		0.32	0.32

These findings suggested that the nitrogen atom was required for both high and broad antifungal activity *in vitro*. In order to determine if this holds true for the intrinsic activity at the enzyme level, analogues of terbinafine, in which the nitrogen atom was replaced by a carbon atom, were tested for sterol biosynthesis inhibition of *C. albicans* under both cellular and cell-free conditions in comparison with the parent compounds **2** and **43**.[28,58]

The results of these experiments (Table 7) demonstrated that the inhibitory activity within the series of analogues **48–52** was dependent on similar steric requirements as for the nitrogen containing derivatives. The most active compound **49**, with the trans double bond and one methyl substituent at the carbon replacing the nitrogen, was the closest analogue of **2**. Furthermore, it could be concluded that the nitrogen atom was not essential for enzymatic activity per se. Comparing the values for inhibition in whole cells and cell-free extracts, it became evident that presence of the amine function is particularly important for penetration of the compounds through the fungal cell envelope. In addition, the findings summarised in Table 7 underlined the absolute requirement of the N-methyl group in **2** for high potency and suggested that the reduced *in vitro* activity of the N-demethyl terbinafine derivative **43** (Table 6) was not only the result of increased polarity of the secondary amine function. The lack of the extra methyl group could also partially explain the weak antifungal potency of ether derivative **47** (Table 6).

Substitution at the α-position of the 1-naphthylmethyl moiety generally led to decrease in potency, in particular against *A. fumigatus* (Table 8). Compound **53**

Table 8 *In vitro* activity of terbinafine analogues substituted at the α-position of the 1-naphthylmethyl residue (* values for **57** from Chen *et al.*[59])

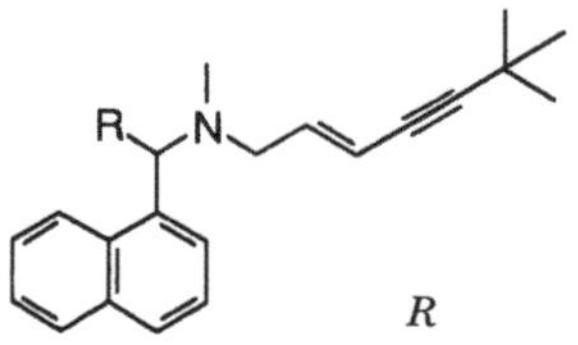

	R	T. ment.	M. can.	A. fum.	Sp. sch.	C. alb.	C. par.
				MIC [mg/L]			
2	–H	0.003	0.006	0.8	0.4	25	0.4
53	–CH$_3$	0.01	0.02	>100	1.6	50	3.1
(S)-53	(S)–CH$_3$	0.01	0.02	>100	3.1	25	1.6
(R)-53	(R)–CH$_3$	0.01	0.02	>100	0.8	50	3.1
54	–CF$_3$	>100	>100	>100	>100	>100	>100
55	=O	1.6	1.6	>100	>100	>100	>100
56	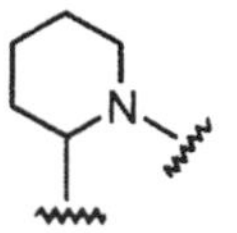	0.05	0.2	>100	3.1	50	25
57*	–CH$_2$–N 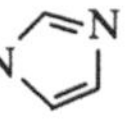	5	5	20	20	20	—

bearing a methyl group at this position was about 3 times less active than **2**. By resolving the racemic **53** into its enantiomers, it could be proven that the newly created stereocentre in **53** had no influence on biological activity. Analogue **54** with a trifluoromethyl instead of a methyl group was completely inactive, possibly due to the electronic impact of the trifluoromethyl substituent. In agreement with the results obtained for the N-formyl analogue **46** (Table 6), the amide derivative **55** was also lacking substantial antifungal potency. Incorporation of the N-methyl substituent into a cyclic structure generated piperidine derivative **56**, which was also less active than **2**.

Chinese researchers reported on the synthesis and antifungal potency of additional terbinafine analogues substituted at the α-position of the 1-naphthylmethyl residue.[59] The most interesting compound was imidazole derivative **57** representing an hybrid structure of the allylamine antimycotics and the azoles, another class of potent synthetic antifungal agents. The reported activity of **57**, however, was quite low. The synthesis of correspondingly substituted benzylamine derivatives, related antimycotics with a benzylamine instead of an allylamine side chain (see Table 14), was also described recently, but no biological data were disclosed for these compounds.[60]

Extensive studies on modifications of the naphthalene residue of **2** were performed. In agreement with the SAR findings for naftifine structures[51] (Figure 3), the 1-naphthyl substitution pattern was found to be essential for high antifungal potency. Thus, the corresponding 2-substituted terbinafine analogue **58** (Table 9) was almost

Table 9 *In vitro* activity of terbinafine analogues with modifications of the naphthalene residue

R =		MIC [mg/L]					
		T.	M. can.	A. fum.	Sp. sch.	C. alb.	C. par.
2	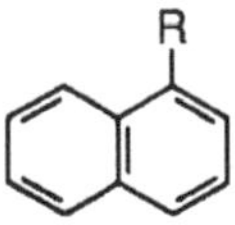	0.003	0.006	0.8	0.4	25	0.4
58	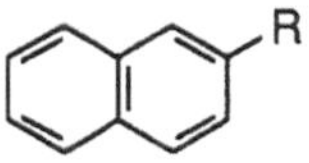	6.2	100	100	100	>100	>100
59	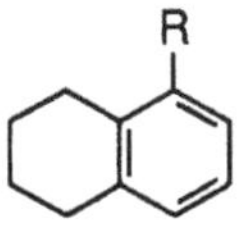	0.02	0.2	>100	1.6	50	6.2
60	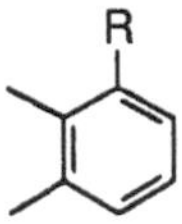	0.2	0.4	25	3.1	>100	25
61	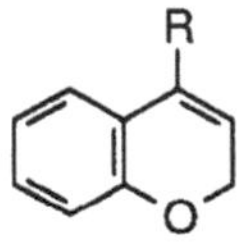	0.01	0.05	6.2	1.6	50	6.2
62	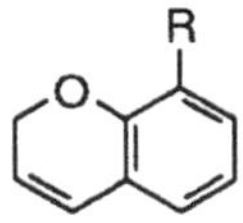	0.8	1.6	>100	>100	>100	>100
63	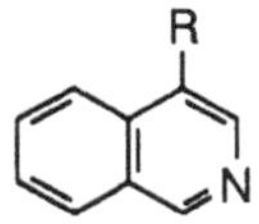	0.02	0.05	6.2	1.6	50	3.1
64	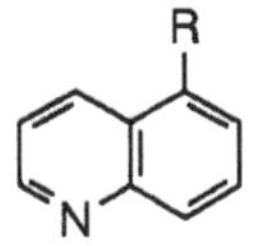	0.05	0.1	6.2	1.6	100	6.2

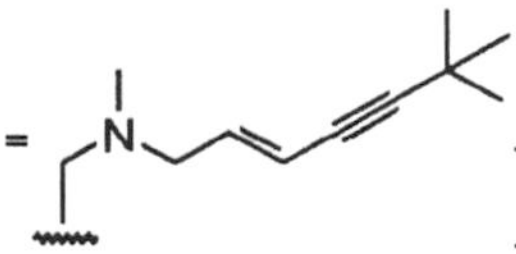

inactive.[12] Furthermore, the moderate results obtained for the tetrahydronaphthalene derivative **59** indicated that a planar bycyclic aromatic ring sytem is required. Consequently, it was not surprising that all attempts to replace the 1-naphthyl group by substituted monocyclic ring structures generally led to compounds with inferior antimycotic properties. This is illustrated by compound **60** (Table 9), in which a 2,3-dimethylphenyl residue mimicks the bycyclic ring system.

In terbinafine analogues **61–64** (Table 9) the naphthyl moiety was replaced by 6,6-membered bycyclic heteroaromatic ring systems. Although they were all less active than **2**, the isoquinoline derivative **63** was of particular interest. Allylamine antimycotics are known to bind non-specifically to several major serum components,[61] probably due to their highly lipophilic nature. This might limit their use in the treatment of systemic mycoses. We intended to solve this problem by synthesising terbinafine analogues with decreased lipophilicity, but nearly identical steric characteristics. From this programme, the isoquinolinyl compound **63** (Table 9) demonstrating comparable potency in inhibiting the hyphal growth of *Candida albicans* both in the presence of serum and under serum-free conditions was identified.[62]

Table 10 summarises the SAR investigations of terbinafine structures containing 5,6-membered bycyclic aromatic ring systems instead of the naphthalene moiety. A detailed study on benzo[b]thienyl allylamine antimycotics revealed that benzo[b]thiophene can act as a bioisostere for naphthalene when the allylamine side chain is linked to position 3, 4, or 7 (Table 10, **65–67**).[58,63] Furthermore, additional substitution at position 3 of the 7-benzo[b]thienyl allylamines led to highly active compounds with substantially enhanced potency against *Candida albicans*. The 3-chloro derivative SDZ 87-469[64] (Table 10, Figure 1, **3**) was found to be the most potent allylamine antimycotic *in vitro* and the most potent inhibitor of fungal squalene epoxidase in enzyme preparations of *C. albicans*.[16] In general, yeasts showed considerably increased susceptibility to **3** relative to **2**. SDZ 87-469 proved to be also highly effective in several animal models and was therefore selected for further development. Toxicological studies, however, indicated undesired side effects resulting in discontinuation of this project. Attempts to replace the naphthalene ring system either by benzo[b]furans (Table 10, **68, 69**) or by indoles (**70, 71**) afforded analogues with moderate to good antifungal activity, but none was as active as **2**. The synthesis of benzo[b]furan and indole analogues of butenafine (Figure 1, **4**) was described recently, but biological results were not reported.[65,66]

The increased activity of the 3-substituted benzo[b]thienyl allylamines against yeasts stimulated further studies in order to investigate whether the 3-substituted thiophene element was essential for this effect. Consequently, terbinafine analogues with various substitution patterns on the naphthalene ring system were prepared.[67]

A set of substituents was introduced at positions 3–7 (substitution at positions 2 and 8 potentially causes a conformational change of the allylamine side chain) of the naphthalene in order to get information on the electronic and steric factors required for optimal activity. An extract of the SAR of these derivatives is listed in Table 11. In general the potency was found to be strongly dependent on the bulkiness of the substituent. Only fluorine could substitute hydrogen at positions 2–4 and 6–8 of the naphthalene moiety without appreciable loss of activity as compared to **2**. In contrast, substituents might be larger in size at position 5, as shown by the high activity of the 5-chloro (**74**), 5-bromo (**75**), and 5-methyl (**76**) analogues. The 5-cyano derivative **77**, however, displayed reduced *in vitro* potency. The fluoro compounds **72**, **73**, and **78** demonstrated enhanced activity against yeasts relative to **2**. This increase in yeast sensitivity could be further improved by introduction of two fluoro substituents at positions 5 and 7 (Table 11, **79**). Compound

Table 10 Antifungal activity of terbinafine analogues with 5,6-membered bicyclic aromatic ring systems

R 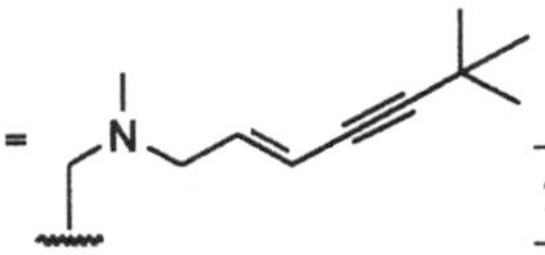 = (structure)		MIC [mg/L]					
		T. ment.	*M. can.*	*A. fum.*	*Sp. sch.*	*C. alb.*	*C. par.*
2	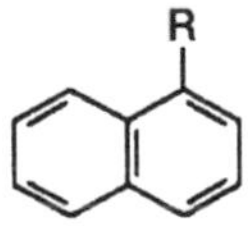	0.003	0.006	0.8	0.4	25	0.4
65	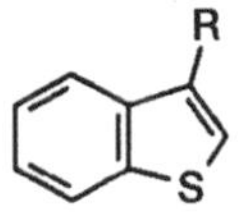	0.01	0.02	3.1	0.8	100	3.1
66	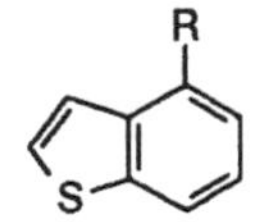	0.006	0.01	0.8	0.4	50	1.6
67	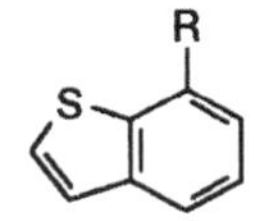	0.006	0.01	0.1	0.2	25	0.8
3	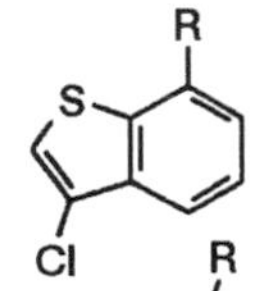	0.002	0.003	0.1	0.2	0.8	0.1
68	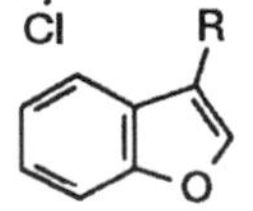	0.05	0.1	>100	3.1	100	12.5
69	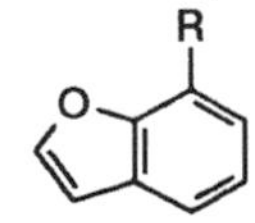	0.2	0.8	100	25	>100	100
70	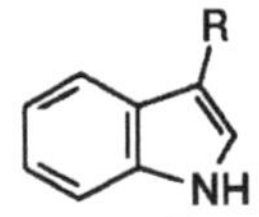	6.2	6.2	>100	25	>100	>100
71	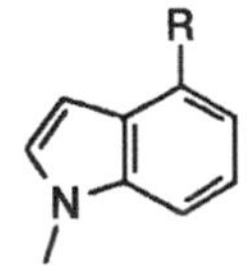	0.2	0.4	>100	3.1	>100	50

79 exhibited 8- to 16-fold enhanced potency against *Aspergillus fumigatus, Candida albicans,* and *C. parapsilosis* in comparison with **2**. The activity spectrum of **79** was found to be very similar to that of the 3-chloro-benzo[b]thienyl analogue **3** (Table 10). Thus, high activity against yeasts is not limited to 3-substituted benzo[b]thienyl

Table 11 *In vitro* activity of naphthalene substituted terbinafine analogues

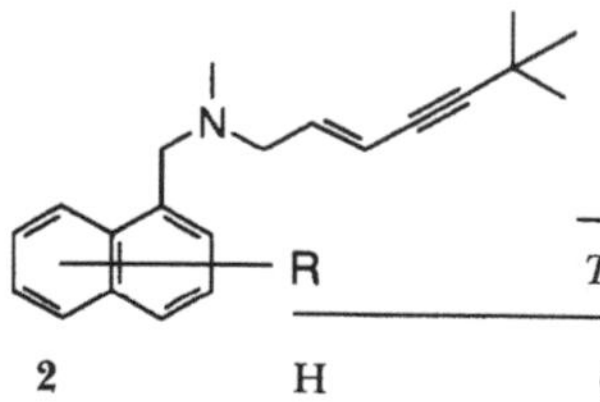

		MIC [mg/L]					
	R	*T. ment.*	*M. can.*	*A. fum.*	*Sp. sch.*	*C. alb.*	*C. par.*
2	H	0.003	0.006	0.8	0.4	25	0.4
72	3-F	0.006	0.006	0.8	0.2	6.2	0.2
73	5-F	0.006	0.006	0.1	0.2	3.1	0.2
74	5-Cl	0.003	0.006	0.1	0.4	1.6	0.2
75	5-Br	0.006	0.01	1.6	0.1	3.1	0.8
76	5-Me	0.006	0.02	0.8	0.4	12.5	1.6
77	5-CN	0.2	0.4	>100	6.2	>100	3.1
78	7-F	0.006	0.006	0.4	0.4	12.5	0.4
79	5,7-diF	0.003	0.003	0.1	0.1	1.6	0.1

allylamines with the side chain at position 7. Furthermore, within the allylamine antimycotics the 5,7-difluoro-1-naphthyl residue and the 3-chloro-7-benzo[b]thienyl element might be considered as bioequivalents displaying similar physicochemical and biological properties.[67]

Unexpectedly, a very similar spectrum of activity was observed for both the *E* and the *Z*-isomer of the 4-fluoronaphthyl compound (Table 12, **80, 81**). Furthermore, the *Z*-isomer **81** displayed lower MIC values against all strains tested than the corresponding unsubstituted derivative **32** (Table 3, 12). In all previous cases of pairs of *cis/trans* isomers, the *E*-isomer was clearly superior to the corresponding *Z*-derivative in the routine *in vitro* screening, as shown for example for terbinafine (**2**) and its *Z*-analogue **32** (Table 12). The surprising finding that (*E*)- and (*Z*)-4-fluoro allylamines **80** and **81** had equivalent potencies was confirmed by investigating additional 4-substituted analogues.

Table 12 lists the results obtained for several additional examples (−CN, −Me, −OMe, −CHO, −SiMe₃). It was found that the effect was independent of the nature of the substituent. All sets of 4-derivatised *E/Z*-isomers demonstrated consistently equivalent antifungal activity, and this observation was not restricted to single strains, but was true for all fungi tested. The translation of this *in vitro* effect to *in vivo* efficacy, however, failed. Whereas oral administration of 40 mg/kg of the (*E*)-4-methyl compound **84** cured all animals, the same dose of the corresponding *Z*-derivative **85** was completely ineffective in the guinea pig trichophytosis model.

Structural modifications of the 1,3-enyne side chain of **2** were studied next. Introduction of a methyl group at each site of the double bond led to diminished activity (Table 13, **92, 93**).[12] The results indicated that within this region steric

Table 12 *In vitro* activity of 4-naphthalene substituted terbinafine analogues

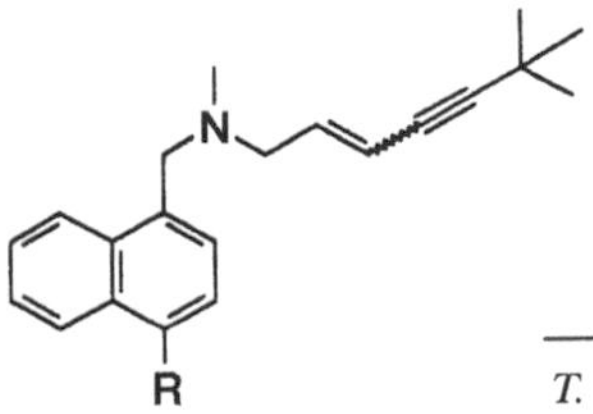

	R		T. ment.	M. can.	A. fum.	Sp. sch.	C. alb.	C. par.
					MIC [mg/L]			
2	H	E	0.003	0.006	0.8	0.4	25	0.4
32		Z	0.1	0.1	>100	3.1	>100	12.5
80	4-F	E	0.01	0.01	3.1	0.4	25	0.8
81		Z	0.02	0.02	3.1	0.8	50	3.1
82	4-CN	E	0.2	0.1	>100	>100	>100	12.5
83		Z	0.02	0.1	>100	12.5	>100	1.6
84	4-Me	E	0.02	0.05	>100	0.8	50	3.1
85		Z	0.02	0.05	12.5	3.1	50	12.5
86	4-OMe	E	0.05	0.05	100	0.8	100	12.5
87		Z	0.05	0.05	>100	1.6	>100	—
88	4-CHO	E	3.1	6.2	>100	6.2	>100	25
89		Z	3.1	3.1	>100	3.1	>100	25
90	4-SiMe$_3$	E	0.8	1.6	>100	50	>100	>100
91		Z	0.8	3.1	>100	100	>100	50

requirements of the side chain are very stringent. Substitution by fluorine at the β-position of the allylamine side chain generated the highly active derivative **94**, whereas the corresponding γ-fluoro analogue **95** showed reduced *in vitro* potency.

Decrease in the antifungal activity was also observed for compounds **96** and **97**, formally resulting from selective reduction of one of the multiple bonds. In addition, variation of the distance between the lipophilic 1-naphthalene moiety and the bulky *tert*-butylacetylene function was intensively studied by using saturated hydrocarbon spacers of different length containing the NMe group.[68] All modifications of the original spacer such as switching the position of the nitrogen atom, shortening and elongation, resulted in decreased potencies with one exception: compounds with the $-CH_2NMeCH_2CH_2-$ group between the two lipophilic domains (1-naphthalene, *tert*-butylacetylene) demonstrated high antifungal activity *in vitro*. Furthermore, the new homopropargylamine antimycotics, represented in Table 13 by compound **98**, are more potent than **2** against *Aspergillus fumigatus*.[68]

These findings suggested that the essential prerequisites for potent allylamine/homopropargylamine antimycotics are two lipophilic domains, one as a bicyclic aromatic ring system (1-naphthalene, 3-, 4-, or 7-benzo[b]thiophene) and the second being a *tert*-alkyl group, linked by a spacer of appropriate length containing

Table 13 Antifungal activity of terbinafine analogues with modified side chains

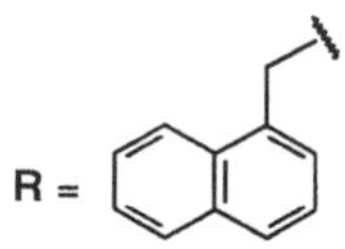

	R =	MIC [mg/L]					
		T. ment.	M. can.	A. fum.	Sp. sch.	C. alb.	C. par.
2	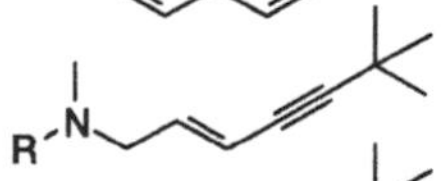	0.003	0.006	0.8	0.4	25	0.4
92	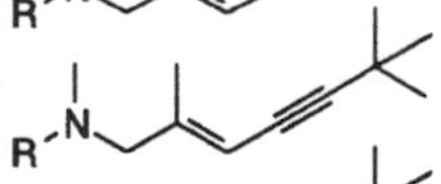	1.6	0.2	>100	6.2	>100	12.5
93	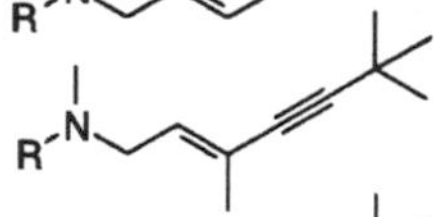	0.006	0.8	>100	6.2	>100	>100
94	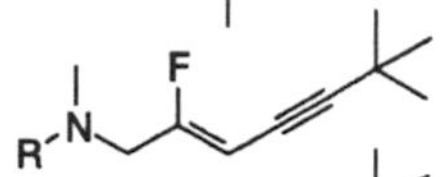	0.006	0.006	0.4	0.8	>100	0.2
95	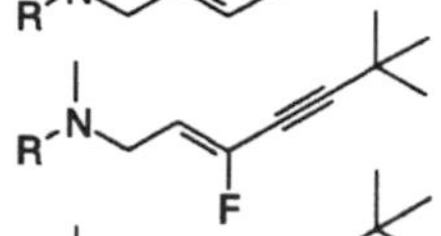	0.05	0.05	50	6.2	100	6.2
96	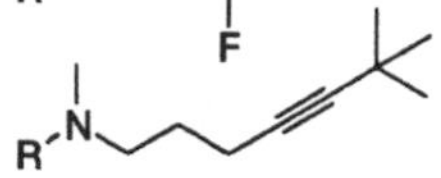	0.05	0.05	50	1.6	100	0.8
97	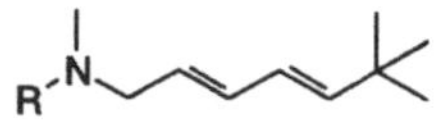	0.01	0.02	25	0.8	50	3.1
98	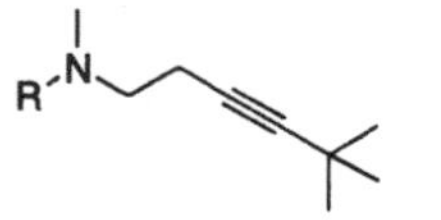	0.006	0.01	0.1	0.1	3.1	1.6

the N-methyl group at a defined position, the latter group being necessary for cell penetration, (Figure 5, page 25).[68]

5.4 SAR of the Benzylamine Antimycotics

The triple bond in terbinafine (**2**) and its homopropargylamine analogue **98** was replaced by a phenyl ring in order to maintain the steric alignment of the adjacent *tert*-butyl group, while modifying the distance between this bulky residue and the polar N-Me function. Compounds **99** and **100** (Table 14), however showed only moderate to weak antifungal activity. Next the (*E*)-enyne structural element of **2** was substituted by a phenyl ring leading to benzylamine analogue **4**.[58] In this structure the allylic double bond is fixed in the *E* configuration, which was proven to be essential for high activity within the allylamine antimycotics. Independently other research groups discovered the antifungal potential of compounds with the benzylamine structure.[69,70] Due to the importance of this novel class of squalene epoxidase inhibitors, the compounds were collectively named benzylamine antimycotics. Compound **4** displayed high antifungal potency[71] resulting in the development of **4** (butenafine) for the topical treatment of mycoses.[72] The reduced activity of the cyclohexenyl analogue **101** proved that the planar aromatic ring system of

Table 14 *In vitro* activity of side chain variations containing a phenyl ring

	R 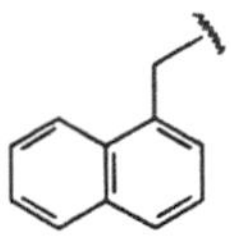 =	MIC [mg/L]					
		T. ment.	M. can.	A. fum.	Sp. sch.	C. alb.	C. par.
2	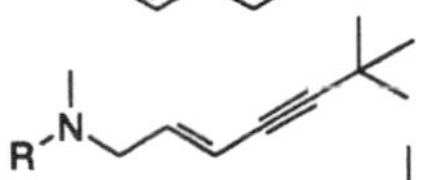	0.003	0.006	0.8	0.4	25	0.4
99	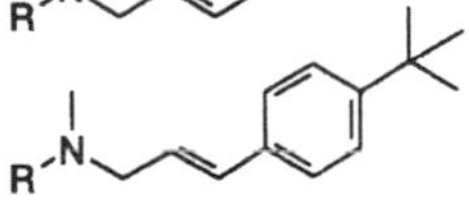	3.1	12.5	>100	>100	100 *	100
100	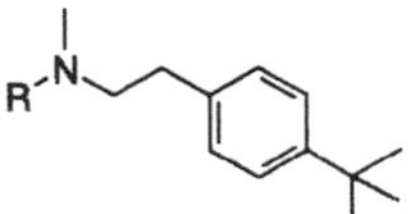	0.2	0.8	>100	6.2	50	25
4	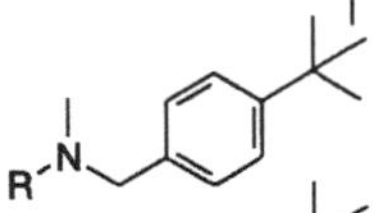	0.006	0.01	0.4	0.2	25	0.8
101	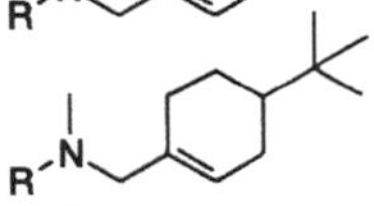	0.1	0.4	>100	3.1	100	12.5
102	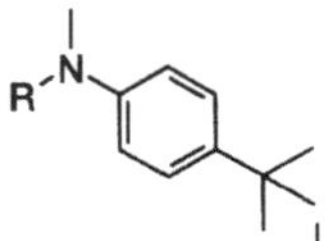	>100	100	>100	>100	>100	>100
103	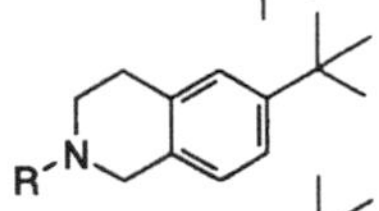	100	100	>100	>100	>100	>100
104	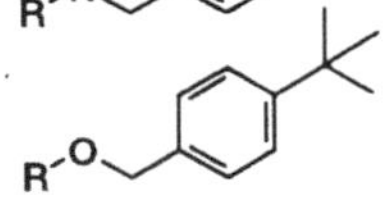	100	100	>100	>100	>100	>100

4 was essential. Direct substitution of the phenyl moiety at the NMe group (**102**) as well as incorporation of the benzylamine side chain into a tetrahydroisoquinoline structure (**103**) caused total loss of activity, indicating that flexibilty of the benzylamine side chain is important. Although the SARs of allylamine and benzylamine antimycotics were found to be rather similar, as confirmed in a combined QSAR investigation,[73] there exist some divergent findings. Thus, the benzyl ether **104** was completely inactive, whereas the corresponding allyl ether analogue (Table 6, **41**) effectively inhibited the growth of dermatophytes.

SAR studies on variation of the substituents at the benzylamine side chain revealed that substitution at the para position is absolutely required for high antifungal potency (Table 15).[74] The unsubstituted benzylamine compound **105** was totally devoid of antimycotic activity.[58] Attachment of a para methyl group (**106**) already led to substantial antifungal potency against several strains, which was further enhanced by increasing the bulkiness of the substituent. The optimum alkyl substituent proved to be the *tert*-butyl group of compound **4**.

Table 15 Antifungal activity of benzylamines **4, 5, 105–111**

	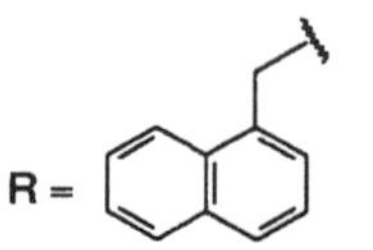	MIC [mg/L]					
R =		T. ment.	M. can.	A. fum.	Sp. sch.	C. alb.	C. par.
4	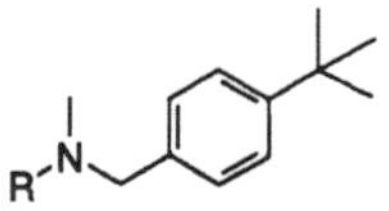	0.006	0.01	0.4	0.2	25	0.8
105	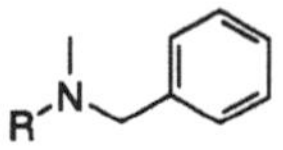	>100	>100	>100	>100	>100	>100
106	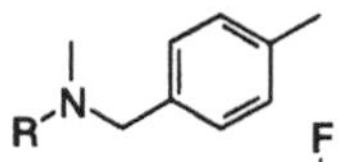	0.8	6.2	>100	6.2	>100	>100
107	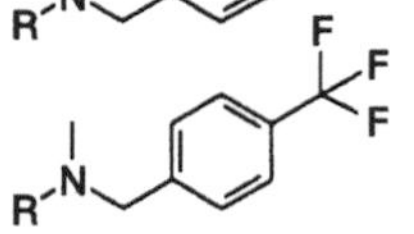	0.02	0.05	3.1	1.6	>100	6.2
108	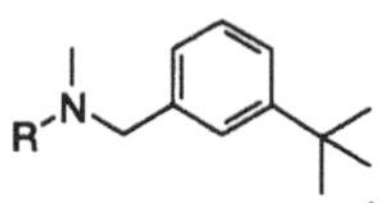	0.4	0.8	>100	1.6	>100	>100
109	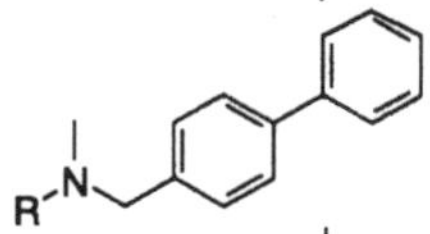	10	>100	>100	>100	>100	>100
5	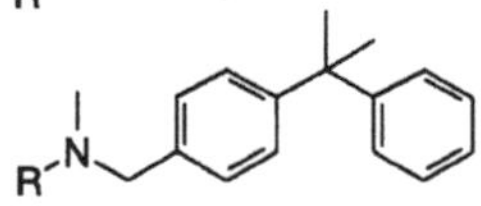	0.006	0.003	0.1	0.2	1.6	0.1
110	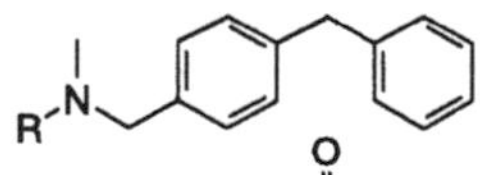	0.05	0.05	>100	1.6	>100	6.2
111	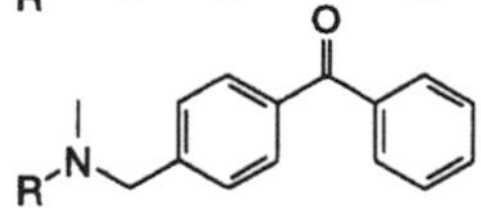	0.05	0.1	>100	>100	>100	50

The high potency of the trifluoromethyl analogue **107**, which is comparable in size to the methyl substituted derivative **106**, indicated that in addition to the steric bulkiness, electronic parameters might also influence the biological activity of benzylamine antimycotics. The poor activity of the meta substituted compound **108** documented the strict requirement for para substitution. The so far most active benzylamine antimycotic **5** (SDZ SBA 586) was identified from a series of compounds containing two phenyl groups linked by various spacers.[75] SDZ SBA 586 was significantly superior to **4** and **2** *in vitro*, in particular against *Aspergillus fumigatus* and *Candida albicans*. In agreement with these findings, **5** was also shown to be superior to **2** as an inhibitor of SE in *C. albicans*.[76]

The high efficacy of **5** was confirmed *in vivo* in several animal models including guinea pig dermatophytoses caused by *T. mentagrophytes*, *T. rubrum* or *M. canis*. In these three models ED_{50} values of 4.6, 8.7, and 4.8 mg/kg were obtained after oral administration of SDZ SBA 586 once daily for 9 days.[77] SDZ SBA 586 also demonstrated high efficacy in the auricular skin temperature test in guinea pigs. Both oral and topical treatment with **5** resulted in reduction of auricular skin temperature back into the normal physiological range.[77] Furthermore, the compound proved to be well absorbed and well tolerated after oral administration. Finally, SDZ SBA 586 was selected as a development candidate for the systemic treatment of mycoses.

The results obtained for analogues **109**, **110**, and **111** (Table 15) underlined that both electronic and steric factors are important within the class of benzylamine antimycotics. In summary, para substitution of the benzylamine side chain by a fully substituted sp^3 carbon atom is required for high antifungal activity.[75]

5.5   Summary of SAR Results

The extensive SAR investigations described above revealed that all highly active allylamine (e.g., compounds **2** and **3**), homopropargylamine (compound **98**) and benzylamine antimycotics (e.g., compounds **4** and **5**) have three structural features in common: (1) a bicyclic aromatic ring system (naphthalene, benzo[*b*]thiophene), connected through (2) a spacer containing the N-methyl function to (3) a bulky, lipophilic residue (*tert*-butyl, 2-phenyl-2-propyl group). This can be summarised by the pharmacophoric model **A** (Figure 5), in which two lipophilic domains L1 and L2 are linked by a spacer of appropriate length containing a polar centre at a defined position.[68] The general structure **A**, describing the structural requirements for all potent antifungal allylamine, homopropargylamine, and benzylamine derivatives, agrees well with a recently suggested hypothesis postulating that the allylamine antimycotics bind with relatively low affinity to two separate sites on the epoxidase (Figure 5).[16] The naphthalene ring system (=L1) might bind to the squalene binding site (=B1), and the bulky end of the side chain (=L2) to an

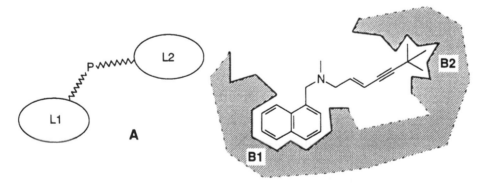

Figure 5   Comparison of the pharmacophoric model **A** for allyl/benzylamine antimycotics extracted from *in vitro* SAR studies and the hypothetical model for inhibition of squalene epoxidase by terbinafine.

Table 16 Inhibition of *Candida albicans* SE by terbinafine fragments

	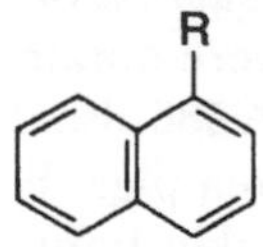 R	*Squalene epoxidase inhibition (%) at*		
		10^{-5} M	10^{-4} M	10^{-3} M
112	—H	11	16	84
113	—CH$_3$	7	0	85
114	—CH$_2$NHCH$_3$	−27	0	26
115	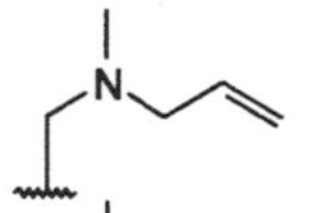	0	0	100
116	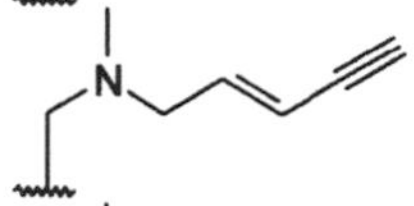	0	37	84
2	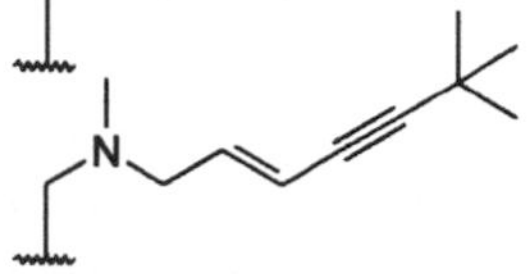	100	100	100

adjacent lipophilic pocket (=B2), resulting in high-affinity entropic binding of the inhibitory molecule to the enzyme. This concept is compatible with all the available data and can be used to interpret the SAR results obtained so far.

5.6 Structural Determination for Squalene Epoxidase Inhibition Using Terbinafine Fragments

The SAR studies presented above were primarily based on *in vitro* results, where compounds are tested for their ability to inhibit the growth of the fungal organisms. This procedure is relevant for the identification of potent antifungal agents and the selection of compounds for further investigations. Although all highly active ally-lamine and benzylamine derivatives were proven to be efficient inhibitors of fungal squalene epoxidase, the cellular antifungal data do not always reflect the intrinsic inhibitory potency of the test compounds against the enzyme. In particular, drug penetration into the fungus and/or metabolic degradation of the drug by the orga-nism may account for divergence of activities between the cellular and the enzymatic test systems.

Therefore, a separate study was performed to determine the minimal structural requirements for inhibition of the *Candida albicans* microsomal squalene epoxidase using fragments of terbinafine (**2**).[28] The (*E*)-1,3-enyne side chain alone had no inhibitory effect on the squalene epoxidase at concentrations up to 1 mM. In contrast, naphthalene (**112**) and its 1-methyl derivative (**113**) showed weak inhibition of the enzyme (Table 16) supporting the concept of entropic binding in which the naphthalene ring structure binds weakly to the squalene binding site of SE.[16]

Addition of the polar methylamino group (**114**) caused loss of this activity, which could be explained by the inability of **114** to bind to the enzyme due to the strong polarisation of the molecule (hydrophobic naphthalene moiety, hydrophilic amine functionality). Weak inhibitory activity at the enzymatic level was regained with the tertiary amine derivatives **115** and **116**. In the routine antifungal screening all these terbinafine fragments (**112–116**) were inactive up to 100 mg/L, except a faint growth inhibitory effect of compound **116** against dermatophytes. Addition of the terminal *tert*-butyl group to produce **2** finally resulted in a 10000-fold increase in potency (IC$_{50}$ for **2** = 0.03 μM). This enormous enhancement in enzymatic activity observed only with the full terbinafine structure again supported the two-site model for epoxidase inhibition by allylamine antimycotics shown in Figure 5.

In general, the SARs for fungal squalene epoxidase inhibition and antifungal activity coincide well. The most divergent results were observed for analogues of terbinafine in which the allylamine nitrogen was replaced by carbon (Tables 6 and 7), probably due to insufficient penetration of the compounds through the fungal cell envelope.[28] Further results on SE inhibition by allylamine antimycotics and selectivity for squalene epoxidases of different origins are discussed in section 9 of this article.

6.   SYNTHETIC ASPECTS OF THE ALLYLAMINE ANTIMYCOTICS

Terbinafine (Figure 1, **2**) was the first pharmaceutical agent containing the (*E*)-1,3-enyne structural element. Hence, several methods have been developed for the preparation of terbinafine type allylamine antimycotics. The diversity of the procedures provided high flexibility for the synthesis of the target molecules. Thus, the most valuable starting material could be introduced at a late stage of the reaction sequence. Since the synthetic methods for the preparation of terbinafine analogues have been reviewed in 1991,[58] only the most interesting and recent results are presented below.

In general, the *E*-isomers of the allylamine derivatives display substantially higher antifungal activity than the corresponding *Z*-isomers (Table 3). Therefore, the (*E*/ *Z*)-product mixtures, which were produced by some synthetic methods, had to be separated into the single isomers. In many cases the purification was successfully performed by chromatography on silica gel and/or by selective crystallisation of the hydrochloride salt of the (*E*)-isomer. In addition several procedures for the selective synthesis of pure (*E*)-products were developed.

6.1   Synthesis of (*E*)-Allylamines from Propargylamines by DiBAH reduction

Diisobutylaluminum hydride (DiBAH) is known to add to alkynes with high *Z*-stereoselectivity producing (*Z*)-alkenes after hydrolysis (Scheme 2). We found that, in contrast, the reduction of tertiary propargylamines with DiBAH selectively yields (*E*)-allylamines.[78,79] The reversed selectivity of the hydroalumination reaction is attributed to a neighbouring effect of the amino function.

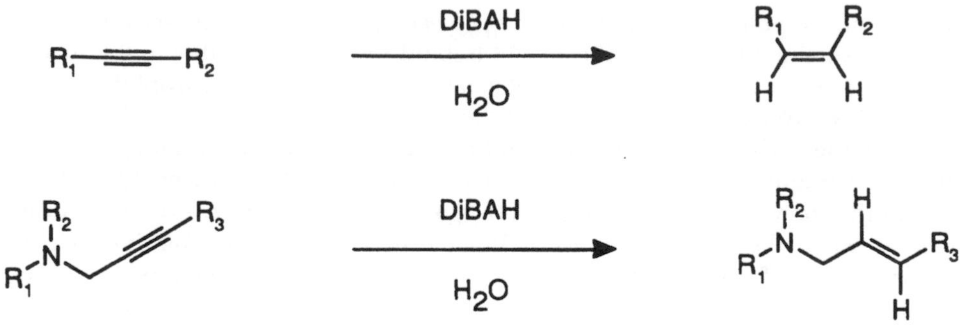

Scheme 2   DiBAH reduction of alkynes and propargylamines.

The results of a study comparing the new method with the known *trans*-reduction of α-hydroxyalkynes to (E)-allylalcohols by lithium aluminum hydride[80] are summarised in Scheme 3. As expected, the treatment of compound **117**, which contained both a propargylamine and a propargylalcohol function, with lithium aluminum hydride resulted in (E)-hydroalumination of the α-hydroxyalkyne moiety only, generating product **118** (Scheme 3, R = 1-naphthalene). In contrast, conversion of compound **117** with DiBAH gave (E)-allylamine **119** (R = 1-naphthalene) in good yield without affecting the propargylalcohol function.[79] Consequently, the *trans*-hydroalumination of tertiary propargylamines represented a novel and selective access to (E)-allylamines. Furthermore, the method has been applied successfully for the preparation of terbinafine type allylamine antimycotics [(E)-2-alken-4-ynylamines] from the corresponding tertiary 2,4-alkadiynylamine precursors.[53]

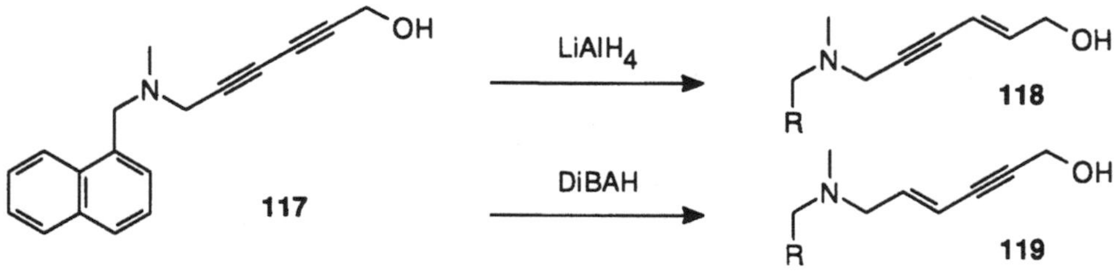

Scheme 3   LiAlH$_4$ versus DiBAH reduction of bifunctionalised compound **117**.

6.2   Selective Synthesis of (E)-2-Alken-4-ynylamines by Pd(0)-Catalysed Coupling Reactions

Pd(0)-catalysed coupling of (E)-vinyl halides with terminal alkynes is a well established method for the synthesis of pure (E)-1,3-enyne products in high yields.[81] This procedure proved to be efficient in particular for the preparation of allylamine derivatives bearing functional groups.[58] As a typical example, the synthesis of ester

42, the precursor in the preparation of the terbinafine metabolites **40** and **41** (Table 5), is described in more detail (Scheme 4).

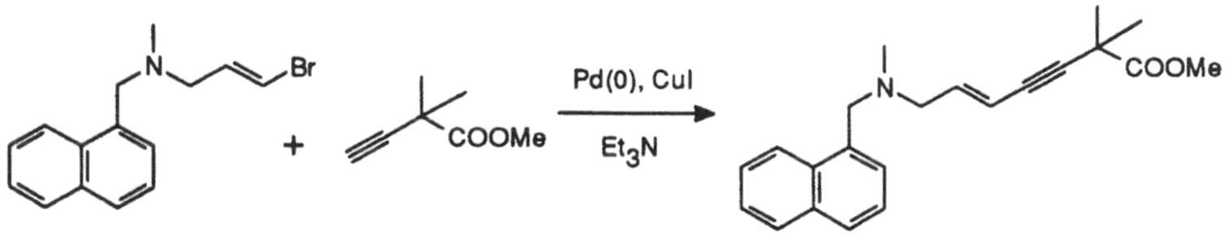

Scheme 4   Synthesis of terbinafine ester derivative **42**.

(*E*)-Vinyl bromide **120** (prepared either by alkylation of N-methyl-1-naphthalene-methanamine (**114**) with 1,3-dibromopropene and separation of the (*E*)-isomer by chromatography, or by hydrozirconation of the corresponding tertiary propargylamine, followed by halo-metal exchange) was allowed to react with methyl 2,2-dimethyl-3-butynoate (**121**) in the presence of a Pd(0)-catalyst, cuprous iodide and triethylamine to give product **42** in high yield with complete retention of the (*E*)-stereochemistry. Starting from (*E*)-3-halo-2-propenol the Pd(0)-catalysed coupling reaction was also used successfully for the synthesis of pure (*E*)-1,3-enyne side chain alcohols, which were subsequently converted into the corresponding bromides or mesylates and reacted with secondary amines to generate allylamine derivatives.[58]

6.3   Selective Preparation of Secondary (*E*)-Allylamines

The allylamine antimycotics with bicyclic aromatic naphthalene equivalents (e.g. benzo[*b*]thiophenes) were generally synthesised by alkylation of the secondary amine **123** (Scheme 5) with the appropriate valuable aromatic bromomethyl compounds.[63] In order to maximise the efficiency of this route, it was desirable to use the pure (*E*)-isomer of the side chain amine. Usually (*E*)-**123** was synthesised by mono-alkylation of methylamine with the easily accessible (*E*/*Z*)-mixture of 1-bromo-6,6-dimethyl-2-hepten-4-yne (**122**) followed by selective crystallisation of its hydrochloride salt. This proceeded only in poor to moderate yields.[63]

Using the procedure shown in Scheme 5, the pure allylamine side chain (*E*)-**123** was prepared in high yields.[82] The secondary amine **124** was treated with allylic bromide **122** (*E*/*Z* = 3/1) to afford a mixture of tertiary amines with the same stereoisomer ratio. The (*E*)-isomer **125** was purified by chromatography on silica gel. The 2,4-dimethoxybenzyl protecting group in **125** was removed by treatment with trifluoroacetic anhydride. The regioselective cleavage of the benzylic C-N bond generated trifluoroacetamide **126**, which was easily converted into the secondary amine (*E*)-**123**. The procedure proved to be a valuable general method for the synthesis of secondary amines, in particular those containing olefinic or acetylenic groups, where deprotection cannot be easily achieved by hydrogenation. Initial

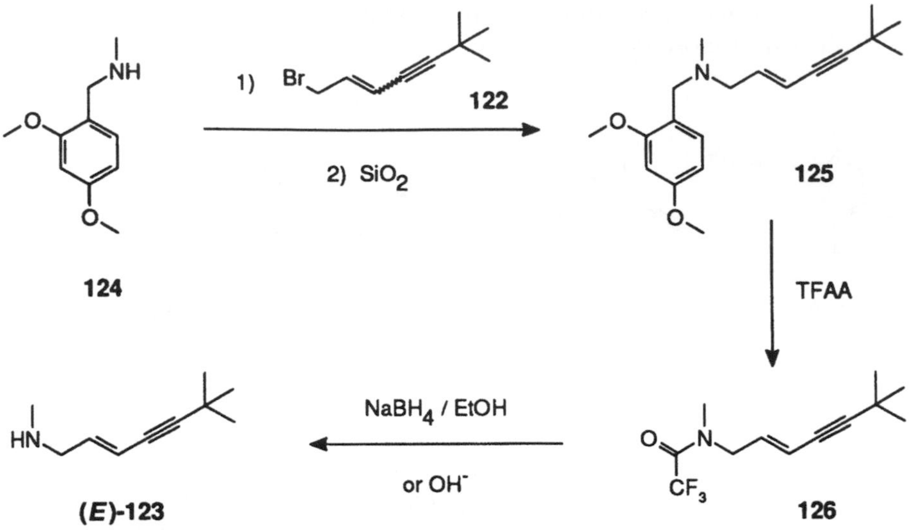

Scheme 5   Selective synthesis of pure allylamine side chain (*E*)-123.

experiments demonstrated also the potential usefulness of the 2,4-di(6-tri)metho-xybenzyl function as new protection group for secondary amines.[82]

6.4   Synthesis of 2-Substituted Homopropargylalcohols

Variations of the spacer between the 1-naphthalene moiety and the *tert*-butylacetylene group of terbinafine led to the discovery of the homopropargylamine antimycotics (Table 13).[68] For the determination of SAR within this type of antifungals, substituted homopropargylalcohols were required as synthetic intermediates.

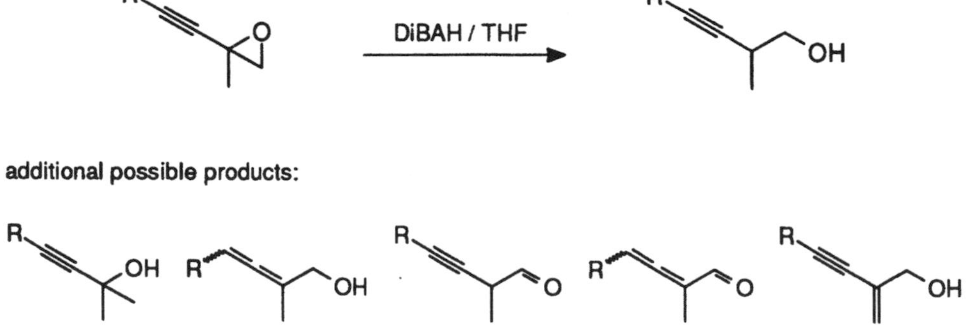

additional possible products:

Scheme 6   Synthesis of 2-substituted homopropargylalcohols by DiBAH reduction of acetylenic epoxides and additional possible products of oxirane treatment with various reducing agents.

For the preparation of 2-substituted homopropargylic alcohols, a new method had to be developed. We envisaged that regioselective reductive ring opening of α-substituted α-alkynyl epoxides could give ready access to the desired intermediates. The use of various combinations of reducing agents and solvents generally resulted in unwanted products (Scheme 6) or mixtures thereof deriving from reversed regioselectivity, propargyl/allene isomerisation and rearrangements. Finally, treatment of the oxiranes with diisobutylaluminum hydride in tetrahydrofuran produced the desired 2-substituted 2-butyn-1-ols in high yields (Scheme 6).[83] The unique high regioselectivity and the low degree of isomerisation to allenic products make our new procedure an efficient method for the synthesis of 2-substituted homopropargylalcohols.

7. SAR OF THIOCARBAMATE ANTIMYCOTICS

The aryl thiocarbamates were obtained as the chemical decomposition products of corresponding aryl thiocarbamoylthiocarbonates, which were synthesised to evaluate their antibacterial activity against pathogenic microorganisms.[84] The thiocarbamates showed no antibacterial potential, but exhibited a strong and selective antifungal activity *in vitro* against *Trichophyton* spp. Further examination revealed that 2-naphthyl-*N*-aryl-methylthiocarbamates, which were named "naphthiomates", had a specific therapeutic effect against experimental trichophytosis in animals.[84]

Although the antifungal potential of the thiocarbamates had already been reported in 1962 and the representative tolnaftate (Figure 1, 6) in the meantime had been used for many years clinically for topical treatment of skin infections, it took a long time until their mode of action was definitively elucidated. Several modes of action were proposed for the antifungal activity of naphthiomates such as inhibition of RNA and DNA synthesis[85] and cell wall formation.[86] But finally, it was found that the thiocarbamate antimycotics block ergosterol biosynthesis in several fungi by inhibiting the squalene epoxidase.[15,39,40,87]

Tolnaftate (Figure 6, 6) was already discovered during initial SAR studies within the antifungal aryl thiocarbamates.[84] Its low toxicity to animals and lack of skin irritation potential led to the development of 6 for the topical treatment of skin mycoses. Detailed SAR investigations[88,89] followed and proved that the original thiocarbamate structural element was essential for high activity against *Trichophyton*. Both the corresponding carbamate (Figure 6, 127) and the dithiocarbamate (128) analogues showed substantially reduced activities *in vitro*. Systematic replacement of the naphthalene moiety in compound 6 by the benzonorbornane system led to the discovery of tolciclate (7),[90] which displayed comparable potency against dermatophytes *in vitro*, but was slightly more effective than 6 in several animal models of dermatophyte infection.[91] Within the antifungal allylamines, the same structural modification caused a drastic drop in activity. SAR studies on tolnaftate and tolciclate analogues furthermore revealed that interchanging the position of oxygen and sulphur (129) and introducing monocyclic aromatic substituents at the oxygen (130) resulted in loss of activity. The N-Me structural element in the

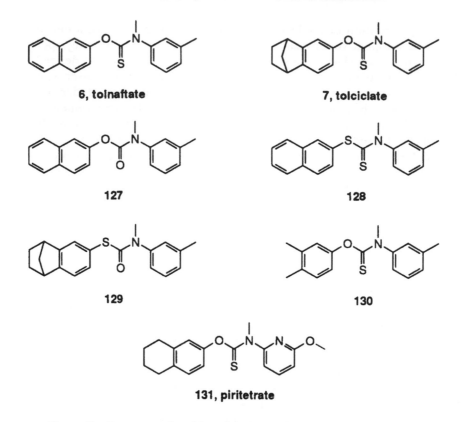

6, tolnaftate

7, tolciclate

127

128

129

130

131, piritetrate

Figure 6   Representative thiocarbamate antimycotics and analogues.

thiocarbamate function was also found to be essential for activity. Analogues with a secondary nitrogen or with higher N-alkyl substituents were substantially less potent or inactive. For a long time the 3-methylphenyl group seemed to be the optimal substituent at the nitrogen for high activity. Replacement by 6-methoxy-2-pyridine, however, resulted in the discovery of piritetrate (Figure 6, **131**), which exhibited *in vitro* up to 10-fold stronger antidermatophytic activity than **6** and showed a broader antifungal spectrum as compared with other antifungal thio-carbamates.[92] The enhanced activity of **131** was confirmed in *in vivo* experiments and is likely due to the higher inhibitory potency of **131** at the enzymatic level.[41] Interestingly, thiocarbamate prototypes **6**, **7**, and **131** were found to inhibit the squalene epoxidase from *Candida albicans* potently, but to be inactive against *Candida* species *in vitro*.[15,40,41] This was explained by a poor drug penetration through the *Candida* cell envelope.

Summarising the SAR findings of the thiocarbamate antimycotics, the following structural features were found to be essential: (1) a bi(tri)cyclic ring system with (2) the ring linked at the β-position to the oxygen of the thiocarbamate element being aromatic, (3) the nitrogen of which has to be substituted by a methyl and a 3-substituted aryl group.

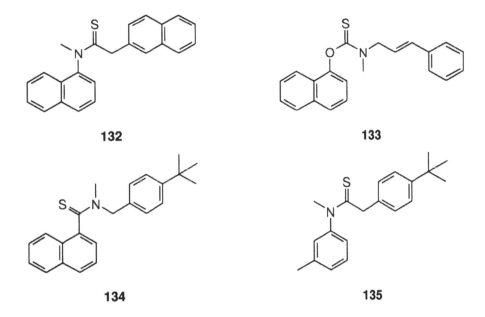

Figure 7 Hybrid structures of allyl/benzylamine and thiocarbamate antimycotics.

Although some structural and SAR related similarities to the allylamine and benzylamine antimycotics could be suspected, so far no common model for both classes of antifungal agents has been advanced. The basic differences between the two compound classes were reflected by the finding that compounds **132–135** (Figure 7), which could be regarded as hybrid structures containing features of both types of antifungal agents, were inactive against all strains tested.

With respect to the biological activity, two major differences between the allyl/benzylamine and thiocarbamate derivatives exist: (1) the allyl/benzylamine antimycotics have a broad spectrum of antifungal activity, whereas the thiocarbamates are almost exclusively antidermatophytic agents (Table 17); (2) within the allyl/benzylamine antimycotics, a number of drugs for systemic treatment of fungal infections have been identified and developed (e.g. terbinafine, SDZ SBA 586), while the thiocarbamates are limited to topical use.

Table 17 *In vitro* antifungal activity (MIC [mg/L]) of thiocarbamates **6** and **7** in comparison with terbinafine **2**

	T. ment.	M. canis	A. fum.	Sp. sch.	C. alb.	C. par
2, terbinafine	0.003	0.006	0.8	0.4	25	0.4
6, tolnaftate	0.05	0.05	>100	>100	>100	>100
7, tolciclate	0.05	0.05	>100	>100	>100	>100

8. MEDICINAL CHEMISTRY OF MAMMALIAN SE INHIBITORS

8.1 SAR of Squalene Analogues

Due to the existence of highly active and selective fungal squalene epoxidase inhibitors (the allylamine, the homopropargylamine, the benzylamine, and the thiocarbamate antimycotics), squalene analogues have been of no importance as inhibitors of the fungal enzyme. In contrast, substrate analogues were the only known inhibitors of the mammalian squalene epoxidase until the recent discovery of a subclass of allylamine derivatives without antifungal activity, but which potently and selectively inhibit this enzyme.[17]

A number of squalene analogues were synthesised by introduction of functional groups in the squalene molecule and tested for their inhibitory potential against squalene epoxidase. Trisnorsqualene alcohol (Table 18, **8**) was found to be a potent inhibitor of vertebrate SE with an IC_{50} of 4 μM for pig liver SE.[93] SAR studies revealed that the primary alcohol function was essential for activity. As shown in Table 18, methylation (**136**) as well as replacement of the hydroxy group by the amino (**137**) or ethylamino function (**138**) caused a drastic decrease in inhibitory potency.[93,94] The absence of classical competitive inhibition, however, indicated a more complex mode of action for compound **8**.[93]

Table 18 Inhibition of mammalian SE by substrate analogues (adapted from Abe *et al.*[6])

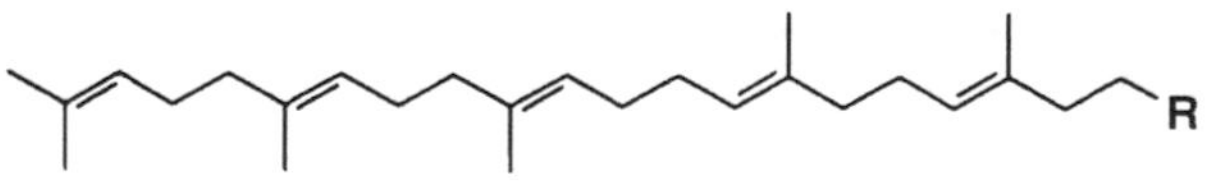

	R	IC_{50} [μM]
8	CH_2OH	4.0
136	CH_2OMe	300.0
137	CH_2NH_2	200.0
138	CH_2NHEt	200.0
9	$CH_2NH\text{-}c\text{-}Pr$	2.0
139	$CH_2NMe\text{-}c\text{-}Pr$	100.0
10	$CH=CF_2$	5.4 (rat liver)

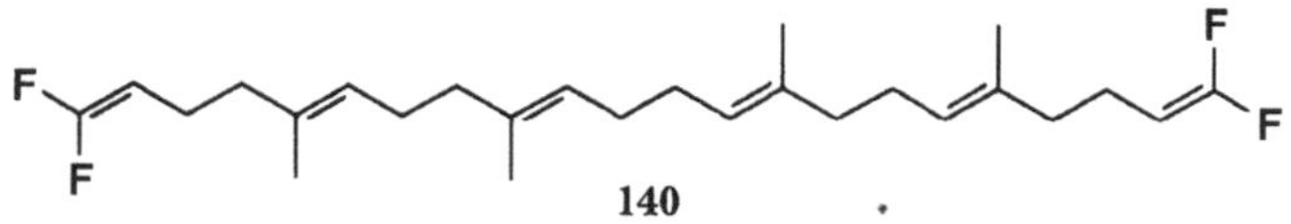

4.5 (rat liver)

140

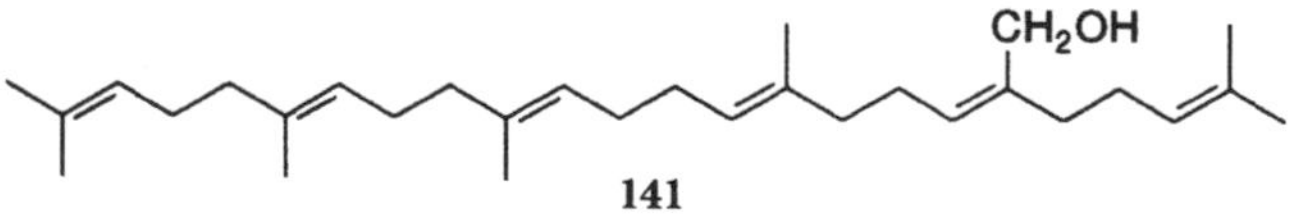

16.0

141

The high sensitivity of the enzyme to minor structural modification is reflected by the results obtained for the *N*-cyclopropylamino derivative of trisnorsqualene (**9**).[94] This compound was the first highly selective, slow tight-binding inhibitor of mammalian SE, but detailed investigations indicated that it was not a mechanism-based inactivator of the enzyme. Chemical derivatisation of compound **9** only yielded analogues with reduced activity, such as derivative **139** with an extra *N*-methyl group. Compound **10** with a terminal difluoroolefin group and its bifunctional analogue **140** also inhibited rat liver SE with IC_{50} values in the μM range.[95] They were shown to be time-dependent inhibitors indicating mechanism-based enzyme inactivation. Analogue **10** was reported to lower cholesterol and lanosterol levels in mice after oral treatment.[96] Concomitant squalene accumulation was observed furnishing evidence that the agent really interfered with the sterol biosynthesis at the SE stage. 2-Fluoromethyl and 2,2-difluoromethyl analogues of squalene were also synthesised, but data on vertebrate SE inhibition were not disclosed.[97]

Further SAR studies included squalene analogues with acetylene, allene, and diene functionalities, but none of these compounds showed substantial potency.[98,99] The first competitive inhibitors of SE were identified from a series of C-26 modified squalene derivatives. 26-Hydroxysqualene (Table 18, **141**) was shown to act as a substrate for SE yielding a 3/1 mixture of 2,3-epoxide and 22,23-epoxide.[100] Further variations at position 26, however, resulted in much weaker inhibitory potency of pig liver SE.[6] It was concluded that at least two of the following characteristics are required for pig liver SE inhibition by squalene analogues: (1) a hydrophobic moiety mimicking the terminal isopropylidene group; (2) an overall size and geometry similar to the substrate; (3) an unpolarised, but reactive unsaturated bond at the position normally epoxidised; (4) a hydroxyl group in the isopropylidene region.[99] In summary, chemical modification of squalene led to two types of mammalian SE inhibitors. The difluoroolefin analogue **10** represents a mechanism-based enzyme inhibitor, whereas 26-hydroxysqualene (**141**) is a competitive inhibitor. *In vivo* efficacy was demonstrated for the first type of SE inhibitors.

8.2 SAR of Allylamine Derivatives

The allylamine antimycotics consistently show high selectivity for the fungal SE with a difference between fungal and mammalian enzyme of several orders of magnitude. Based on the antifungal allylamine terbinafine (**2**), a research group at Banyu Pharmaceutical Co. discovered analogues with inhibitory potential for mammalian SE.[17] The first potent inhibitor of this type was compound **142** (Table 19) in which a 3-benzyloxyphenyl group replaces the naphthalene moiety in **2**.[101] Whereas terbinafine showed an IC_{50} of 130 μM for rat SE in their test system, compound **142** demonstrated about 100-fold higher activity. Moreover, comparable potency of **142** was found in both the enzymatic and the cellular test system using Hep G2 cells.[101] Table 19 summarises the results from initial SAR investigations on structurally related substances.

Variation of the position of the benzyloxy substituent led to complete loss of activity as shown with compound **143**. The same negative effect was observed when

Table 19 Enzymatic and cellular mammalian SE inhibition by allylamine derivatives (adapted from Iwasawa et al.[101])

		Enzyme, rat IC_{50} [μM]	Cell, Hep G2 IC_{50} [μM]
2	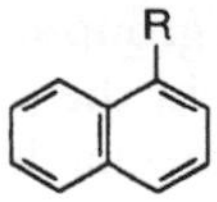	130	—
142	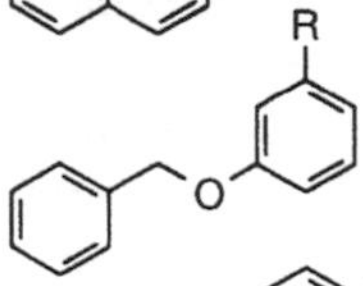	1.4	1.5
143	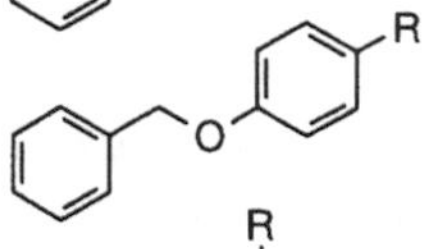	198	>30
144	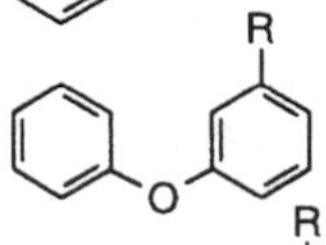	129	>30
145	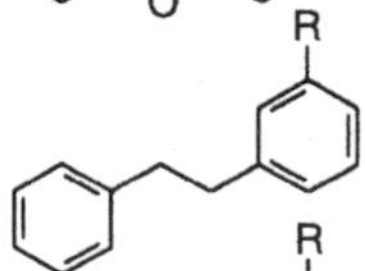	3.6	1.8
146	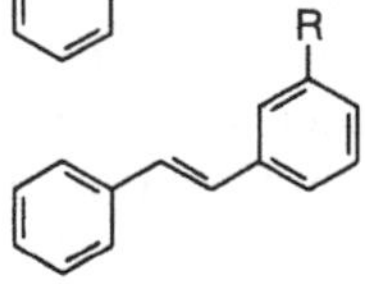	0.32	0.32

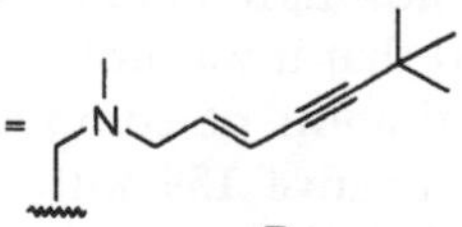

the benzyloxy substituent in **142** was replaced by the phenoxy group yielding the diphenyl ether derivative **144**. These findings indicated that the appropriate alignment of and distance between the two phenyl rings in this region were very important for high potency, which was confirmed by the results obtained for compounds **145** and **146**. These carbon-analogues of the 3-benzyloxyphenyl allylamine **142** displayed good activity in the enzymatic and the cellular test system. The (E)-stilbene derivative **146** was the most active compound of this series, whereas its (Z)-isomer was reported to be much less effective.[101]

SAR studies on analogues additionally substituted at the benzyloxy group revealed that at position 2 only small substituents such as the methyl group (**147**, Table 20) were tolerated, whereas substitution at position 4 generally led to decreased activity (data not shown). Introduction of a methyl or methoxy group at position 3 (**148**, **149**) did not substantially affect the affinity for the enzyme, but a tendency for lower cellular compared to enzymatic activity was observed. The first substance showing substantially increased potency relative to **142** was the 3-cyanobenzyloxy derivative **150** (Table 20). Extensive investigations disclosed that the activity could be further

Table 20 Activity of substituted 3-benzyloxyphenyl allylamine derivatives (adapted from Iwasawa et al.[101])

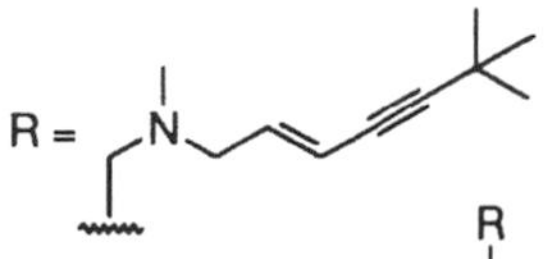

		Enzyme, rat IC_{50} [μM]	Cell, Hep G2 IC_{50} [μM]
142	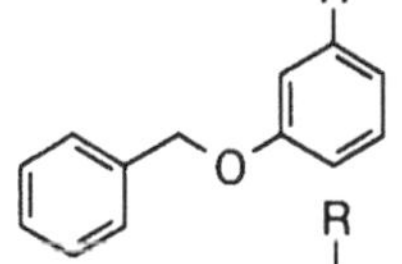	1.4	1.5
147	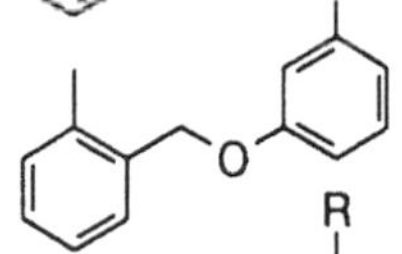	0.56	0.3
148	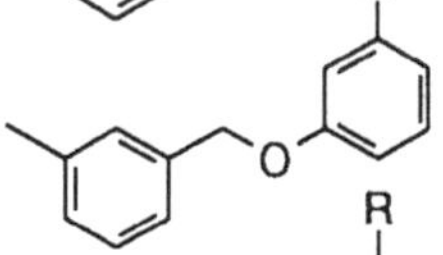	0.66	3.6
149	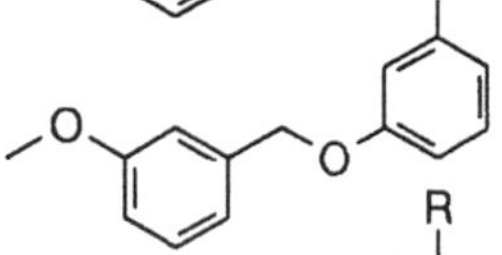	1.4	3.2
150	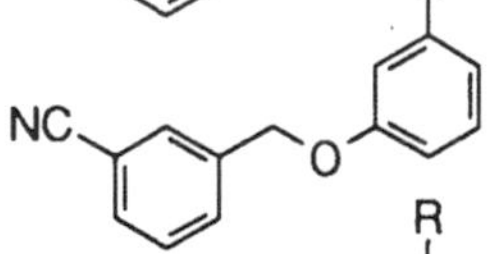	0.5	0.1
151	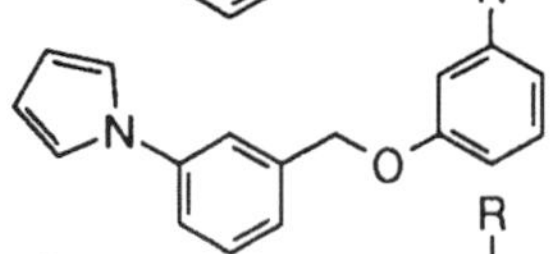	0.016	0.07
152	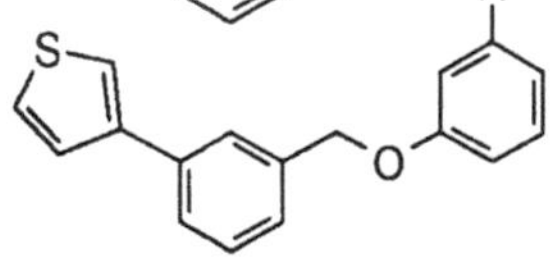	0.011	0.003

improved by addition of 5-membered heterocyclic ring systems at position 3 of the benzyloxy residue. For example, the pyrrolyl compound **151** showed about 100-fold higher inhibitory potency for rat liver SE and 20-fold increased activity in the cellular assay relative to the unsubstituted lead compound **142**. Out of this series of heterocyclic substituents, the 3-thienyl group produced the highest gain in cellular activity (**152**, Table 20).[101]

Studies of modifications of the central amino group demonstrated that the tertiary amino group was necessary for high activity. The corresponding secondary amines were only weak inhibitors of rat SE, whereas ether derivatives exhibited no significant activity in the enzymatic assay (data not shown). Variation of the N-alkyl residue revealed, unlike methyl for the antifungal allylamines, ethyl and propyl to

Table 21 Activity of potent mammalian SE inhibitors **11** and **153** (adapted from Iwasawa *et al.*[101])

		Enzyme, rat IC_{50} [μM]	Cell, Hep G2 IC_{50} [μM]
153	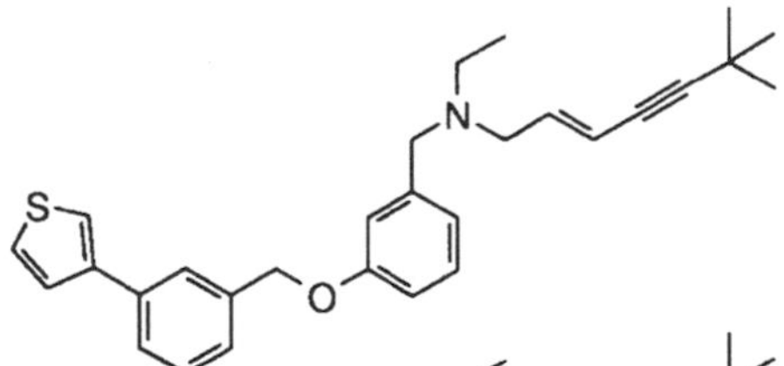	0.0078	0.002
11	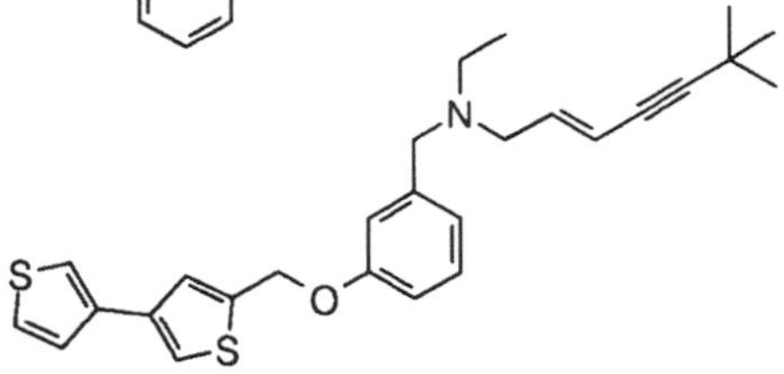	0.0044	0.00075

be the best substituents. Thus, compound **153** (Table 21) was even slightly superior to the corresponding *N*-methyl derivative **152** (Table 20). Finally, testing of some bis-heterocyclic analogues resulted in the discovery of NB-598 (**11**, Table 21). This compound featuring a 3,3-bithienyl moiety was selected for further evaluation[101] as the most effective mammalian SE completely lacking antifungal activity.

NB-598 (**11**) was shown to be a competitive inhibitor of SE from Hep G2 cells and to effectively reduce cholesterol synthesis from [^{14}C]-acetate, causing intracellular accumulation of [^{14}C]-squalene in those cells.[17] The synthesis of free fatty acids, phospholipids and triacylglycerol was not affected. This profile of activity was confirmed in animal experiments with rats and dogs.[17,102] Multiple administration of oral doses to dogs significantly decreased serum total and low density lipoprotein cholesterol levels. Furthermore, NB-598 produced an increase of both HMG-CoA reductase activity and LDL receptor activity in Hep G2 cells.[18] Studies on lipid metabolism in Hep G2 cells suggested that NB-598 inhibits not only cholesterol synthesis, but also the secretion of lipids.[103]

Banyu researchers reported that replacement of the 6,6-dimethyl-2-hepten-4-ynyl side chain in compound **142** by the cinnamyl (naftifine-type) or the 4-*tert*-butylbenzyl group (butenafine-type) resulted in drastic loss of activity.[101] SAR studies on NB-598 analogues performed at the Sandoz Research Institutes in East Hanover and Vienna revealed that this finding is not generally valid for all allylamine-type mammalian SE inhibitors. As shown in Table 22, the corresponding 4-*tert*-butylbenzylamine (**155**, **158**) and 4-benzylbenzylamine (**156**, **159**) analogues of the potent oxazolyl and pyrrolyl substituted mammalian SE inhibitors **154** and **157** indeed did not show relevant activity against Hep G2 microsomal squalene epoxidase. In contrast, the benzylamine derivatives **160** and **12** (SDZ 281-915)[104] containing the 3,3-bithienyl structural feature of NB-598 showed strong inhibition of Hep G2 SE. Their IC_{50} values of 42 and 45 nM, respectively, were almost as low as that determined for NB-598 (**11**, Table 22).

Table 22 Comparison of highly active allylamine SE inhibitors with their benzyl- and benzylbenzylamine analogues

Enzyme, Hep G2 IC_{50} [μM] (compound)	R 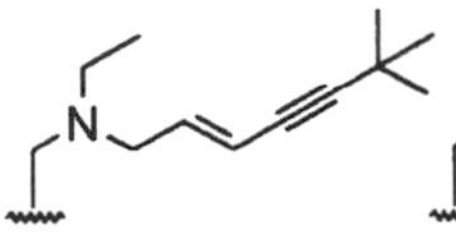 =	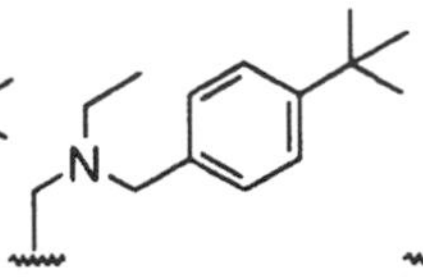	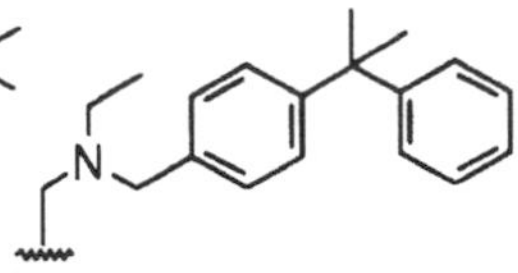
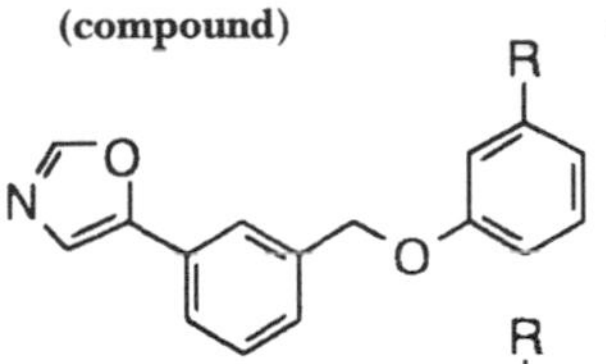 (oxazole structure)	0.38 (154)	7.9 (155)	14.8 (156)
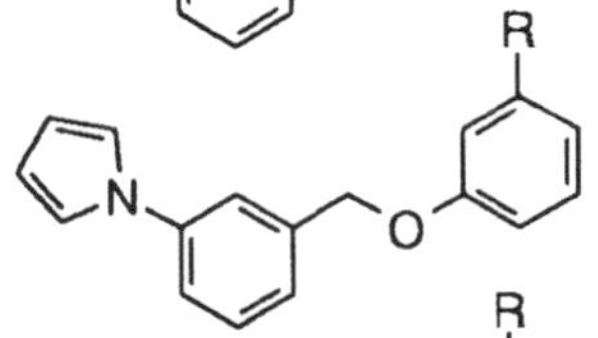 (pyrrole structure)	0.043 (157)	>10 (158)	>10 (159)
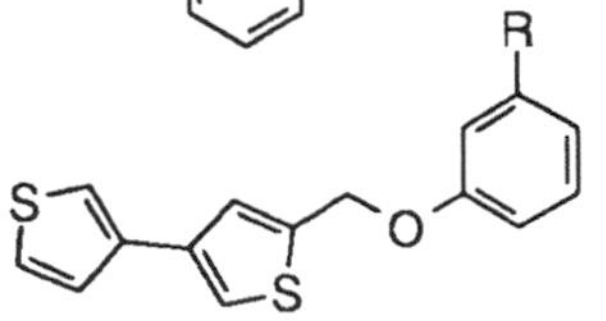 (bithiophene structure)	0.017 (11)	0.042 (160)	0.045 (12)

Additional SAR studies demonstrated that the high mammalian SE inhibitory potency of NB-598 related structures could be maintained when the internal aromatic groups were replaced by other spacers generating compounds with altered physicochemical properties. Due to the extremely high lipophilicity of NB-598 and SDZ 281-915 analogues, this new generation of "hydrophilic" mammalian SE inhibitors might provide improved biological profiles as reported for compound **161** (Figure 8, FW-1045) by Banyu.[105] Recently, compound **162**[106] was disclosed by

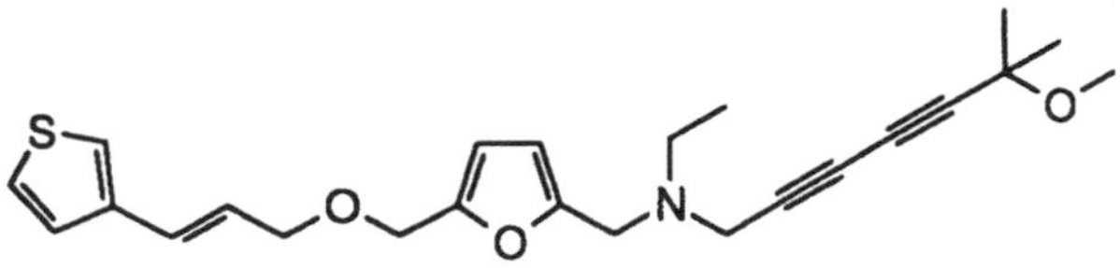

161, FW-1045

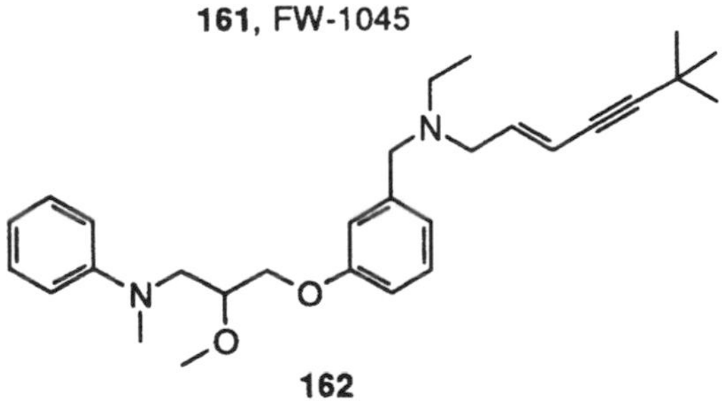

162

Figure 8 Structures of recently disclosed mammalian SE inhibitors

Yamanouchi claiming slightly superior potency in rat liver SE inhibition, relative to NB-598.

9.   SELECTIVITY OF SQUALENE EPOXIDASE INHIBITORS

Selectivity of drug action is a very broadly understood term including selectivity within a superfamily of enzymes, enzyme subtype selectivity, species selectivity, tissue selectivity, etc. It is a common finding that drugs usually have more than one action and are selective rather than specific.

In the case of SE inhibitors, the primary questions were: Is SE the sole target of the compounds, or do they also affect other steps of the sterol biosynthesis pathway? Can selectivity for SE of different species be achieved? Answering the first question, the allylamine antimycotics were shown to be specific inhibitors of fungal SE by using a range of cell-free tests using different labelled sterol precursors.[30,36] The second question was particularly important for the antifungal SE inhibitors, because high selectivity for the fungal enzyme relative to the human SE was required for safe systemic treatment. Species selectivity of inhibitors in cellular and *in vivo* systems, however, cannot be attributed only to selectivity at the enzymatic level, as other parameters, such as drug penetration or pharmacokinetic behaviour, might strongly influence the activity. For example, the thiocarbamate antimycotics tolnaftate and tolciclate (Figure 6, **6** and **7**) are potent inhibitors of microsomal SE from *Candida albicans*,[15,87] but are inactive against *Candida* species *in vitro* (Table 17). Thus, the generally selective activity of the thiocarbamates against dermatophytes was found to be the consequence of their poor penetration through the cell envelope of other fungi.[15,29]

Both the thiocarbamate and the allylamine antimycotics proved to be highly selective for the fungal enzyme as well as for inhibition of ergosterol versus cholesterol biosynthesis (Table 23). However, in a different therapeutic area, there is also a medical need for cholesterol lowering agents to treat hypercholesterolemia in

Table 23   Selective inhibition of fungal and mammalian squalene epoxidase and *C. albicans* sterol biosynthesis (cell free test) (IC_{50} [µM], n.t. = not tested, * values from Iwasawa *et al.*[101])

Compound	Candida albicans		Rat Liver epoxidase	Hep G2 epoxidase
	epoxidase	cell-free		
1 naftifine	1.1	0.59	144	n.t.
2 terbinafine	0.03	0.03	77	32
3 SDZ 87-469	0.01	0.007	43	n.t.
6 tolnaftate	1.04	0.42	215	n.t.
7 tolciclate	0.12	0.03	145	n.t.
11 NB-598	n.t.	34	0.0044*	0.00075*

patients. Inhibition of SE seemed to be an attractive target to interfere with human cholesterol biosynthesis, since side effects resulting from alteration of the production of other key biological substances like dolichol and ubiquinone could be avoided. Based on the antifungal allylamine SE inhibitors, a medicinal chemistry programme led to the identification of allylamine analogues with high inhibitory potency against mammalian SE. These compounds (e.g. NB-598; Figure 2, **11**), derived from the antimycotic terbinafine (Figure 1, **2**) by modification of the bicyclic aromatic ring system, completely lack antifungal activity and display a selectivity profile which is exactly the reverse of that of the antimycotic allylamines. Examples of the divergent activities between antifungal and cholesterol-lowering SE inhibitors in microsomal enzyme preparations of different species are shown in Table 23.

Thus, medicinal chemistry of allylamine derivatives led to the identification of highly active and selective inhibitors of both the fungal and the mammalian SE.

10. CONCLUSION

Allylamine derivatives constitute the most important class of squalene epoxidase inhibitors. A first group of compounds is highly selective for the fungal enzyme and used safely and successfully in the topical and systemic therapy of mycoses. A second group selectively inhibits mammalian squalene epoxidase, and appears to have a promising potential for the treatment of hypercholesterolemia. Comparison of the SAR of the two types of allylamine derivatives reveals almost identical structural requirements of the allylamine side chains, while basic structural differences at the distal aromatic region of the compounds are responsible for potency and selectivity between fungal or mammalian SE. These findings are compatible with a two-binding site model for inhibition of squalene epoxidase by allylamine derivatives (Figure 5), on both the fungal and the human enzyme. One is postulated to be a lipophilic side chain binding site, the other to be an aromatic ring system binding site. The high degree of selectivity achieved for fungal versus mammalian SE inhibition can be

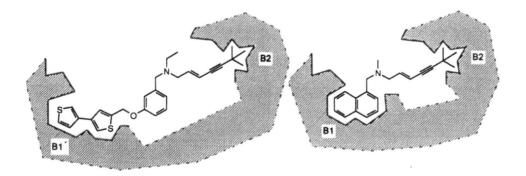

Figure 9 Hypothetical two binding site models for mammalian (NB-598) vs fungal SE inhibitors (terbinafine)

explained either by different relative positions of these two sites on the enzymes[101] or by variation of the aromatic ring binding site. One of these sites is thought to overlap with the squalene binding site.[16,28]

In summary, the accidental discovery of a lead compound and the elucidation of its mode of action enabled the search for more potent squalene epoxidase inhibitors. Medicinal chemistry programmes led to the establishment of detailed SARs and the identification of highly active and selective drugs. Furthermore, the medicinal chemistry of SE inhibitors has provided evidence that intrinsic differences between squalene epoxidases from various origins exist, which can be used to design drugs displaying high species selectivity.

11. REFERENCES

1. Ryder, N.S. (1990) Squalene epoxidase: enzymology and inhibition. In *Biochemistry of Cell Walls and Membranes in Fungi*, pp. 189–203, edited by Kuhn, P.J., Trinci, A.P.J., Jung, M.J., Goosey, M.W. and Copping, L.G. Berlin: Springer-Verlag.

2. Oehlschlager, A.C. and Czyzewska, E. (1992) Rationally designed inhibitors of sterol biosynthesis. In *Emerging targets in antibacterial and antifungal chemotherapy*, pp. 437–475, edited by Sutcliffe, J.A. and Georgopapadakou, N.H. New York: Chapman and Hall.

3. Georgopapadakou, N.H. and Walsh, T.J. (1994) Human mycoses: drugs and targets for emerging pathogens. *Science*, **264**, 371–373.

4. Schulz, U. and Scheinpflug, H. (1988) Sterol biosynthesis inhibiting fungicides: antifungal properties and application in cereals. In *Sterol Biosynthesis Inhibitors*, pp. 211–261, edited by Berg, D. and Plempel, M. Weinheim, Chichester: VCH, Ellis Horwood Ltd.

5. Baldwin, B.C. (1990) Inhibitors of ergosterol biosynthesis as crop protection agents. *Biochem. Soc. Trans.*, **18**, 61–62.

6. Abe, I., Tomesch, J.C., Wattanasin, S. and Prestwich, G.D. (1994) Inhibitors of squalene biosynthesis and metabolism. *Nat. Prod. Rep.* (in press).

7. Suckling, K.E. (1993) Emerging strategies for the treatment of atherosclerosis as seen from the patent literature. *Biochem. Soc. Trans.*, **21**, 660–662.

8. Mercer, E.I. (1993) Inhibitors of sterol biosynthesis and their applications. *Prog. Lipid Res.*, **32** (4), 357–416.

9. Barrett-Bee, K. and Ryder, N.S. (1992) Biochemical aspects of ergosterol biosynthesis inhibition. In *Emerging Targets for Antibacterial and Antifungal Chemotherapy*, pp. 410–436, edited by Sutcliffe, J.A. and Georgopapadakou, N.H. New York: Chapman and Hall.

10. Paltauf, F., Daum, G., Zuder, G., Hoegenauer, G., Schulz, G. and Seidl, G. (1982) Squalene and ergosterol biosynthesis in fungi treated with naftifine, a new antimycotic agent. *Biochim. Biophys. Acta*, **712**, 268–273.

11. Ryder, N.S., Seidl, G. and Troke, P.F. (1984) Effect of the antimycotic drug naftifine on growth of and sterol biosynthesis in *Candida albicans*. *Antimicrob. Agents Chemother.*, **25**, 483–487.

12. Stütz, A. (1988) Synthesis and structure-activity correlations within allylamine antimycotics. *Ann. N.Y. Acad. Sci.*, **544**, 46–62.

13. Ryder, N.S. and Dupont, M.-C. (1985) Inhibition of squalene epoxidase by allylamine antimycotic compounds: a comparative study of the fungal and mammalian enzymes. *Biochem. J.*, **230**, 765–770.

14. Ryder, N.S. (1988) Mechanism of action and biochemical selectivity of allylamine antimycotic agents. *Ann. N.Y. Acad. Sci.*, **544**, 208–220.

15. Ryder, N.S., Frank, I. and Dupont, M.-C. (1986) Ergosterol biosynthesis inhibition by the thiocarbamate antifungal agents tolnaftate and tolciclate. *Antimicrob. Agents Chemother.*, **29**, 858–860.

16. Ryder, N.S. (1990) Inhibition of squalene epoxidase and sterol side-chain methylation by allylamines. *Biochem. Soc. Trans.*, **18**, 45–46.

17. Horie, M., Tsuchiya, Y., Hayashi, M., Iida, Y., Iwasawa, Y., Nagata, Y., Sawasaki, Y., Fukuzumi, H., Kitani, K. and Kamei, T. (1990) NB-598: a potent competitive inhibitor of squalene epoxidase. *J. Biol. Chem.*, **265**, 18075–18078.

18. Hidaka, Y., Hotta, H., Nagata, Y., Iwasawa, Y., Horie, M. and Kamei, T. (1991) Effect of a novel squalene epoxidase inhibitor, NB-598, on the regulation of cholesterol metabolism in Hep G2 cells. *J. Biol. Chem.*, **266**, 13171–13177.

19. Yamamoto, S. and Bloch, K. (1970) Studies on squalene epoxidase of rat liver. *J. Biol. Chem.*, **245**, 1670–1674.

20. Ono, T., Nakazono, K. and Kosaka, H. (1982) Purification and partial characterization of squalene epoxidase from rat liver microsomes. *Biochim. Biophys. Acta*, **709**, 84–90.

21. Ryder, N.S. and Dupont, M.-C. (1984) Properties of a particulate squalene epoxidase from *Candida albicans*. *Biochim. Biophys. Acta*, **794**, 466–471.

22. Ryder, N.S. (1991) Squalene epoxidase as a target for the allylamines. *Biochem. Soc. Trans.*, **19**, 774–776.

23. Jahnke, L. and Klein, H.P. (1983) Oxygen requirements for formation and activity of the squalene epoxidase in Saccharomyces cerevisiae. *J. Bacteriol.*, **155**, 488–492.

24. M'Baya, B. and Karst, F. (1987) *In vitro* assay of squalene epoxidase of Saccharomyces cerevisiae. *Biochem. Biophys. Res. Comm.*, **147**, 556–564.

25. Jandrositz, A., Turnowsky, F. and Hoegenauer, G. (1991) The gene encoding squalene epoxidase from Saccharomyces cerevisiae: cloning and characterization. *Gene*, **107**, 155–160.

26. Satoh, T., Horie, M., Watanabe, H., Tsuchiya, Y. and Kamei, T. (1993) Enzymatic properties of squalene epoxidase from Saccharomyces cerevisiae. *Biol. Pharm. Bull.*, **16**, 349–352.

27. Petranyi, G., Ryder, N.S. and Stütz, A. (1984) Allylamine derivatives: new class of synthetic antifungal agents inhibiting fungal squalene epoxidase. *Science*, **224**, 1239–1241.

28. Ryder, N.S., Stütz, A. and Nussbaumer, P. (1992) Squalene epoxidase inhibitors: structural determinants for activity and selectivity of allylamines and related compounds. In *Regulation of Isopentenoid Metabolism*, pp. 192–204, edited by Nes, W.D., Parish, E.J. and Trzaskos, J.M. Washington D.C.: American Chemical Society.

29. Ryder, N.S., Stütz, A. and Nussbaumer, P. (1994) Inhibitors of squalene epoxidase. In *Design of enzyme inhibitors as drugs: Vol. II*, pp. 378–395, edited by Sandler, M. and Smith, H.J. Oxford: Oxford University Press.

30. Ryder, N.S. (1985) Specific inhibition of fungal sterol biosynthesis by SF 86-327, a new allylamine antimycotic agent. *Antimicrob. Agents Chemother.*, **27**, 252–256.

31. Georgopoulos, A.G., Petranyi, G., Mieth, H. and Drews, J. (1981) *In vitro* activity of naftifine, a new antifungal agent. *Antimicrob. Agents Chemother.*, **19**, 386–389.

32. Petranyi, G., Meingassner, J.G. and Mieth, H. (1987) Antifungal activity of the allylamine derivative terbinafine *in vitro*. *Antimicrob. Agents Chemother.*, **31**, 1365–1368.

33. Mieth, H. (1993) *In vitro* and *in vivo* activities of terbinafine. In *Cutaneous Antifungal Agents*, pp. 137–149, edited by Rippon, J.W. and Fromtling, R.A. New York: Marcel Dekker, Inc.

34. Ryder, N.S. and Mieth, H. (1992) Allylamine antifungal drugs. In *Current Topics in Medical Mycology Vol. 4*, pp. 158–188, edited by Borgers, M., Hay, R. and Rinaldi, M.G. New York: Springer-Verlag.

35. Ryder, N.S. (1985) Effect of allylamine antimycotic agents on fungal sterol biosynthesis measured by sterol side-chain methylation. *J. Gen. Microbiol.*, **131**, 1595–1602.

36. Ryder, N.S. (1988) Mode of action of allylamines. In *Sterol Biosynthesis Inhibitors. Pharmaceutical and Agrochemical Aspects*, pp. 151–167, edited by Berg, D. and Plempel, M. Chichester: Ellis Horwood Ltd.

37. Ryder, N.S. (1992) Terbinafine: mode of action and properties of the squalene epoxidase inhibition. *Br. J. Dermatol.*, **126** (Suppl. 39), 2–7.

38. Gnamusch, E., Ryder, N.S. and Paltauf, F. (1992) Effect of squalene on the structure and functions of fungal membranes. *J. Dermatol. Treat.*, **3** (Suppl. 1), 9–13.

39. Morita, T. and Nozawa, Y. (1985) Effects of antifungal agents on ergosterol biosynthesis in *Candida albicans* and *Trichophyton mentagrophytes*: differential inhibitory sites of naphthiomates and miconazole. *J. Invest. Dermatol.*, **85**, 434–437.

40. Barrett-Bee, K., Lane, A.C. and Turner, R.W. (1986) The mode of antifungal action of tolnaftate. *J. Med. Vet. Mycol.*, **24**, 155–160.

41. Morita, T., Iwata, K. and Nozawa, Y. (1989) Inhibitory effect of a new mycotic agent, piritetrate on ergosterol biosynthesis in pathogenic fungi. *J. Med. Vet. Mycol.*, **27**, 17–25.

42. Ryder, N.S. (1987) Squalene epoxidase as the target of antifungal allylamines. *Pestic. Sci.*, **21**, 281–288.

43. Schuster, I. (1985) The interaction of representative members from two classes of antimycotics – the azoles and the allylamines – with cytochromes P-450 in steroidogenic tissues and liver. *Xenobiotica*, **15**, 529–546.

44. Schuster, I. (1987) Potential of allylamines to inhibit cytochrome P-450. In *Recent Trends in the Discovery, Development and Evaluation of Antifungal Agents*, pp. 471–478, edited by Fromtling, R.A. Barcelona: J.R. Prous Science Publishers.

45. Schaude, M., Ackerbauer, H. and Mieth, H. (1987) Inhibitory effect of antifungal agents on germ tube formation in *Candida albicans*. *Mykosen*, **30**, 281–287.

46. Petranyi, G., Meingassner, J.G. and Mieth, H. (1987) Activity of terbinafine in experimental fungal infections of laboratory animals. *Antimicrob. Agents Chemother.*, **31**, 1558–1561.

47. Mieth, H. and Petranyi, G. (1989) Preclinical evaluation of terbinafine *in vivo*. *Clin. Exp. Dermatol.*, **13**, 104–107.

48. Berney, D. and Schuh, K. (1978) Heterocyclic spiro-naphthalenones. Part I: synthesis and reactions of some spiro[(1*H*-naphthalenone-1,3'-piperidines]. *Helv. Chim. Acta*, **61**, 1262–1273.

49. Clayton, Y.M., Meinhof, W. and Seeliger, H. (1985) In *Mykosen*, **28** (Suppl. 1), Berlin: Grosse Verlag.

50. Braeutigam, M. and Weidinger, G. (1993) Clinical experience with naftifine. In *Cutaneous Antifungal Agents*, pp. 99–115, edited by Rippon, J.W. and Fromtling, R.A. New York: Marcel Dekker, Inc.

51. Stütz, A., Georgopoulos, A.G., Granitzer, W., Petranyi, G. and Berney, D. (1986) Synthesis and structure-activity relationships of naftifine-related allylamine antimycotics. *J. Med. Chem.*, **29**, 112–125.

52. Stütz, A. (1987) Allylamine derivatives – a new class of active substances in antifungal chemotherapy. *Angew. Chem. Int. Ed. Engl.*, **26**, 320–328.

53. Stütz, A. and Petranyi, G. (1984) Synthesis and antifungal activity of (*E*)-*N*-(6,6-dimethyl-2-hepten-4-ynyl)-*N*-methyl-1-naphthalenemethanamine (SF 86-327) and related allylamine derivatives with enhanced oral activity. *J. Med. Chem.*, **27**, 1539–1543.

54. Baudraz-Rosselet, F., Rakosi, T., Wili, P.B. and Kenzelmann, R. (1992) Treatment of onychomycosis with terbinafine. *Br. J. Dermatol.*, **126** (Suppl. 39), 40–46.

55. Goodfield, M.J.D. (1992) Short-duration therapy with terbinafine for dermatophyte onychomycosis: a multicentre trial. *Br. J. Dermatol.*, **126** (Suppl. 39), 33–35.

56. Villars, V.V. and Jones, T.C. (1992) Special features of the clinical use of oral terbinafine in the treatment of fungal diseases. *Br. J. Dermatol.*, **126** (Suppl. 39), 61–69.

57. Villars, V.V. and Jones, T.C. (1993) Clinical use of oral and topical terbinafine in the treatment of fungal diseases. In *Cuteanous Antifungal Agents*, pp. 151–168. edited by Rippon, J.W. and Fromtling, R.A. New York: Marcel Dekker, Inc.

58. Nussbaumer, P., Ryder, N.S. and Stütz, A. (1991) Allylamine antimycotics: recent trends in structure-activity relationships and syntheses. *Pestic. Sci.*, **31**, 437–455.

59. Chen, W.P., Liu, L.L. and Yang, J.Q. (1989) Synthesis and antifungal activity of *N*-(6,6-dimethyl-2-hepten-4-ynyl)-*N*-methyl-α-substituted-1-(4-substituted)naphthalenemethanamines. *Yaoxue Xuebao*, **24** (12), 895–905.

60. Stanetty, P. and Wallner, H. (1993) Synthesis of potential inhibitors of squalenepoxidase with conformational fixation of the structural elements of butenafine. *Arch. Pharm.*, **326**, 341–350.

61. Ryder, N.S. and Frank, I. (1992) Interaction of terbinafine with human serum and serum proteins. *J. Med. Vet. Mycol.*, **30**, 451–460.

62. Nussbaumer, P., Leitner, I., Mraz, K., Ryder, N.S. and Stütz, A. (1994) Quinoline and isoquinoline allylamine antimycotics. *Med. Chem. Res.*, **3**, 517–522.

63. Nussbaumer, P., Petranyi, G. and Stütz, A. (1991) Synthesis and structure-activity relationships of benzo[*b*]thienylallylamine antimycotics. *J. Med. Chem.*, **34**, 65–73.

64. Stütz, A. and Nussbaumer, P. (1989) SDZ 87-469. *Drugs Fut.*, **14**, 639–642.

65. Stanetty, P., Koller, H., Puerstinger, G. and Grubner, S. (1993) Synthesis of new 7-benzofuran-methanamines as heterocyclic analogues of the squalene epoxidase inhibitor butenafine. *Arch. Pharm.*, **326**, 351–358.

66. Stanetty, P. and Koller, H. (1992) Synthesis of new indolemethanamines using the Leimgruber-Batcho reaction. *Arch. Pharm.*, **325**, 433–437.
67. Nussbaumer, P., Dorfstatter, G., Leitner, I., Mraz, K., Vyplel, H. and Stütz, A. (1993) Synthesis and structure-activity relationships of naphthalene-substituted derivatives of the allylamine antimycotic terbinafine. *J. Med. Chem.*, **36**, 2810–2816.
68. Nussbaumer, P., Leitner, I. and Stütz, A. (1994) Synthesis and structure-activity relationships of the novel homopropargylamine antimycotics. *J. Med. Chem.*, **37**, 610–615.
69. Maeda, T., Yamamoto, T., Takase, M., Sasaki, K., Arika, T., Yokoo, M., Hashimoto, R., Amemiya, K. and Koshikawa, S. (1985) *Eur. Pat. Appl.*, EP 164697.
70. Arai, K., Arita, M., Sekino, T., Komoto, N. and Hirose, S. (1987) *Eur. Pat. Appl.*, EP 221781.
71. Arika, T., Yokoo, M., Hase, T., Maeda, T., Amemiya, K. and Yamaguchi, H. (1990) Effects of butenafine hydrochloride, a new benzylamine derivative, on experimental dermatophytosis in guinea pigs. *Antimicrob. Agents Chemother.*, **34**, 2250–2253.
72. Fukushiro, R., Urabe, H., Kagawa, S., Watanabe, S., Takahashi, H., Takahashi, S. and Nakajima, H. (1992) Butenafine hydrochloride, a new antifungal agent: clinical and experimental study. In *Recent Progress in Antifungal Chemotherapy*, pp. 147–157, edited by Yamaguchi, H., Kobayashi, G.S. and Takahashi, H. New York: Marcel Dekker, Inc.
73. Hecht, P., Vyplel, H., Nussbaumer, P. and Berner, H. (1992) A combined use of quantum chemical parameters, hydrophobic and geometrical descriptors to establish QSARs of allylamine antimycotics. *Quant. Struct-Act. Relat.*, **11**, 339–347.
74. Maeda, T., Takase, M., Ishibashi, A., Yamamoto, T., Sasaki, K., Arika, T., Yokoo, M. and Amemiya, K. (1991) Synthesis and antifungal activity of butenafine hydrochloride (KP-363), a new benzylamine antifungal agent. *Yakugaku Zasshi*, **111** (2), 126–137.
75. Nussbaumer, P., Dorfstaetter, G., Grassberger, M.A., Leitner, I., Meingassner, J.G., Thirring, K. and Stütz, A. (1993) Synthesis and structure-activity relationships of phenyl-substituted benzylamine antimycotics: a novel benzylbenzylamine antifungal agent for systemic treatment. *J. Med. Chem.*, **36**, 2115–2120.
76. Ryder, N.S., Favre, B. and Leitner, I. (1994) Biochemical mechanism of action of SDZ SBA 5 86, a new orally active antimycotic. Oral presentation at the *XIIth ISHAM Congress*, Adelaide, Australia. (Abstracts book P 7.1).
77. Nussbaumer, P., Dorfstaetter, G., Leitner, I., Meingassner, J.G., Ryder, N.S. and Stütz, A. (1993) SDZ SBA 586: a novel benzylamine antifungal agent for systemic and topical treatment. Oral presentation at the *International Summit on Cutaneous Antifungal Therapy*, San Francisco.
78. Granitzer, W. and Stütz, A. (1979) Stereoselective trans-reduction of tert. Propargylamines to (*E*)-allylamines using DiBAH. *Tetrahedron Lett.*, 3145–3148.
79. Stütz, A., Granitzer, W. and Roth, S. (1985) Diisobutylaluminum hydride for the stereoselective synthesis of tertiary (*E*)-2-alkenylamines, (*E*)-2-alken-4-ynylamines and (2*E*,4*Z*)-alkadienylamines. *Tetrahedron*, **41** (23), 5685–5696.
80. Corey, E.J., Katzenellenbogen, J.A. and Posner, H.G. (1967) A new stereospecific synthesis of trisubstituted olefins. Stereospecific synthesis of farnesol. *J. Am. Chem. Soc.*, **89**, 4245–4247.
81. Ratovelomana, V. and Linstrumelle, G. (1981) New synthesis of the sex pheromone of the egyptian cotton leaf-worm, Spodoptera littoralis. *Synth. Commun.*, **11** (11), 917–923.
82. Nussbaumer, P., Baumann, K., Dechat, T. and Harasek, M. (1991) Highly selective TFAA-cleavage of tertiary 2,4-dimethoxybenzylamines and its use in the synthesis of secondary amines. *Tetrahedron*, **47** (26), 4591–4602.
83. Nussbaumer, P. and Stütz, A. (1992) Regioselective ring opening of α-substituted α-alkynyl oxiranes to 2-substituted 3-butyn-1-ols. *Tetrahedron Lett.*, **33** (49), 7507–7508.
84. Noguchi, T., Kaji, A., Igarashi, Y., Shigematsu, A. and Taniguchi, K. (1962) Antitrichophyton activity of naphthiomates. *Antimicrob. Agents Chemother.*, 259–267.
85. Nishino, T., Okano, Y., Isogawa, Y., Koshi, T. and Tanino, T. (1981) Antimycotic studies on tolciclate. *Chemotherapy (Tokyo)*, **29**, 1304–1317.
86. Hiratani, T., Yamaguchi, H., Uchida, E., Osumi, Y., Watanabe, S., Okozumi, K., Yamani, H., Tanaka, N., Yamamoto, Y. and Iwata, K. (1983) Mechanism of action of antifungal agents – biochemical study. *Jpn. J. Med. Mycol.*, **24**, 194–204.

87. Nozawa, Y. and Morita, T. (1992) Biochemical aspects of squalene epoxidase inhibition by a thiocarbamate derivative, naphthiomate-T. In *Recent progress in antifungal chemotherapy*, pp. 53–64, edited by Yamaguchi, H., Kobayashi, G.S. and Takahashi, H. New York: Marcel Dekker, Inc.

88. Noguchi, T., Hashimoto, Y., Miyazaki, K. and Kaji, A. (1968) Studies on the selective toxicity II. Relationship between chemical structure and selective antimicrobial activities of aryl thiocarbamates. *Yakugaku Zasshi*, **88**, 335–343.

89. Noguchi, T., Hashimoto, Y., Kosaka, S., Kikuchi, M., Miyazaki, K., Sakimoto, R. and Kaji, A. (1968) Studies on the selective toxicity III. Some consideration of antitrichophyton activity of 2-naphthyl-*N*-methyl-*N*-arylthiocarbamtes. *Yakugaku Zasshi*, **88**, 344–352.

90. Melloni, P., Metelli, R., Vecchietti, V., Logemann, W., Castellino, S., Monti, G. and deCarneri, I. (1974) New antifungal agents. *Eur. J. Med. Chem.*, **9**, 26–31.

91. deCarneri, I., Monti, G., Bianchi, A., Castellino, S., Meinardi, G. and Mandelli, V. (1976) Tolciclate against dermatophytes. *Drug Res.*, **26** (5), 769–772.

92. Iwata, K., Yamashita, T. and Uehara, H. (1989) *In vitro* and *in vivo* activities of piritetrate (M-732), a new antidermatophytic thiocarbamate. *Antimicrob. Agents Chemother.*, **33** (12), 2118–2125.

93. Sen, S.E. and Prestwich, G.D. (1989) Trisnorsqualene alcohol, a potent inhibitor of vertebrate squalene epoxidase. *J. Am. Chem. Soc.*, **111**, 1508–1510.

94. Sen, S.E. and Prestwich, G.D. (1989) Trisnorsqualene cyclopropylamine: a reversible tight-binding inhibitor of squalene epoxidase. *J. Am. Chem. Soc.*, **111**, 8761–8763.

95. Moore, W.R., Schatzman, G.L., Jarvi, E.T., Gross, R.S. and McCarthy, J.R. (1992) Terminal difluoro olefin analogues of squalene are time-dependent inhibitors of squalene epoxidase. *J. Am. Chem. Soc.*, **114**, 360–361.

96. Jarvi, E.T., McCarthy, J.R. and Edwards, M.L. (1991) *Eur. Pat. Appl.*, EP 0448934A2.

97. Mann, J. and Smith, G.P. (1991) Synthesis of potential inhibitors of squalene epoxidase. *J. Chem. Soc. Perkin Trans. 1*, 2884–2885.

98. Ceruti, M., Viola, F., Grosa, G., Balliano, G., Delprino, L. and Cattel, L. (1988) Synthesis of squalenois acetylenes and allenes, as inhibitors of squalene epoxidase. *J. Chem. Res. (S)*, 18–19.

99. Sen, S.E. and Prestwich, G.D. (1989) Squalene analogues containing isopropylidine mimics as potential inhibitors of pig liver squalene epoxidase and oxidosqualene cyclase. *J. Med. Chem.*, **32**, 2152–2158.

100. Bai, M., Xiao, X. and Prestwich, G.D. (1991) 26-Hydroxysqualene and derivatives: substrates and inhibitors for squalene epoxidase. *Bioorg. Med. Chem. Lett.*, **1**, 227–232.

101. Iwasawa, Y. and Horie, M. (1993) Mammalian squalene epoxidase inhibitors and structure-activity relationships. *Drugs Fut.*, **18** (10), 911–918.

102. Horie, M., Sawasaki, Y., Fukuzumi, H., Watanabe, K., Iizuka, Y., Tsuchiya, Y. and Kamei, T. (1991) Hypolipidemic effects of NB-598 in dogs. *Atherosclerosis*, **88**, 183–192.

103. Nagata, Y., Horie, M., Hidaka, Y., Yonemoto, M., Hayashi, M., Watanabe, H., Ishida, F. and Kamei, T. (1992) Effect of an inhibitor of squalene epoxidase, NB-598, on lipid metabolsim in Hep G2 cells. *Chem. Pharm. Bull.*, **40** (2), 436–440.

104. Tomesch, J.C., Nussbaumer, P., Haupt, E.M., Mraz, K., Kathawala, F., Stütz, A., Wareing, J.R., Boettcher, B. and Cornell, S. (1993) Mammalian species selectivity of SDZ 281-915, a potent inhibitor of human squalene epoxidase. Poster presentation at the *17th RSC-SCI Medicinal Chemistry Symposium*, Cambridge, UK.

105. Iwasawa, Y., Tsuchiya, Y., Iida, Y., Hayashi, M., Masaki, H., Nomoto, T., Sakuma, Y., Kato, M., Watanabe, Y., Takezawa, H., Hosoi, M., Sawasaki, Y., Kitani, K. and Kamei, T. (1992) Synthesis and structure-activity relationships of a novel squalene epoxidase inhibitor, NB-598 and its related compounds. Oral presentation at the *XIIth International Symposium on Medicinal Chemistry*, Basel.

106. Matsuda, K., Harada, H., Tsuzuki, R., Morihira, K., Ito, N., Kakuta, H. and Iizumi, Y. (1993) *World Patent*, WO 9312069A1.

2. DUAL INHIBITORS OF 5-LIPOXYGENASE AND CYCLOOXYGENASE

DAVID T. CONNOR and DIANE H. BOSCHELLI

Department of Chemistry, Parke-Davis, Pharmaceutical Research,
Division of Warner-Lambert, 2800 Plymouth Road, Ann Arbor, MI

1. INTRODUCTION

Arachidonic acid (AA) is oxidatively metabolised to the eicosanoids, which include the prostaglandins and leukotrienes, both potent mediators of inflammation (see Scheme 1). AA is released from phospholipase stores via the action of enzymes such as phospholipase A_2 (PLA_2). Various stimuli then generate the production of eicosanoids from AA. The rate limiting steps in the biosynthesis of proinflammatory prostaglandins and leukotrienes are catalyzed by cyclooxygenase (CO, prostaglandin synthetase, PGHS) and 5-lipoxygenase (5-LO). Inhibitors of these enzymes should therefore downregulate an inflammatory response.[1]

The prostaglandins were first identified in 1936.[2] Subsequent work led to the elucidation of their structures[3,4] and biosynthetic transformations.[5] CO converts AA to prostaglandin G_2 (PGG_2) the precursor to both the thromboxanes and additional prostaglandins, including PGE_2. In 1971, Vane proposed that the mechanism of action of nonsteroidal antiinflammatory drugs (NSAIDs) was their ability to inhibit CO.[6] These drugs were originally developed based on their activity in *in vivo* animal models of inflammation. Aspirin (**1**), ibuprofen (**2**), naproxen (**3**), meclofenamic acid (**4**), diclofenac (**5**), indomethacin (**6**), sulindac (**7**) and piroxicam (**8**) are examples of commercially successful NSAIDs, which greatly improve the quality of life for millions of patients including those suffering from osteoarthritis (OA) and rheumatoid arthritis (RA).

While prostaglandins are proinflammatory they are also cytoprotective.[7] Thus, inhibition of prostaglandin biosynthesis provides a mechanism for the gastrointestinal (GI) side effects, commonly seen with NSAID use.[8] In fact, the production of gastric ulcers associated with NSAIDs can be prevented by coadministration of misoprostol, a synthetic PGE_2 analog.[9]

In 1992, it was discovered that mammalian cells contain two isozymes of CO.[10–12] Cyclooxygenase-1 (COX-1) is a constitutively expressed enzyme responsible for gastrointestinal, vascular and renal homeostasis. The recently discovered isozyme, cyclooxygenase-2 (COX-2, prostaglandin synthetase-2, PGHS-2) appears to be expressed in inflamed tissue and is upregulated by stimulation with cytokines, growth factors, or other mediators of inflammation. Marketed NSAIDs inhibit both isozymes,[13] and it is attractive to speculate that a selective inhibitor of COX-2 would be a superior antiinflammatory agent with reduced side effects.[14] Therefore, selective COX-2 inhibitors have become a major focus for drug discovery. NS-398 (**9**) was recently described to be a selective COX-2 inhibitor *in vivo*.[15,16]

DAVID T. CONNOR and DIANE H. BOSCHELLI

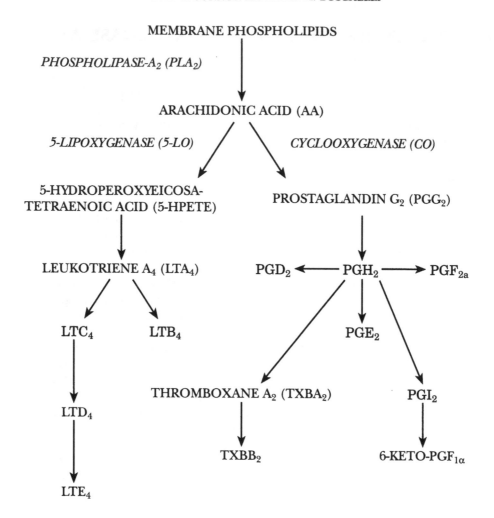

Scheme 1   Metabolism of arachidonic acid by 5-lipoxygenase and cyclooxygenase.

The leukotrienes were discovered several years after the CO pathway was delineated.[17-22] Like prostaglandins, the leukotrienes are implicated in inflammatory processes, and in addition they promote mucosal damage. 5-LO, 12-LO and 15-LO convert AA into 5-hydroperoxyeicosatetraenoic acid (5-HPETE), 12-HPETE and 15-HPETE, respectively. 5-HPETE is then dehydrated to form unstable leukotriene A_4 (LTA$_4$), which is the source of the peptidoleukotrienes (LTC$_4$, LTD$_4$ and LTE$_4$) and also LTB$_4$. LTC$_4$, LTD$_4$ and LTE$_4$, the components of slow reacting substance of anaphylaxis (SRS-A),[23,24] are potent bronchoconstrictors in airway smooth muscle.[25] LTB$_4$ can induce an inflammatory response in human volunteers.[26] LTB$_4$ levels are elevated in peripheral blood[27] from RA patients and in synovial fluid obtained from RA[28] and OA patients.[29]

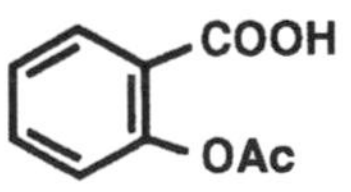
COOH
OAc

1

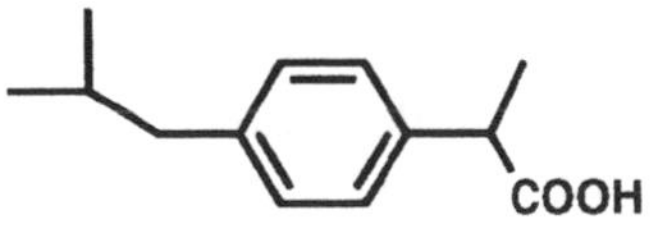
COOH

2

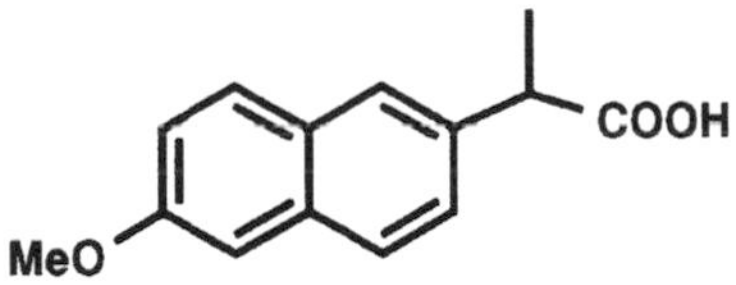
COOH
MeO

3

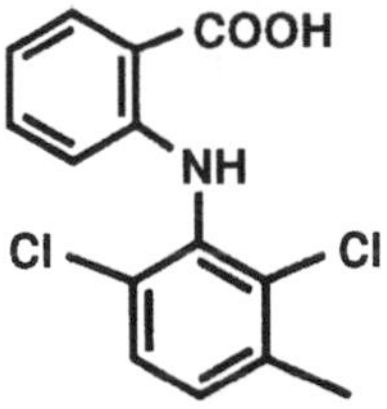
COOH
NH
Cl
Cl

4

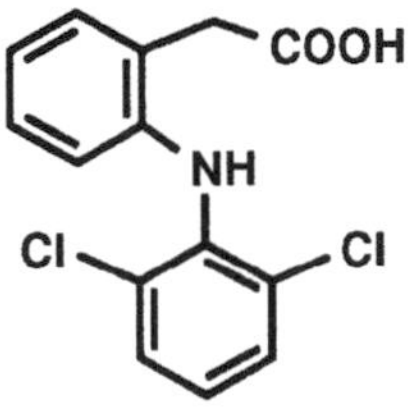
COOH
NH
Cl
Cl

5

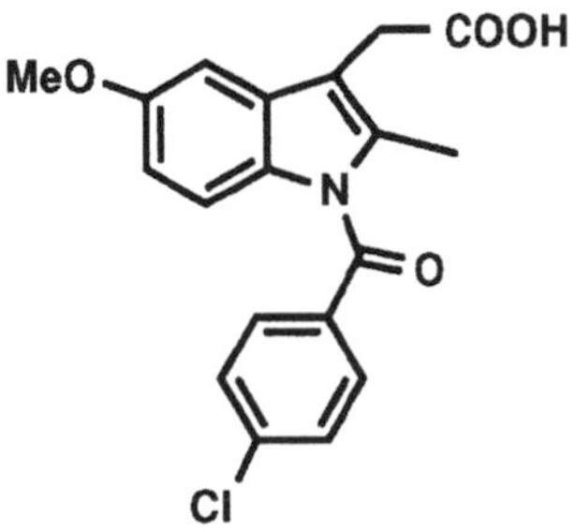
COOH
MeO
N
O
Cl

6

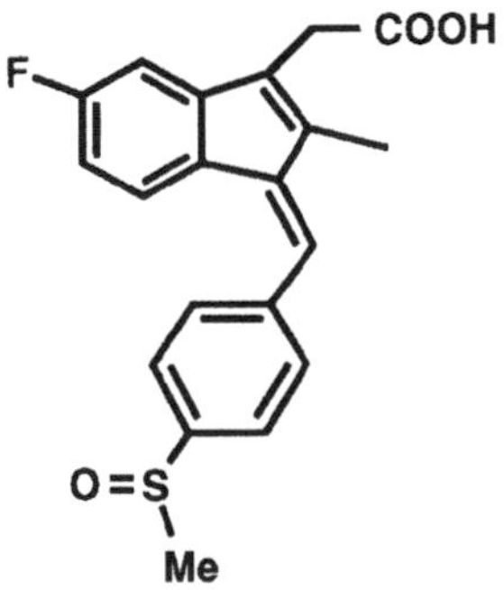
COOH
F
O=S
Me

7

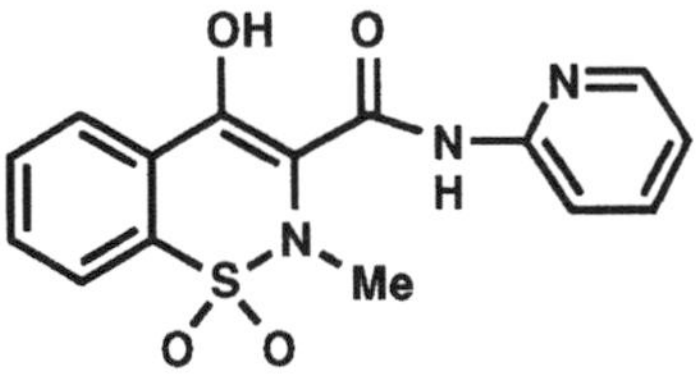
OH
O
N
N
H
S
N
Me
O
O

8

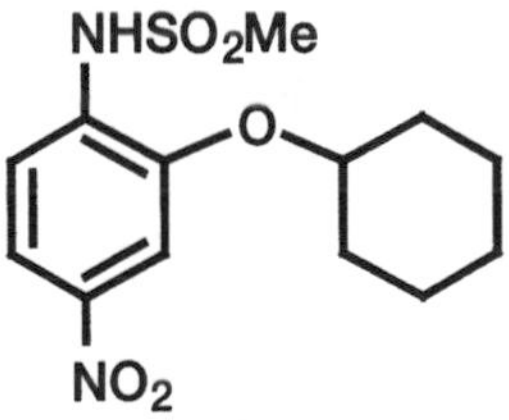
NHSO2Me
O
NO2

9

LTB$_4$ is also a potent chemokinetic and chemotactic factor for neutrophils and may be a key player in NSAID induced GI lesions via neutrophil chemotaxis.[30] Indomethacin induced ulceration in the rat stomach[31] is reduced by prior treatment with L 663536 (10), a 5-LO inhibitor, and SC 41930 (11), an LTB$_4$ antagonist.[32] In addition, the selective 5-LO inhibitors, PF-5901 (12), NDGA (13), zileuton (14), and L-689037 (15) show a dose related inhibition of LTB$_4$ synthesis and a parallel reduction in the severity of indomethacin induced gastric injury.[33] Although this data points to a role for LTB$_4$ in NSAID induced GI damage, other studies have indicated that chronic leukotriene inhibition in the rat fails to modify indomethacin induced gastroenteropathy.[34]

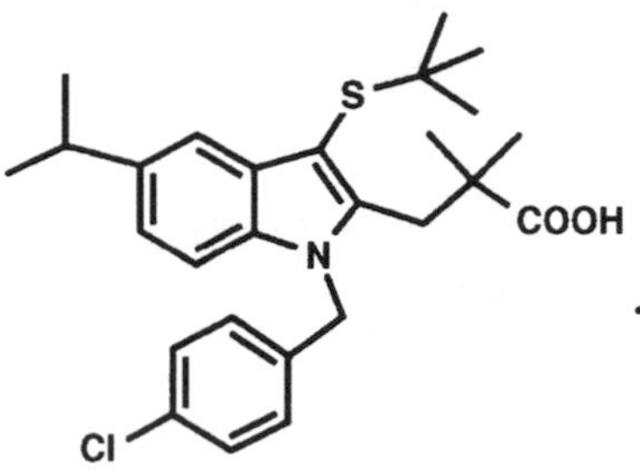

10

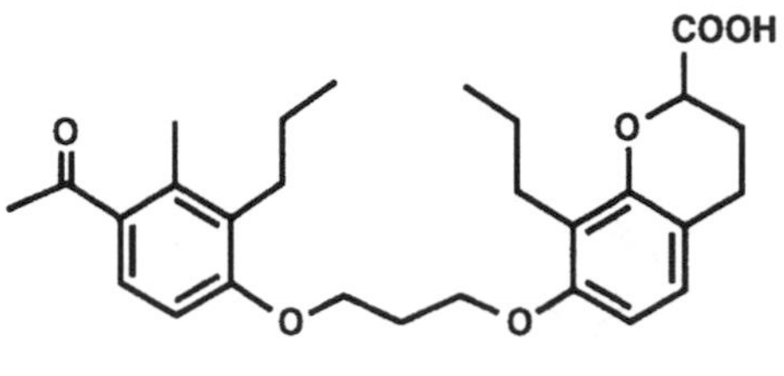

11

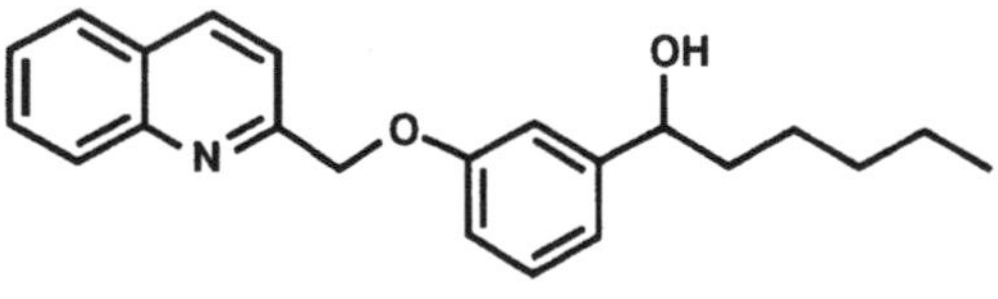

12

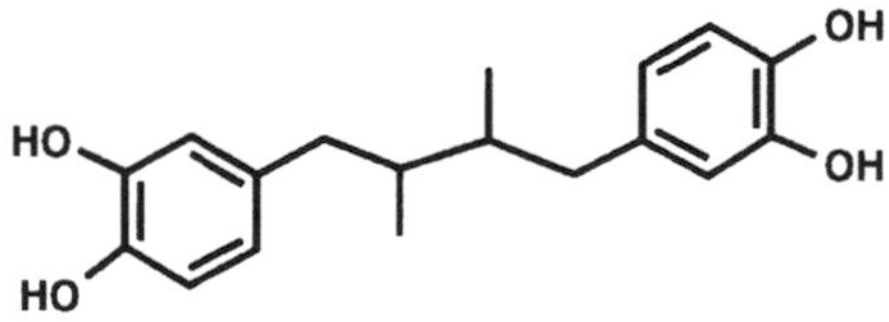

13

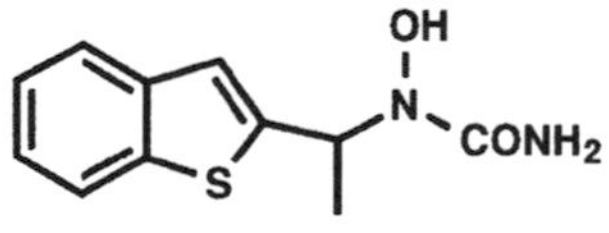

14

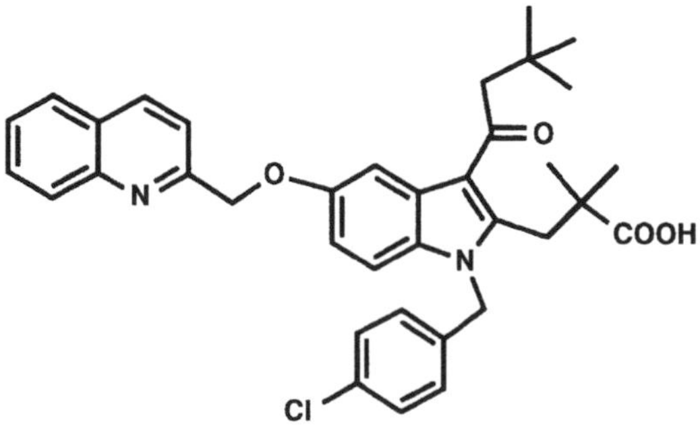

15

In addition to CO, other current targets for drug discovery based on the AA cascade, include PLA$_2$,[35] LTB$_4$,[36] LTD$_4$,[37,38] 5-LO[39–42] and 5-lipoxygenase activating protein (FLAP).[43] 5-Lipoxygenase inhibitors currently in clinical trials include D 2138 (**16**), zileuton, A-78773 (**17**), MK 591 (**18**), and BAY x1005 (**19**). These compounds are primarily targeted as antiasthma drugs, but could also be developed as cytoprotective agents. The demonstrated effectiveness of CO inhibitors and the promising initial results observed with 5-LO inhibitors, suggest that a dual inhibitor might provide an antiinflammatory agent with improved efficacy, reduced GI side effects and a cytoprotective component.

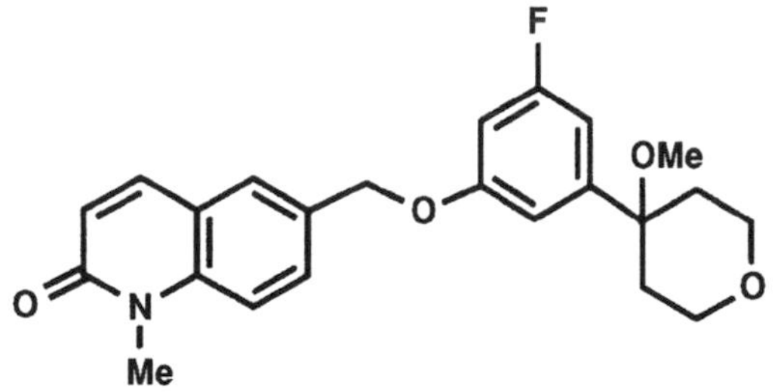

16

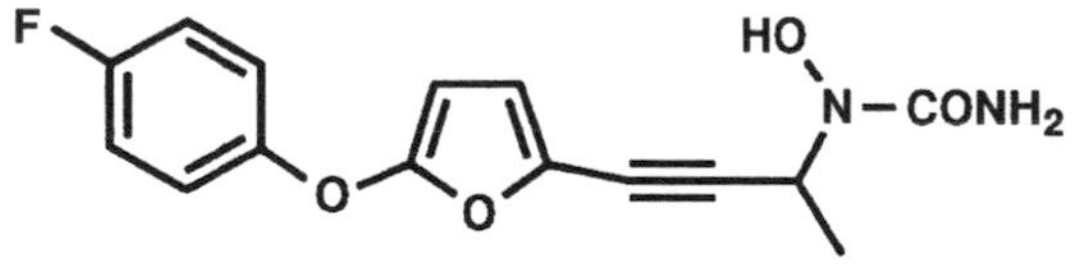

17

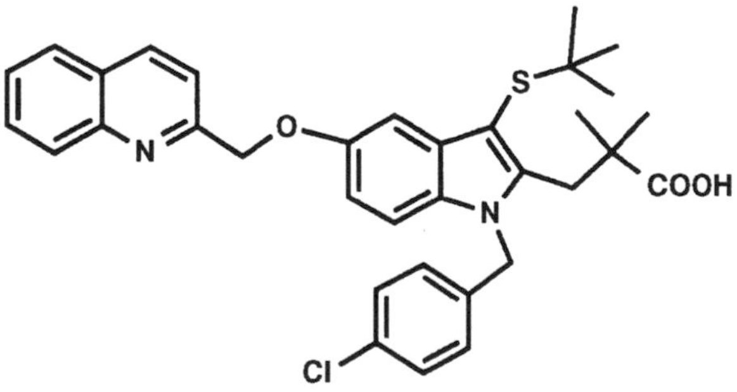

18

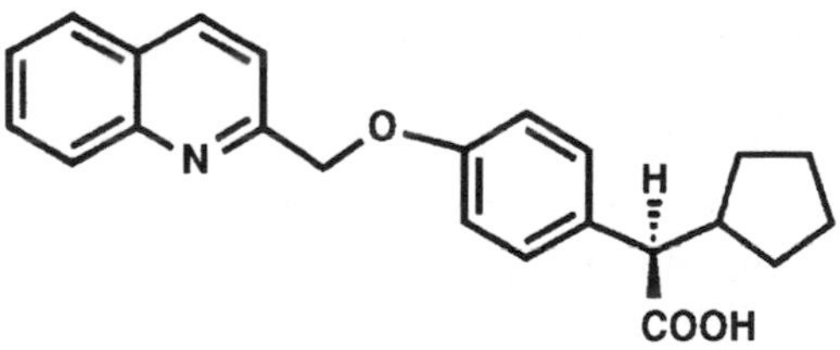

19

2. DUAL INHIBITORS OF CYCLOOXYGENASE AND 5-LIPOXYGENASE NON-DI-*tert*-BUTYLPHENOLS

2.1 Introduction

Over the past several years a major focus of several pharmaceutical companies had been the identification and development of dual inhibitors of CO and 5-LO. Inhibition has been assessed using a variety of enzyme preparations from different cell types and species. 5-LO from human leukocytes was not purified until 1985 and the most significant discrepancies are with this enzyme.[44] The test systems used to determine inhibition of 5-LO and CO are therefore stated along with the IC_{50} for the agent of interest. Several companies have evaluated analogs of di-*tert*-butyl-phenol as dual inhibitors and these will be discussed in Section 3.

2.2 Phenidone and BW 755C

The first dual inhibitor structurally unrelated to arachidonic acid was 1-phenyl-3-pyrazolidone (**20**), phenidone, reported by Wellcome in 1978.[45,46] Phenidone inhibited LO and CO from horse platelets with IC_{50}s of 50 μg/ml and 63 μg/ml respectively. Several analogs of phenidone that inhibited CO, including phenylbutazone, antipyrine and dipyrone, had no effect on LO. However, a structural analog of phenidone, 3-amino-1-[*m*-(trifluoromethyl)phenyl]-2-pyrazoline (**21**), BW 755C, was a more potent inhibitor of both LO ($IC_{50} = 1.7$ μg/ml) and CO ($IC_{50} = 0.72$ μg/ml).[47] In addition BW 755C showed *in vivo* activity in carrageenin foot pad edema (CFE), a model of acute inflammation in the rat that is dependent on CO inhibition.

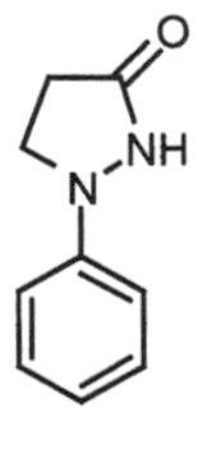

20 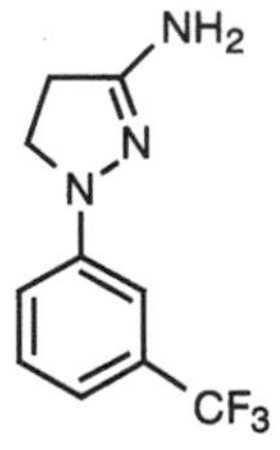   21

Wellcome later reported that N-alkyl analogs of BW 755C had increased *in vitro* activity against LO.[48] The N-methyl analog (**22**), BW 540C, showed a 3:1 preference for LO over CO. The IC_{50} values for the inhibition of LO and CO from rabbit leukocytes were 2.9 μM and 3.1 μM for BW 755C and 0.4 μM and 1.2 μM for BW 540C. Abbott published a study of the activity of various aromatic substituted analogs of BW 755C in the reverse passive Arthus reaction, an *in vivo* model of inflammation.[49] No LO data was reported for these compounds. Several analogs of BW 755C were prepared including a carbonyl linker between the two rings and substitution of a pyrrole for the amino group and of a pyridine for the phenyl ring.[50] 4,5-Dihydro-3-(1H-pyrrol-1-yl)-3-(trifluoromethyl)phenyl-1H-pyrazole (**23**) was the most potent compound against 15-LO ($IC_{50} = 0.08$ μM) from soybeans and against CO ($IC_{50} = 35$

μM) from bovine seminal vesicle microsomes (BSVM). Most of these compounds were selective for 15-LO over CO, including BW 755C (15-LO: $IC_{50} = 0.2$ μM, CO: $IC_{50} = 77$ μM).

Abbott has disclosed that a one carbon homolog of phenidone, tetrahydro-1-phenyl-3(2H)-pyridazinone (**24**), A-53612, inhibited 5-LO ($IC_{50} = 4.0$ μM) from rat basophilic leukemia (RBL-1) cells with no effect on CO.[51] Sterling has reported a related series of 4-substituted 1-phenyl-3-pyrazolidinones as inhibitors of 5-LO from RBL-1 cells.[52] The most active analog was **25**, with a thioethyl substituent at C4 ($IC_{50} = 0.06$ μM).

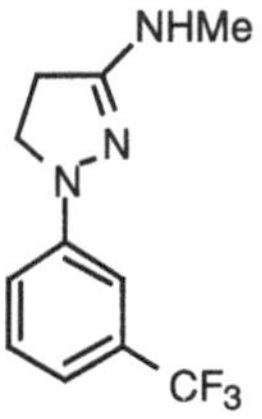

22 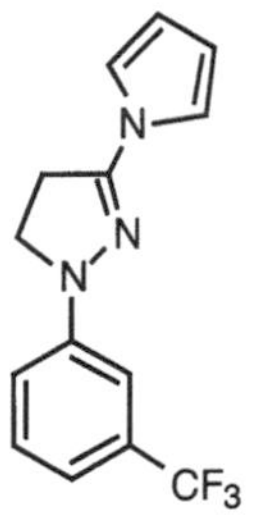   23 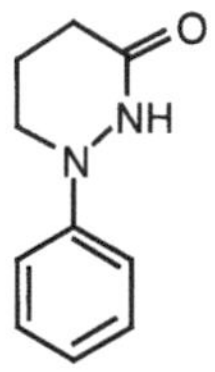   24 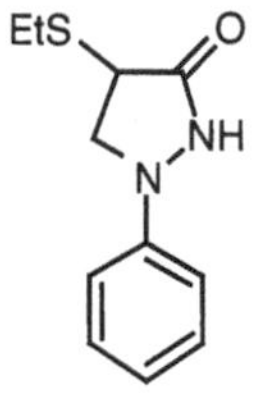   25

2.3   Timegadine

Leo Pharmaceuticals disclosed a series of N,N',N''-trisubstituted guanidines that inhibited CO and were active in CFE and adjuvant arthritis models of inflammation.[53] One of these compounds, N-cyclohexyl-N''-4-[2-methyl-quinolyl]-N'-2-thiazolylguanidine, (**26**), timegadine, gave 100% inhibition of CO from BSVM at a concentration of 10 μM. In washed rabbit platelets, timegadine inhibited LO ($IC_{50} = 0.15$ mM), albeit at higher concentrations than those required to inhibit CO ($IC_{50} = 5$ nM).[54] Timegadine gave a similar inhibition of 5-LO ($IC_{50} = 0.41$ mM) when intact rat neutrophils were used.[55] A study of the effect of timegadine on the production of PGE_2 and LTB_4 from human neutrophils found a 2000 fold ratio in the IC_{50}s for CO vs 5-LO inhibitory activity.[56] Timegadine was active in antipyretic and delayed type hypersensitivity models in the rat[57] and was also efficacious in rheumatoid arthritis clinical trials.[58–60] Due to its impressive inhibition of CO, the *in vivo* potency of this compound may be unrelated to its inhibition of 5-LO.

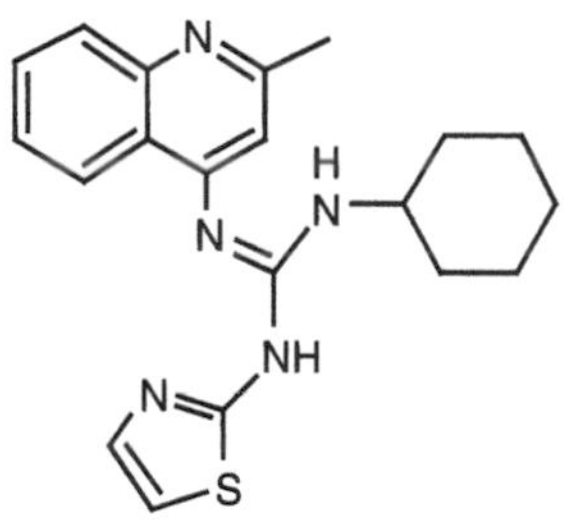

26

2.4 L-652,343

Merck has identified 3-hydroxy-5-trifluoromethyl-N-(2-(2-thienyl)-2-phenyl-ethenyl)-benzo[b]thiophene-2-carboxamide (**27**), L-652,343 as inhibiting the production of LTB$_4$ (IC$_{50}$=0.62 µM) and PGE$_2$ (IC$_{50}$=0.001 µM) from rat neutrophils.[61] L-652,343 was active in CO dependent models of inflammation and showed decreased gastric irritation when compared to standard NSAIDs.[62] While no SAR studies have been reported, it was claimed that 5-LO inhibition was due to the 3-hydroxybenzo[b] thiophene moiety while CO inhibition was a result of the enamide side chain.[61] Merck further stated that the trifluoromethyl group was necessary for *in vivo* activity.

In an initial human safety study, L-652,343 effectively reduced the serum concentration of PGF$_{2\alpha}$, but had no effect on *ex vivo* LTB$_4$ production.[63] This result was confirmed in a second study using skin samples from patients with psoriasis.[64] It was proposed that this inactivity of L-652,343 *in vivo* may be due to non-specific protein binding.

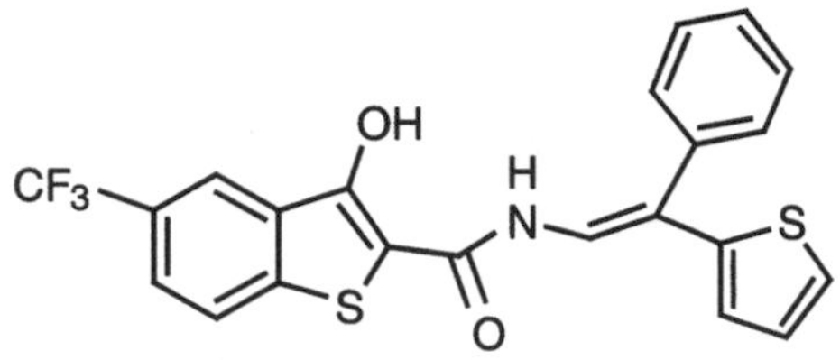

27

2.5 SKF-86002, SKF-104493, SKF-105809

Smith Kline has reported a series of 5,6-diaryl-2,3-dihydroimidazo[2,1-b]thiazoles containing either a 4-pyridyl or 4-substituted phenyl group at both the 5 and 6 positions.[65] Only those isomers with a 5-(4-pyridyl) and 6-(4-substituted phenyl) were active in a model of adjuvant arthritis. The most potent was 5-(4-pyridyl)-6-(4-fluorophenyl)-2,3,-dihydroimidazo[2,1-b]thiazole (**28**), SKF-86002. Two subsequent routes to SKF-86002 avoid the formation of isomers that occurred in the original route.[66,67] It was later disclosed by Smith Kline that SKF-86002 was a dual inhibitor. In a cell free RBL sonicate, SKF-86002 inhibited CO with an IC$_{50}$=100 µM and LO with an IC$_{50}$ = 75 µM.[68] SKF-86002 inhibited prostanoid production by human monocytes (IC$_{50}$=1.0 µM) and LTB$_4$ production by human neutrophils (IC$_{50}$=20 µM). SKF-86002 showed oral activity in CO dependent models of inflammation, such as CFE, and also demonstrated analgesic properties.[69] In addition, SKF-86002 reduced the neutrophil infiltration and edema in AA induced inflammation in the mouse ear, a CO independent model.[70]

SKF-86002 inhibited the production of interleukin-1 (IL-1) (IC$_{50}$=1.3 µM) in lipopolysaccharide (LPS) stimulated human monocytes.[71] This inhibition of IL-1 biosynthesis was independent of stimulus and independent of SKF-86002's effects on arachidonic acid metabolism.[72] SKF-86002 also inhibited the production of

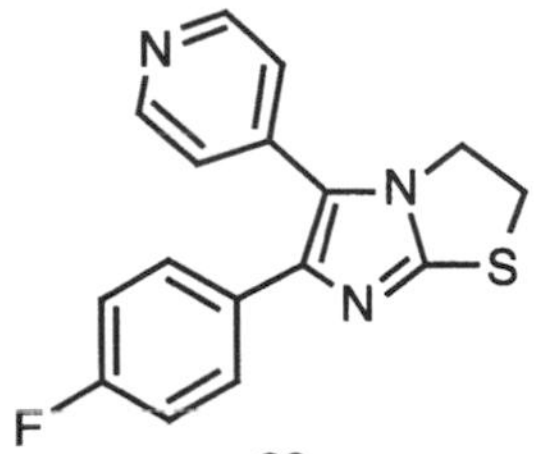

28

tumor necrosis factor alpha (TNF-α) (IC$_{50}$=5–8 µM) while less of an effect was seen against alpha interferon, interferon beta-2 and granulocyte colony stimulating factor (all IC$_{50}$s > 20 µM).[73] In addition to its effects on monocytes, SKF-86002 also reduced TNF-α production in two murine cell lines.[74] The inhibition of TNF-α and IL-1 is thought to occur at the level of translation.[75] Both SKF-86002 and antibodies to TNF-α were efficacious in an animal model of endotoxin shock.[76] Several of SKF-86002's effects on helper T-cell function, including inhibition of proliferation, are attributed to its ability to block IL-1 production.[77]

Dual inhibitor activity similar to that of SKF-86002 has been reported for an analog, 6,7-dihydro-2-(4-methoxyphenyl)-3-(4-pyridyl)-5[H]-pyrrolo[1,2-a]imidazole (29), SKF-104493.[78] SKF-104493 was active in CO independent models of inflammation as was another analog 6,7-dihydro-2-[4-(methylsulfinyl)phenyl]-3-(4-pyridyl)-5[H]-pyrrolo-[1,2-a]imidazole (30), SKF-105809.[79] SKF-105809 did not inhibit CO or 5-LO *in vitro* although it did inhibit LTB$_4$ and prostanoid synthesis *in vivo*.[80] It was determined that SKF-105809 was a prodrug for the corresponding sulfide, 31, SKF-105561, which did inhibit 5-LO (IC$_{50}$ = 3µM) and CO (IC$_{50}$ = 3 µM) *in vitro*. Interestingly, the sulfone metabolite, 32, SKF 105942 had no effect on CO or 5-LO.[81]

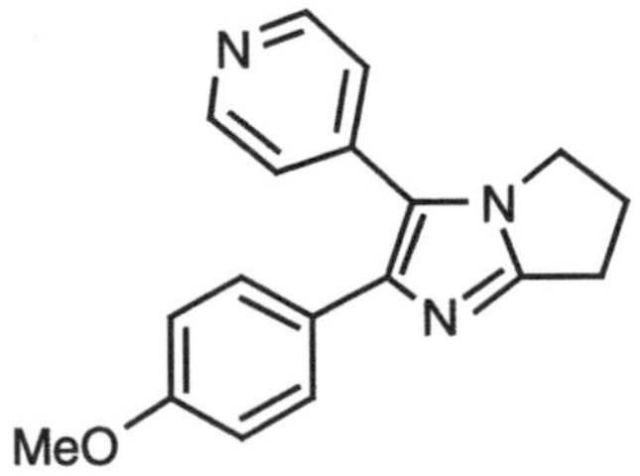

29

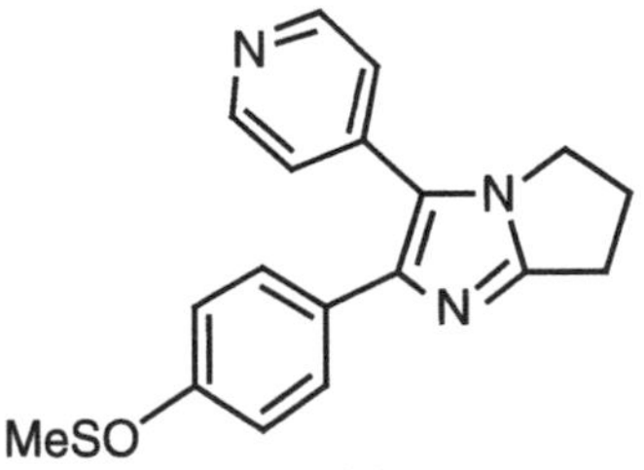

30

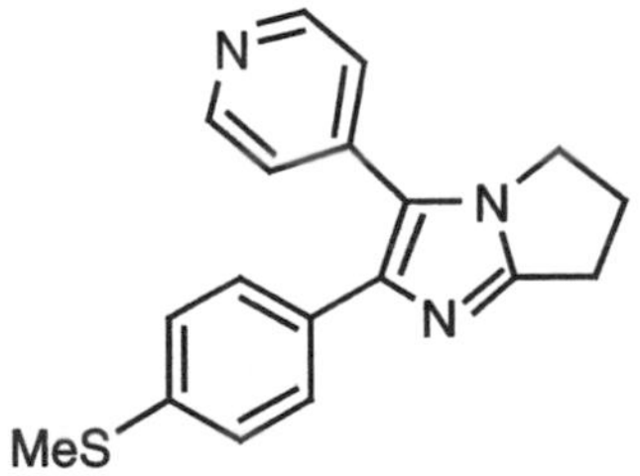

31

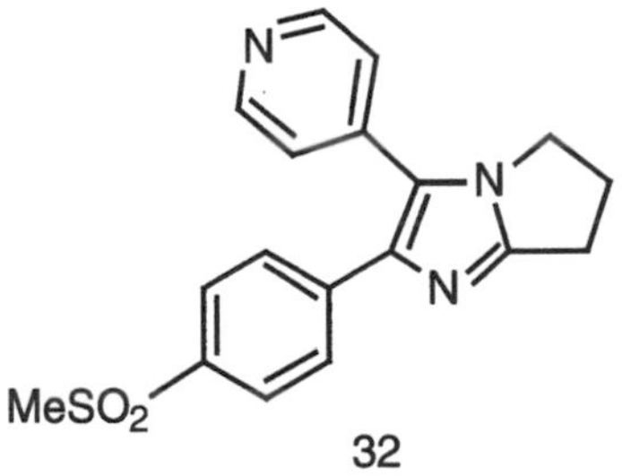

32

Like SKF-86002, SKF-105561 also inhibited the production of IL-1 and TNF-α. To distinguish this new class of compounds from NSAIDs, Smith Kline has coined the term CSAIDs for cytokine supressive antiinflammatory drugs.

2.6 Tepoxalin

5-(4-Chlorophenyl)-N-hydroxy-1-(4-methoxyphenyl)-N-methyl-1H-pyrazole-3-propanamide (**33**), tepoxalin, RWJ 20485, is a dual inhibitor from the R.W. Johnson Institute (formerly Ortho Pharmaceuticals).[82] Tepoxalin inhibited CO and 5-LO from both RBL cell lysates (IC$_{50}$=2.8 µM and 0.15 µM) and intact RBL cells (IC$_{50}$ =4.2 µM and 1.7 µM). Tepoxalin showed topical activity in a mouse ear model of inflammation without any evidence of toxicity,[83–85] and oral activity in adjuvant arthritis without significant ulcerogenicity.[86,87] Unlike standard NSAIDs such as indomethacin, low doses of tepoxalin had no effect on gastric CO.[88] It is not known if tepoxalin is selective for COX-2. Additional studies have shown that tepoxalin blocks the adherence of leukocytes to vascular endothelium, an initial step in gastric ulcer formation. This compound also inhibited the production of PGE$_2$ and LTB$_4$ in both adjuvant arthritic rats and the knee joints of dogs challenged with injections of sodium urate or immune complexes.[89]

The activity of several analogs of tepoxalin has been reported.[82] The data included inhibition of 5-LO from RBL cell lysates and *in vivo* activity in a model of adjuvant arthritis. Data for *in vitro* CO inhibition was not presented. The 1,3-diaryl isomer of tepoxalin, **34**, did not show antiinflammatory activity, although it did inhibit 5-LO. Only the 1,5-isomers showed *in vivo* antiinflammatory activity. The hydroxamic acid functionality of tepoxalin was required for 5-LO inhibition. Reversed hydroxamic acids retained 5-LO activity, but their *in vivo* effectiveness decreased. The major metabolite of tepoxalin results, as predicted, from hydrolysis of the hydroxamic acid to the carboxylic acid. Two synthetic routes to tepoxalin have appeared in the literature.[90,91] This new class of dual inhibitors has been given the name SRADs for steroid replacing antiinflammatory drugs.

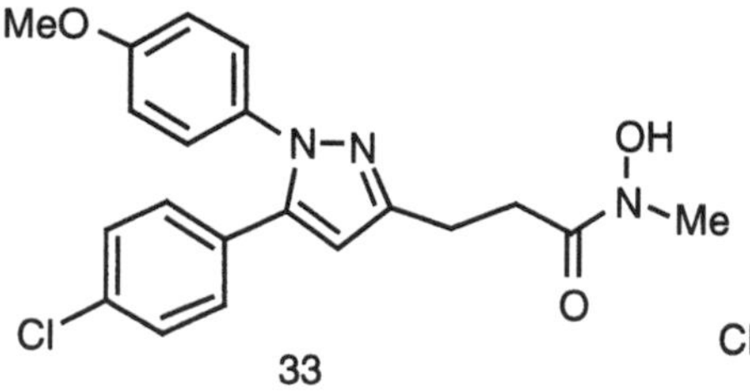

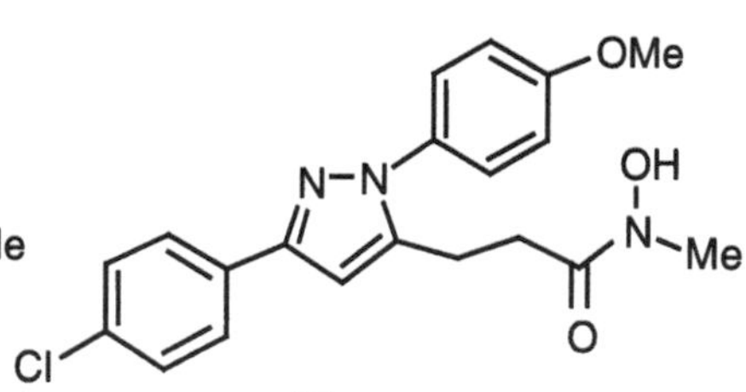

2.7 Tenidap

Pfizer has reported that 5-chloro-2-hydroxy-3-(2-thienylcarbonyl)-indole-1-carboxamide (**35**), CP-66,248, tenidap, inhibited both 5-LO (IC$_{50}$=9.6 µM) and CO (IC$_{50}$ =0.7 µM) in RBL cells.[92] Tenidap also blocked the production of interleukin-1[93,94] and was active in an RA clinical trial.[95] Patients that received tenidap had decreased levels of IL-1 in their synovial fluid.[96] In human PMNLs, tenidap was much more potent against CO ($IC_{50}=32$ nM) than against 5-LO ($IC_{50}=13$ μM).[97] This result was confirmed by others who reported that 100 fold higher concentrations of tenidap were required to inhibit 5-LO than to inhibit CO and that in the presence of plasma protein, tenidap did not inhibit 5-LO.[98] Other reports claim that tenidap blocked LTB_4 production in the synovial fluid of RA patients.[99,100] There is an additional report that tenidap inhibits PLA_2, which would result in decreased production of both prostaglandins and leukotrienes.[101]

The activities currently being emphasized for tenidap are its abilities to inhibit prostaglandins and cytokines. In a detailed study of tenidap's effects on IL-1, a concentration of 30 μM gave 50% inhibition of IL-1 production from both LPS and zymosan stimulated macrophages.[102] Tenidap also inhibited the production of IL-1 and the additional cytokines, IL-6 and TNF-α from human monocytes.[103] Several other activities have been reported for tenidap including inhibition of collagenase release from activated neutrophils[104] and lowering levels of serum C-reactive protein, a marker of arthritic disease.[105]

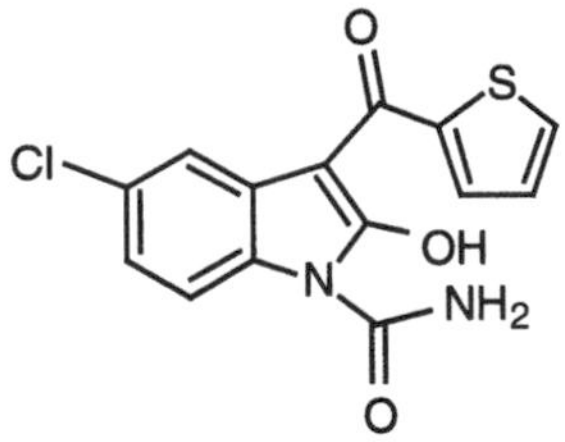

35

Although no SAR study of tenidap has been reported, Pfizer has published a SAR study of a series of related tetracyclic compounds that are also dual inhibitors of 5-LO and CO.[106] Both this series, exemplified by 36, and tenidap, resulted from the finding that 37, an earlier antiinflammatory compound from Pfizer, was a dual inhibitor.[107] 36 inhibited 5-LO ($IC_{50}=1.4$ μM) and CO ($IC_{50}=0.7$ μM) in RBL-1 cells and was also active in rat models of inflammation and analgesia. Another structurally related enolamide 38, RU 43,526 was reported to be a dual inhibitor by Roussel Uclaf.[108] 38 inhibited 5-LO from rat neutrophils ($IC_{50}=3.0$ μM) and BSV CO ($IC_{50}=2.1$ μM) and was also active in CFE and adjuvant arthritis in the rat.

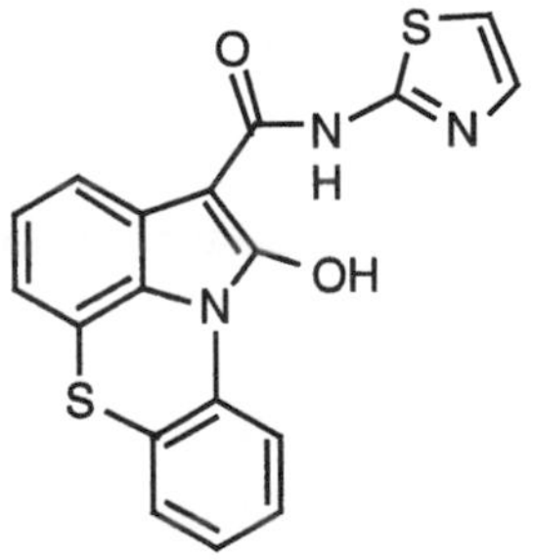

36

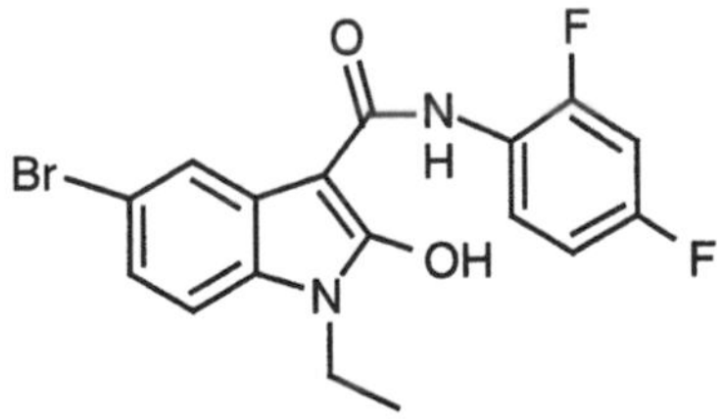

37

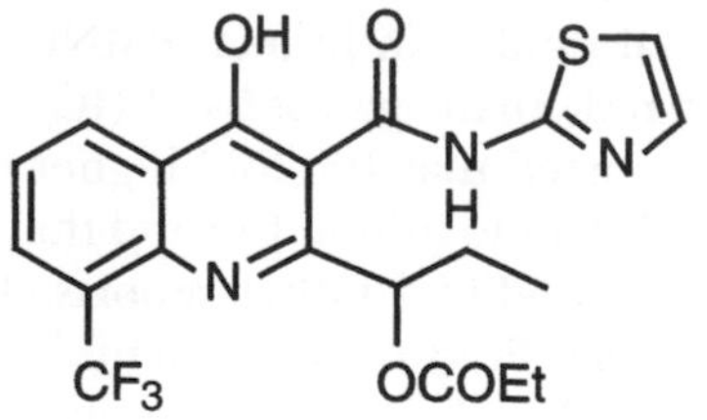

38

2.8   DuP 654

Du Pont has described a series of 2-substituted-1-napthols as potent 5-LO inhibitors.[109] These compounds also inhibited CO but in most cases were selective for 5-LO. One analog, DuP 654, 2-benzyl-1-napthol (**39**), inhibited 5-LO from RBL-1 cells (IC_{50} = 0.019 μM) and BSVM CO (IC_{50} = 3.4 μM). DuP 654 was effective in the mouse AA induced ear edema model, when applied topically. DuP 654 was also active when the inflammation was induced with a phorbol ester, a CO dependent model of inflammation.[110]

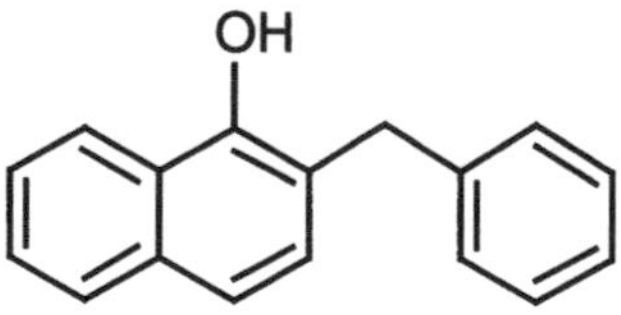

39

3.   DUAL INHIBITORS OF CYCLOOXYGENASE AND 5-LIPOXYGENASE DI-*tert*-BUTYLPHENOLS

3.1   Introduction

Since CO and 5-LO are oxidative enyzmes that convert AA into pro-inflammatory mediators, antioxidants have been investigated as possible dual inhibitors and antiinflammatory agents.[111,112] Derivatives of butylated hydroxytoluene (BHT), a potent antioxidant, were studied extensively, and from this work evolved the 2,6-di-*tert*-butylphenol class of antiinflammatory agents.[113] R-830, 2,6-di-*tert*-butyl-4-(2'-thenoyl)-phenol (**40**), reported by Riker in 1982 was an early example of this class.[114] R-830 inhibited guinea pig lung LO (IC_{50} = 20 μM) and BSVM CO (IC_{50} = 0.5 μM) and was active in several animal models of inflammation including CFE and adjuvant arthritis. The desirable *in vitro* and *in vivo* profile of R-830 generated increased interest in this class, resulting in the preparation of diverse 2,6-di-*tert*-butylphenols.

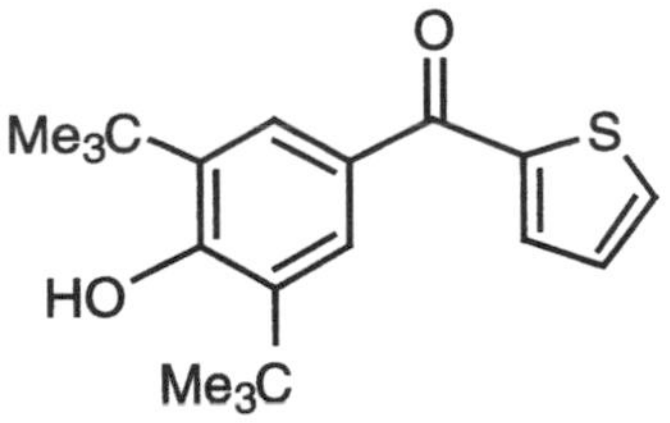

40

3.2 KME-4

α-(3,5-Di-*tert*-butyl-4-hydroxy-benzylidene)-γ-butyrolactone (**41**), KME-4, is a balanced dual inhibitor from Kanegafuchi.[115] In RBL cells, KME-4 inhibited 5-LO ($IC_{50} =$ 1.3 μM) and CO ($IC_{50} = 0.74$ μM) and was also active against BSVM CO. In a later report, KME-4 inhibited 5-LO in guinea pig peritoneal neutrophils ($IC_{50} = 11.5$ μM) and CO isolated from rabbit platelets ($IC_{50} = 0.44$ μM) but had no effect on platelet 12-LO.[116]

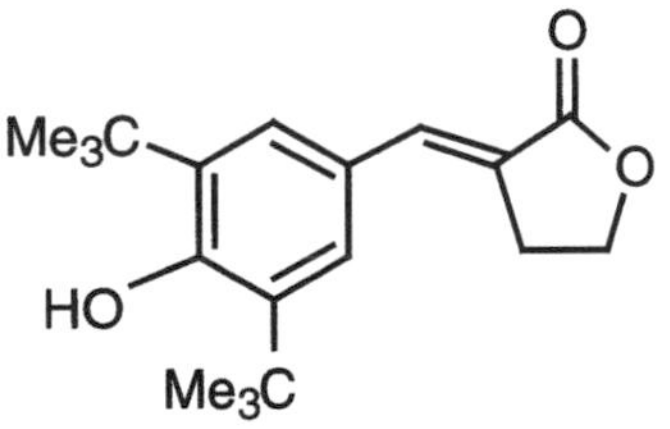

41

KME-4 was active in models of inflammation, including CFE, and possessed anti-pyretic and anti-platelet aggregation activities.[115] In addition, KME-4 inhibited carrageenan induced pleurisy in the rat.[117] In a model of established adjuvant arthritis, a 10 mg/kg dose of KME-4 was as effective as a 2 mg/kg dose of indomethacin.[118] While doses of 3 mg/kg of indomethacin are reported to be toxic in this model, only 10% mortality was observed with a 200 mg/kg dose of KME-4. An additional study reported that although KME-4 caused gastric lesions, its therapeutic index was superior to standard NSAIDs.[119]

There are two published SAR studies on KME-4. In the first of these the γ-butyrolactone portion was held constant and the substitution on the aromatic ring was varied.[120] Using CFE as the test protocol, the requirements for antiinflammatory activity were a *tert*-butyl group at C-3, an oxygen at C-4 and an alkyl group at C-5. Interestingly the 4-methoxy analog of KME-4, **42**, gave good inhibition of CO (IC_{50} = 1.2 μM) but was ineffective against 5-LO ($IC_{50} > 100$ μM). The most potent inhibitor of 5-LO, after KME-4, was the 5-methyl analog, **43**, ($IC_{50} = 10.7$ μM) which also inhibited CO ($IC_{50} = 0.89$ μM).

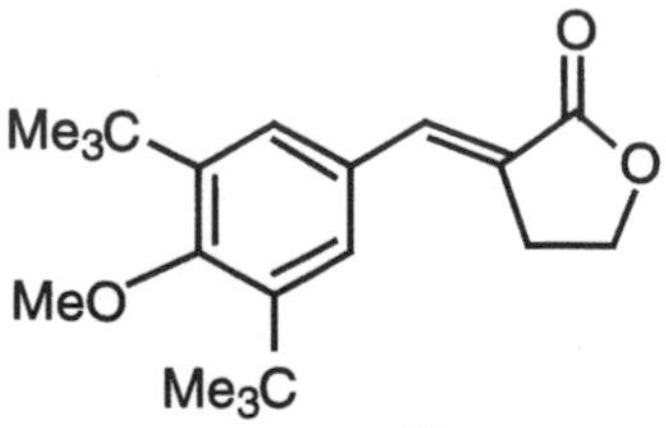

42

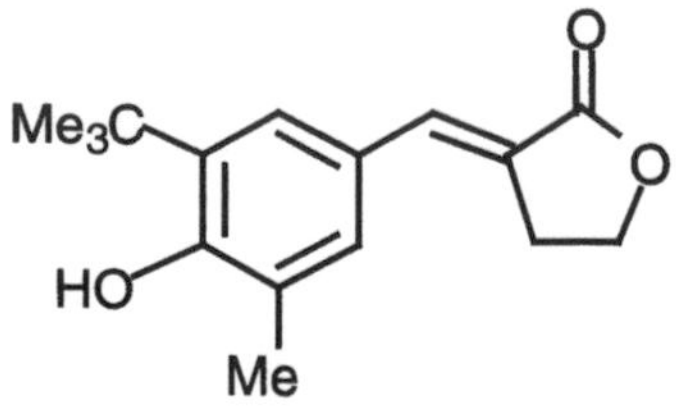

43

The second SAR study involved various 3,5-di-*tert*-butyl-4-hydroxystyrenes.[121] Several of these analogs inhibited CO and/or had antiinflammatory activity. However there was no correlation between potency in the *in vitro* assays and activity in the *in vivo* models. Only three derivatives, **44** ($IC_{50}=1.34\ \mu M$), **45** ($IC_{50}=1.16\ \mu M$) and **46** ($IC_{50} =1.37\ \mu M$), gave inhibition of 5-LO similar to KME-4.

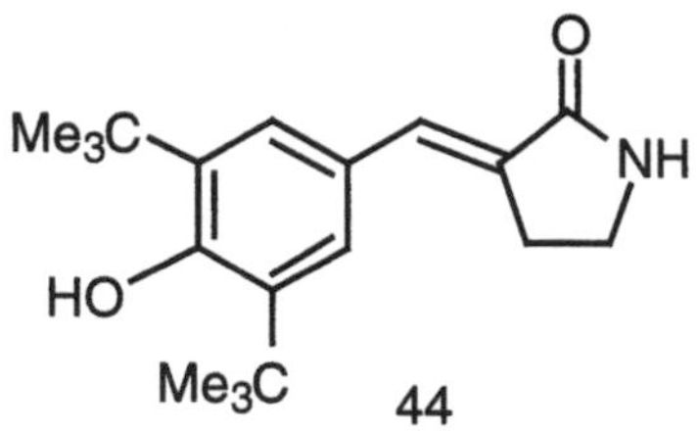

44

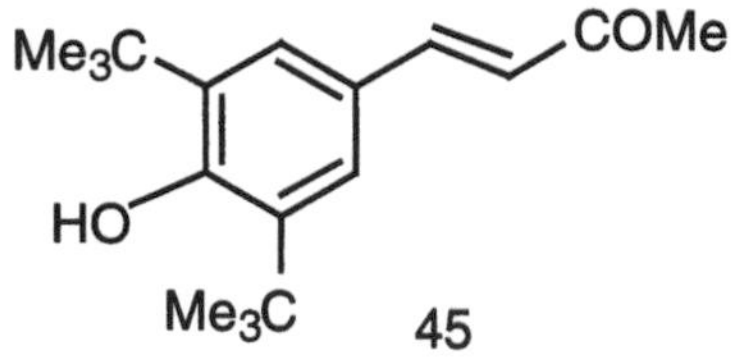

45

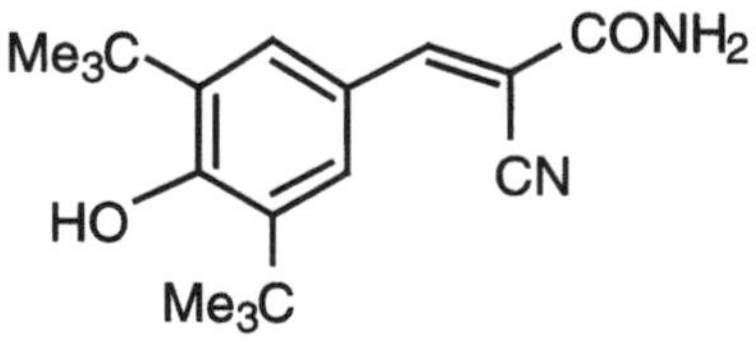

46

3.3 E-5110

N-Methoxy-3-(3,5-di-*tert*-butyl-4-hydroxy-benzylidene)-2-pyrrolidone (**47**), E-5110, was disclosed by Eisai to be a dual inhibitor.[122] E-5110 inhibited 5-LO from RBL cells ($IC_{50} = 0.90\ \mu M$) and CO from sheep seminal vesicles ($IC_{50} = 0.02\ \mu M$). E-5110

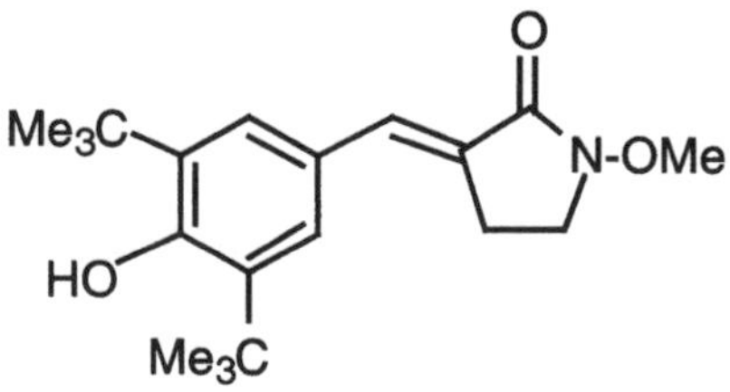

47

blocked production of PGE_2 from rat synovial cells ($IC_{50}=0.026$ μM) and LTB_4 from human neutrophils ($IC_{50}=0.20$ μM). The compound also inhibited the release of superoxide from zymosan stimulated human neutrophils.[123] E-5110 inhibited the production of IL-1 from human monocytes independent of stimulus.[124] When the stimulus was LPS, an IC_{50} of 1.21 μM was obtained.

Upon evaluation in several animal models of inflammation, including CFE, adjuvant-induced arthritis, established adjuvant arthritis and type II collagen induced arthritis, E-5110 was equipotent with indomethacin.[125,126] E-5110 also had analgesic effects similar to indomethacin. Although E-5110 caused gastric lesions, its therapeutic index was superior to indomethacin and piroxicam.

A brief SAR study of N-substituted E-5110 analogs including alkyl, substituted alkyl or acetyl, but no other alkoxy analogs, was reported.[127] While the unsubstituted, the N-methyl and the N-ethyl analogs had *in vitro* activity as dual inhibitors equivalent to E-5110, only the unsubstituted analog, **44**, was equipotent to E-5110 *in vivo*.

3.4 LY 178002, LY 256548

Lilly has identified 5-(3,5-di-*tert*-butyl-4-hydroxyphenyl)-methylene-4-thiazolidinone (**48**), LY 178002 and its N-methyl derivative (**49**), LY 256548 as dual inhibitors.[128] LY 178002 and LY 256548 inhibited 5-LO from human neutrophils ($IC_{50}=0.6$ μM and 1.6 μM) and CO from human platelets ($IC_{50}=4.2$ μM and 3.3 μM). These compounds also inhibited PLA_2 and LTB_4 production from human neutrophils. Both LY 178002 and LY 256548 were active in a rat model of adjuvant arthritis.

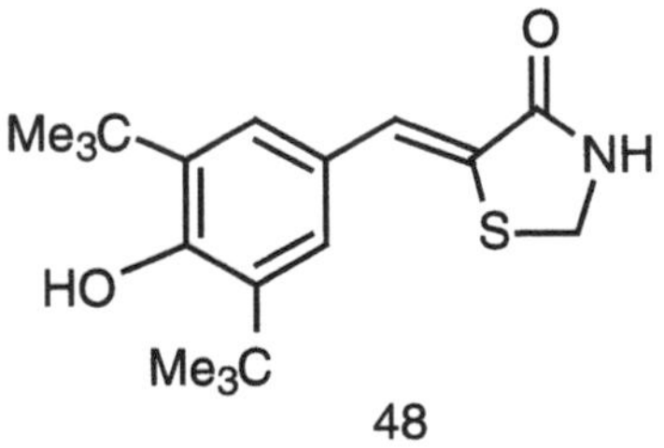

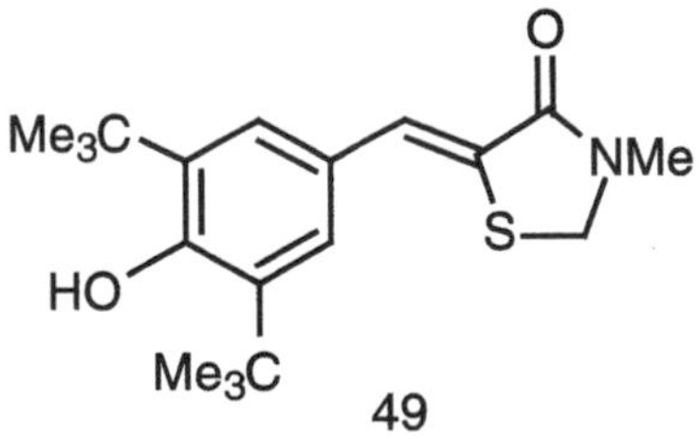

3.5 BI-L-93

Boehringer Ingelheim has reported a series of dual inhibitors containing a 2,6-di-*tert*-butyl-4-phenol linked by a double bond to various aromatic groups.[129] Of this series, (E)-2,6-di-*tert*-butyl-4-[2-(3-pyridinyl)ethenyl]phenol (**50**), BI-L-93, inhibited 5-LO from human leukocytes ($IC_{50}=2.7$ μM) and CO from human platelets ($IC_{50}=0.67$ μM). BI-L-93 had ED_{50}s of 2.1, 4.0 and 16.5 mg/kg in developing adjuvant arthritis, established adjuvant arthritis and CFE, respectively. Replacing one of the *tert*-butyl groups with a methyl or hydrogen retained *in vitro* potency, but the compounds lost *in vivo* activity. Reducing the size of the *tert*-butyl groups to *iso*-propyl gave increased activity against 5-LO. Expanding on this finding, Boehringer prepared a series of 2,6-di-methyl-4-phenol analogs. 2,6-Di-methyl-4-[2-(2-thienyl)ethenyl] phenol (**51**), BI-L-226 and 2,6-di-methyl-4-[2-(4-fluoro-phenyl)ethenyl]phenol (**52**), BI-L-239, were 10-fold more potent against 5-LO than CO.[130] The succinate ester of BI-L-226, BI-L-357, is under investigation as an anti-asthmatic agent.[131]

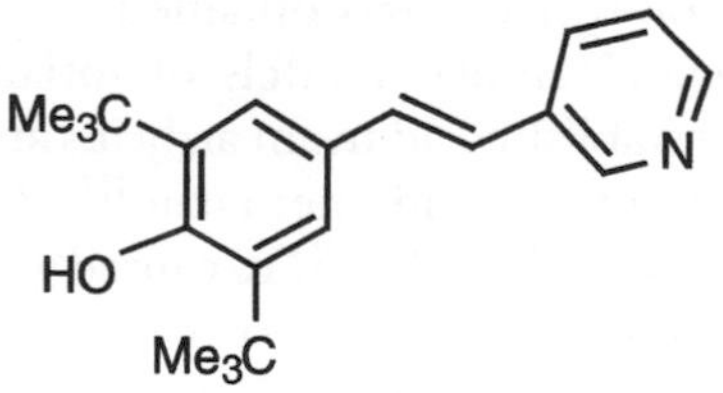

50

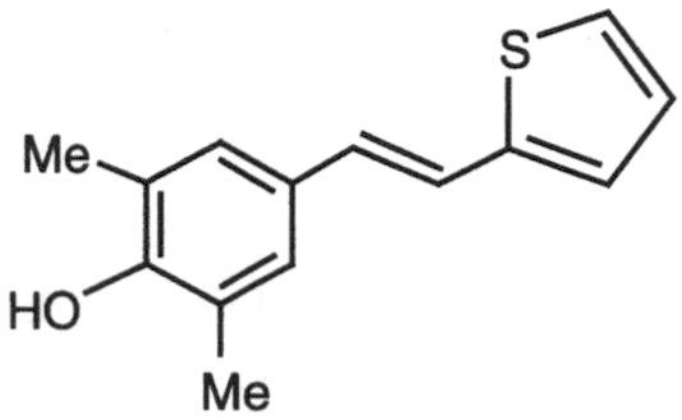

51

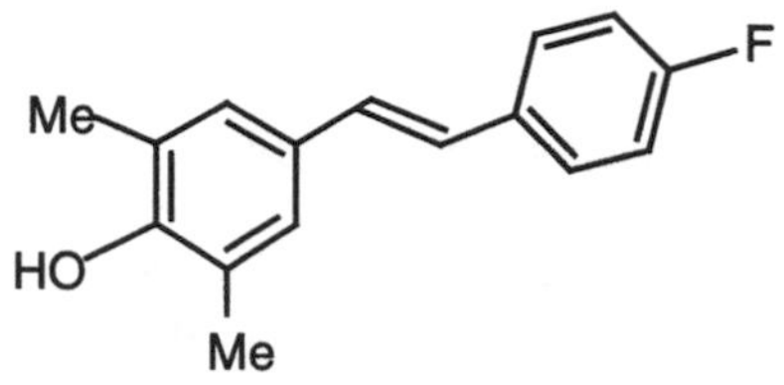

52

3.6   Tebufelone

Procter and Gamble's tebufelone, 1-(3,5-di-*tert*-butyl-4-hydroxyphenyl)-5-hexyn-1-one (**53**), NE 11740, inhibited CO from human monocytes by 80% at 2.5 μM and at 12.5 μM gave over 50% inhibition of 5-LO.[132] At the levels of tebufelone that caused inhibition of 5-LO, the production of IL-1 and TNF was increased. This effect was not observed at the lower concentrations of tebufelone that inhibited CO. Tebufelone also blocked the production of LTB_4 in ionophore stimulated human whole blood (ED_{50}=22 μM).[133] In a clinical trial with 120 patients, the anti-pyretic effect of 60 mg of tebufelone was equivalent to that of 650 mg of aspirin.[134]

A pharmacokinetic study with labeled tebufelone demonstrated that at a dose of 2 mg/kg the compound is 100% bioavailable.[135] However, tebufelone can be enzymatically hydroxylated on one of its *tert*-butyl groups providing the metabolites **54** and **55**.[136] A "metabolism-resistant" analog of tebufelone, **56**, has been prepared where the *tert*-butyl substituents have been replaced by trifluoromethyl groups.[137] No biological activity has been reported for this compound.

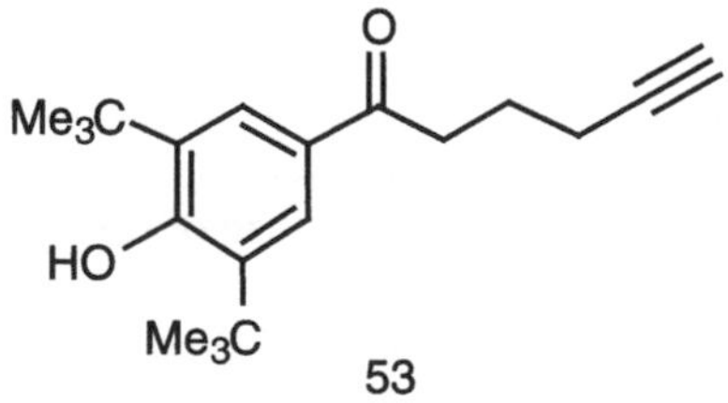

53

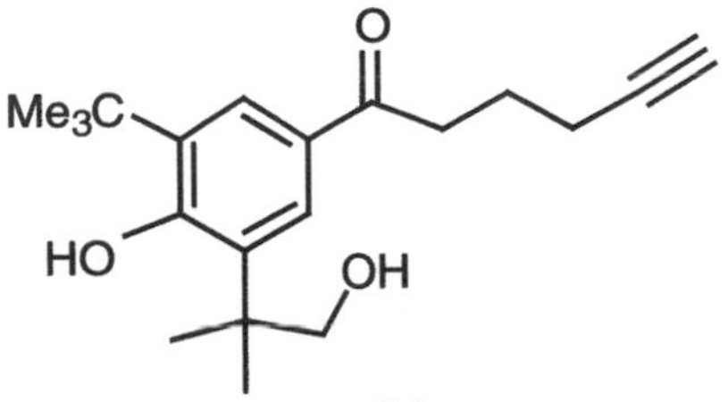

54 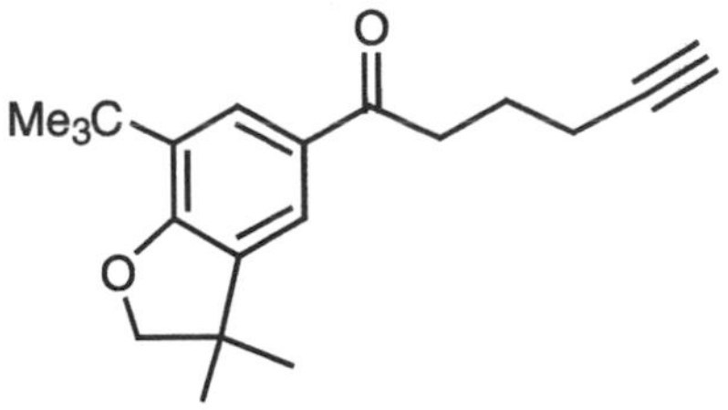   55

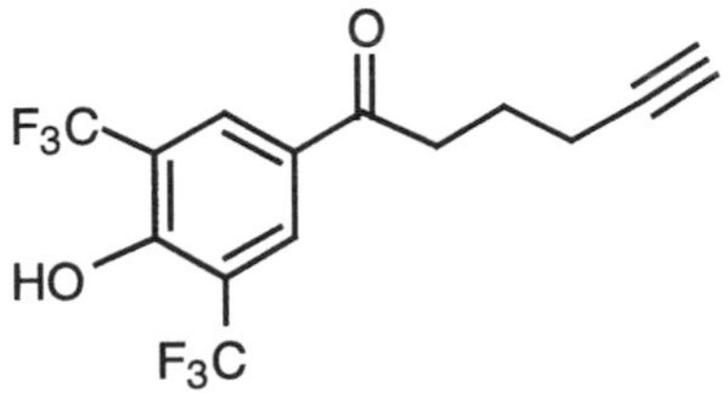

56

3.7 BF-389

Dihydro-4-(3,5-di-*tert*-butyl-4-hydroxybenzylidene)-2-methyl-2H-1,2-oxazin-3(4H)-one
(**57**), BF 389, from Biofor inhibited 5-LO ($IC_{50}=3.65$ µM) and CO ($IC_{50}=0.84$ µM)
from human leukocytes and was orally active in models of established and developing
adjuvant arthritis.[138] BF-389 also showed analgesic properties and was efficacious in
type II collagen arthritis and in a 21 day model of lipoidalamine induced arthritis
(LA).[139] Interestingly, BF 389 did not inhibit LTB_4 production in the animals from
the LA model. A recent report claims that BF 389 is a potent and selective inhibitor
of COX-2.[140]

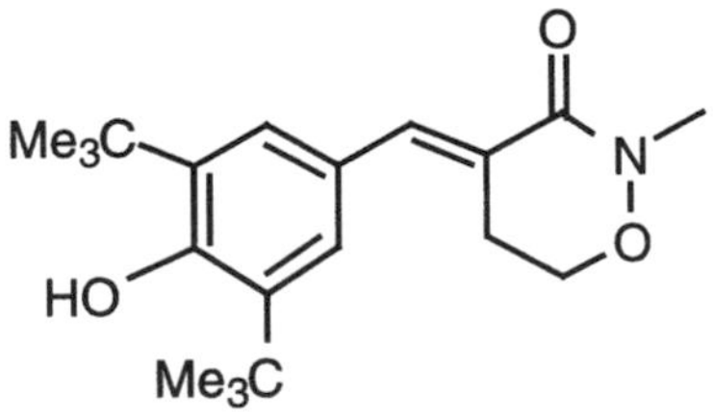

57

4. PARKE-DAVIS DUAL INHIBITOR STRATEGIES

4.1 Introduction

In this section several strategies for the production of dual inhibitors of 5-LO and
CO are described. Our major areas of emphasis were 1) modification of NSAIDs
by introducing a 5-LO inhibiting pharmacophore, 2) synthesis of a wide variety of

di-*tert*-butylphenols, and 3) synthesis of di-*tert*-butylhydroxypyrimidine analogs of key di-*tert*-butylphenols.

4.2 Modified NSAIDs

An efficient and elegant method for obtaining molecules expressing dual activities is to add a pharmacophore, which introduces the additional desired activity while retaining the initial activity, to compounds showing potent activity against one enzyme. The marketed NSAIDs were an obvious choice as CO inhibitors that could be converted to dual inhibitors of CO and 5-LO. Two major types of compounds were prepared using this strategy; 1) those where the NSAID's CO pharmacophore (a carboxylic acid in many cases) is modified or 2) the CO pharmacophore is retained and a 5-LO inhibitory pharmacophore is added at another part of the molecule. NSAIDs are readily available starting materials and provided easy access to modifed pharmacophore type compounds. For compounds containing both pharmacophores, new synthetic routes, often including functional group protection, were required.

Our initial strategy was to structurally modify the carboxylic acid group of various NSAIDs. The hydroxamic acid derivative of AA had been described as having 5-LO inhibiting activity.[141] Thus a series of hydroxamic acid derivatives of marketed NSAIDs including indomethacin (6) and the fenamate derivatives: meclofenamic acid (4), flufenamic acid (58) and mefenamic acid (59) were prepared.[142–144] This approach led to the series listed in Tables 1 and 2. Both the fenamate and indomethacin derivatives retained the CO inhibiting activity of their parent, and a 5-LO inhibiting component had been added. These compounds and all subsequent ones were assayed using RBL 5-LO and CO, with the data reported as the IC_{50} (μM) or the percent inhibition at the stated dose. Fenamate analogs, 60, 61, and 62, were

Table 1

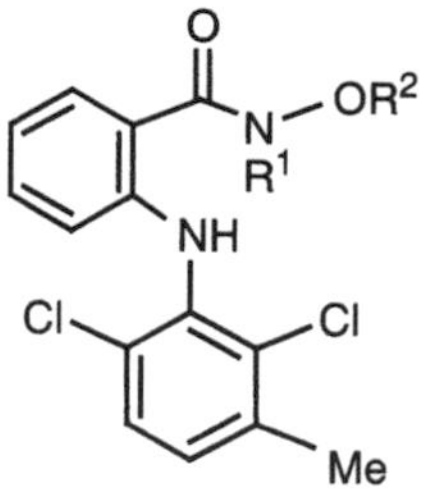

Compound	R^1	R^2	CO	5-LO	Ratio
4	Meclofenamic acid		0.10	24	240
60	H	H	1.1	3.9	3.5
61	H	Me	0.55	16	29
62	Me	H	15	1.5	0.1

Table 2

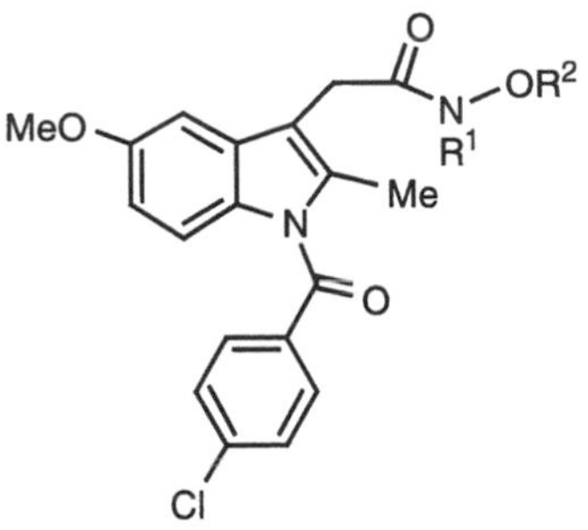

Compound	R^1	R^2	CO	5-LO	Ratio
6	Indomethacin		0.50	>100	200
63	H	H	1.1	7.5	6.8
64	Me	H	5.2	1.4	0.27
65	i-Pr	H	2.7	0.9	0.33
66	H	Me	<0.2	24	>120

balanced dual inhibitors compared to meclofenamic acid (ratio 240), and indomethacin analogs, **63** (ratio 6.8), **64** (ratio 0.27) and **65** (ratio 0.33) were balanced dual inhibitors compared to indomethacin (ratio >200). The activity of **66** (ratio >120), was comparable to that of indomethacin. These results indicated that potent CO inhibitors could be modified to give dual inhibitors.

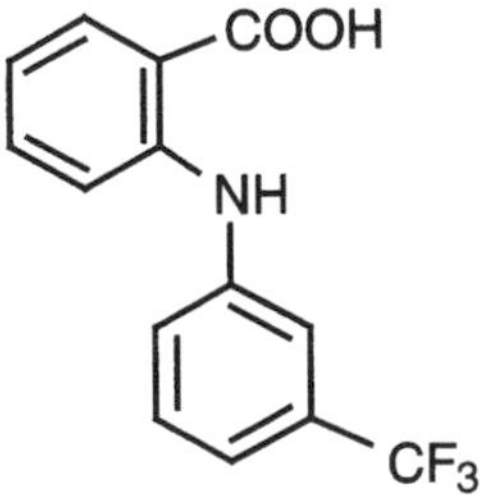

58

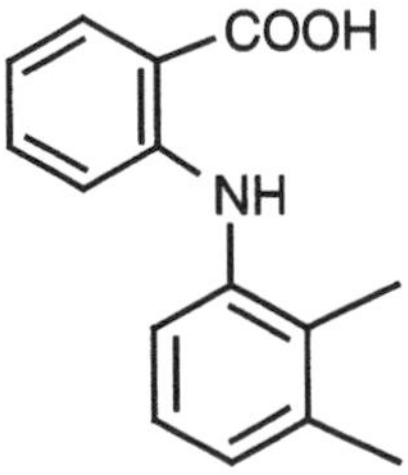

59

This series showed *in vivo* activity in CFE and also in mycobacterium footpad edema (MFE), another model predictive of clinical efficacy for NSAIDs. Compounds **61** (ED_{40}=1.3 mg/kg) and **62** (ED_{40}=0.2 mg/kg) showed comparable potency to meclofenamic acid (ED_{40}=8.2 mg/kg) in CFE, and **65** (CFE: ED_{40}=27 mg/kg; MFE: ED_{40}=0.82 mg/kg) showed the same range of potency as indomethacin (CFE: ED_{40} =36 mg/kg; MFE: ED_{40}=0.21 mg/kg). In a rat model of ulcerogenic potential, the dual inhibitors were less ulcerogenic than sodium meclofenamate (50% incidence of ulcers at an oral dose of 36 mg/kg) or indomethacin (50% incidence of ulcers at an oral dose of 5.4 mg/kg). In fact compounds **60**, **61**, **62**, and **65** showed a lack of ulcerogenicity in this model. This *in vivo* data strengthens the concept that dual inhibitors of CO and 5-LO will provide the antiinflammatory efficacy of NSAIDs with reduced potential for GI side-effects. These results, together with the prospects for additional efficacy by blocking both enzymes, confirmed dual inhibition as being a very attractive strategy for novel antiinflammatory drugs. As is documented for other hydroxamic acid analogs,[145,146] which were metabolised to the corresponding carboxylic acids, additional *in vivo* studies with these compounds indicated metabolism to the parent NSAID had occurred.

Although the concept of a dual inhibitor that is metabolised to a selective CO inhibitor is a potential alternate approach to novel antiinflammatory agents, it seemed less attractive than our original concept. Thus the next step was to modify the above series in a way to retain activity and to block metabolism. In the case of selective inhibitors of 5-LO this problem was solved with the reverse hydroxamates.[147,148] We adopted this approach, which involves linking the nitrogen to the aralkyl system directly rather than the carbonyl group i.e. the positions of the nitrogen and carbonyl groups are reversed, to prepare the class of modified NSAID dual inhibitors shown in Tables 3 and 4.[149-151] In the meclofenamic acid series, compounds lacking an α-methyl group **67**, **69–71** were potent dual inhibitors, whereas **68** with an α-methyl group in the side-chain was a selective 5-LO inhibitor. *In vivo* studies indicated the series lacking the α-methyl group underwent metabolism back to the parent NSAIDs. This same metabolism involving oxidation of the carbon adjacent to the hydroxamate, followed by cleavage, occurs with selective 5-LO inhibitors containing analogous side chains.[38] In the indomethacin series, compounds **72**, **73** and **74** were 5-LO inhibitors, whereas the compounds of the α-methyl series **75** and **76** were potent dual inhibitors, that were resistant to metabolism.

Table 3

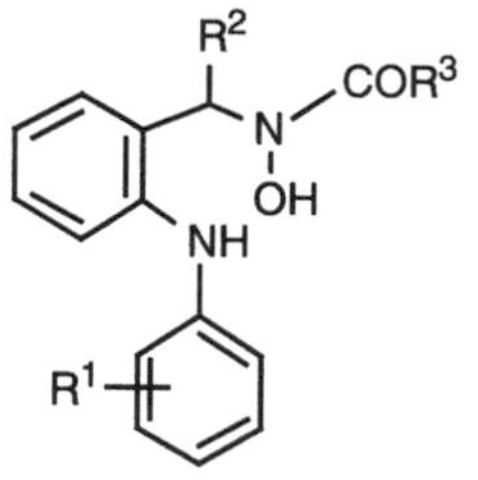

Compound	R^1	R^2	R^3	CO	5-LO
67	2,6-di-Cl, 3-Me	H	Me	5.0	0.18
68	2,6-di-Cl, 3-Me	Me	Me	N @ 32	1.8
69	3-CF$_3$	H	Me	5.2	0.28
70	3-CF$_3$	H	NHMe	2.2	0.12
71	3-CF$_3$	H	OEt	5.9	0.58

Table 4

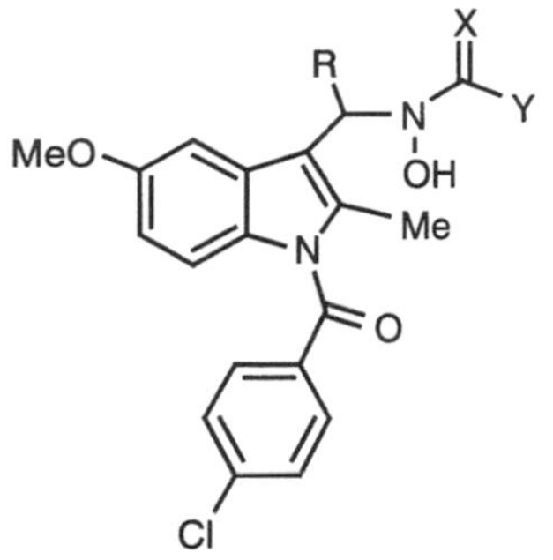

Compound	R	X	Y	CO	5-LO
72	H	O	Me	15	1.4
73	H	O	NH$_2$	>20	0.34
74	H	S	NHMe	7.1	0.40
75	Me	O	Me	2.4	0.83
76	Me	O	NH$_2$	0.89	1.4
77	Me	S	NHMe	7.0	0.57

The tetrazole group is a well known bioisoster for the carboxylic acid group, and the tetrazole analog of meclofenamic acid had been shown to be an antiinflammatory agent.[152] This led to the preparation of the compounds, **78–88** listed in Table 5, in which the carboxylic acid of meclofenamic acid was replaced with various heterocyclic systems.[153–157] The 1,3,4-oxadiazole-2-thione **79** and 1,3,4-thiadiazole-2-thione **80** were dual inhibitors, which were active in CFE (**79**: ED$_{50}$=8.5 mg/kg; **80**: ED$_{50}$=4.7 mg/kg) but not MFE, possibly due to facile metabolism at the thio group. The corresponding 1,2,4-triazole-3-thione was inactive. Replacing the 2-thio group of **79** and **80** with the acidic aminonitrile group provided **85** and **86** that retained *in vitro* activity, but were inactive in CFE. Replacement of the 2-thio group with a basic guanidine group gave **87** and **88**, both selective 5-LO inhibitors. This series provided another example of a successful modification of NSAIDs that lead to dual inhibitors, and also identified different functional groups that could be used to control the type of biological activity desired, and provided access to a wide variety of salt forms.

A second class of modified NSAIDs, that we investigated, retained the CO inhibiting pharmacophore, the arylcarboxylic acid, and a 5-LO inhibiting pharmacophore was introduced into the second aromatic ring, to provide **89–95**. These compounds have the advantage of having both pharmacophores, so each can be fine tuned. A set of fenamates of this type is listed in Table 6.[158] The intermediate esters showed good 5-LO inhibitory activity as expected, and **89**, **90** and **91** were dual inhibitors. Conversion to the target carboxylic acids resulted in **94** and **95**, both selective CO inhibitors. This was a rare situation since compounds with a reversed hydroxamate

Table 5

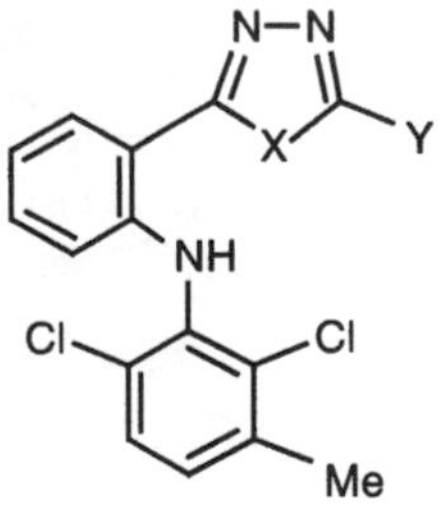

Compound	X	Y	CO	5-LO
78	O	OH	N @ 16	N @ 16
79	O	SH	0.70	0.74
80	S	SH	1.7	1.4
81	O	NH_2	N @ 16	45% @ 16
82	S	NH_2	6.1	0.69
83	NH	SH	N @ 16	N @ 16
84	NMe	SH	N @ 16	86% @ 16
85	O	NHCN	0.25	0.89
86	S	NHCN	1.9	0.29
87	O	$NHC(NH)NH_2$	N @ 16	46% @ 16
88	S	$NHC(NH)NH_2$	N @ 16	100% @ 16

Table 6

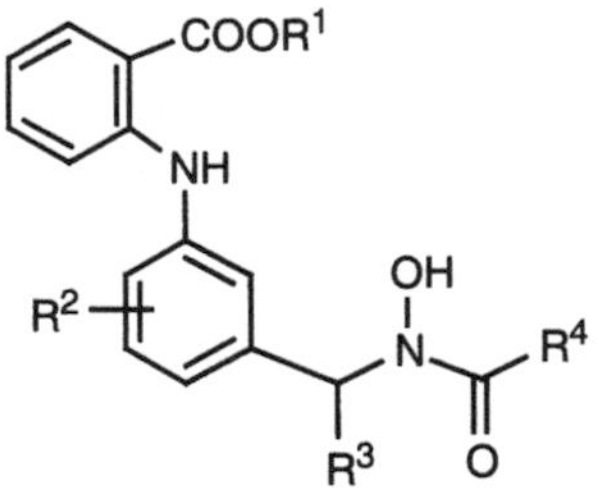

Compound	R^1	R^2	R^3	R^4	CO	5-LO
89	Me	2-Cl	H	Me	4.3	0.58
90	Me	2-Cl	H	OEt	0.67	0.21
91	Me	2,6-di-Cl	H	Me	6.0	0.36
92	Me	2-Cl	Me	Me	N @ 10	100% @ 10
93	Me	2,6-di-Cl	Me	Me	N @ 10	100% @ 10
94	H	2-Cl	H	Me	79% @ 10	N @ 10
95	H	2-Cl	Me	Me	53% @ 10	N @ 10

Table 7

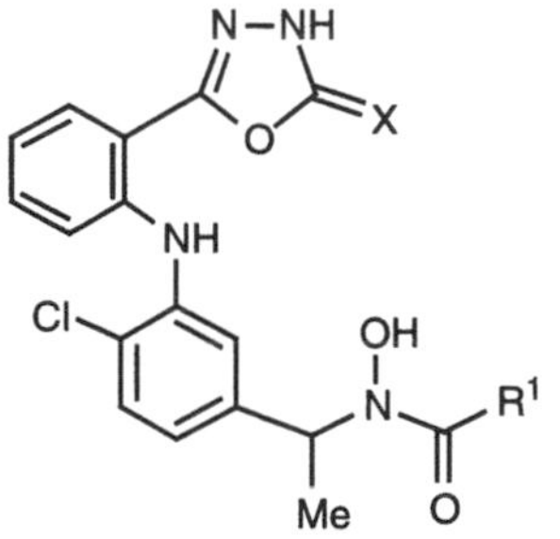

Compound	X	R^1	CO	5-LO
96	O	Me	1.9	0.68
97	O	OEt	0.86	1.1
98	O	NHMe	0.36	1.9
99	S	Me	1.1	9.6
100	S	OEt	0.91	1.1
101	NH	Me	N @ 10	100% @ 10

side chain attached to an aromatic nucleus are usually 5-LO inhibitors. The α-methylesters **92** and **93** lacked CO inhibitory activity.

To circumvent the loss of 5-LO activity in the above dual pharmacophore targets, we replaced the carboxylic acid group with heterocycles.[159] A set of fenamates containing these modifications is shown in Table 7. In contrast to the previous α-methyl series **92** and **93**, these compounds were dual inhibitors **96–100**, with the exception of 2-amino-1,3,4-oxadiazole **101**. 1,3,4-Oxadiazole-2-thione **100** was active in CFE (41% inhibition @ 30 mg/kg), but was inactive in MFE probably due once again to metabolism at the thiol group. Thus, replacing the carboxylic acid with heterocycles in this series yielded modified NSAID dual inhibitors which were not metabolised to NSAIDs and have the potential to be developed as drugs.

4.3   Di-*tert*-butylphenols

As discussed in Section 3, several 2,6-di-*tert*-butylphenol analogs have been reported to be dual inhibitors. Those compounds which were potent CO inhibitors showed antiinflammatory activity, whereas compounds with a potent antioxidant component, but no CO inhibitory component were inactive *in vivo*. There were also many problems associated with the antioxidant class of selective 5-LO inhibitors, including lack of specificity, methemoglobin formation, insolubility, poor absorption, rapid metabolism, and lack of biochemical efficacy *in vivo*. We undertook extensive investigations of various classes of di-*tert*-butylphenols to assess the potential for obtaining dual inhibitors of 5-LO and CO, without the deficiencies described above.

Table 8

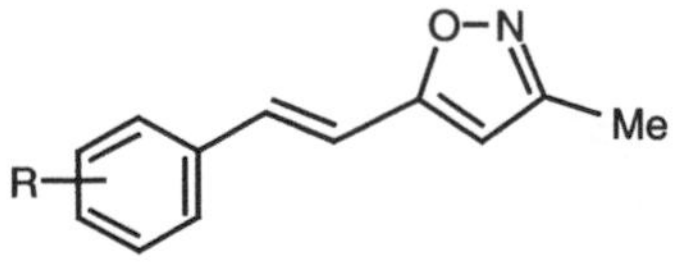

Compound	R	CO	5-LO
102	H	>30	>30
103	4-OH	4.0	20
104	3-OMe, 4-OH	12	5.5
105	3,5-di-OMe, 4-OH	22	1.4
106	3,5-di-Me, 4-OH	1.2	4.5
107	3,5-di-t-Bu, 4-OH	1.5	2.4

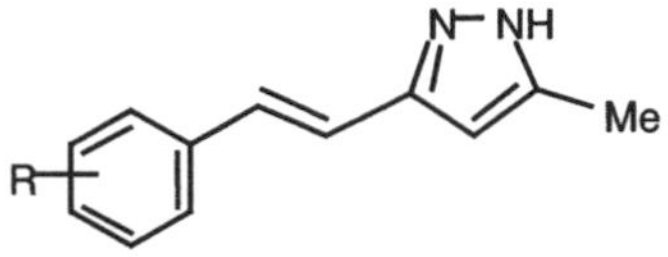

Compound	R	CO	5-LO
108	3,5-di-Me, 4-OH	4.6	0.80
109	3,5-di-t-Bu, 4-OH	0.90	3.0

Initially we investigated a variety of phenols linked to oxazoles and pyrazoles via styryl groups. A selection of compounds **102–109** from this work is shown in Table 8.[160–162] In this series a phenolic group was required for potent activity against both enzymes. Electron releasing groups such as methoxy lead to enhanced selectivity for 5-LO inhibition **104** and **105**, whereas alkyl groups gave dual inhibition **106–109**. The di-$tert$-butylphenol dual inhibitors, **107** (CFE: ID_{25}=24 mg/kg) and **109** (CFE: ID_{25}=15 mg/kg) showed *in vivo* antiinflammatory activity, and were not ulcerogenic up to 200 mg/kg in a rat model. Compounds such as **106**, a potent CO inhibitor, lacked *in vivo* antiinflammatory activity presumably due to rapid metabolism via glucoronidation of the hydroxyl group. From the biological data obtained with this series, it was demonstrated that hydroxy and alkyl groups were required for potent dual inhibition, and the 2,6-di-$tert$-butylphenol configuration is required to prevent rapid metabolism at the hydroxy group.

Further evaluation of **107** revealed major deficiencies: low oral bioavailability due to poor systemic absorption and long plasma half-life due to lack of metabolism. Even so, the superior oral activity and lack of ulcerogenicity shown by **107** and **109** made the di-$tert$-butylphenols a favored class of compounds for development. Our objective became to build a more desirable pharmacokinetic profile into these compounds.

Table 9

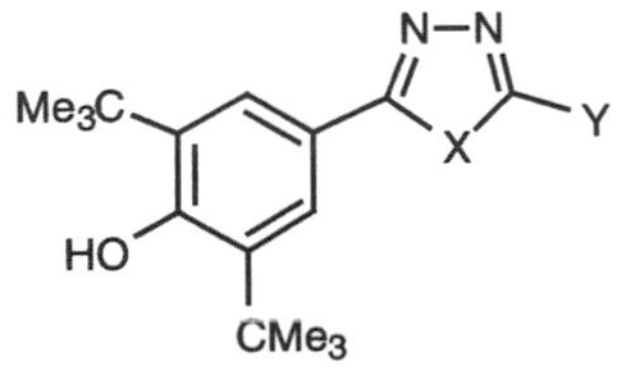

Compound	X	Y	CO	5-LO
110	S	SH	0.80	2.8
111	S	OH	>16	5.9
112	S	NH_2	0.13	1.4
113	O	SH	3.7	4.8
114	O	OH	2.5	1.4
115	O	NH_2	2.4	0.70
116	NH	SH	0.78	11
117	NH	OH	5.5	4.5
118	NH	NH_2	>30	5.1

A major challenge was the high lipophilicity of the compounds caused by the two *tert*-butyl groups. This lipophilicity was reduced by removing the styryl linker group and adding an additional hetero atom to the heterocyclic systems. The di-*tert*-butylphenol 1,3,4-oxadiazoles, 1,3,4-thiadiazoles and 1,2,4-triazoles **110–118** (Table 9) were the result of this approach.[163–166] The compounds were dual inhibitors with the exception of **111** and **118**, which were selective 5-LO inhibitors. 1,3,4-Thiadiazoles **110** (tested as its choline salt) and **112** were especially potent *in vitro* and showed *in vivo* activity (**110**: CFE: $ID_{40}=1.1$ mg/kg; MFE: $ID_{40}=7.7$ mg/kg; AIP: $ID_{40}=7.2$ mg/kg; **112**: CFE: $ID_{40}=1.9$ mg/kg; MFE: $ID_{40}=2.3$ mg/kg) on a par with naproxen (CFE: $ID_{40}=0.7$ mg/kg; MFE: $ID_{40}=6.3$ mg/kg; AIP: $ID_{40}=3.8$ mg/kg). 1,3,4-Thiadiazole-2-thione **110** caused no GI lesions at oral doses of 200 mg/kg in acute and chronic studies.[167] **110** ($ID_{50}=0.23$ mg/kg) also showed an analgesic effect equivalent to indomethacin ($ID_{50}=0.87$ mg/kg).

To expand this series further, and to obtain a wide variety of functional groups that would allow for the formation of acid and base salts, an investigation of the SAR at the 2 position was carried out. A selection of the more potent compounds **119–125** is shown in Table 10. The thiol group at 2 was replaced with an acidic aminonitrile, basic guanidine or neutral thiomethyl or hydrazide group. The 5-LO and CO inhibitory potency was retained in each case, thus providing a broad range of di-*tert*-butylphenol dual inhibitors.

The effect of the linking group on activity was investigated **126–130** (Table 11), and a dramatic effect was observed in contrast to the results of changes at the 2 position.[168–170] A comparison of the potency of these compounds with that of **110** indicates a preference toward a single bond for obtaining the highest level of

Table 10

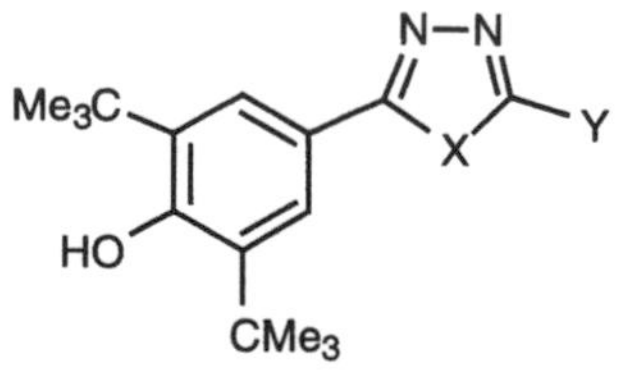

Compound	X	Y	CO	5-LO
119	S	NHCN	4.4	1.5
120	S	NH(C=NH)NH$_2$	2.4	0.78
121	S	NHNH$_2$	0.70	0.89
122	S	SMe	0.41	1.2
123	O	NHCN	0.34	1.2
124	O	NH(C=NH)NH$_2$	0.06	1.8
125	O	SMe	0.03	0.30

CO inhibition. Compounds with a double bond **126** or sulfur **127** linker retained 5-LO inhibition, but all the linking groups investigated resulted in a decrease of CO inhibition. Especially dramatic was the complete loss of activity in the case of **130**, a compound with an amide linker.

The directly linked isomeric 1,2,4-oxadiazoles and 1,2,4-thiadiazoles **131–135** (Table 12) were also potent dual inhibitors, further broadening the scope of the heterocyclic systems that could be attached to the di-*tert*-butyl pharmacophore.[171–173] 1,2,4-Oxadiazole **131** was a very strong CO inhibitor, showing a ratio of 68:1 for inhibition of CO over 5-LO.

Table 11

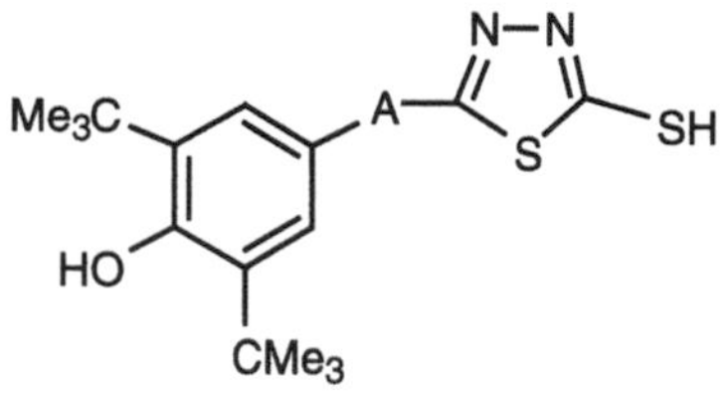

Compound	A	CO	5-LO
126	CH=CH	5.2	1.8
127	S	3.6	1.1
128	O	63% @ 10	73% @ 10
129	CH$_2$CH$_2$	N @ 10	4.5
130	CONH	N @ 10	N @ 10

Table 12

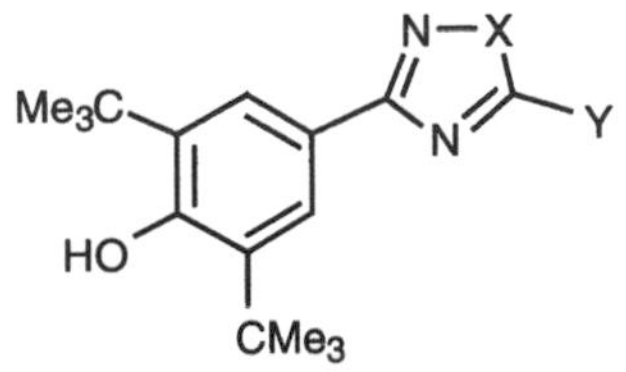

Compound	X	Y	CO	5-LO
131	O	SH	0.05	3.4
132	O	OH	4.7	N @ 10
133	O	NH$_2$	3.0	2.5
134	S	NH$_2$	2.2	0.33

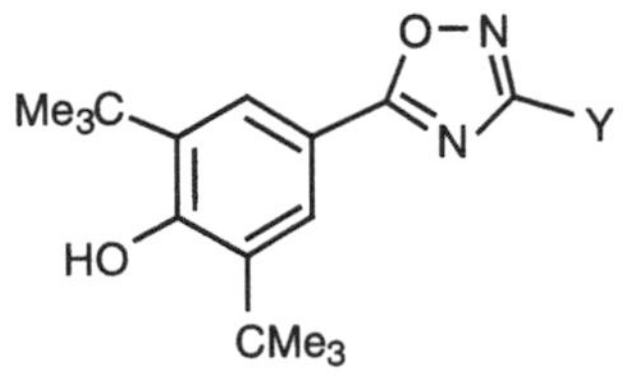

Compound	Y	CO	5-LO
135	NH$_2$	1.9	3.6

KME-4 , E-5110, LY 178002 and BF-389 are examples of the benzylidene class of di-*tert*-butylphenol antiinflammatory agents, which are structurally distinct from the types described above. The thiazolones and oxazolones, listed in Table 13, which were readily accessible from 3,5-di-*tert*-butyl-4-hydroxybenzaldehyde, are other examples of this class. This series contains a one carbon linker between the two rings, and the key substituents from the 1,3,4-thiadiazole series (Tables 9 and 10) were introduced in the 2-position. This strategy produced a broad range of compounds **136–147** which were potent dual inhibitors, and was the most fruitful in providing potential development candidates.[174–179] Both the choline salt of **137**, **CI-987** (CFE: ID$_{40}$=0.5 mg/kg; MFE: ID$_{40}$=2.0 mg/kg) and the methanesulfonate salt of **138**, **CI-1004** (CFE: ID$_{40}$=1.5 mg/kg; MFE: ID$_{40}$=0.6 mg/kg) showed good *in vivo* activity after oral administration. These salts were also orally active in AIP (**137**: ID$_{40}$=2.2 mg/kg; **138**: ID$_{40}$=1.7 mg/kg), and were as efficacious as sodium meclofenamate (ID$_{40}$=1.3 mg/kg) in this model. Neither **137** or **138** caused ulcers upon oral administration at 200 mg/kg to fasted rats. The hydrochloride salt of guanidine, **140** also showed *in vivo* activity (CFE: 34% inhibition @ 10 mg/kg; MFE: ID$_{40}$=2.2 mg/kg) and lack of ulcerogenicity at 200 mg/kg. **140** is a rare example of a potent basic CO inhibitor. This series provided an excellent illustration of the utility of heterocyclic systems to balance the lipophilicity of the di-*tert*-butyl pharmacophore and the use of key substituents to allow for salt formation and

DAVID T. CONNOR and DIANE H. BOSCHELLI

Table 13

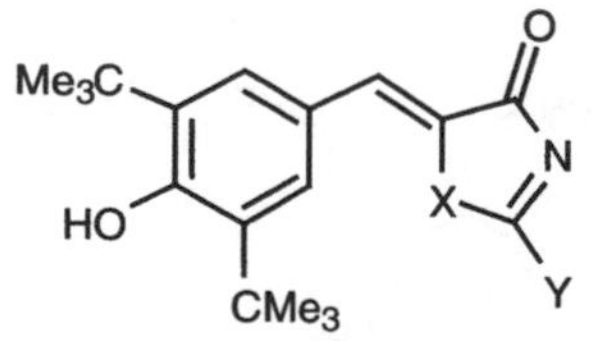

Compound		X	Y	CO	5-LO
136		S	SH	0.012	0.38
137	CI-987	S	OH	0.35	1.4
138	CI-1004	S	NH_2	2.6	1.8
139		S	NHCN	0.16	0.63
140		S	$NH(C=NH)NH_2$	0.08	0.90
141		O	SH	1.7	0.84
142		O	OH	1.0	1.7
143		O	NHCN	1.1	2.6
144		O	$NH(C=NH)NH_2$	0.34	1.2
145		NH	SH	0.77	0.092
146		CH_2	SH	83% @ 10	94% @ 10
147		NMe	OH	4.5	0.78

provided two antiinflammatory agents **137** and **138**, which appear to have significant advantages over standard selective CO inhibitors.

4.4 Di-*tert*-butylhydroxypyrimidines

The introduction of nitrogens into the phenolic ring is another approach to balance the lipophilicity of the *tert*-butyl groups, and potentially improve the compound's absorption/elimination profiles. These nitrogens provide additional polar groups and handles for salt formation. After examining routes to various nitrogen containing rings, the symmetrical 5-hydroxy-1,3-pyrimidines were selected for evaluation. An advantage of the 1,3-pyrimidines was that the symmetry eliminates many of the difficulties associated with the synthesis of various unsymmetrical nitrogen containing structures. Also the 4,6-di-*tert*-butyl-5-hydroxy-1,3-pyrimidines were not described in the literature and potential routes to these compounds had broad applicability for the synthesis of pyrimidine versions of di-*tert*-butylphenol dual inhibitors, which had been widely exploited by many pharmaceutical companies.

4,6-Di-*tert*-butyl-5-hydroxy-1,3-pyrimidine analogs of our key di-*tert*-butylphenols, and of compounds taken to the clinic by other labs including KME-4, E- 5110,

Table 14

Compound	CO	5-LO
148	74% @ 10	N @ 10
149	1.4	1.1
150	0.19	5.8
151	0.021	5.5
152	0.003	81% @ 10
153	0.014	2.5
154	0.0026	0.16
155	0.33	0.16
156	56% @ 10	N @ 10
157	2.0	6.0
158	0.022	100% @ 10
159	85% @ 10	46% @ 10
160	0.16	7.9

tebufelone, BI-L-93 and BF-389 were prepared and evaluated (Table 14). Pyrimidine analogs of the directly attached di-*tert*-butylphenol series (Table 9), tended to be less potent and lack a 5-LO inhibiting component. A typical example is 1,3,4-thiadiazole-2-thione, **148**, which was a selective CO inhibitor in contrast to the corresponding

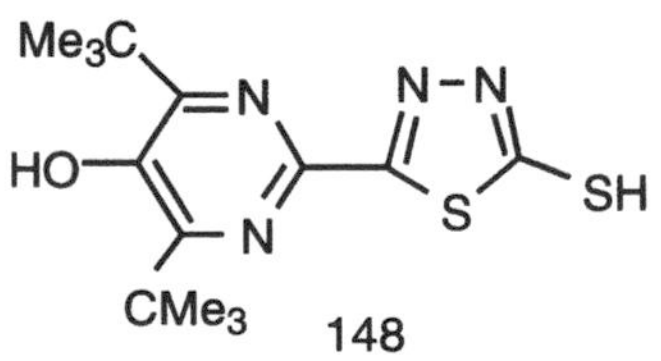

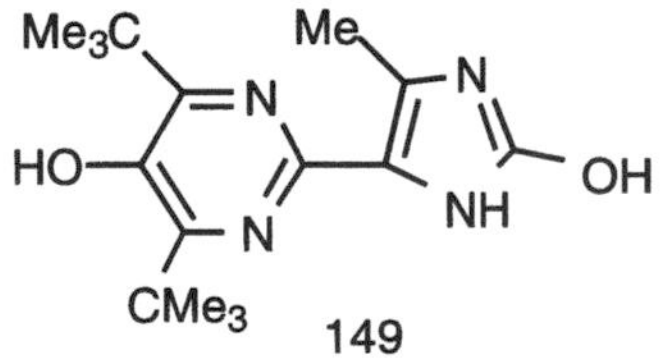

di-*tert*-butylphenol, **110**. An exception to this trend was imidazole, **149**, which was a good inhibitor of both enzymes, and was selected for further study.[180] Inserting a styryl linker between the two rings provided a series of dual inhibitors.[181] Typical examples were the compounds **150** and **151**, which still followed a trend towards selective inhibition of CO. This trend was continued with the carbonyl linked

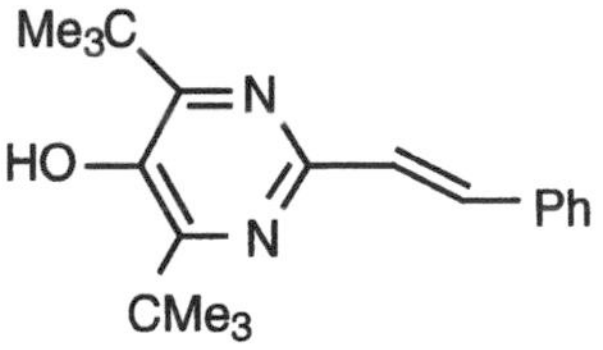

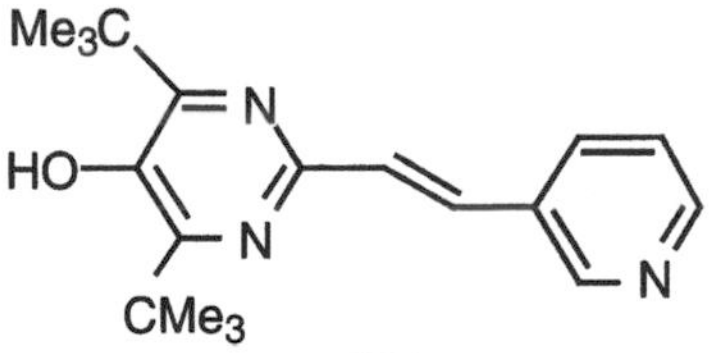

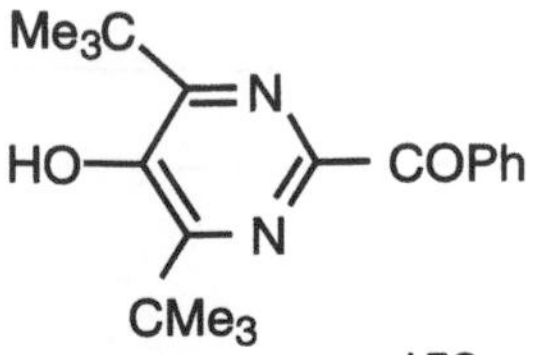

152

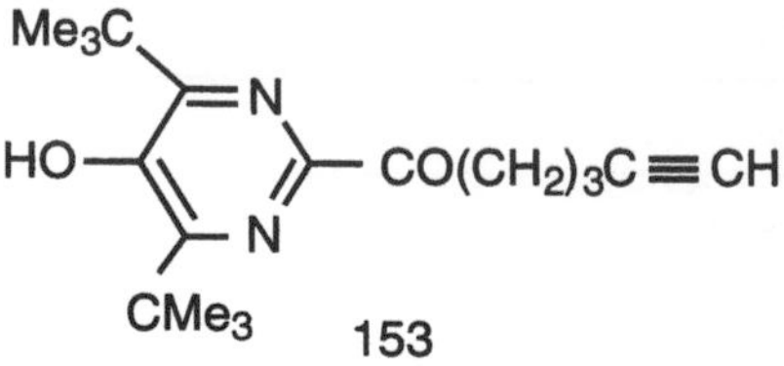

153

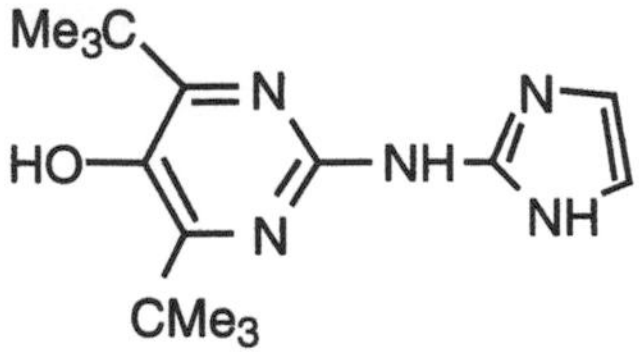

154

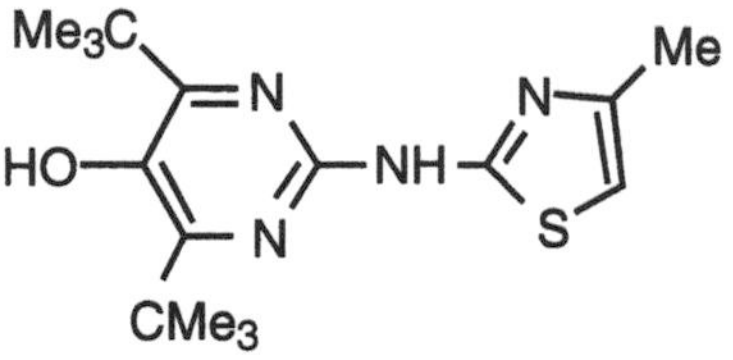

155

compounds.[182] Carbonyl linked pyrimidines, **152** and **153**, were extremely potent selective inhibitors of CO, whereas their di-*tert*-butylphenol counterparts were balanced dual inhibitors. Amino linked pyrimidines, **154** and **155**, the best 5-LO inhibitors of the pyrimidines, were also strong inhibitors of CO.[183]

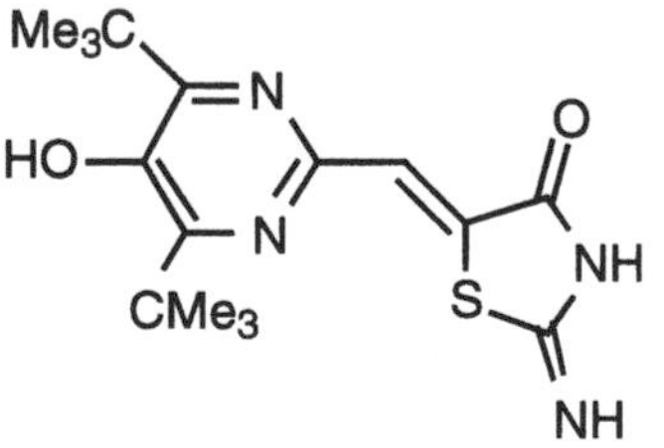

156

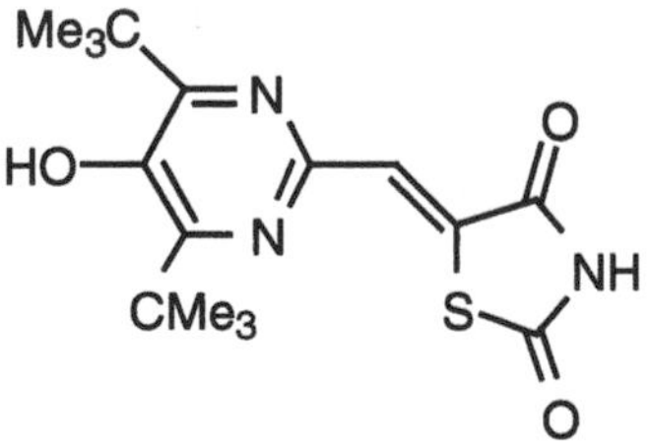

157

Compounds **156** and **157** are di-*tert*-butylhydroxypyrimidine analogs of two of our di-*tert*-butylphenols that were taken into development, and again they showed selectivity for CO inhibition.[184] Compounds, **158**,[184] **159**,[184] and **160**,[185] pyrimidine analogs of KME-4, E-5110, and BF-389 respectively, were dual inhibitors, but with a much stronger CO inhibiting component. The series of 4,6-di-*tert*-butyl-5-hydroxy-1,3-pyrimidines all showed a trend towards greater inhibitory potency against CO and this was a general property of the class compared to the corresponding di-*tert*-butylphenols.

5. SUMMARY

There is an urgent demand for innovative therapeutic agents to treat inflammatory diseases including OA and RA. Current first-line therapy with NSAIDs provides symptomatic relief but does not halt the progression of the disease, and subjects

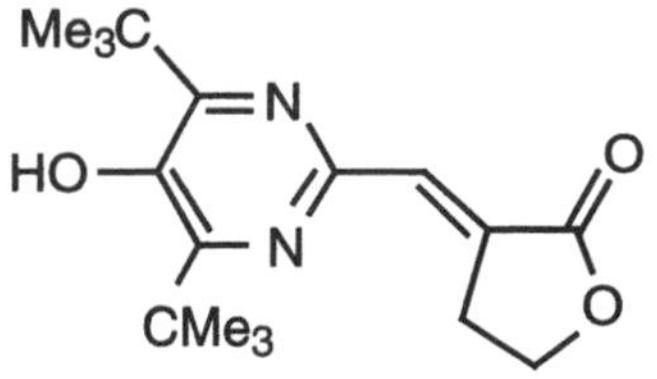

158

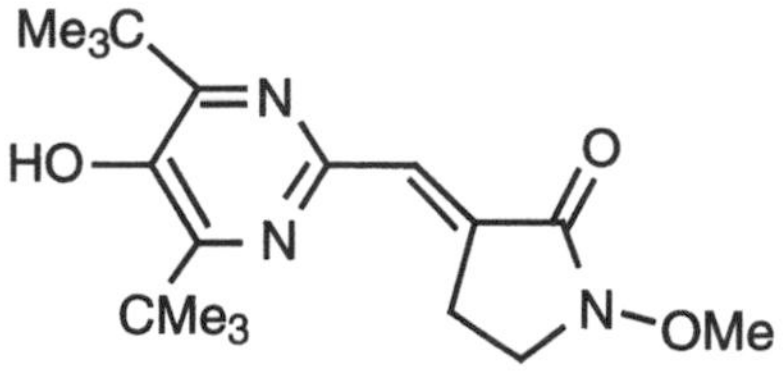

159

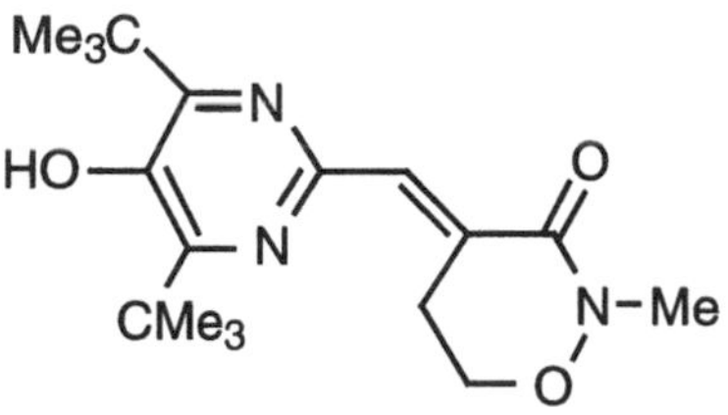

160

patients, especially the elderly, to serious side-effects. Dual inhibitors hold the promise of being more efficacious, with reduced side-effects and may also be disease modifying. Upon closer examination, many dual inhibitors of CO and 5-LO are actually highly selective for one enzyme and/or have poor bioavailabilty. The fact that some dual inhibitors also inhibit cytokine production could add to their clinical efficacy, but would make interpretation of the contributions of the various components difficult. Although authoritative clinical data for a well defined dual inhibitor has not been published, the concept of a dual inhibitor being a superior anti-inflammatory drug to either standard NSAIDs or the developing class of 5-LO inhibitors, appears to be valid. Future clinical studies should determine whether dual inhibitors will be a major advance in the treatment of inflammatory diseases.

6. REFERENCES

1. Salmon, J.A. and Higgs, G.A. (1987) Prostaglandins and leukotrienes as inflammatory mediators. Br. Med. Bull., 43, 285–296.
2. von Euler, U.S. (1936) Specific vasodilating and plain muscle stimulating substances from accessory genital glands in man and certain animals. J. Physiol. (Lond.), 88, 213–215.
3. Bergstrom, S. and Sjovall, J. (1957) The isolation of prostaglandins. Acta Chem. Scand., 11, 1086.
4. Samuelsson, B. (1965) The prostaglandins. Angew. Chem. Int. Ed. Eng., 4, 410–416.
5. Van Dorp, D.A., Beerthius, R.K., Nugteren, D.H. and Vonkeman H. (1964) The biosynthesis of prostaglandins. Biochim. Biophys. Acta, 90, 204–207.
6. Vane, J.R. (1971) Inhibition of prostaglandin synthesis as a mechanism of action of aspirin like drugs. Nature New Biol., 231, 232–235.
7. Miller, T.A. (1983) Protective effects of prostaglandins against gastric mucosal damage: current knowledge and proposed mechanisms. Am. J. Physiol., 245, G601–G623.
8. Soll, A.H., Weinstein, W.M., Kurata, J. and McCarthy, D. (1991) Nonsteroidal anti-inflammatory drugs and peptic ulcer. Ann. Int. Med., 114, 307–319.
9. Graham, D.Y., Agrawal, N.M. and Roth, S.H. (1988) Prevention of NSAID-induced gastric ulcer with misoprostol: multicentre, double-blind, placebo-controlled trial. Lancet, 1277.

10. O'Banion, M.K., Winn, V.D. and Young, D.A. (1992) cDNA cloning and functional activity of a glucocorticoid-regulated inflammatory cyclooxygenase. *Proc. Natl. Acad. Sci. U.S.A.*, **89**, 4888–4892.

11. Hla, T. and Neilson, K. (1992) Human cyclooxygenase-2 cDNA. *Proc. Natl. Acad. Sci. U.S.A.*, **89**, 7384–7388.

12. Xie, W., Robertson, D.L. and Simmons D.L. (1992) Mitogen inducible prostaglandin G/H synthase: a new target for nonsteroidal antiinflammatory drugs. *Drug Dev. Res.*, **25**, 249–265.

13. Meade, E.A., Smith, W.L. and DeWitt, D.L. (1993) Differential inhibition of prostaglandin endoperoxide synthase (cyclooxygenase) isozymes by aspirin and other non-steroidal antiinflammatory drugs. *J. Bio. Chem.*, **268**, 6610–6614.

14. DeWitt, D.L., Meade, E.A. and Smith, W.L. (1993) PGH synthase isoenzyme selectivity: the potential for safer nonsteroidal antiinflammatory drugs. *Am. J. Medicine*, **95**, (suppl. 2A), 40S–44S.

15. Arai, I., Hamasaka, Y., Futaki, N., Takahashi, S., Yoshikawa, K., Higuchi, S. and Otomo, S. (1993) Effect of NS-398, a new nonsteroidal antiinflammatory agent, on gastric ulceration and acid secretion in rats. *Res. Comm. Chem. Pathol. Pharmacol.*, **81**, 259–270.

16. Futaki, N., Takahashi, S., Yokoyama, M., Arai, I., Higuchi S. and Otomo, S. (1994) NS-398, a new anti-inflammatory agent, selectively inhibits prostaglandin G/H synthase/cyclooxygenase (COX-2) activity *in vitro*. *Prostaglandins*, **47**, 55–59.

17. Borgeat, P. and Samuelsson, B. (1979) Arachidonic acid metabolism in polymorphonuclear leukocytes: effects of ionophore A23187. *Proc. Nat. Acad. Sci. U.S.A.*, **76**, 2148–2152.

18. Murphy, R.C., Hammarstrom, S. and Samuelsson, B. (1979) Leukotriene C: a slow reacting substance from murine mastocyte cells. *Proc. Nat. Acad. Sci. U.S.A.*, **76**, 4275–4279.

19. Morris, H.R., Taylor, G.W., Piper, P.J. and Tippins, J.R. (1980) Structure of slow reacting substance of anaphalaxis from guinea pig lung. *Nature*, **285**, 104–106.

20. Anderson, M.E., Allison, R.D. and Meister, A. (1982) Interconversion of leukotrienes catalysed by purified glutamyltranspeptidase: Concomitant formation of leukotriene D4 and glutamylamino acids. *Proc. Nat. Acad. Sci. U.S.A.*, **79**, 1088–1091.

21. Samuelsson, B. (1981) Leukotrienes: Mediators of allergic reactions and inflammation. *Int. Arch. Allergy Appl. Immunol.*, **66(s1)**, 98–106.

22. Samuelsson, B. (1983) Leukotrienes: Mediators of immediate hypersensitivity reactions and inflammation. *Science*, **220**, 568–575.

23. Feldberg, W. and Kellaway, C.H. (1938) Liberation of histamine and formation of lysocithin-like substance by cobra venom. *J. Physiol.*, **94**, 187–197.

24. Kellaway, C.H. and Trethewie, E.R. (1940) The liberation of slow reacting smooth muscle–stimulating substance in anaphalaxis. *Q. J. Exp. Physiol.*, **38**, 121–145.

25. Drazen, J.M. (1986) Inhalation challenge with sulfidopeptide leukotrienes in human subjects. *Chest*, **89**, 414–419.

26. Camp, R.D.R., Mallet, A.I., Wollard, P.M., Brain, S.D., Black, A.K. and Greaves, M.W. (1983) The identification of hydroxy fatty acids in psoriatic skin. *Prostaglandins*, **26**, 431–448.

27. Belch, J.J., O'Dowd, A., Ansell, D. and Sturrock, R.D. (1989) Leukotriene B_4 production by peripheral blood neutrophils in rheumatoid arthritis. *Scand. J. Rheumatol.*, **18**, 213–219.

28. Davidson, E.M., Rae, S.A. and Smith, M.J.H. (1983) Leukotriene B_4, a mediator of inflammation present in synovial fluid in rheumatoid arthritis. *Ann. Rheumatol. Dis.*, **42**, 677–679.

29. Atik, O.S. (1990) Leukotriene B_4 and prostaglandin E_2 like activity in synovial fluid in osteoarthritis. *Prostagland., Leuk., Essen. Fatty Acids*, **39**, 253–254.

30. Bray, M.A. (1983) The pharmacology and pathophysiology of leukotriene B_4. *Br. Med. Bull.*, **39**, 249–254.

31. Wallace, J.L. and Granger, D.N. (1992) Pathogenesis of NSAID gastropathy: are neutrophils the culprits. *Trends Pharmacol. Sci.*, **13**, 129–131.

32. Asako, H., Kubes, P., Wallace, J.L., Gaginella, T., Wolf, R.E. and Granger, D.N. (1992) Indomethacin-induced leukocyte adhesion in mesenteric venules: role of lipoxygenase products. *Am. J. Physiol.*, **262**, G903–908.

33. Vaananen, P.M., Keenan, C.M., Grisham, M.B. and Wallace, J.L. (1992) Pharmacological investigations of the role of leukotrienes in the pathogenesis of experimental NSAID gastropathy. *Inflammation*, **16**, 227–240.

34. Ford-Hutchinson, A.W., Tagari, P., Ching, S.V., Anderson, J.B., Coleman, J.B. and Peter, C.P. (1993) Chronic leukotriene inhibition in the rat fails to modify the toxicological effects of a cyclooxygenase inhibitor. *Can. J. Physiol. Pharmacol.*, **71**, 806–810.

35. Connolly, S. and Robinson, D.H. (1993) A new phospholipase A_2 comes to the surface. *Drug News Perspect.*, **6**, 584–590.

36. Djuric, S.W., Fretland, D.J. and Penning, T.D. (1992) The leukotriene B_4 receptor antagonists – a most discriminating class of antiinflammatory agents. *Drugs Fut.*, **17**, 819–830.

37. Sawyer, J.S. and Saussy, D.L. (1993) Cysteinyl leukotriene receptors. *Drug News Perspect.*, **6**, 139–149.

38. Salmon, J.A. and Garland, L.G. (1991) Leukotriene antagonists and inhibitors of leukotriene biosynthesis as potential therapeutic agents. *Drug Research*, **37**, 9–90.

39. McMillan, R.M. and Walker, E.R.H. (1992) Designing therapeutically effective 5-lipoxygenase inhibitors. *Trends Pharmacol. Sci.*, **13**, 323–330.

40. Summers, J.B. (1990) Inhibitors of leukotriene biosynthesis. *Drug News Perspect.*, **3**, 517–526.

41. Musser, J.H. and Kreft, A.F. (1992) 5-Lipoxygenase: properties, pharmacology, and the quinolyl (bridged)aryl class of inhibitors. *J. Med. Chem.*, **35**, 2501–2524.

42. Batt, D.G. (1992) 5-Lipoxygenase inhibitors and their antiinflammatory activities. *Prog. in Med. Chem.*, **29**, 1–63.

43. Young, R.N., Gillard J.W., Hutchinson, J.H., Leger, S. and Prasit P. (1993) Discovery of inhibitors of the 5-lipoxygenase activating protein (flap). *J. Lipid Mediators*, **6**, 233–238.

44. Rouzer, C.A. and Samuelsson, B. (1985) On the nature of the 5-lipoxygenase reaction in human leukocytes: Enzyme purification and requirement for multiple stimulatory factors. *Proc. Nat. Acad. Sci. U.S.A.*, **82**, 6040–6044.

45. Blackwell, G.J. and Flower, R.J. (1978) 1-Phenyl-3-pyrazolidone: an inhibitor of arachidonate oxidation in lung and platelets. *Br. J. Pharmacol.*, **63**, 360P.

46. Blackwell, G.J. and Flower, R.J. (1978) 1-Phenyl-3-pyrazolidone: an inhibitor of cyclooxygenase and lipoxygenase in lung and platelets. *Prostaglandins*, **16**, 417–425.

47. Higgs, G.A., Flower, R.J. and Vane, J.R. (1979) A new approach to antiinflammatory drugs. *Biochem. Pharmacol.*, **28**, 1959–1961.

48. Copp, F.C., Islip, P.J. and Tateson, J.E. (1984) 3-N-Substituted-amino-1-[3-(trifluoromethyl)phenyl]-2-pyrazolines have enhanced activity against arachidonate 5-lipoxygenase and cyclooxygenase. *Biochem. Pharmacol.*, **33**, 339–340.

49. Kim, H.K., Martin, Y.C., Norris, B., Young, P.R., Carter, G.W., Haviv, F. and Walters, R.L. (1990). Quantitative structure-activity relationships of inhibitors of immune complex-induced inflammation: 1-phenyl-3-amino-pyrazoline derivatives. *J. Pharm. Sci.*, **79**, 609–613.

50. Frigola, J., Columbo, A., Pares, J., Martinez, L., Sagarra, R. and Roser, R. (1989) Synthesis, structure and inhibitory effects on cyclooxygenase, lipoxygenase, thromboxane synthetase and platelet aggregation of 3-amino-4,5-dihydro-1H-pyrazole derivatives. *Eur. J. Med. Chem.*, **24**, 435–445.

51. Albert, D.H., Dyer, R., Young, P.R., Barlow, J., Bornemeier, D., Bouska, J., Roberts, E., Sonsalla, J., Brooks, D.W. and Carter, G.W. (1988) Inhibition of leukotriene biosynthesis with A-53612, a selective and orally active inhibitor of 5-lipoxygenase (5-LO). *FASEB J.*, **2**, A369.

52. Hlasta, D.J., Casey, F.B., Ferguson, E.W., Gangell, S.J., Heimann, M.R., Jaeger, E.P., Kullnig, R.K. and Gordon, R.J. (1991) 5-Lipoxygenase inhibitors: the synthesis and structure-activity relationships of a series of 1-phenyl-3-pyrazolidinones. *J. Med. Chem.*, **34**, 1560–1570.

53. Rachlin, S., Bramm, E., Ahnfelt-Ronne, I. and Arrigoni-Martelli, E. (1980) Basic antiinflammatory compounds. N,N',N"-Trisubstituted guanidines. *J. Med. Chem.*, **23**, 13–20.

54. Ahnfelt-Ronne, I. and Arrigoni-Martelli, E. (1980) A new anti-inflammatory compound, timegadine (N-cyclohexyl-N"-4-[2-methylquinolyl]-N'-2-thiazolyl-guanidine), which inhibits both prostaglandin and 12-hydroxy-eicosatetraenoic acid (12-HETE) formation. *Biochem. Pharmacol.*, **29**, 3265–3269.

55. Ahnfelt-Ronne, I. and Arrigoni-Martelli, E. (1982) Multiple effects of a new anti-inflammatory agent, timegadine, on arachidonic acid release and metabolism in neutrophils and platelets. *Biochem. Pharmacol.*, **31**, 2619–2624.

56. Moilanen, E., Alanko, J., Seppala, E. and Vapaatalo, H. (1988) Effects of antirheumatic drugs on leukotriene B_4 and prostanoid synthesis in human polymorphonuclear leukocytes *in vitro*. *Agents Actions*, **24**, 387–394.

57. Bramm, E., Binderup, L. and Arrigoni-Martelli, E. (1981) An unusual profile of activity of a new basic anti-inflammatory drug, timegadine. *Agents Actions*, **11**, 402–409.

58. Berry, H., Bloom, B., Fernandes, L. and Morris, M. (1983) Comparison of timegadine and naproxen in rheumatoid arthritis. A placebo controlled trial. *Clin. Rheumatol.*, **2**, 357–361.

59. O'Sullivan, M. and Molloy, M.G. (1985) Comparison of timegadin versus naproxen in rheumatoid arthritis. *Clin. Rheumatol.*, **4**, 362–363.

60. Egsmose, C., Lund, B. and Andersen, R.B. (1988) Timegadine: more than a non-steroidal for the treatment of rheumatoid arthritis. *Scand. J. Rheum.*, **17**, 103–111.

61. Tischer, A., Bailey, P., Dallob, A., Witzel, B., Durette, P.L., Rupprecht, K., Allison, D., Dougherty, H., Humes, J., Ham, E., Bonney, R., Egan, R., Gallagher, T., Miller, D. and Goldenberg, M. (1986) L-652,343: A novel dual 5-lipoxygenase/cyclooxygenase inhibitor. *Ad. Prost., Throm., Leuk. Res.*, **16** edited by U. Zor *et al*, 63–66.

62. Bailey, P.J., Dallob, A.L., Allison, D.L., Anderson, R.L., Bach, T., Durette, P., Hand, K.M., Hopple, S.L., Luell, S., Meurer, R., Rosa, R., Tischer, A.N., Witzel, B.E. and Goldenberg, M.M. (1988) Pharmacology of the dual inhibitor of cyclooxygenase and 5-lipoxygenase 3-hydroxy-5-trifluoromethyl-N-(2-(2-thienyl)-2-phenyl-ethenyl)-benzo(b)thiophene-2-carboxamide. *Arzneim.-Forch./Drug Res.*, **38**, 372–378.

63. De Schepper, P.J., Van Hecken, A., De Lepeleire, I., Gresele, P., Arnout, J. and Vermylen, J. (1986) First human studies with L-652,343, a dual 5-lipoxygenase/cyclooxygenase inhibitor. *Acta Pharm. Tox.*, **59 S V**, 164.

64. Barr, R.M., Kobza Black, A., Dowd, P.M., Koro, O., Mistry, K., Issacs, J.L. and Greaves, M.W. The *in vitro* 5-lipoxygenase and cyclooxygenase inhibitor L-652,343 does not inhibit 5-lipoxygenase *in vivo* in human skin. *Br. J. Clin. Pharmacol.*, **25**, 23–26.

65. Lantos, I., Bender, P.E., Razgaitis, K.A., Sutton, B.M., DiMartino, M.J., Griswold, D.E. and Walz, D.T. (1984) Antiinflammatory activity of 5,6-diaryl-2,3,-dihydroimidazo[2,1-b]thiazoles. Isomeric 4-pyridyl and 4-substituted phenyl derivatives. *J. Med. Chem.*, **27**, 72–75.

66. Lantos, I., Gombatz, K., McGuire, M., Pridgen, L., Remich, J. and Shilcrat, S. (1988) Synthetic and mechanistic studies on the preparation of pyridyl-substituted imidazothiazoles. *J. Org. Chem.*, **53**, 4223–4227.

67. Hayes, J.F., Mitchell, M.B. and Procter, G. (1994) A novel and efficient synthesis of a tetra-substituted imidazole. *Tetrahedron Lett.*, **35**, 273–274.

68. Griswold, D.E., Marshall, P.J., Webb, E.F., Godfrey, R., Newton Jr., J., DiMartino, M.J., Sarau, H.M., Gleason, J.G., Poste, G. and Hanna, N. (1987) SK&F 86002: A structurally novel anti-inflammatory agent that inhibits lipoxygenase- and cyclooxygenase-mediated metabolism of arachidonic acid. *Biochem. Pharmacol.*, **36**, 3463–3470.

69. DiMartino, M.J., Griswold, D.E., Berkowitz, B.A., Poste, G. and Hanna, N. (1987) Pharmacologic characterization of the antiinflammatory properties of a new dual inhibitor of lipoxygenase and cyclooxygenase. *Agents Actions*, **20**, 113–123.

70. Griswold, D.E., Webb, E., Schwartz, L. and Hanna, N. (1987) Arachidonic acid-induced inflammation: Inhibition by dual inhibitor of arachidonic acid metabolism, SK&F 86002. *Inflammation*, **11**, 189–199.

71. Lee, J.C., Griswold, D.E., Votta, B. and Hanna, N. (1987) Inhibition of monocyte IL-1 production by the anti-inflammatory compound, SK&F 86002. *Int. J. Immunopharmacol.*, **10**, 835–843.

72. Lee, J.C., Votta, B., Dalton, B.J., Griswold, D.E., Bender, P.E. and Hanna, N. (1990) Inhibition of human monocyte IL-1 production by SK&F 86002. *Int. J. Immunother.*, **61**, 1–12.

73. Lee, J.C., Rebar, L. and Laydon, J.T. (1989) Effect of SK&F 86002 on cytokine production by human monocytes. *Agents Actions*, **27**, 277–279.

74. Olivera, D.L., Laydon, J.T., Hillegass, L., Badger, A.M. and Lee, J.C. (1993) Effects of pyridinyl imidazole compounds on murine TNF-α production. *Agents Actions*, **39**, Special Conference Issue, C55–C57.

75. Young, P., McDonnell, P., Dunnington, D., Hand, A., Laydon, J. and Lee, J. (1993) Pyridinyl imidazoles inhibit IL-1 and TNF production at the protein level. *Agents Actions*, **39**, Special Conference Issue, C67–C69.

76. Badger, A.M., Olivera, D., Talmadge, J.E. and Hanna, N. (1989) Protective effect of SK&F 86002, a novel dual inhibitor of arachidonic acid metabolism, in murine models of endotoxin shock. *Circ. Shock*, **27**, 51–61.

77. Barton, B.E. and James, L.C. (1990) The effect of dual inhibitor, SK&F 86002, on helper T cell functions. *Immunopharmacol. Immunotoxicol.*, **12**, 105–121.

78. Griswold, D.E., Hoffstein, S., Marshall, P.J., Webb, E.F., Hillegass, L., Bender, P.E. and Hanna, N. (1989) Inhibition of inflammatory cell infiltration by bicyclic imidazoles, SK&F 86002 and SK&F 104493. *Inflammation*, **13**, 727–739.

79. Hanna, N., Marshall, P.J., Newton, Jr., J., Schwartz, L., Kirsh, R., DiMartino, M.J., Adams, J., Bender, P. and Griswold, D.E. (1990) Pharmacological profile of SK&F 105809, a dual inhibitor of arachidonic acid metabolism. *Drugs Exptl. Clin. Res.*, **16**, 137–147.

80. Griswold, D.E., Marshall, P.J., Lee, J.C., Webb, E.F., Hillegass, L.M., Wartell, J., Newton, Jr., J. and Hanna, N. (1991) Pharmacology of the pyrroloimidazole, SK&F 105809-II. Antiinflammatory activity and inhibition of mediator production *in vivo. Biochem. Pharmacol.*, **42**, 825–831.

81. Marshall, P.J., Griswold, D.E., Breton, J., Webb, E.F., Hillegass, L.M., Sarau, H.M., Newton, Jr., J., Lee, J.C., Bender, P.E. and Hanna, N. (1991) Pharmacology of the pyrroloimidazole, SK&F 105809-I. Inhibition of inflammatory cytokine production and of 5-lipoxygenase- and cyclo-oxygenase-mediated metabolism of arachidonic acid. *Biochem. Pharmacol.*, **42**, 813–824.

82. Ferro, M.P., Adams, R.E., Mart, A.J., Wachler, M.P., Argentieri, D.C. and Levinson, S.L. (1993) Tepoxalin, an orally active dual cyclooxygenase/5-lipoxygenase inhibitor. *205th National Meeting of the American Chemical Society, Denver*, Abstr. MED 138.

83. Argentieri, D.C., Ritchie, D.M., Tolman, E.L., Ferro, M.P., Wachter, M.P., Mezick, J.A. and Capetola, R.J. (1988) Topical and *in vitro* pharmacology of tepoxalin (ORF 20485) — a new dual cyclooxygenase (CO)/lipoxygenase (LO) inhibitor. *FASEB J.*, **2**, Abstr. 427.

84. Mezick, J.A., Bishop, C.M., Argentieri, D.C., Ritchie, D.M., Thorne, L.G., Rosenthale, M.E. and Capetola, R.J. (1988) Evaluation of topical tepoxalin (ORF 20485) — a novel dual cyclooxygenase (CO)/lipoxygenase (LO) inhibitor, for dermal irritation in rabbits and contact sensitization in guinea pigs. *Clin. Res.*, **36**, 675A.

85. Capetola, R.J., Kirchner, T., Argentieri, D.C., Meeks, A., Ferro, M.P., Wachter, M.P., Rosenthale, M.E. and Ritchie, D.M. (1988) Pulmonary and antiinflammatory profile of tepoxalin (ORF 20485) — a dual cyclooxygenase/lipoxygenase inhibitor. *FASEB J.*, **2**, Abstr. 428.

86. Anderson, D.W., Argentieri, D.C., Ritchie, D.M., Katz, L.B., Shriver, D.A., Rosenthale, M.E. and Capetola, R.J. (1990) Gastrointestinal (GI) profile of tepoxalin (TX) — an orally active dual cyclooxygenase (CO)/lipoxygenase(LO) inhibitor with potent antiinflammatory activity. *FASEB J.*, **4**, Abstr. 4973.

87. Wallace, J.L., Cirino, G., Cicala, C., Anderson, D.W., Argentieri, D. and Capetola, R.J. (1991) Comparison of the ulcerogenic properties of tepoxalin with those of non-steroidal anti-inflammatory drugs (NSAIDs). *Agents Actions*, **34**, 247–250.

88. Wallace, J.L., McCafferty, D.M., Carter, L., McKnight, W. and Argentieri, D. (1993) Tissue-selective inhibition of prostaglandin synthesis in rat by tepoxalin: anti-inflammatory without gastropathy? *Gastroenterology*, **105**, 1630–1636.

89. Argentieri, D.C., Anderson, D.W., Ritchie, D.M., Rosenthale, M.E. and Capetola, R.J. (1990) Tepoxalin (RWJ 20485) inhibits prostaglandin (PG) and leukotriene (LT) production in adjuvant arthritic rats and in dog knee joints challenged with sodium urate and immune complexes. *FASEB J.*, **4**, Abstr. 4974.

90. Murray, W., Wachter, M., Barton, D. and Forero-Kelly, Y. (1991) The regioselective synthesis of tepoxalin, 3-[5-(4-chlorophenyl)-1-(4-methoxyphenyl)-3-pyrazolyl]-N-hydroxy-N-methylpropanamide and related 1,5-diarylpyrazole antiinflammatory agents. *Synthesis*, 18–20.

91. Murray, W.V. and Hadden, S.K. (1992) A facile synthesis of tepoxalin, 5-(4-chlorophenyl)-N-hydroxy-1-(4-methoxyphenyl)-N-methyl-1H-pyrazole-3-propanamide. *J. Org. Chem.*, **57**, 6662–6663.

92. Carty, T.J., Showell, H.J., Loose, L.D. and Kadin, S.B. (1988) Inhibition of both 5-lipoxygenase (5-LO) and cyclooxygenase (CO) pathways of arachidonic acid (AA) metabolism by CP-66,248. *Arthritis Rheum.*, **31**, Suppl. 4, C54.

93. Otterness, I.G., Bliven, M.L., Downs, J.T. and Hanson, D.C. (1988) Effects of CP-66,248 on IL-1 synthesis by murine peritoneal macrophages. *Arthritis Rheum.*, **31**, Suppl. 4, C55.

94. McDonald, B., Loose, L. and Rosenwasser, L.J. (1988) The influence of a novel arachidonate inhibitor, CP 66,248 on the production and activity of human monocyte IL-1. *Arthritis Rheum.*, **31**, Suppl. 4, 32.

95. Katz, P., Borger, A.P. and Loose, L.D. (1988) Evaluation of CP-66,248 [5-chloro-2,3,-dihydro-2-oxo-3-(2-thienylcarbonyl)-indole-1-carboxamide] in rheumatoid arthritis. *Arthritis Rheum.*, **31**, Suppl. 4, A87.

96. McDonald, B., Loose, L.D. and Rosenwasser, L.J. (1988) Synovial fluid IL-1 in RA patients receiving a novel arachidonate inhibitor CP-66,248. *Arthritis Rheum.*, **31**, Suppl. 4, A88.

97. Moilanen, E., Alanko, J., Asmawi, M.Z. and Vapaatalo, H. (1988) CP-66,248, a new anti-inflammatory agent, is a potent inhibitor of leukotriene B_4 and prostanoid synthesis in human polymorphonuclear leucocytes *in vitro*. *Eicosanoids*, **1**, 35–39.

98. Proudman, K.E. and McMillan, R.M. (1991) Are tolfenamic acid and tenidap dual inhibitors of 5-lipoxygenase and cyclooxygenase? *Agents Actions*, **34**, 121–124.

99. Blackburn, Jr., W.D., Heck, L.W., Loose, L.D., Eskra, J.D. and Carty, T.J. (1991) Inhibition and 5-lipoxygenase product formation and polymorphonuclear cell degranulation by tenidap sodium in patients with rheumatoid arthritis. *Arthritis Rheum.*, **34**, 204–210.

100. Blackburn, Jr., W.D., Heck L.W. and Loose, L.D. (1990) Inhibition of PMN degranulation and synovial fluid lipoxygenase (LO) and cyclooxygenase (CO) activity by tenidap [5-chloro-2,3-dihydro-2-oxo-3-(2-thienyl-carbonyl)-indole-1-carboxamide]. *Arthritis Rheum.*, **33**, Suppl. 5, P9S.

101. Smith, D.M., Johnson, J.A., Loeser, R. and Turner, R.A. (1990) Evaluation of tenidap (CP-66,248) on human neutrophil arachidonic acid metabolism, chemotactic potential and clinical efficacy in the treatment of rheumatoid arthritis. *Agents Actions*, **31**, 102–109.

102. Otterness, I.G., Bliven, M.L, Downs, J.T., Natoli, E.J. and Hanson, D.C. (1991) Inhibition of interleukin 1 synthesis by tenidap: a new drug for arthritis. *Cytokine*, **3**, 277–283.

103. Sipe, J.D., Bartle, L.M. and Loose, L.D. (1992) Modification of proinflammatory cytokine production by the antirheumatic agents tenidap and naproxen. *J. Immun.*, **148**, 480–484.

104. Blackburn, Jr., W.D., Loose, L.D., Heck, L.W. and Chatham, W.W. (1991) Tenidap, in contrast to several available nonsteroidal antiinflammatory drugs, potently inhibits the release of activated neutrophil collagenase. *Arthritis Rheum.*, **34**, 211–216.

105. Otterness, I.G., Pazoles, P.P., Morre, P.F. and Pepys, M.B. (1991) C-reactive protein as an index of disease activity. Comparison of tenidap, cyclophosphamide and dexamethasone in rat adjuvant arthritis. *J. Rheumatol.*, **18**, 505–511.

106. Mylari, B.L., Carty, T.J., Moore, P.F. and Zembrowski, W.J. (1990) 1,2-Dihydro-1-oxopyrrolo[3,2,1-kl]phenothiazine-2-carboxamides and congeners, dual cyclooxygenase/5-lipoxygenase inhibitors with anti-inflammatory activity. *J. Med. Chem.*, **33**, 2019–2024.

107. Wiseman, E.H., Chiaini, J. and McManus, J.M. (1973) Studies with antiinflammatory oxindole-carboxanilides. *J. Med. Chem.*, **16**, 131–134.

108. Clemence, F., Le Martret, O., Delevallee, F., Benzoni, J., Jouanen, A., Jouquey, S., Mouren, M. and Deraedt, R. (1988) 4-Hydroxy-3-quinolinecarboxamides with antiarthritic and analgesic activities. *J. Med. Chem.*, **31**, 1453–1462.

109. Batt, D.G., Maynard, G.D., Petraitis, J.J., Shaw, J.E., Galbraith, W. and Harris, R.R. (1990) 2-Substituted-1-naphthols as potent 5-lipoxygenase inhibitors with topical antiinflammatory activity. *J. Med. Chem.*, **33**, 360–370.

110. Harris, R.R., Batt, D.G., Galbraith, W. and Ackerman, N.R. (1989) Topical anti-inflammatory activity of DuP 654, a 2-substituted 1-naphthol. *Agents Actions*, **27**, 297–299.

111. Vanderhoek, J.Y. and Lands, W.E.M. (1973) The inhibition of the fatty acid oxygenase of sheep vesicular gland by antioxidants. *Biochim. Biophys. Acta*, **296**, 382–385.

112. Thody, V.E., Buckle, D.R. and Foster, K.A. (1987) Studies on the antioxidant activity of 5-lipoxygenase inhibitors. *Biochem. Soc. Trans.*, **15**, 416–417.

113. Swingle, K.F., Bell, R.L and Moore, G.G.I. (1985) Anti-inflammatory activity of antioxidants. *Antiinflammatory and Antirheumatic Drugs-Volume 3*, pp. 105–126, ed. by Rainsford, K.D. CRC Press.

114. Moore, G.G.I. and Swingle, K.F. (1982) 2,6-Di-tert-butyl-4-(2'-thenoyl)phenol (R-830): a novel non-steroidal anti-inflammatory agent with antioxidant properties. *Agents Actions*, **12**, 674–683.

115. Hidaka, T., Hosoe, K., Ariki, Y., Takeo, K., Yamashita, T., Katsumi, I., Kondo, H., Yamashita, K. and Watanabe, K. (1984) Pharmacological properties of a new anti-inflammatory compound, α-(3,5-di-tert-butyl-4-hydroxybenzylidene)-γ-butyrolactone (KME-4), and its inhibitory effects on prostaglandin synthetase and 5-lipoxygenase. *Japan J. Pharmacol.*, **36**, 77–85.

116. Hidaka, T., Takeo, K., Hosoe, K., Katsumi, I., Yamashita, T. and Watanabe, K. (1985) Inhibition of polymorphonuclear leukocyte 5-lipoxygenase and platelet cyclooxygenase by α-(3,5-di-tert-butyl-4-hydroxybenzylidene)-γ-butyrolactone (KME-4), a new anti-inflammatory drug. *Japan J. Pharmacol.*, **38**, 267–272.

117. Hidaka, T., Hosoe, K., Katsumi, I., Yamashita, T. and Watanabe, K. (1986) The effect of α-(3,5-di-tert-butyl-4-hydroxybenzylidene)-γ-butyrolactone (KME-4), a new anti-inflammatory drug, on leucocyte migration in rat carrageenan pleurisy. *J. Pharm. Pharmacol.*, **38**, 242–245.

118. Hidaka, T., Hosoe, K., Yamashita, T. and Watanabe, K. (1986) Effect of α-(3,5-di-tert-butyl-4-hydroxybenzylidene)-γ-butyrolactone (KME-4), a new anti-inflammatory drug, on the established adjuvant arthritis in rats. *Japan J. Pharmacol.*, **42**, 181–187.

119. Hidaka, T., Hosoe, K., Yamashita, T., Watanabe, K., Hiramatsu, Y. and Fujimura, H. (1986) Analgesic and anti-inflammatory activities in rats of α-(3,5-di-tert-butyl-4-hydroxybenzylidene)-γ-butyrolactone (KME-4), and its intestinal damage. *J. Pharm. Pharmacol.*, **38**, 748–753.

120. Katsumi, I., Kondo, H., Yamashita, K., Hidaka, T., Hosoe, K., Yamashita, T. and Watanabe, K. (1986) Studies on styrene derivatives. I. Synthesis and antiinflammatory activities of α-benzylidene-γ-butyrolactone derivatives. *Chem. Pharm. Bull.*, **34**, 121–129.

121. Katsumi, I., Kondo, H., Fuse, Y., Yamashita, K., Hidaka, T., Hosoe, K., Takeo, K., Yamashita, T. and Watanabe, K. (1986) Studies on styrene derivatives. II. Synthesis and antiinflammatory activities of 3,5-di-tert-butyl-4-hydroxystyrenes. *Chem. Pharm. Bull.*, **34**, 1619–1627.

122. Katayama, K., Shirota, H., Kobayashi, S., Terato, K., Ikuta, H. and Yamatsu, I. (1987) *In vitro* effect of N-methoxy-3-(3,5-di-tert-butyl-4-hydroxybenzylidene)-2-pyrrolidone (E-5110), a novel nonsteroidal anti-inflammatory agent, on generation of some inflammatory mediators. *Agents Actions*, **21**, 269–271.

123. Shirota, H., Kobayashi, S., Terato, K., Sakuma, Y., Yamada, K., Ikuta, H., Yamagishi, Y., Yamatsu, I. and Katayama, K. (1987) Effect of the novel nonsteroidal anti-inflammatory agent N-methoxy-3-(3,5-di-tert-butyl-4-hydroxybenzylidene)pyrrolidin-2-one on *in vitro* generation of some inflammatory mediators. *Arzneim.-Forsch./Drug Res.*, **37**, 936–940.

124. Shirota, H., Goto, M., Hashida, R., Yamatsu, I. and Katayama, K. (1989) Inhibitory effects of E-5110 on interleukin-1 generation from human monocytes. *Agents Actions*, **27**, 322–324.

125. Shirota, H., Katayama, K., Ono, H., Chiba, K., Kobayashi, S., Terato, K., Ikuta, H. and Yamatsu, I. (1987) Pharmacological properties of N-methoxy-3-(3,5-di-tert-butyl-4-hydroxybenzylidene)-2-pyrrolidone (E-5110), a novel nonsteroidal antiinflammatory agent. *Agents Actions*, **21**, 250–252.

126. Shirota, H., Chiba, K., Ono, H., Yamamoto, H., Kobayashi, S., Terato, K., Ikuta, H., Yamatsu, I. and Katayama, K. (1987) Pharmacological properties of the novel nonsteroidal antiinflammatory agent N-methoxy-3-(3,5-di-tert-butyl-4-hydroxybenzylidene)pyrrolidin-2-one. *Arzneim.-Forsch./Drug Res.*, **37**, 930–936.

127. Ikuta, H., Shirota, H., Kobayashi, S., Yamagishi, Y., Yamada, K., Yamatsu, I. and Katayama, K. (1987) Synthesis and antiinflammatory activities of 3-(3,5-di-tert-butyl-4-hydroxybenzylidene)pyrrolidin-2-ones. *J. Med. Chem.*, **30**, 1995–1998.

128. Panetta, J.A., Benslay, D.N., Phillips, M.L., Towner, R.D., Bertsch, B., Wang, L. and Ho, P.P.K. (1989) The antiinflammatory effects of LY178002 and LY256548. *Agents Actions*, **27**, 300–302.

129. Lazer, E.S., Wong, H.-C., Possanza, G.J., Graham, A.G. and Farina, P.R. (1989) Antiinflammatory 2,6-di-tert-butyl-4-(2-aryl-ethenyl)phenols. *J. Med. Chem.*, **32**, 100–104.

130. Lazer, E.S., Wong, H.-C., Wegner, C.D., Graham, A.G. and Farina, P.R. (1990) Effect of structure on potency and selectivity in 2,6-di-substituted-4-(2-arylethenyl)phenol lipoxygenase inhibitors. *J. Med. Chem.*, **33**, 1892–1898.

131. Farina, P.R., Graham, A.G., Homon, C.A., Lazer, E.S., Hattox, S.E., Riska, P.S., Gundel, R.H. and Wegner, C.D. (1993) A prodrug of a 2,6-di-substituted 4-(2-arylethenyl)phenol is a selective and orally active 5-lipoxygenase inhibitor. *J. Pharmacol. Exp. Ther.*, **265**, 483–489.

132. Sirko, S.P., Schindler, R., Doyle, M.J., Weisman, S.M. and Dinarello, C.A. (1991) Transcription, translation and secretion of interleukin 1 and tumor necrosis factor: effects of tebufelone, a dual cyclooxygenase/5-lipoxygenase inhibitor. *Eur. J. Immunol.*, **21**, 243–250.

133. Doyle, M.J., Eichhold, T.H., Hynd, B.A. and Weisman, S.M. (1990) Determination of leukotriene B_4 in human plasma by gas chromatography using a mass selective detector and a stable isotope labelled internal standard. Effect of NE-11740 on arachidonic acid metabolism. *J. Pharm. Biomed. Anal.*, **8**, 137–142.

134. Powell, J.H., Meredith, M.P., Vargas, R., McMahon, F.G. and Jain, A.K. (1991) Antipyretic activity of tebufelone (NE-11740) in man. *Agents Action*, **Suppl. 32**, 45–49.

135. Sietsema, W.K., Kelm, G.R., Deibel, R.M., Doyle, M.J., Loomans, M.E., Smyth, R.E., Kinnett, G.O., Eichhold, T.H. and Farmer, R.W. (1993) Absorption, bioavailability, and pharmacokinetics of tebufelone in the rat. *J. Pharm. Sci.*, **82**, 610–612.

136. Miller, J.A. and Matthews, R.S. (1992) A short, efficient synthesis of tert-butyl-hydroxylated di-tert-butylphenols. *J. Org. Chem.*, **57**, 2514–2516.

137. Miller, J.A., Coleman, M.C. and Matthews, R.S. (1993) Synthesis of 2,6-bis(trifluoromethyl)phenol and its elaboration into "metabolism-resistant" analogs of tebufelone. *J. Org. Chem.*, **58**, 2637–2639.

138. Wong, S., Lee, S.J., Frierson, III, M.R., Proch, J., Miskowski, T.A., Rigby, B.S., Schmolka, S.J., Naismith, R.W., Kreutzer, D.C. and Lindquist, R. (1992) Antiarthritic profile of BF-389 – a novel anti-inflammatory agent with low ulcerogenic liability. *Agents Actions*, **37**, 90–98.

139. Bendele, A.M., Benslay, D.N., Hom, J.T., Spaethe, S.M., Ruterbories, K.J., Lindstrom, T.D., Lee, S.J. and Naismith, R.W. (1992) Anti-inflammatory activity of BF-389, a di-t-butylphenol, in animal models of arthritis. *J. Pharmacol. Exp. Ther.*, **260**, 300–305.

140. Mitchell, J.A., Akarasereenont, P., Thiemermann, C., Flower, R.J. and Vane, J.R. (1993) Selectivity of nonsteroidal antiinflammatory drugs as inhibitors of constitutive and inducible cyclooxygenase. *Proc. Nat. Acad. Sci. U.S.A.*, **90**, 11693–11697.

141. Corey, E.J., Cashman, J.R., Kantner, S.S. and Wright, S.W. (1984) Rationally designed, potent competitive inhibitors of leukotriene biosynthesis. *J. Amer. Chem. Soc.*, **106**, 1503–1504.

142. Flynn, D.L., Capiris, T., Cetenko, W.A., Connor, D.T., Dyer, R.D., Kostlan, C.R., Nies, D.E., Schrier, D.J. and Sircar, J.C. (1990) Nonsteroidal antiinflammatory drug hydroxamic acids. Dual inhibitors of both cyclooxygenase and 5-lipoxygenase. *J. Med. Chem.*, **33**, 2070–2072.

143. Sircar, J.C., Capiris, T., Cetenko, W.A., Connor, D.T., Dyer, R.D., Flynn, D.L., Kostlan, C.R., Nies, D.E. and Schrier, D.J. (1990) Nonsteroidal antiinflammatory drug hydroxamic acids as potent dual inhibitors of cyclooxygenase and 5-lipoxygenase. *ICOI*, Barcelona, Spain, June 17–22.

144. Cetenko, W.A., Connor, D.T., Flynn, D.L. and Sircar, J.C. (1990) Hydroxamate derivatives of selected nonsteroidal antiinflammatory acyl residues and their use for cyclooxygenase and 5-lipoxygenase inhibition. US 4,943,587.

145. Summers, J.B., Mazdiyasni, H., Holms, J.H., Ratajczyk, J.D., Dyer, R.D., and Carter, G.W. (1987) Hydroxamic acid inhibitors of 5-lipoxygenase. *J. Med. Chem.*, **30**, 574–578.

146. Summers, J.B., Gunn, B.P., Mazdiyasni, H., Goetze, A.M., Young, P.R., Bouska, J.B., Dyer, R.D., Brooks, D.W. and Carter, G.W. (1987) *In vivo* characterization of hydroxamic acid inhibitors of 5-lipoxygenase. *J. Med. Chem.*, **30**, 2121–2126.

147. Summers, J.B., Gunn, B.P., Martin, J.G., Mazdiyasni, J.D., Young, P.R., Goetze, A.M., Bouska, J.B., Dyer, R.D., Brooks, D.W. and Carter, G.W. (1988) Orally active hydroxamic acid inhibitors of 5-lipoxygenase. *J. Med. Chem.*, **31**, 3–5.

148. Jackson, W.P., Islip, P.J., Kneen, G., Pugh, A. and Wates, P.J. (1988) Acetohydroxamic acids as potent, selective orally active 5-lipoxygenase inhibitors. *J. Med. Chem.*, **31**, 499–500.

149. Belliotti,T.R., Cetenko, W.A., Connor, D.T., Flynn, D.L., Kostlan, C.R., Kramer, J.B. and Sircar, J.C. (1991) N-Hydroxyamide, N-hydroxy-thioamide, N-hydroxy-urea, and N-hydroxythiourea derivatives of selected NSAIDs as antiinflammatory agents. US 4,981,865.

150. Kostlan, C.R., Belliotti,T.R., Bornemeier, D., Connor, D.T., Dyer, R.D., Flynn, D.L., Nies, D.E., Okonkwo, G.C., Schrier, D.J. and Sircar, J.C. (1990) Synthesis and biological evaluation of hydroxamate analogues of fenamates as dual inhibitors of 5-lipoxygenase and cyclooxygenase. *22nd National Medicinal Chemistry Symposium*, Austin, Texas, July 30–August 2.

151. Kramer, J.B., Boschelli, D.H., Connor, D.T., Kostlan, C.R., Flynn, D.L., Dyer, R.D., Bornemeier, D.A., Kennedy, J.A., Wright, C.D. and Kuipers, P.J. (1992) Synthesis of reversed hydroxamic acids of indomethacin: Dual inhibitors of cyclooxygenase and 5-lipoxygenase. *Bio. Med. Chem. Lett.*, **2**, 1655–1660.

152. Juby, P.F., Hudyma, T.W. and Brown, M. (1968) Preparation and antiinflammatory properties of some 5-(2-anilinophenyl)tetrazoles. *J. Med. Chem.*, **11**, 111–117.
153. Boschelli, D.H. and Connor, D.T. (1991) Preparation of heterocyclic analogs of the fenamates. *13th International Congress of Heterocyclic Chemistry*, Corvallis, Oregon, August 11–16, Abstract PO3-150.
154. Boschelli, D.H., Connor, D.T., Bornemeier, D.A., Dyer, R.D., Kuipers, P.J. and Wright, C.D. (1992) Novel 1,2,4-oxadiazoles and 1,2,4-thiadiazoles as dual inhibitors of 5-lipoxygenase and cyclooxygenase. *203rd National Meeting of the American Chemical Society*, San Francisco, California, April 5–10, MEDI 122.
155. Boschelli, D.H., Connor, D.T., Hoefle, M., Bornemeier, D.A. and Dyer, R.D. (1992) Conversion of NSAIDS into balanced dual inhibitors of cyclooxygenase and 5-lipoxygenase. *Bio. Med. Chem. Lett.*, **2**, 69–72.
156. Boschelli, D.H., Connor, D.T., Bornemeier, D.A., Dyer, R.D., Kennedy, J.A., Kuipers, P.J., Okonkwo, G.C., Schrier, D.J. and Wright, C.D. (1993) 1,3,4-Oxadiazole, 1,3,4-thiadiazole and 1,2,4-triazole analogs of the fenamates: *In vitro* inhibition of cyclooxygenase and 5-lipoxygenase activities. *J. Med. Chem.*, **36**, 1802–1810.
157. Boschelli, D.H., Connor, D.T., Hoefle, M., Flynn, D.L. and Sircar, J.C. (1990) Triazole derivatives of fenamates as antiinflammatory agents. US 4,962,119.
158. Belliotti, T.R., Kostlan, C.R., Connor, D.T., Flynn, D.L., Nies, D.E., Sircar, J.C., Dyer, R.D., Bornemeier, D., Wright, C.D., Kuipers, P.J., Kennedy, J.A. and Schrier, D.J. (1991) Novel hydroxamic acid fenamates as dual inhibitors of cyclooxygenase and 5-lipoxygenase. (1991) *202nd National Meeting of the American Chemical Society*, New York, New York, August 25–30, MEDI 177.
159. Belliotti T.R., Connor, D.T. and Kostlan, C.R. (1993) Thiadiazole or oxadiazole analogs of fenamic acids containing substituted hydroxamate side-chains as antiinflammatory agents. US 5,212,189.
160. Flynn, D.L., Belliotti, T.R., Boctor, A.M., Connor, D.T., Kostlan, C.R., Nies, D.E., Ortwine, D.F., Schrier, D.J. and Sircar, J.C. (1988) Design and synthesis of novel styrylheterocycles as *in vitro* and *in vivo* inhibitors of 5-lipoxygenase and cyclooxygenase. *196th National Meeting of the American Chemical Society*, Los Angeles, California, September 25–30, Abstr. MEDI 124.
161. Flynn, D.L., Belliotti, T.R., Boctor, A.M., Connor, D.T., Kostlan, C.R., Nies, D.E., Ortwine, D.F., Schrier, D.J. and Sircar, J.C. (1991) Strylpyrazoles, styrylisoxazoles, and styrylisothiazoles. Novel 5-lipoxygenase and cyclooxygenase inhibitors. *J. Med. Chem.*, **34**, 518–525.
162. Belliotti, T.R., Connor, D.T., Flynn, D.L., Kostlan, C.R. and Nies, D.E. (1993) Styryl pyrazoles, isoxazoles and analogs thereof having activity as 5-lipoxygenase inhibitors, pharmaceutical compositions and methods of use thereof. US 5,208,251.
163. Wilson, M.W., Mullican, M.D., Kostlan, C.R. and Connor, DT. (1991) Synthesis and physical chemical properties of 3,5-di-t-butyl-4-hydroxyphenyl-1,3,4-thiadiazoles, 1,3,4-oxadiazoles and 1,2,4-triazoles. *13th International Congress of Heterocyclic Chemistry*, Corvallis, Oregon, August 11–16, Abstract PO3-158.
164. Mullican, M.D., Wilson, M.W., Connor, D.T., Kostlan, C.R., Schrier, D.J. and Dyer, R.D. (1992) 5-(3,5-Di-t-butyl-4-hydroxyphenyl)-1,3,4-thiadiazoles, -1,3,4-oxadiazoles, and -1,2,4-triazoles as nonulcerogenic antiinflammatory Agents: Discovery of CI-986. *203rd National Meeting of the American Chemical Society*, San Francisco, California, April 5–10, MEDI 121.
165. Mullican, M.D., Wilson, M.W., Connor, D.T., Kostlan, C.R., Schrier, D.J. and Dyer, R.D. (1993) Design of 5-(3,5-di-tert-butyl-4-hydroxyphenyl)-1,3,4-thiadiazoles, -1,3,4-oxadiazoles, and -1,2,4-triazoles as orally-active, nonulcerogenic antiinflammatory agents. *J. Med. Chem.*, **36**, 1090–1099.
166. Connor, D.T., Flynn, D.L., Kostlan, C.R., Mullican, M.D., Shrum, G.P., Unangst, P.C. and Wilson, M.W. (1992) 3,5-Di-tert-butyl-4-hydroxy-phenyl-1,3,4-thiadiazole and oxadiazoles, and 3,5-di-t-butyl-4-hydroxy-phenyl-1,2,4-thiadiazoles, oxadiazoles and triazoles as antiinflammatory agents. US 5,155,122.
167. Schrier, D.J., Baragi, V.M., Connor, D.T., Dyer, R.D., Jordan, J.H., Imre, K.M., Lesch, M.E., Mullican, M.D., Okonkwo, G.C.N. and Conroy, M.C. (1994) The pharmacologic effects of 5-[3,5-bis(1,1-dimethylethyl)-4-hydroxyphenyl]-1,3,4-thiadiazole-2(3H)-thione, choline salt (CI-986), a novel inhibitor of arachidonic acid metabolism in models of inflammation, analgesia and gastric irritation. *Prostaglandins*, **47**, 17–30.

168. Kramer, J.B., Connor, D.T., Boschelli, D.H. and Kostlan, C.R. (1992) Synthesis of sulfur and ethylene linked di-t-butylphenol-1,3,4-thiadiazoles, -1,3,4-oxadiazoles, and -1,2,4-triazoles. *International Conference on Organic Synthesis*, Montreal, Canada, June 28–July 2.

169. Kramer, J.B., Boschelli, D.H., Connor, D.T., Kostlan, C.R., Kuipers, P.J., Kennedy, J.A., Wright C.D., Bornemeier, D.A. and Dyer, R.D. (1993) Cyclooxygenase and 5-lipoxygenase inhibitory activity of 2,6-di-t-butylphenols linked via a sulfur atom to 1,3,4-thiadiazoles and 1,3,4-oxadiazoles. *Bio. Med. Chem. Lett.*, **3**, 2827–2830.

170. Kramer, J.B., Boschelli, D.H., Connor, D.T., Kuipers, P.J., Kennedy, J.A., Wright, C.D., Bornemeier, D.A. and Dyer, R.D. (1994) Cyclooxygenase and 5-lipoxygenase inhibitory activity of 2,6 di-t-butylphenols linked to 1,3,4-diazoles and 1,3,4-oxadiazoles. *207th National Meeting of the American Chemical Society*, San Diego, California, March 13–17, MEDI 122.

171. Unangst, P.C., Shrum, G.P. and Connor, D.T. (1991) Synthesis and transformation of 5-[3,5-bis(1,1-dimethylethyl)-4-hydroxyphenyl]-1,2,4-oxadiazoles. *13th International Congress of Heterocyclic Chemistry*, Corvallis, Oregon, August 11–16, Abstract PO2–94.

172. Unangst, P.C., Shrum, G.P., Connor, D.T., Schrier, D.J. and Dyer, R.D. (1992) Novel 1,2,4-oxadiazoles and 1,2,4-thiadiazoles as dual inhibitors of 5-lipoxygenase and cyclooxygenase. *203rd National Meeting of the American Chemical Society*, San Francisco, California, April 5–10, MEDI 122.

173. Unangst, P.C., Shrum, G.P., Connor, D.T., Dyer, R.D. and Schrier, D.J. (1992) Novel 1,2,4-oxadiazoles and 1,2,4-thiadiazoles as dual 5-lipoxygenase and cyclooxygenase inhibitors. *J. Med. Chem.*, **35**, 3691–3698.

174. Kostlan, C.R., Cetenko, W.A., Connor, D.T., Sorenson, R.J., Sircar, J.C., Bornemeier, D.A., Dyer, R.D., Kuipers, P.J., Okonkwo, G.C., Schrier, D.J. and Wright, C.D. (1992) Structure-activity relationship of thiazolidinone derivatives as dual inhibitors of 5-lipoxygenase and cyclooxygenase. *6th International Conference of the Inflammation Research Association*, White Haven, Pennsylvania, September 20–24.

175. Sorenson, R.J., Cetenko, W.A., Connor, D.T., Dyer, R.D., Bornemeier, D.A., Kuipers, P.J. and Wright, C.D. (1993) Benzylidene-thiazolinones as dual inhibitors of cyclooxygenase and 5-lipoxygenase: Potential antiinflammatory agents. *205th National Meeting of the American Chemical Society*, Denver, Colorado, March 28–April 2, Abstr. MEDI 141.

176. Unangst, P.C., Connor, D.T., Cetenko, W.A., Sorenson, R.J., Sircar, J.C., Wright, C.D., Schrier, D.J. and Dyer, R.D. (1993) Oxazole, thiazole, and imidazole derivatives of 2,6-di-tert-butylphenol as dual 5-lipoxygenase and cyclooxygenase inhibitors. *206th National Meeting of the American Chemical Society*, Chicago, Illinois, Aug 22–27, Abstr. MEDI 37.

177. Unangst, P.C., Connor, D.T., Cetenko, W.A., Sorenson, R.J., Sircar, J.C., Wright, C.D., Schrier, D.J. and Dyer, R.D. (1993) Oxazole, thiazole, and imidazole derivatives of 2,6-di-t-butylphenols as dual 5-lipoxygenase and cyclooxygenase inhibitors. *Bio. Med. Chem. Lett.*, **3**, 1729–1734.

178. Unangst, P.C., Connor, D.T., Cetenko, W.A., Sorenson, R.J., Kostlan, C.R., Sircar, J.C., Wright, C.D., Schrier, D.J. and Dyer, R.D. (1994) Synthesis and biological evaluation of 5-[[3,5-bis(1,1-dimethylethyl)-4-hydroxyphenyl]-methylene]oxazoles, -thiazoles, and -imidazoles: Novel dual 5-lipoxygenase and cyclooxygenase inhibitors with antiinflammatory activity. *J. Med. Chem.*, **37**, 322–328.

179. Cetenko, W.A., Connor, D.T., Sircar, J.C., Sorenson, R.J. and Unangst, P.C. (1992) 3,5-Di-tertiarybutyl-4-hydroxyphenylmethylene derivatives of 2-substituted thiazolidinones, oxazolidinones, imidazolidinones as antiinflammatory agents. US 5,143,928.

180. Connor, D.T., Kostlan, C.R. and Unangst, P.C. (1993) 2-Heterocyclic-5 hydroxy-1,3-pyrimidines, useful as antiinflammatory agents. US 5,240,929.

181. Connor, D.T. and Kostlan, C.R. (1993) 2-substituted-4,6-di-tertiarybutyl 5-hydroxy-1,3-pyrimidines useful as anti-inflammatory agents. US 5,177,079.

182. Belliotti, T.R., Connor, D.T., Kostlan, C.R. and Miller, S.R. (1993) 2-Carbonyl substituted-5-hydroxyl-1,3-pyrimidines as antiinflammatory agents. US 5,187,175.

183. Belliotti, T.R., Connor, D.T. and Kostlan, C.R. (1993) 2-Substituted amino-4,6-di-tertiarybutyl-5-hydroxy-1,3-pyrimidines as antiinflammatory agents. US 5,196,431.

184. Belliotti, T.R., Connor, D.T. and Kostlan, C.R. (1993) 5-Hydroxy-2-pyrimidinylmethylene derivatives useful as antiinflammatory agents. US 5,270,319.

185. Connor, D.T., Kostlan, C.R., Shrum, G.P. and Unangst, P.C. (1993) 5-Hydroxy-2-pyrimidinylmethene oxaza heterocycles. US 5,215,986.

3. INHIBITORS OF CHOLESTEROL BIOSYNTHESIS

DRAGO R. SLISKOVIC and BRUCE D. ROTH

Department of Chemistry, Parke-Davis Pharmaceutical Research,
Division of Warner-Lambert Company, 2800 Plymouth Road, Ann Arbor, MI 48105, USA

Before the discovery of the mevinic acids as potent inhibitors of HMG-CoA reductase, the rate limiting enzyme of the cholesterol biosynthetic pathway, there was no rational approach to the design and synthesis of potent and efficacious hypocholesterolemic agents. Since this discovery considerable effort has been devoted to the design and synthesis of totally or partially synthetic inhibitors of this enzyme. Today, inhibition of other steps of the biosynthetic pathway is being investigated for the potential to yield novel cholesterol lowering agents. This review will chronicle the efforts made to identify inhibitors of each step in the cholesterol biosynthetic pathway.

Atherosclerosis is an insidious disease of the arteries which can begin in the first decade of life with the deposition of lipids within the arterial intima. The resulting fatty streaks, whose most prominent component is the cholesteryl ester-enriched macrophage foam cell, eventually progress to form more complicated, fibrous lesions which can obstruct blood flow, creating turbulence which often results in laceration of the atheroma and emboli formation. The resulting myocardial infarctions are the leading cause of death in the industrialized nations. For example, in the USA, coronary heart disease accounts for 25–30%, of the annual deaths or 600,000 deaths each year.[1] Early hypotheses of the pathogenesis of this disease by Virchow[2] (the "lipid" hypothesis) and von Rokitansky[3] (the "incrustation" hypothesis) were confirmed by the landmark studies of N.N. Anitschkow, who fed rabbits egg yolks and produced a high level of cholesterol (1) in the blood as well as atherosclerosis in the aorta and coronary arteries.[4] Since these experiments were performed, there have been many clinical studies which have established hypercholesterolemia as a definite risk factor for the development of atherosclerosis in man.

Atherosclerosis is initiated or aggravated by a variety of environmental and genetic factors. The Expert Panel of the National Cholesterol Education Program (NCEP) in the US has recognized 10 such risk factors, including smoking, hypertension, severe obesity, gender, diabetes mellitus, etc.[5] However, among these risk factors, only an elevated serum total or, more specifically, low density lipoprotein-cholesterol (LDL-C) level has been shown to be an unequivocal <u>independent</u> risk factor for increased morbidity due to myocardial infarction.[6] Thus, the lowering of plasma cholesterol by dietary or pharmacologic intervention has become an accepted preventative measure against coronary heart disease (CHD). However, it was not until the results of the Lipid Research Clinic Coronary Primary Prevention Trial (LRC-CPPT) became available in 1984 that a widespread acceptance was established among clinicians about the desirability of treating of asymptomatic hypercholesterolemia.[7] This study demonstrated that long term treatment (seven years) with the bile acid sequestrant, cholestyramine (2), resulted in a 12.5% decrease in LDL-C which was associated with a 19% decrease in the incidence of myocardial infarction and CHD death. The Helsinki Heart Study showed that treatment with gemfibrozil (3) led to an 11% decrease in LDL-C and an 11% increase in high-density lipoprotein cholesterol (HDL-C), these changes were associated with a 34% decrease in the CHD end points of myocardial infarction and/or deaths.[8] The Cholesterol Lowering Atherosclerosis Study (CLAS) demonstrated that aggressive therapy with niacin (4) and cholestipol (5) resulted in a 43% reduction in LDL-C and a 37% elevation in HDL-C. Repeat angiography demonstrated significant reductions in both the progression and regression of atherosclerotic lesions.[9]

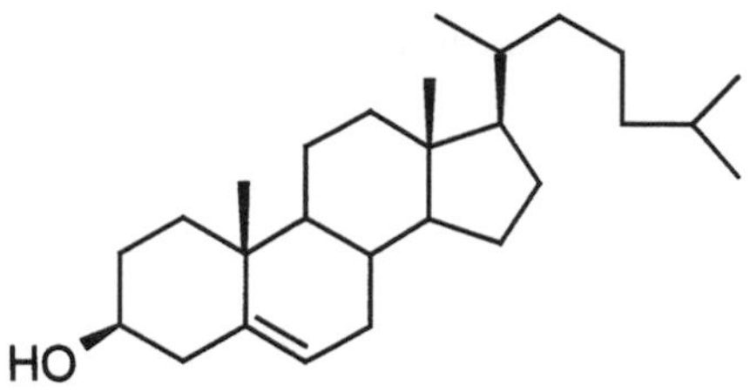

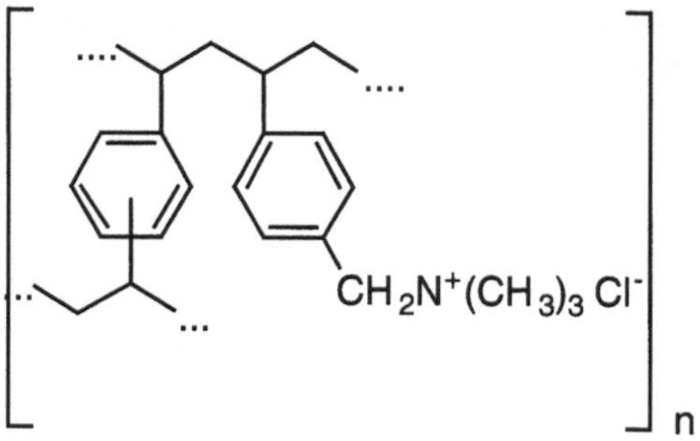

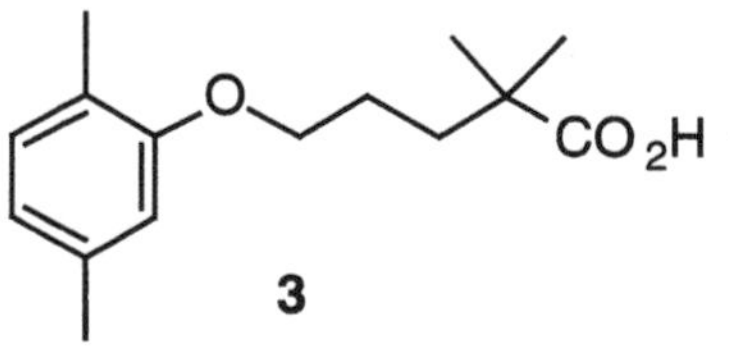

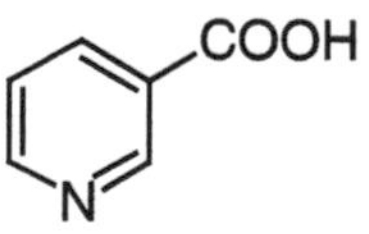

However, in none of these trials was a statistically significant difference in overall mortality demonstrated. The POSCH (Program on the Surgical Control of the Hyperlipidemias) trial has shown that lipid reduction, through partial ileal bypass surgery, significantly reduced CHD end points.[10]

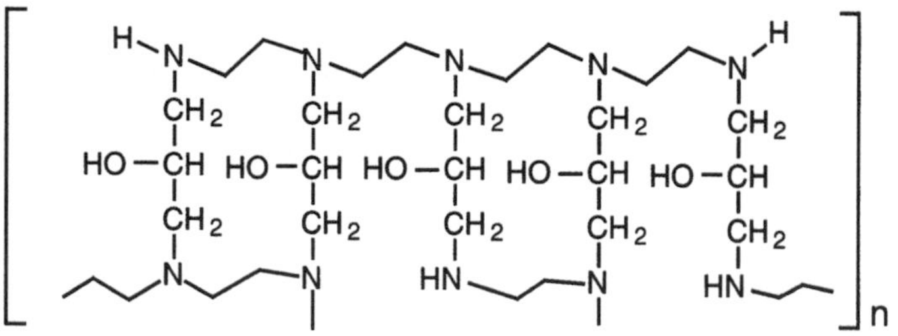

Most of the drugs involved in these trials exhibit some undesirable side-effects (e.g., gastrointestinal discomfort and compliance problems with cholestyramine and colestipol, increases in hepatocellular enzymes and creatinine phosphokinase (CPK) levels with gemfibrozil, and flushing with niacin). Thus researchers continue to search for safer and more effective lipid lowering agents.[11]

Cholesterol (1) is an essential component of all eukaryotic cells. It is utilized as a structural component in the membranes of cells, as well as in the production of steroid hormones, bile acids, lipoproteins and certain vitamins. Cholesterol within the body originates from two sources, either by absorption from the diet (exogenous source), which accounts for 300–500 mg/day in humans, or by endogenous biosynthesis within the tissues of the body (700–900 mg/day).[12] Studies in squirrel monkeys have indicated that 97% of whole body cholesterol synthesis can be accounted for by two tissues: liver and gastrointestinal tract. Furthermore, it was found that the rate of cholesterol synthesis in liver was decreased dramatically by feeding a high-cholesterol diet, whereas other tissues were essentially unaffected.[13] Cells possess an enzymatic pathway to synthesize the cholesterol they need. In this remarkable series of enzyme catalyzed reactions, acetyl-CoA is transformed into cholesterol. These reactions take place both in the cytosolic compartment and on specific membranes.[14] The rate limiting step in this sequence is the conversion of hydroxymethylglutaryl-CoA (HMG-CoA) into mevalonic acid catalyzed by the enzyme, HMG-CoA reductase (HMGR).[15] Inhibition of any of the steps in this pathway would, in theory, lower the amount of cholesterol produced by the body.

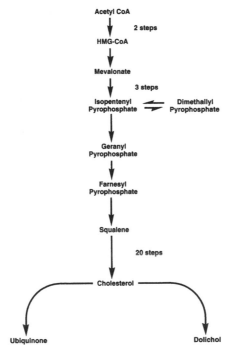

CHOLESTEROL BIOSYNTHETIC PATHWAY

Acetyl CoA

2 steps

HMG-CoA

Mevalonate

3 steps

Isopentenyl Pyrophosphate ⇄ Dimethallyl Pyrophosphate

Geranyl Pyrophosphate

Farnesyl Pyrophosphate

Squalene

20 steps

Cholesterol

Ubiquinone   Dolichol

Early searches for inhibitors of cholesterogenesis yielded only general inhibitors. Curran and Azarnoff outlined the desirable characteristics of a inhibitor of cholesterol biosyntheis. They identified the preferred site of action to be post acetoacetate, since before that crucial intermediates are required for other cellular processes, and pre-squalene, because it was thought that steroid precursors of cholesterol were themselves atherogenic.[16] The earliest known inhibitor was shown to be 2-phenylbutyric acid (6), which was shown to inhibit the formation of acetyl-CoA.[17] Even though this compound was shown to be hypocholesterolemic in rats,[18] it had no effect in humans.[19]

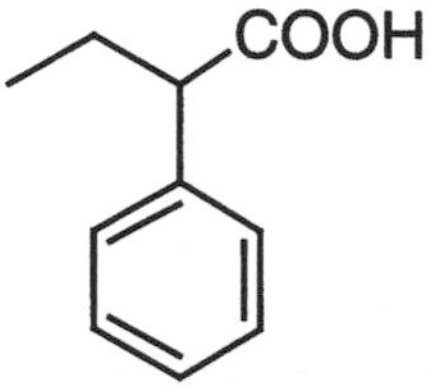

6

Other early inhibitors were based on the observation that pre-menopausal women are less susceptible to atherosclerosis than men.[20] This was thought to be due to the protective effects of estrogen, so a search for a nonestrogenic estrogen was undertaken. Structure-activity studies on chlorotrianisene (7) a synthetic estrogen, yielded triparanol (MER-29, (8)).[21] This compound was shown to be hypocholesterolemic in rats[22] and was eventually approved for human use.[23] However, it was eventually withdrawn from use due to the development of a number of side effects including cataract formation,[24] alopechia and icthyosis.[25] It was later shown that the compound inhibited the biosynthetic pathway at the penultimate step, i.e., the conversion of desmosterol to cholesterol and that this steroid intermediate accumulated in the lenses of patients.[26] The fears of Curran and Azarnoff thus were realized.[16] More recently, tamoxifen (9), a synthetic antiestrogen, has been shown to be hypocholesterolemic in humans by virtue of its ability to inhibit Δ^8-isomerase, one of the enzymes responsible for the conversion of Δ^8-cholestanol to cholesterol.[27]

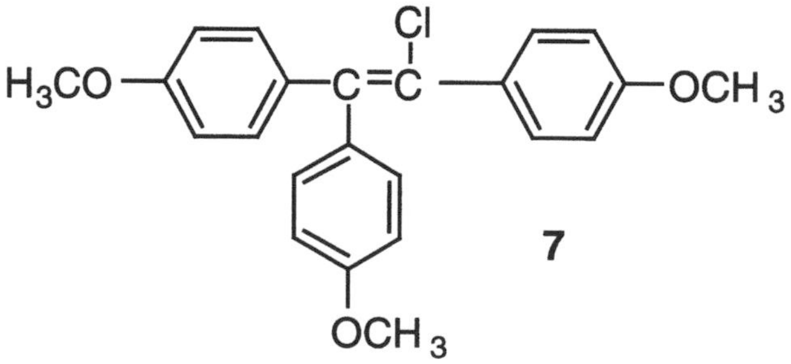

7

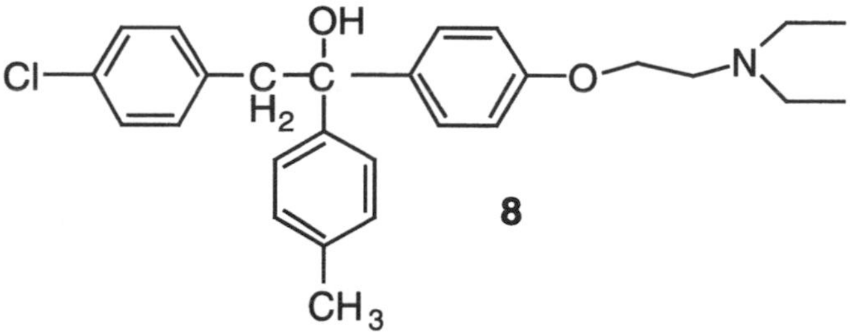

8

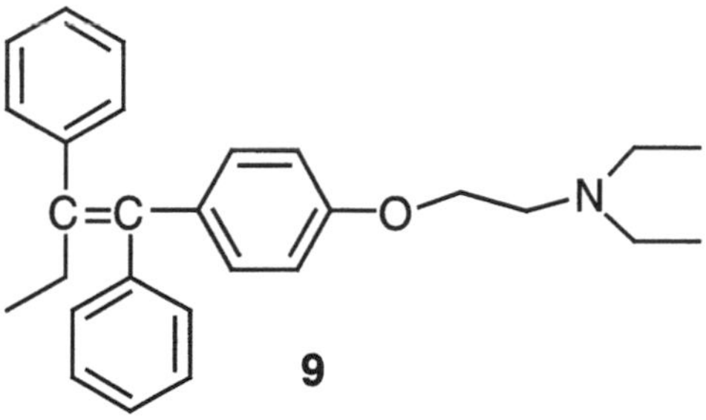

9

The search for inhibitors of cholesterol biosynthesis has continued for nearly four decades, but it was not until the discovery of the mevinic acids in 1976 that the full potential of this therapy in the treatment of hypercholesterolemia was realized.[28,29] This review will document the development of cholesterol biosynthesis inhibitors targeted at different enzymes in the cholesterol biosynthetic pathway. The search for inhibitors has concentrated mainly on three areas;

(1) inhibitors of steps before the formation of mevalonate,
(2) inhibitors of HMGR (natural products, semisynthetic, totally synthetic),
(3) inhibitors of steps after the formation of mevalonate.

Progress in each area will be reviewed.

1. INHIBITORS OF STEPS PRIOR TO MEVALONATE FORMATION

All of the carbon atoms found in cholesterol are derived from acetyl-CoA, which is the primary building block not only for all steroids, but for fatty acids and ketone bodies, as well. The first step in the cholesterol biosynthetic pathway is the condensation of two molecules of acetyl-CoA, with loss of CoA-SH, to produce acetoacetyl-CoA.

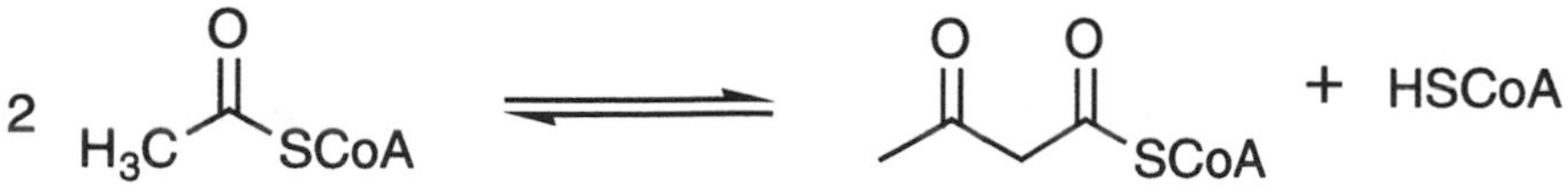

This reversible reaction is catalyzed by the enzyme **acetoacetyl-CoA thiolase** (acyl-CoA-AcCoA-C-acetyltransferase, EC 2.3.1.16).[30] This enzyme exists in at least three isoforms, a mitochondrial isozyme (EC 2.3.1.9) which is specific for acetoacetyl-CoA and is involved solely in the synthesis of ketone bodies, a second form which has broader specificity for acyl-CoA substrates and is involved in the β-oxidation of fatty acids, and the cytosolic isozyme, which is also specific for acetoacetyl-CoA, with a role restricted to cholesterol biosynthesis.[31] Studies with porcine heart thiolase demonstrated that the enzyme exists as a tetramer of four (probably) identical subunits, each comprising approximately 400 amino acids, including one catalytically active cysteine.[32] Based on the results of mechanistic studies with this enzyme, it has been concluded that the enzyme acts by a "ping-pong" mechanism, wherein an acetyl group is transferred to a critical cysteine in the active site, followed by attack by another acetyl-CoA.[32,33] Advantage has been taken of the understanding of the mechanism of this reaction to design affinity labels and irreversible inhibitors of this enzyme by alkylation of the active site thiol with an α-haloketone derived from a fatty acid[34,35] or by an alkylalkane thiosulfonate.[36] In the case of the latter agents, it was demonstrated with a series of polymethylene bismethane thiosulfonates (**10**) that inhibitors of this enzyme lower plasma total cholesterol in chow fed rats and rats on high fat, low carbohydrate and low fat, high carbohydrate diets. Plasma triglycerides were also reduced in this study, however, no correlation was found between *in vitro* potency and the reductions in either plasma total cholesterol (TC) or triglycerides, suggesting that these inhibitors may have poor distribution properties or may not be specific for thiolase.[36] More recently, a triyne carbonate (**11**) from an *Actinomycete* culture has also been found to inhibit this enzyme with an IC_{50} of 0.01 μM, without inhibiting HMG-CoA synthase, HMG-CoA reductase or fatty acid synthase. Compound **11** also inhibited ^{14}C-acetate and ^{14}C-octanoate incorporation into sterols in Hep-G2 cells in culture by 56% and 48%, respectively, at a concentration of 3 μM.[38]

$$CH_3SO_2S(CH_2)_nSSO_2CH_3$$

10

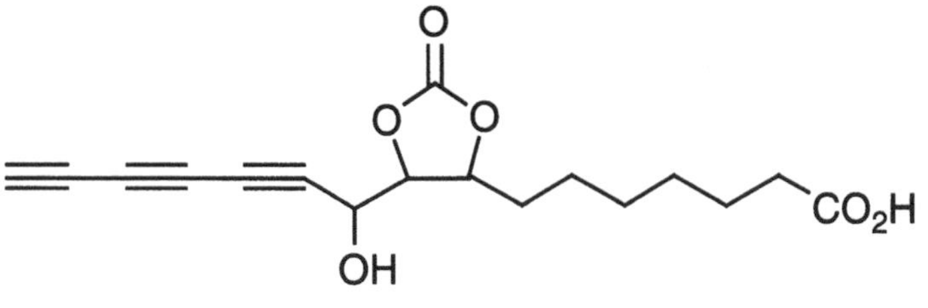

11

The second step of the biosynthetic pathway is the condensation of the acetoacetyl-CoA produced by thiolase with another molecule of acetyl-CoA to form 3-hydroxy-3-methylglutaryl-CoA (HMG-CoA). This reaction is catalyzed by the enzyme **HMG-CoA synthase** (HMGS, EC 4.1.3.5).

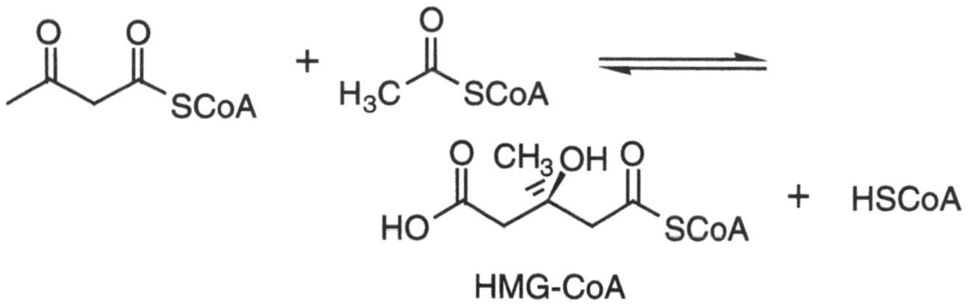

HMG-CoA

This enzyme has been found to be a dimer, also possessing an active site cysteine, which reacts with acetyl-CoA to form an Ac-S-enzyme complex which can reverse in the presence of added acetoacetyl-CoA or CoA-SH.[38] Like thiolase, HMGS can be deactivated by methyl methanethiosulfonates and other alkylating agents.[38] As seen with HMGR, HMGS activity is down-regulated by oxysterols.[39]

Screening for active metabolites in microbial cultures from *Fusarium* sp.[40] and *Scopulariopis* sp.[41] led to the near simultaneous identification of the β-lactone antibiotic, 1233A (12) as a potent (IC_{50} = 0.1 μM), specific and irreversible inhibitor of HMGS *in vitro*,[42] with a rate of inactivation nearly equal to the rate of catalysis.[43] In fact, utilizing a radiolabeled derivative of 1233A (^{3}H-L-668,411), a stable enzyme-inhibitor complex could be isolated. In cell culture and in rats *in vivo*, however, inhibition was found to be reversible on removal of the inhibitor.[42] In order to better understand the mechanism of inhibition, a structure-activity study was undertaken to determine the effect of changes proximate to the β-lactone on inhibition kinetics. In this study, it was found that the β-lactone was essential for inhibition, however, not all β-lactones were found to be irreversible inhibitors. Thus, compound 13 (IC_{50} = 0.1 μM), which lacks the α-hydroxymethyl group was found to exhibit kinetics characteristic of reversible inhibition or rapid turnover of enzyme. Rapid turnover and synthesis of new enzyme may also explain the apparent reversibility of inhibition with 12 in cell culture and *in vivo*.[43]

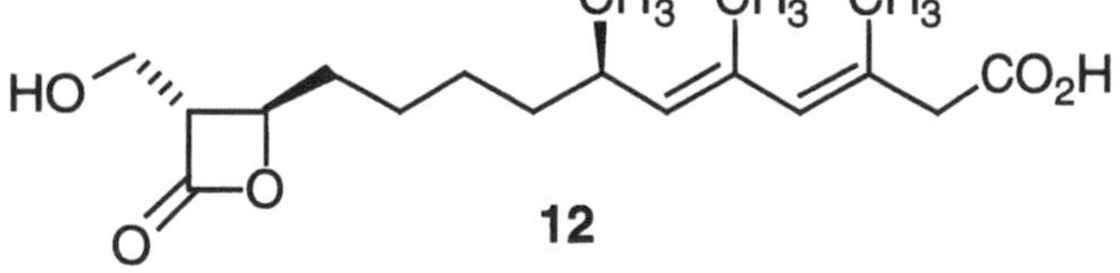

12

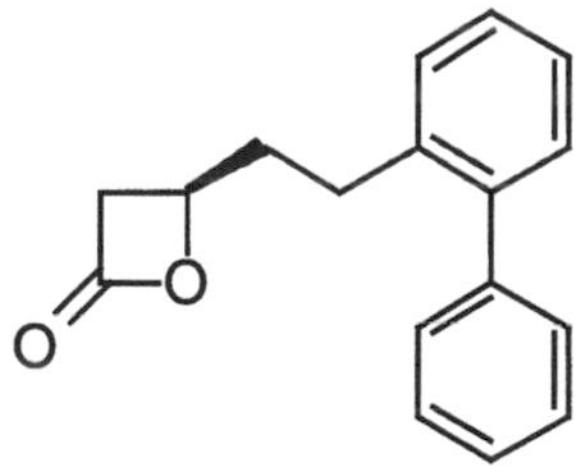

13

2.   INHIBITORS OF HMG-CoA REDUCTASE

2.1   Natural Products and Semi-Synthetic Analogs

The enzyme, **HMG-CoA reductase** (HMGR, EC 1.1.1.34), is the rate limiting enzyme in the cholesterol biosynthetic pathway and its product, mevalonic acid is an essential component of several cellular constituents, such as dolichol, ubiquinone, and cholesterol. In order to maintain cholesterol homeostasis within the cell, negative feedback mechanisms, involving cholesterol or an oxygenated metabolite, suppress the activities of both HMGS and HMGR.[44] In states of cholesterol deprivation, cells maintain high activities of both of these enzymes, thereby satisfying the cells need for cholesterol and other mevalonate-derived products. When the cell is replete with cholesterol, the activities of both enzymes decrease by more than 90% and the cells produce only the small amounts of mevalonate needed for the production of dolichol, ubiquinone, etc.[15] Also when cellular sterols rise, the LDL receptor (LDL-R) gene is repressed, avoiding furthur overaccumulation of cholesterol.[45] Conversely, when intracellular sterol levels drop, the cell responds by increasing cholesterol synthetic rates and the number of LDL-R's. Thus, the mechanism by which inhibition of HMGR lowers plasma LDL levels is by upregulating LDL-R activity and thus increasing the clearance of plasma LDL.

HMGR catalyzes the irreversible reduction of HMG-CoA to mevalonic acid. The reaction takes place in two sequential reductive steps. After the binding of HMG-CoA to the enzyme, reduction, utilising the first equivelent of NADPH, gives a putative hemithioacetal. The second NADPH then releases the mevaldate from the hemitioacetal, and this is then reduced to mevalonate. The optimum for the reaction is pH 7 and there is an absolute requirement for NADPH.[46] Microsomal HMGR is the only non-soluble enzyme in the cholesterol biosynthetic pathway prior to the formation of squalene.

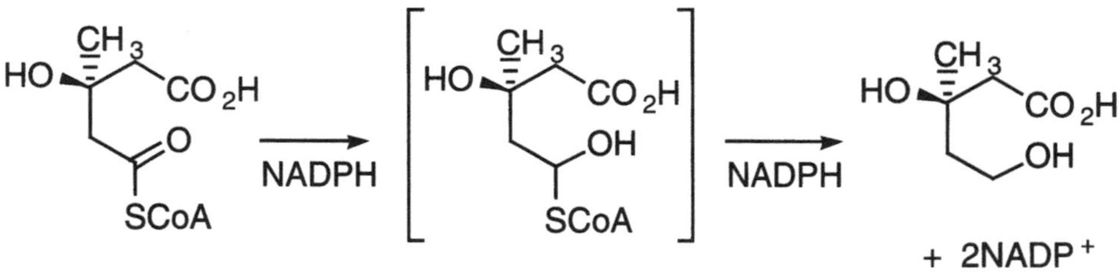

It was recognized early in the development of HMGR inhibitors that either substrate or product analogs could possibly act as competitive inhibitors. 3-Hydroxy-3-methylglutaric acid has been shown to lower plasma cholesterol in man and shown to inhibit HMGR, however, additional mechanisms of action could not be ruled out.[47]

The seminal finding in the quest for inhibitors of HMGR was the discovery of the fungal metabolites, mevastatin (**15**) and lovastatin (**18**). Endo and co-workers first described the discovery of three inhibitors of cholesterol biosynthesis (designated

ML-236A (**14**), ML-236B (**15**) and ML-236C (**16**)), in the culture broth of the fungus, *Penicillium citrinum*.[48] ML-236B (**15**, mevastatin) was shown to be identical to a compound (called compactin) isolated by Brown *et al.* from cultures of *Penicillium brevicompactum*.[49] These discoveries intiated a new search for more potent inhibitors. A closely related analog of mevastatin, called monacolin K, was isolated from cultures of the fungus, *Monascus ruber*.[50] The same compound, **18**, lovastatin, originally called mevinolin, was isolated, independantly by workers at Merck, from *Aspegillus terreus*.[51] A number of other metabolites were isolated from these fungii, dihydrocompactin (**21**) from *P.citrinum*,[52] dihydromevinolin (**22**) from *A. terreus*,[53] monacolin J (**17**) and L (**19**) from *M. ruber*,[54] and monacolin X (**20**) and dihydromonacolin L (**23**) from a mutant strain of *M. ruber*.[55]

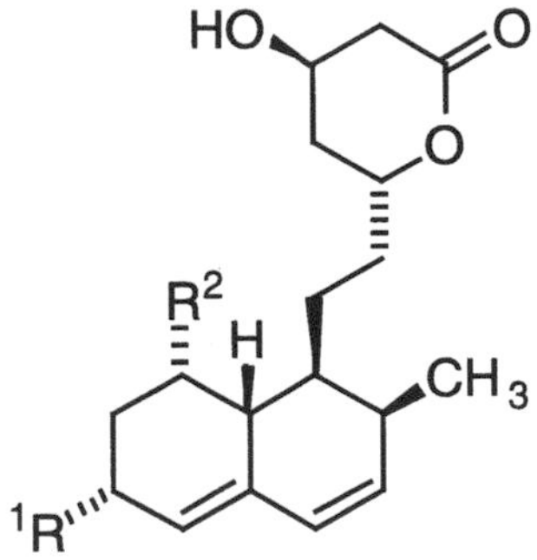

	R[1]	R[2]
ML-236A (**14**)	H	OH
ML-236B,Compactin (**15**)	H	$CH_3CH_2(CH_3)CHCO_2$
ML-236C (**16**)	H	H
Monacolin J (**17**)	CH_3	OH
Monacolin K,Lovastatin (**18**)	CH_3	$CH_3CH_2(CH_3)CHCO_2$
Monacolin L (**19**)	CH_3	H
Monacolin X (**20**)	CH_3	$CH_3CO(CH_3)CHCO_2$

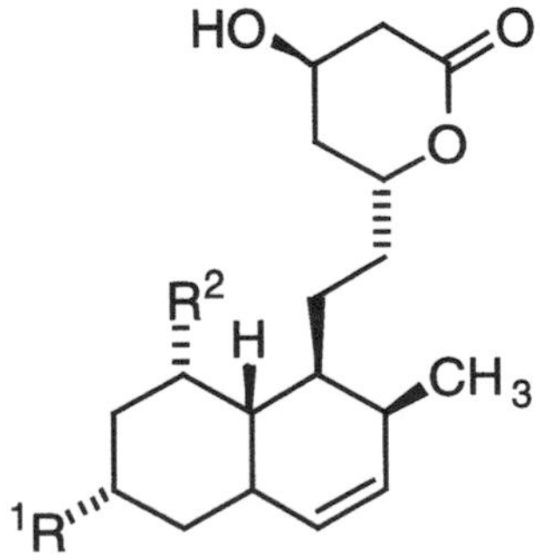

	R[1]	R[2]
Dihydrocompactin (**21**)	H	$CH_3CH_2(CH_3)CHCO_2$
Dihydromevinolin (**22**)	CH_3	$CH_3CH_2(CH_3)CHCO_2$
Dihydromonacolin L (**23**)	CH_3	H

Several active compounds have been derived from either compactin (**15**) or lovastatin (**18**) by microbial conversion. The most noteworthy of these compounds is pravastatin (**24**, CS-514)), the 6β-hydroxy open acid form of compactin. This compound was originally identified as a urinary metabolite of compactin in the dog,[56] however, it can be produced by microbial transformation of compactin using *Norcardia autotrophica*.[57]

The active form of the naturally occuring fungal metabolites is the ring-opened dihydroxy acid form. Inhibition of HMGR by these compounds is reversible and competitive with respect to HMG-CoA and non-competitive with respect to NADPH. The K_i values for the acid forms of **15**, **18** and **24** are 1.0,[58] 0.6[51] and 2.3 nM,[59] respectively. Under the same conditions the K_m value for HMG-CoA is 10 µM. Thus, these agents have binding affinities for HMGR up to 20000 times higher than the binding affinity of the substrate, HMG-CoA.[58]

The ability of these compounds to inhibit cholesterol biosynthesis has been evaluated *in vitro* in cultured cells and *in vivo* in the rat. All compounds were shown to inhibit cholesterol biosynthesis at concentrations in the low nanomolar range.[60] All of these compounds are ineffective in lowering plasma cholesterol in normo-lipidemic rats, due to dramatically increased synthesis of the enzyme in response to the competitive inhibition of HMGR.[61]

The best model for demonstrating plasma cholesterol lowering with these agents is the beagle dog. Efficacy can be demonstrated when the compounds are either dosed alone[62] or in combination with a bile acid sequestrant, such as cholestyramine or colestipol.[63] For example, in one study, 18 was dosed to dogs at 8 mg/kg for 34 days and produced a 27% lowering in plasma TC levels.[51] Combining lovastatin (1–8 mg/kg/day) with cholestyramine (12 g/day) led to a lowering of TC by 14–49%.[64] Similar results have been obtained with both compactin (15)[65] and pravastatin (24).[66]

The mechanism by whch this plasma cholesterol lowering is achieved is best understood by recognising that cellular cholesterol comes from two sources: *de novo* endogenous synthesis and from plasma LDL via the LDL-R. The rate of synthesis of the LDL-R is inversely related to the amount of cellular cholesterol. When synthesis is inhibited, the hepatic cholesterol pool is decreased and the LDL-R is upregulated so that the number of cell surface receptors is increased and more cholesterol enters the cell by this route. In fact, HMGR inhibitors increase the mRNA for the LDL-R in the liver and enhance cell surface expression of the LDL-R on other hepatocytes and other cell types. This increased LDL-R activity enhances the clearance of cholesterol from the plasma and thus lowers plasma TC. In homozygous familial hypercholesterolemia (FH), a condition in which patients have no LDL-receptors, HMGR inhibitors are ineffective.[67]

Compactin (15), lovastatin (18) and pravastatin (24) have all been shown to be effective hypocholesterolemic agents in man. Both lovastatin (Mevacor®) and pravastatin (Pravachol®) have been approved by regulatory authorities around the world for the treatment of hypercholesterolemia. Compactin was discontinued in clinical trials due to toxicity in dogs.[68]

These agents have made a profound impact on the treatment of heterozygous familial hypercholesterolemia (het. FH), a condition where high levels of LDL are found, due to the fact that these patients possess only 50% of the normal LDL-R number.[68] In patients with this disorder, 15, at doses ranging from 60–100 mg/day, produced a 27% decrease in TC after 4–8 weeks treatment.[69] The addition of a bile acid sequestrant produced much greater lowering.[70] When given in doses of 20 mg twice daily, 18 reduced LDL by 25–30%. Dosages of 40 mg twice daily reduced levels by 35–40%.[71,72] At a dose of 40 mg/day, over a 12 month period, 24 reduced LDL by 26%.[73] These agents are also effective in the treatment of other hypercholes-terolemic states such as, primary moderate hypercholesterolemia (non FH), familial dysbetalipoproteinemia and familial combined hyperlipidemia.[67]

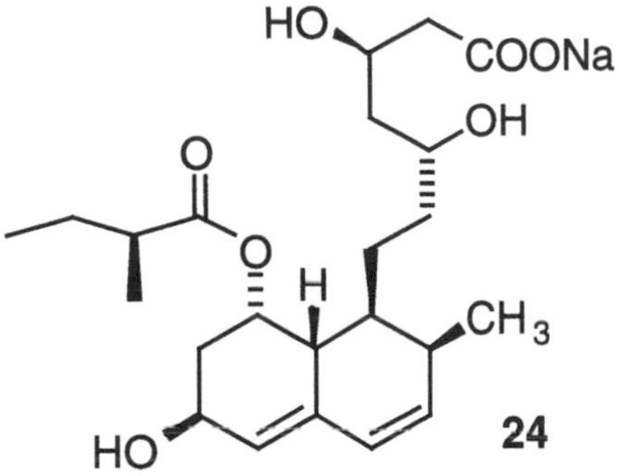

Recently, the MARS study (Monitored Atherosclerosis Regression Study) has shown that treatment with **18** (80 mg/day) plus a cholesterol lowering diet slows the rate of progression and increases the frequency of regression in coronary artery lesions. Over the study period ($\approx$2 years), TC levels were reduced by 32% and LDL levels by 38%.[74]

Two types of toxicity were expected from treatment with these agents, 1) inherent toxicity from the drug itself and 2) toxicity due to the mechanism of action, i.e., arising from a reduction of the key metabolites of mevalonic acid. There has been little convincing data that this class of inhibitors affect the synthesis of either dolichol or the ubiquinones, however, there has been data which showed that **18** did decrease coenzyme Q in tissues isolated from rats[75] and in the plasma of humans treated with **18**.[76] The most notable adverse reaction is an elevation of serum transaminases in a small number of patients, however, in general this class of drugs is remarkably well tolerated.

The discovery of the fungal metabolites has led to a number of SAR studies designed to identify the pharmacophore necessary for HMGR inhibition and to use this information in the design of semisynthetic and totally synthetic inhibitors with improved biological activity. Workers at Merck have pioneered the efforts in these directions.[77] In considering the structure of these compounds, these workers made a number of interesting observations. The relative and absolute stereochemistry of the lactone moiety was shown to be critically important for activity since the corresponding 3,5-dihydroxy acid moiety must mimic the HMG portion of HMGR and bind to the active site of the enzyme. Inversion of one or both asymmetric centers in the lactone moiety leads to a marked loss of activity.[78a] Replacement of the axial β-hydroxy in the lactone moiety of **18** with the acetylthio group leads to a tenfold reduction in activity.[79] Activity is also totally ablated when a methylene group is inserted between the carboxyl and β-carbinol group.[80] Interestingly, Heathcock has shown that oxidation of the 5-carbinol groups in the dihydroxy acid forms of **15** or **22** to give the 5-keto derivatives **25** and **26** leads to no loss in activity. The authors speculate that these compounds are in fact HMGR substrates and undergo reduction, *in situ*, to the active 5-carbinol.[78b]

With regards to the structural requirements in the hexahydronaphthalene ring, the *trans*-fused 4a,5-dihydro compounds (**21**, **22**) retain activity comparable to both

15 and 18.[81] Further reduction of the 3,4-double bonds in these compounds, to give the corresponding *trans*-fused decalins minimally reduces activity. Reduction of 15 and 18 to give the 3,4-dihydro derivatives, 27 and 28, also has little effect on activity. The cis fused analogs, 29 and 30, are much less active than 15 and 18. In total these results illustrate the need for a relatively planar ring system.[60,77,82] Some modifications have been made at the 6 position of the hexahydronaphthalene ring system. Homologation of the 6α-methyl group of 18 to the corresponding ethyl analog has been shown to retain activity,[83] and epimerization of the asymmetric center at position 6 and methyl group oxidation gave the hydroxymethyl analog, 31, which was also shown to be active.[84] The monocyclic compactin analog, 32, retains all the oxygen functionality and the optimal relative and absolute stereochemistry, however, it inhibits HMGR with an $IC_{50} = 320$ μM compared to a corresponding value of 32 nM for 25.[85] Compound 33 was shown to have comparable potency to 18. SAR studies on this compound showed that all four substituents around the cyclohexyl ring are required for potent HMGR inhibition and the C2 position of these inhibitors is highly sensitive and has strict structural requirements.[86]

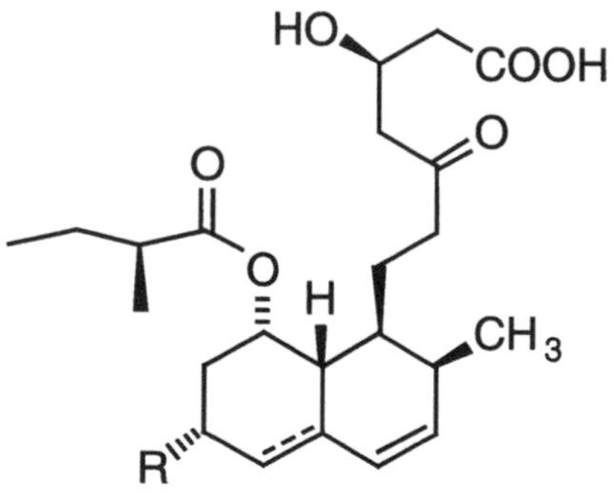

25, R=H (double bond)
26, R=CH₃ (single bond)

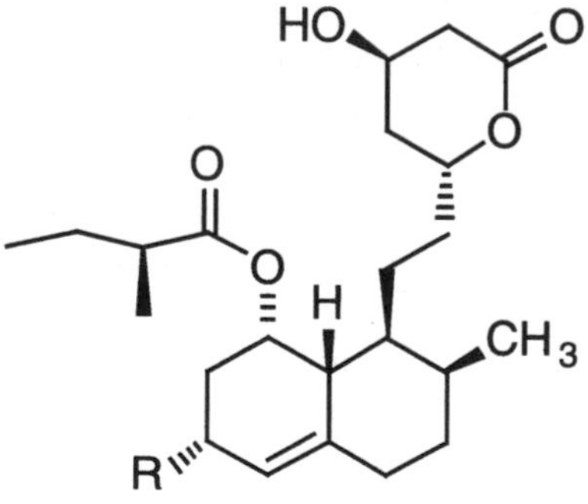

27, R=H
28, R=CH₃

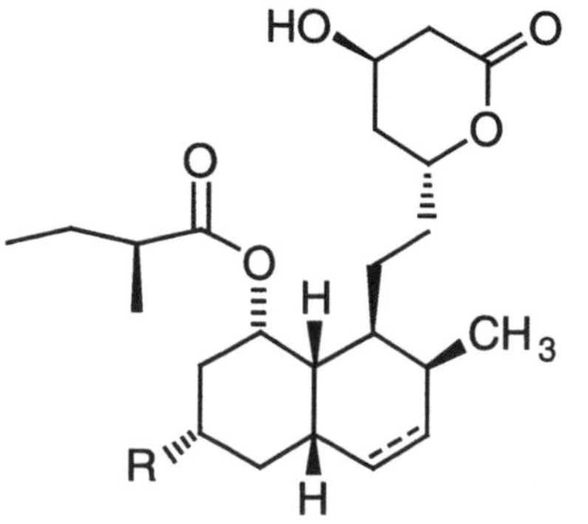

29, R=H (single bond)
30, R=CH₃ (double bond)

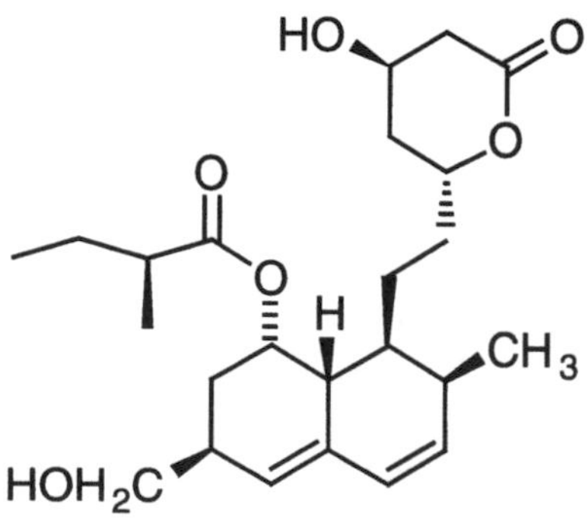

31

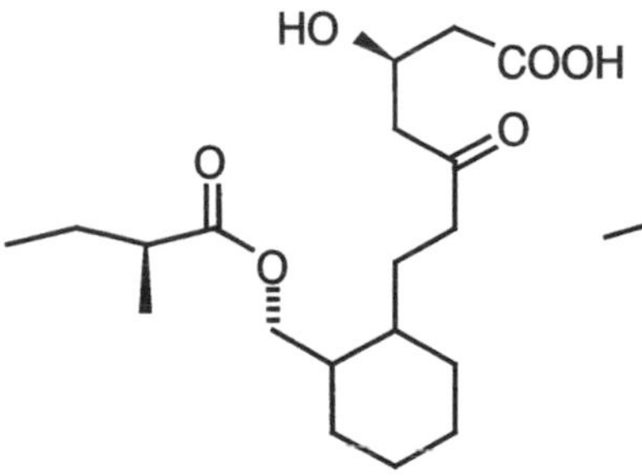

32

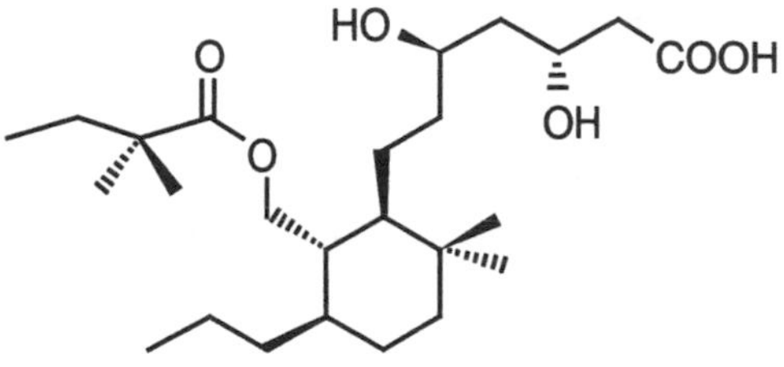

33

The modification of the sidechain ester moiety has proved to be one of the most fruitful areas of SAR development. Saponification of the sidechain ester moiety to the corresponding 8α-alcohol led to a marked loss of activity.

The sidechain epimer was also equipotent to **18**.[87] Such observations led to an intense research effort to delineate the SAR trends in this region of the molecule. It was found that the stereochemistry in the sidechain was not important, the presence of the acyl moiety was essential for good activity, and branching at the acyl α-carbon increases potency.[88] Sidechain ether analogs were also shown to be an order of magnitude less active than their ester counterparts.[89] A recent QSAR study showed that the activity of the sidechain ester analogs correlated best with the molecular size of the acyl moiety and not hydrophobicity.[90] These studies led to the identification of simvastatin (**34**), the 2,2-dimethylbutyrate analog of **18**. The ring opened form of this compound was shown to be about 2.5 fold more potent than **18**, with a K_i value of 0.2 nM. It is also a potent inhibitor of cholesterol biosynthesis in both cellular assays and *in vivo*.[91] Simvastatin has also been shown to be effective in lowering plasma TC in dogs and in humans.[92] Simvastatin (Zocor®) has been approved for use in humans throughout the world. It is more effective than either **18** or **24** and has the potential advantage of once a day dosing at a low dose of 10 mg/kg.[92]

Recent SAR studies have identified a potent series of 5-oxygenated analogs of **34**. The 5 α-hydroxy analog (**35**) and its 6β-methyl epimer (**36**) were shown to be potent

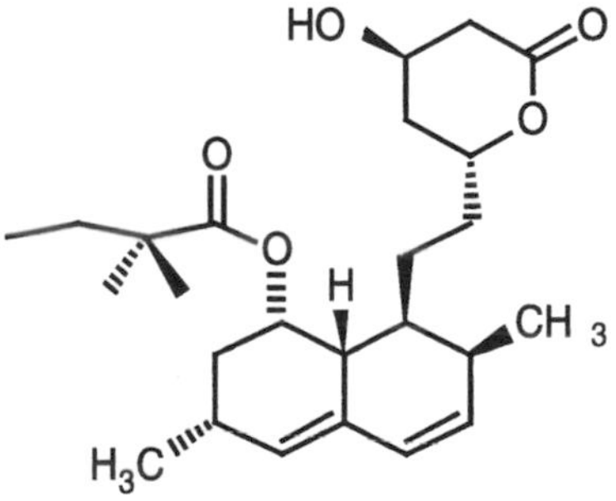

34

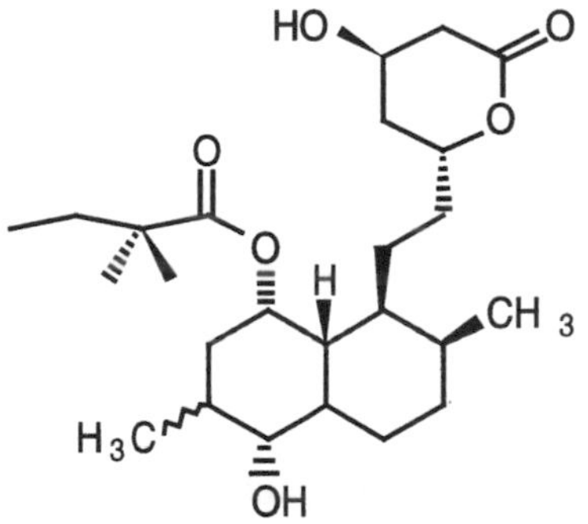

35 , 6 α-methyl epimer
36 , 6 β-methyl epimer

inhibitors of HMGR with IC_{50}'s of 49 and 41 nM, respectively. They were also shown to inhibit cholesterol biosynthesis in rats administered ^{14}C-acetate and were as effective as **18** at lowering plasma TC in hypercholesterolemic dogs.[93]

The issue of tissue selectivity for these agents has evoked considerable controversy in the literature as each pharmaceutical company vies for any marketing edge with regards to side effect profiles.[94] This controversy arose from studies by Tsujita *et al.* who showed that **18** (lovastatin) and **24** (pravastatin) were equipotent as inhibitors of cholesterol biosynthesis in freshly isolated hepatocytes, however, **24** was 100 fold less active than **18** in inhibiting biosynthesis in cultured human skin fibroblasts. Acute experiments *ex vivo* in rats indicated that *de novo* biosynthesis was inhibited solely in the liver and intestine by **24**, but that **15** and **18** also inhibited synthesis in kidney, lung, spleen, prostate, and testes.[66] In marked contrast to these studies, a measurement of drug distribution in peripheral tissues relative to liver following the oral administration of **18**, **24** and **34** in rats and dogs demonstrated that **18** and **34** are more hepatoselective than **24**. Both **18** and **34** are administered as lactone pro-drugs, and undergo first pass metabolism, hepatic sequestration and hydrolysis to their active ring-opened forms. This confines the activity to the liver and minimizes peripheral exposure to active drug.[95] Neither **18**, **24** or other synthetic HMGR inhibitors display hepatoselectivity at the cellular level when assessed in Hep G2 cells (liver model) or human skin fibroblasts (peripheral model).[96]

Hybrid bile acid-HMGR inhibitors (**37**, **38**) were designed utilizing the affinity that bile acids, such as cholic acid, have for enterohepatic circulation. These compounds incorporate the functionality necessary for bile acid transport in the intestine and liver together with the dihydroxy acid moiety necessary for HMGR inhibition. When compared to **18**, these compounds are weak inhibitors of HMGR (e.g., **37** ($IC_{50} = 39.2$ μM), **38** ($IC_{50} = 12.3$ μM), **18** ($IC_{50} = 3.7$ nM)).[97]

Whether tissue selectivity for any compound imparts any improved side effect profile is still an unanswered question since all the HMGR inhibitors currently available are remarkably free from serious adverse reactions.[94]

A series of compounds related to dihydromevinolin (**22**) were synthesized. Compound **39** was an extremely potent inhibitor of HMGR ($IC_{50} = 3$ nM), **18** ($IC_{50} = 11$ nM). Tissue selectivity was assessed *ex vivo* and it was found that this compound showed greater inhibition in the liver than in the spleen, testes and kidney. These studies continue to map the active site of HMGR and provide some evidence for additional binding sites at the C7 methyl group site and at the C3 alkyl position.[98]

The first synthesis of dihydroeptastatin (**40**) was recently described. It was shown to be equipotent to **18** at inhibiting HMGR ($IC_{50} = 14$ nM vs $IC_{50} = 13$ nM for **18**).[99]

2.2 Totally Synthetic Analogs

The earliest synthetic compounds reported to inhibit HMGR were a series of 1-(4-biphenylyl)-1-pentyl monoesters of succinic (**41**) and glutaric acids (**44**).[100] Structure-activity studies led to a series of monoesters of β-hydroxy-β-alkylglutaric acids, the most potent of which, **44**, was the 1-(4-biphenylyl)-1-pentyl ester of the natural substrate, β-hydroxy-β-methylglutaric acid. This compound was found to be a

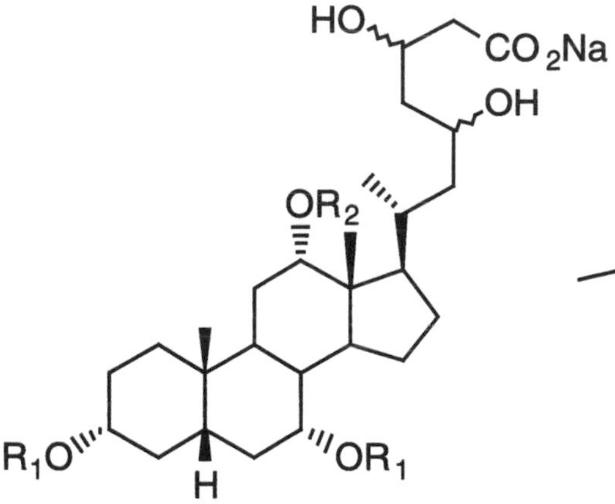

37, R$_1$=R$_2$=Ac
38, R$_1$=H, R$_2$=COtBu

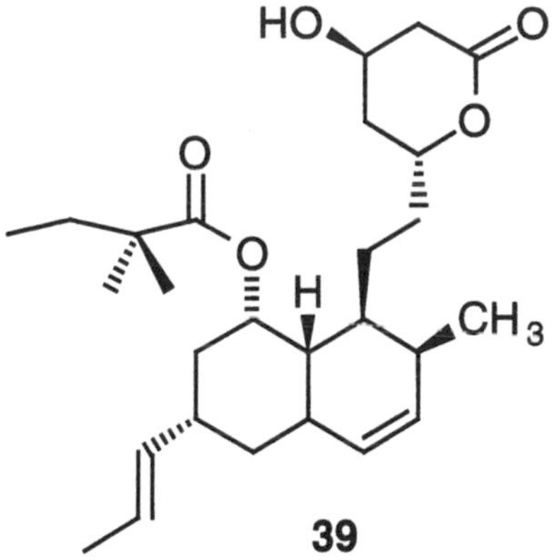

39

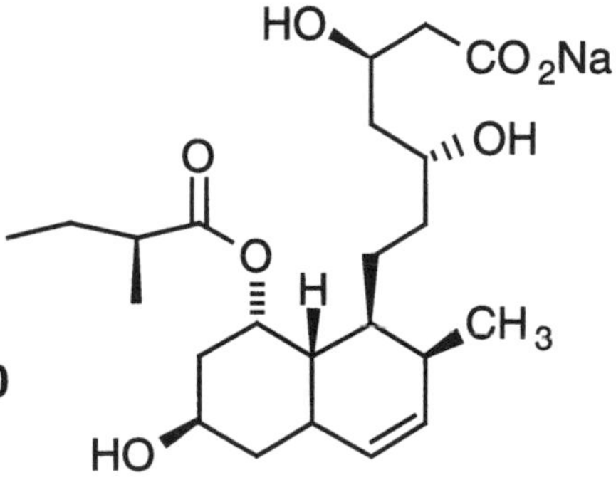

40

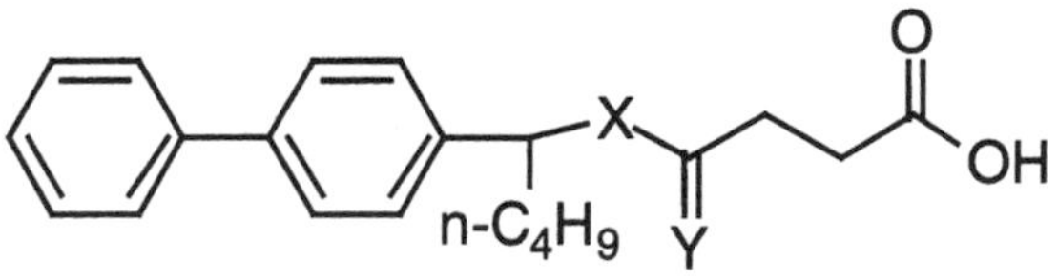

41, X= Y = O
42, X= O, Y = H
43, X= NH, Y = O

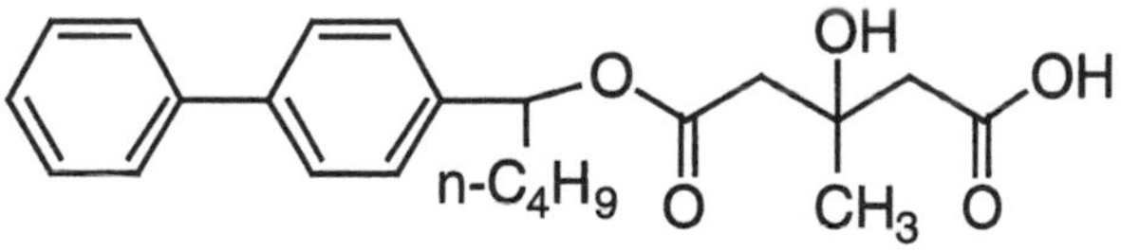

44

reversible, but noncompetitive inhibitor with respect to both substrate (HMG-CoA) and NADPH (50% inhibition at an inhibitor/substrate ratio of 1.5).[101] The introduction of other β-alkyl groups in this series led to reduced inhibitory potency.[102] Replacement of the ester in **41** by various isosteres, including ether (**42**) and the all carbon analog retained activity in this series, but replacement by an amide (**43**) abolished activity.[103] Other simple inhibitors derived from mevalonic acid have also been identified.[104]

Because of the success of the fungal metabolite HMGR inhibitors at lowering plasma TC and LDL-C in animals and man, major efforts to elucidate the key structural features needed for potent inhibition were initiated at several pharmaceutical and academic research institutions. The first reports of totally synthetic inhibitors, incorporating the mevalonate moiety found in the naturally occuring fungal metabolites, originated from the Sankyo[105] and Merck laboratories,[106–110] the same organizations which discovered the utility of the fungal metabolites. The Sankyo scientists prepared a series of β-hydroxy-β-methylglutaric acid analogs, the most potent of which, **45**, displayed an ID_{50} of 15 µM for inhibiting cholesterol biosynthesis in mouse L cells.[105] The work from Merck proved seminal. Utilizing a 2,4-dichlorophenyl group as a surrogate for the hexahydronaphthalene ring system of the fungal metabolites, it was demonstrated that all of the biological activity was contained in the dextrorotatory mevalonolactone side-chain isomer (+)-**46** ($IC_{50} =$ 10.8 µM, 0.1% of the potency of compactin),[106] which was subsequently demonstrated to correspond to the specific 4R, 6R stereochemistry found in both HMG-CoA and the fungal metabolites (4R, 6S isomer in the case where the 6-substituent is unsaturated).[107] Introduction of the 4-methyl group found in HMG-CoA into the lactone of these inhibitors led to a significant reduction in inhibitory potency, as did elimination or oxidation (to the ketone) of the 4-hydroxyl, conversion of the 6-hydroxyl to a methyl ether, or conversion of the lactone to the lactol methyl ether. A preference for a two atom spacer (preferably the E-olefin, although good activity was also found with the dimethylene and the oxymethylene spacers) between the mevalonolactone and the 2,4-dichlorophenyl ring was also found. The most potent inhibitor found in this study inhibited HMGR with an IC_{50} of 0.89 µM, equivalent to approximately 1% of the inhibitory potency found with compactin.[106] Continued structure-activity studies in this series revealed that the introduction of a bulky 6-substituent in the 2,4-dichlorophenyl ring led to significant increases in potency *in vitro*. Thus, compound (+)-**47** possessed 50% of the *in vitro* potency of compactin.[108] Further definition of the optimal inhibitor topology for high *in vitro* potency suggested that exceptional potency could be obtained with an orthobiphenyl connected to the mevalonolactone by an E-olefin. Systematic examination of the ortho aryl ring in these compounds led to the identification of the 4-fluoro-3-methyl analog (**48**) as the most potent inhibitor, with an $IC_{50} = 7.6$ nM, approximately three times more potent than compactin.[107] When the orthobiphenyl ring system was tied together to form a fluorenyl ring (**50**) activity was reduced ($IC_{50} = 85$ nM vs. 24 nM for **49**), suggesting the need for these rings to be orthogonal for highest potency.[109] Potent inhibitors were also found when the mevalonolactone was appended to a 2-naphthyl system (**51**, $IC_{50} = 33$ nM).[110]

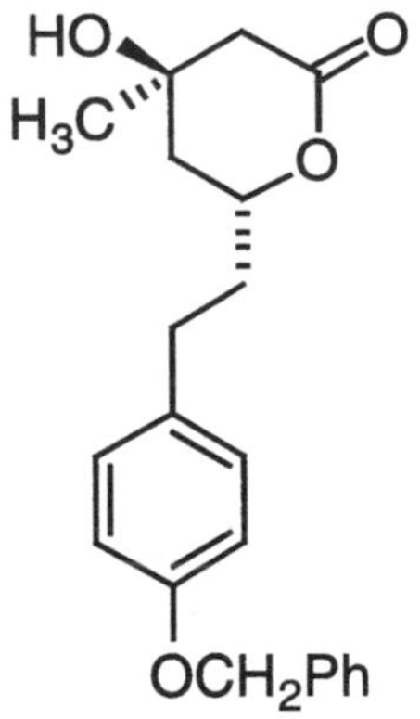

45

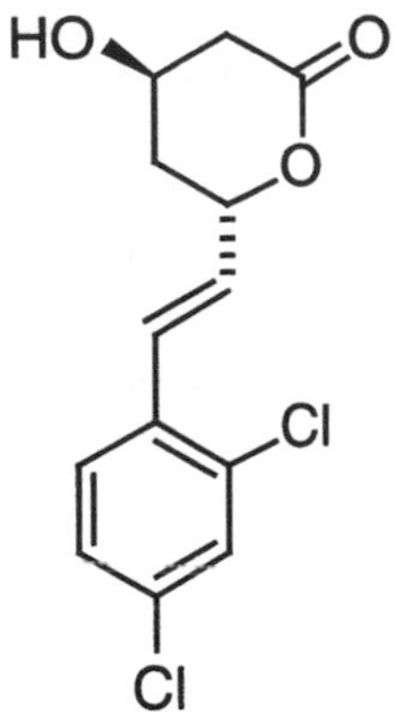

(+)-46

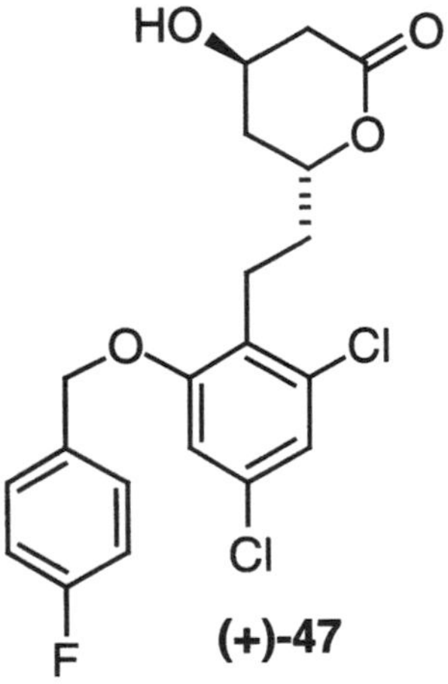

(+)-47

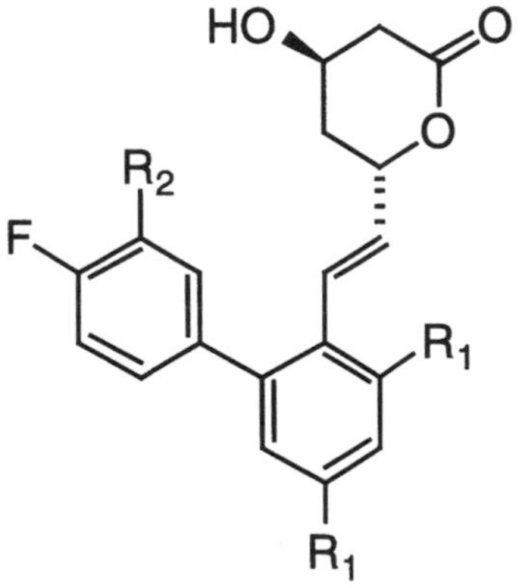

48, $R_1 = R_2 = CH_3$
49, $R_1 = Cl$, $R_2 = H$

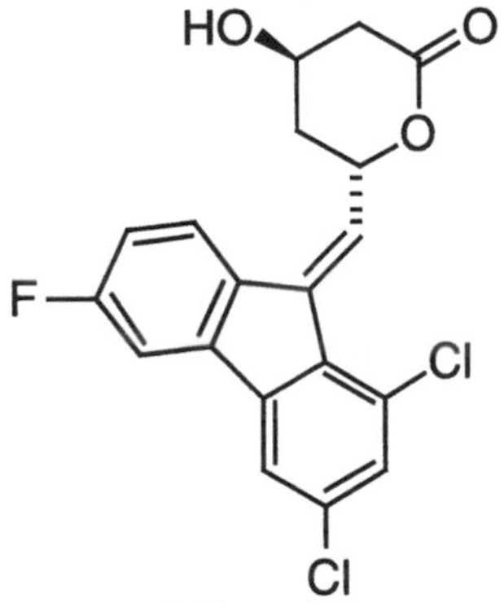

50

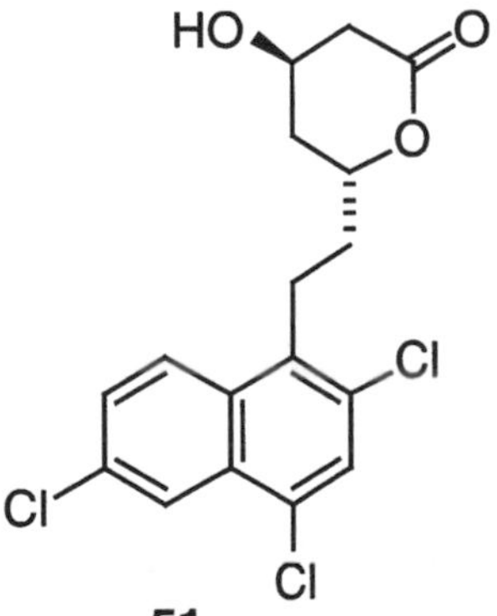

51

With the key topological features for potent inhibition defined, it was left to others to conclude that the actual nature of the center piece (see Figure 1) was unimportant and served only as a template to hold the lactone and a lipophilic moiety in the proper spatial arrangement and provide sufficient lipophilicity for tight binding to the enzyme.[111] Because of this observation, and the seemingly essential nature of the mevalonolactone moiety, most of the effort in the synthetic inhibitors has focused on variations of the central template. This has led to a host of inhibitors containing a wide variety of templates, including indole,[112] pyrrole,[111,113–115] imidazole,[112,116] pyrazole,[112,118] pyridine,[118] pyrimidine,[118] quinoline,[119] cyclohexene,[120] phenyl ether[121] and tetrahydroindazoles.[122]

The stepwise development of the pyrrole series is illustrative of the evolution of synthetic inhibitors. Thus, the early SAR in this series involved the definition of the optimal substituents in the 2- and 5-positions of a pyrrole appended to the mevalonolactone by a saturated 2 carbon spacer at the pyrrole 1-position.[111] In this study, it was found that the 2-position was optimally substituted by a 4-F-phenyl group (although other bulky aryl and cycloalkyl groups also produced potent inhibitors). At the 5-position, it was found that potency increased with increasing substituent size, with a maximum at isopropyl (52, $IC_{50}=230$ nM). Larger substituents led to reduced activity and the cyclohexyl substituted analog was inactive. This led to the conclusion that the overall width of the molecule (across the two pyrrole substituents) must be less than 10Å. Substitution at the 2-position by electron withdrawing groups, such as CF_3, was also favorable (53, $IC_{50}=630$ nM). Examination of substitution at the 3- and 4-positions revealed that although introduction of a bulky substituent into either the 3 or 4-position led to an improvement in potency *in vitro*, potency equivalent to the fungal metabolites was obtained only when both positions were substituted by bulky lipophilic groups (e.g., pentasubstitution).[113] The best activity was found in compound 54 (CI-981, $IC_{50}=7$ nM), which was found to be 5 times more potent than compactin *in vitro*.

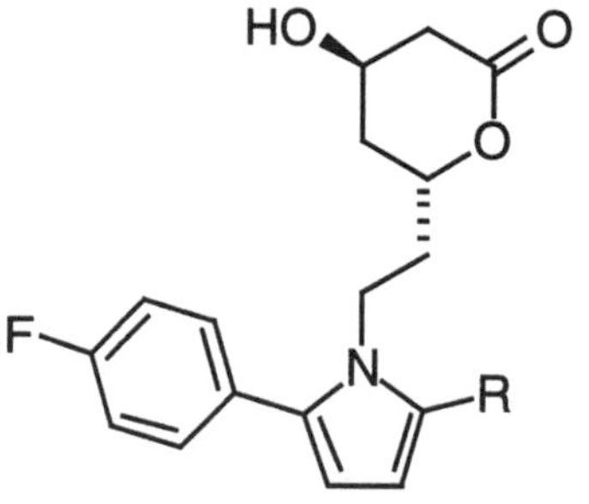

52, R =i-Pr
53, R =CF₃

Maximum width of 10A

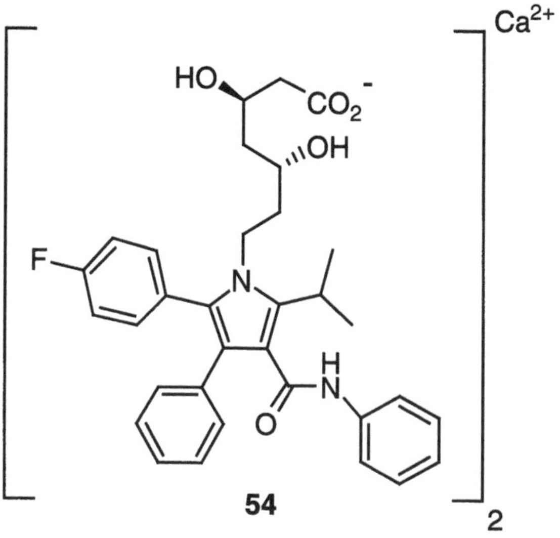

54

Further independent investigations into the pyrrole series suggested that activity in this series could be further improved by replacing the saturated two carbon spacer with an E-olefin.[115] With this change, a fifth pyrrole substituent was not required to obtain inhibitors with potency *in vitro* equivalent to or better than the fungal metabolites and compound **55** possessed an IC_{50} of 0.5 nM (although this could be further improved to 0.3 nM if a bromine was introduced into the 4-position to produce the pentasubstituted analog **56**). Pyrroles connected to the mevalonolactone by a 2 carbon spacer attached to the pyrrole 2-position also produced potent inhibitors.[114] The SARs in this series confirmed the trends seen with the 1-substituted isomers, although potency was not affected significantly by the nature of the two atom spacer or the presence or absence of a fifth substituent on the pyrrole ring (**57**, **58**, IC_{50}'s = 6.0 and 2.5 nM, respectively). These compounds were also shown to be more effective hypocholesterolemic agents than lovastatin in endogenously hypercholesterolemic rabbits and cholestyramine-primed dogs, although the compounds containing the saturated two carbon spacers appeared to be preferred due to some acid instability found with the unsaturated analogs.[114] Interestingly, compounds containing the E-olefin in the 1-position of the pyrrole nucleus (**55**, **56**) were reported to have "remarkable" acid stability. The authors postulated that this might be due to the lack of conjugation of the pyrrole ring with the double bond, something demonstrated in the imidazole series by X-ray crystallography.[116]

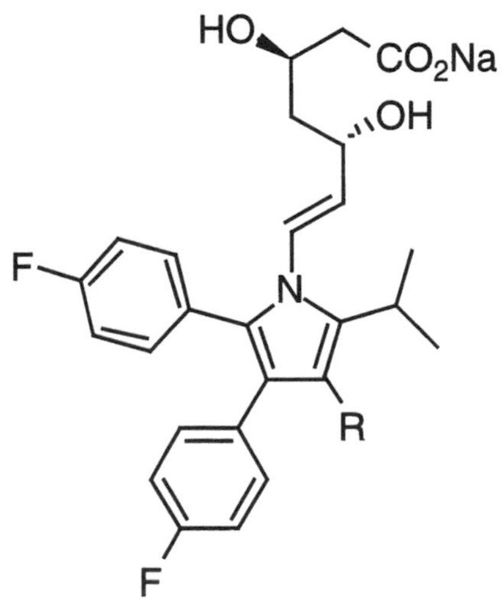

55, R = H
56, R = Br

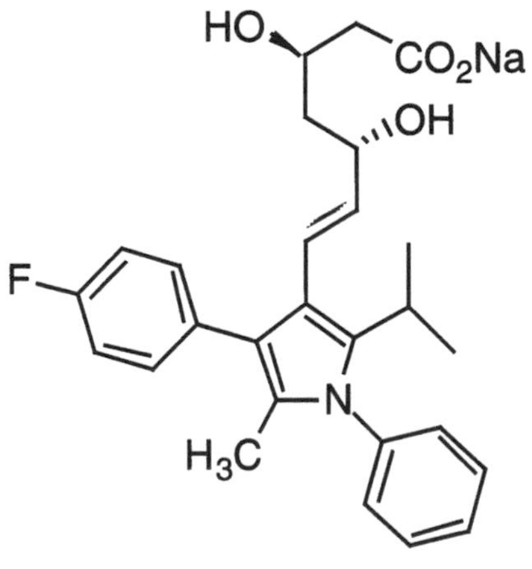

57, 6,7-unsaturated
58, 6,7-saturated

A similar acid catalyzed degradation has been reported for the indole fluvastatin (**59**).[123]

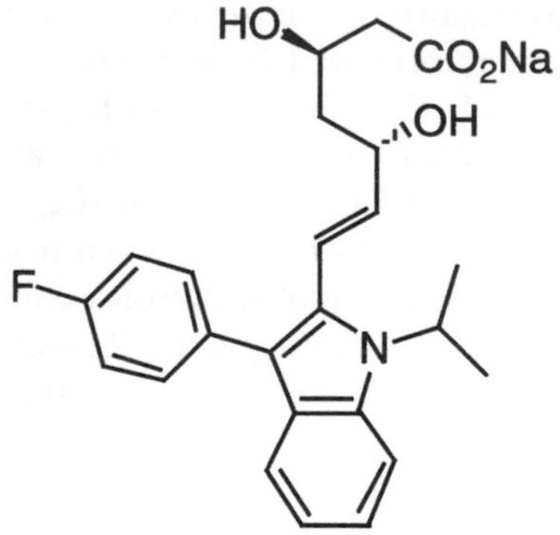

59

Thus, based on the SARs of the many inhibitor classes reported, a composite of all of the inhibitors can be developed (Figure 1). In general, the overall topology of the inhibitors matches well with the fungal metabolites. Thus, the mevalonolactone (or ring-opened dihydroxyacid) is connected at the 6-position by a two carbon spacer to the lipophilic moiety, with highest potency generally found with the trans-olefin spacer.[115] The central template serves to hold (at least) three lipophilic moieties in the proper spatial arrangement to obtain highest potency. Two of these groups occupy the "ortho" positions flanking the spacer connecting the template and the mevalonolactone. These groups are optimally a 4-fluoro or 4-fluoro-3-methylphenyl (occupying the same region in space as the isobutyryl group of the

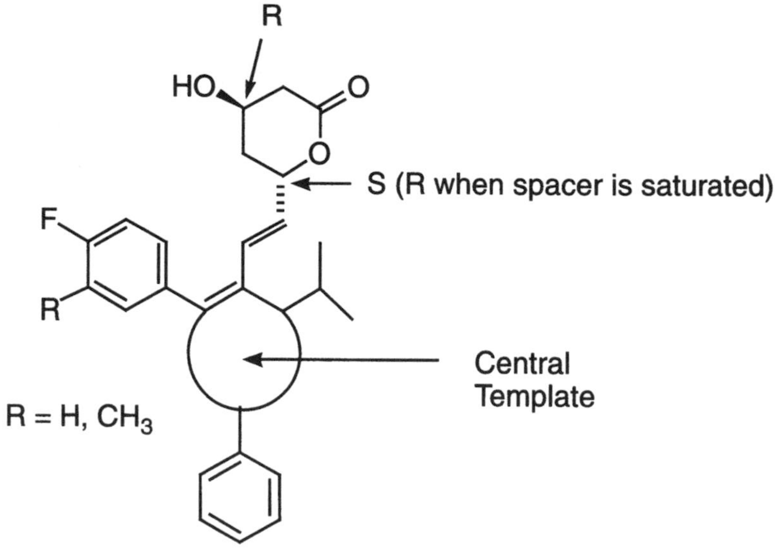

Figure 1

fungal metabolites) on one side and a group approximately the size of an isopropyl on the other (presumably occupying the same region as the 2-methyl group of the fungal metabolites).[112,113] Generally, at least one other lipophilic moiety is needed for potency equivalent to or exceeding that of the fungal metabolites. This group (often a phenyl) is optimally located in a position approximately "para" to the spacer.[118] In the 5-membered ring heterocycles, this requirement has been satisfied by a fused ring in this location or by bulky substituents in either the 3 or 4 positions, or preferably, substituents in both the 3 and 4 positions.[112,113] Others have developed similar models by comparing computer derived molecular models of lovastatin and synthetic inhibitors with a model of HMG-CoA derived from the X-ray crystal structure of CoA bound to citrate synthase.[124] Recently, the incorporation of a fluorine onto the *E*-olefin in the pyrimidine series has also been shown to result in potent inhibitors.[125]

A more recent and novel variation has been to replace the central ring with an olefin.[126–129] Optimum potency is achieved in these compounds when the olefin is tetrasubstituted, with the terminal carbon (9 position) substituted by two aryl groups and the adjacent (8-position) substituted by a group approximately the size of an isopropyl (**60**, $IC_{50} = 14$ nM, one-half the *in vitro* potency of lovastatin), thus approximating the spatial arrangement found in the compounds containing a ring as central template. Removal of either of the olefin substituents or inclusion of these substituents in fused ring systems led to significant reductions in potency *in vitro*.[129,130]

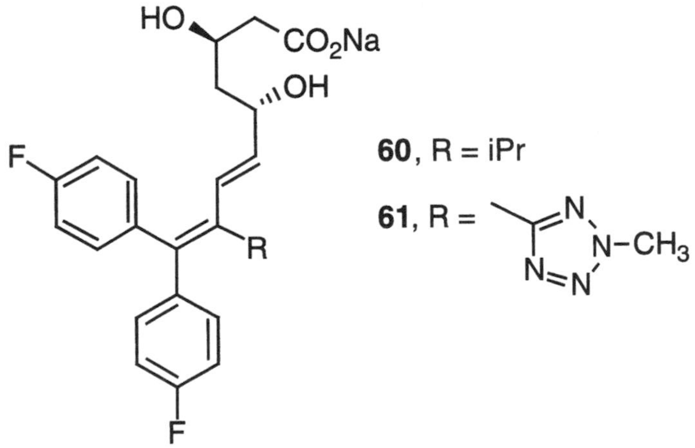

The observation that wide variations in structure and lipophilicity in the central template did not seem to significantly affect inhibitor potency (given that they satisfied the requirements outlined above), has been taken advantage of in attempts to design "tissue selective" inhibitors, i.e., inhibitors which affect cholesterol biosynthesis primarily in the liver, the principal site of *de novo* synthesis.[127,128,131] The relationship between lipophilicity and tissue selectivity has been systematically studied in tissue cubes,[132] tissue culture[96,127,138] and in whole animals.[133] In tissue cubes, a linear relationship was found between lipophilicity and tissue selectivity, suggesting that tissue selectivity is directly related to lipophilicity. When activity was plotted against inhibitor lipophilicity (as estimated by clogP) in each tissue examined (spleen, testes and adrenal tissue cubes), an optimal lipophilicity for penetration into peripheral tissue for the ring opened dihydroxyacids of a clogP of 2–3 was found.[128,132] Penetration into liver seemed to be unaffected by the inhibitor lipophilicity.[132] This phenomenon has recently been explained by observation that the hydrophilic inhibitor pravastatin is taken up by liver through the bile acid transporter system, which is unavailable to other tissues.[134] Despite the good correlation between lipophilicity and tissue selectivity *in vitro*, no clear relationship has been demonstrated *in vivo*, presumably due to differences between inhibitors in absorption and metabolism. In fact, one of the most lipophilic inhibitors, CI-981 (**54**, atorvastatin) has been found to be one of the more tissue selective inhibitors *in vivo*, apparently due to very high first pass metabolism.[133]

An alternate strategy for introducing tissue selectivity has been based on the observation that the 5-hydroxyl of the dihydroxy acid can be replaced by a phosphinic acid, thus, mimicking the transition state for reduction of HMG-CoA to mevalonic acid.[135] This modification has been combined with several of the "hydrophobic anchors" previously connected to the mevalonolactone with no loss, and often an improvement in potency *in vitro*. Unlike the mevalonolactone containing inhibitors, these compounds are somewhat less sensitive to the nature of the two carbon spacer and excellent potency is obtained with acetylene (**62**), *E*-olefin (**63**) and saturated two carbon (**64**) spacers (IC$_{50}$'s=6, 26 and 24 nM, respectively). Oxymethylenes also retained good activity. These compounds also show good selectivity for inhibition of cholesterol biosynthesis in hepatocytes, but not in fibroblasts.[135] The only other lactone modification which has retained good inhibitory potency has been the replacement of the 5-hydroxyl with a sulfoxide.[136] Other modifications, such as conversion of the lactone to a piperidinone, have resulted in the complete loss of activity.[137] Other attempts to remove the dependency on the specific stereochemistry in the mevalonolactone, such as the preparation of symmetrical 3-alkyl-3-hydroxy-glutaric acids, has also resulted in weak inhibitors.[138]

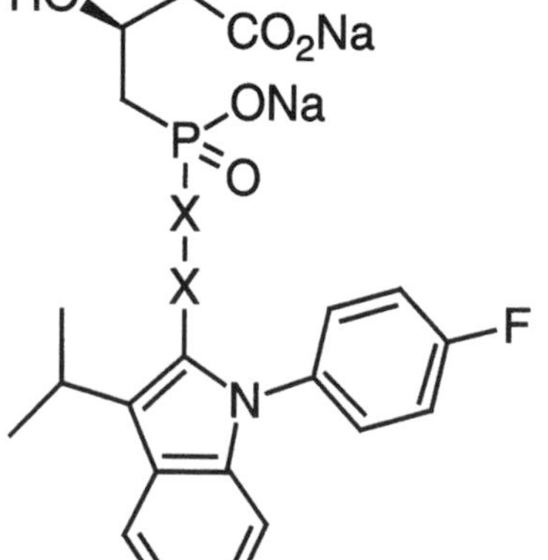

62,	X—X =	C≡C
63,	X—X =	(E)-CH=CH
64,	X—X =	CH$_2$-CH$_2$

In order to better understand the binding of the synthetic and natural inhibitors to HMGR, several kinetic studies have been performed examining the effects of HMG-CoA, coenzyme A (CoASH) and NADPH on the binding of compactin and the synthetic compounds to HMGR. Because of the differential effects of CoASH on inhibitor binding, it was concluded that compactin and the synthetic inhibitors studied bound to HMGR in different manners. Thus, compactin was concluded to occupy both the HMG and CoA binding sites, whereas the synthetic inhibitors (**65** and **66**) occupied the HMG site and an adjacent hydrophobic site distinct from the CoA binding site. It was concluded that the high affinity of these inhibitors for HMGR was due to the simultaneous binding to two separate binding sites.[139] More recent kinetic studies have suggested that the more potent competitive HMGR inhibitors (including lovastatin, fluvastatin and **49**) are slow binding inhibitors, while other less potent inhibitors are not. It was proposed based on this kinetic analysis that binding of the potent inhibitors is biphasic and that the lipophilic moieties distal to the lactone may play an important role in determining the mechanism of inhibitor binding to HMGR.[124]

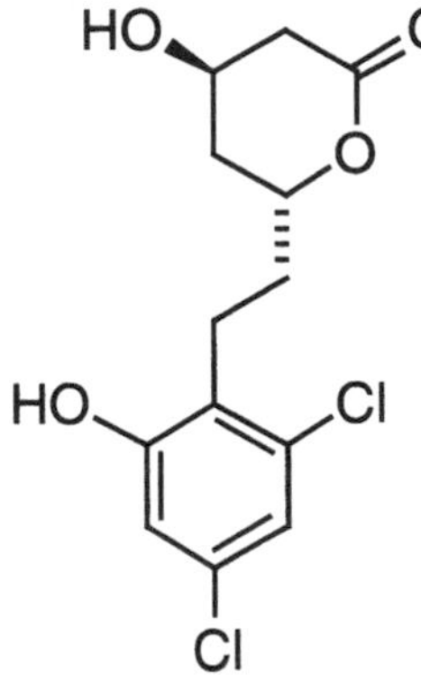

65 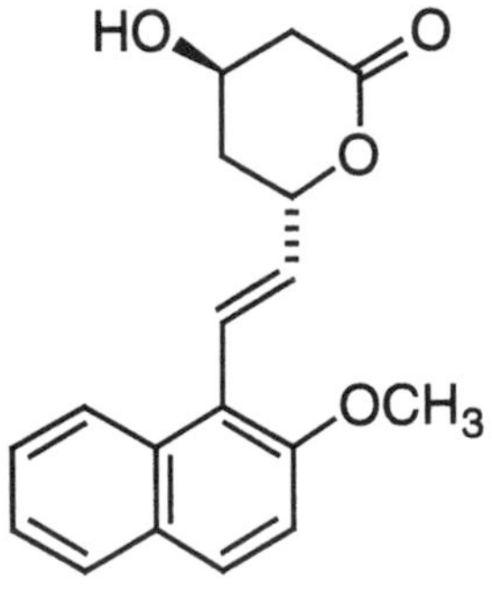   **66**

Several compounds from the inhibitor design strategy outlined above, have been selected for evaluation in clinical trials. Thus, from the indole series, fluvastatin (**59**, XU-62-320) when dosed to hypercholesterolemic patients for 6 weeks at a mean daily dose of 23.1 mg reduced TC and LDL-C by 19 and 25%, respectively, comparable to the changes seen with the fungal metabolites.[140] Dose-response studies with **59** at doses of 5 to 40 mg QPM resulted in reductions in TC and LDL-C of 11–21% and 15–28%, respectively. Fluvastatin was well tolerated at all doses in this study.[112] Fluvastatin has recently become the first totally synthetic HMGR inhibitor approved for use in the United States. Dalvastatin (**67**), from the cyclohexene derived inhibitors, has also progressed into clinical trials. In a dose-response study in 30 men with non-familial hypercholesterolemia, dalvastatin at doses of 5 to 160 mg/day, produced 10–25% and 14–32% reductions in TC and LDL-C, respectively, suggesting that dalvastatin may be somewhat less potent than other HMGR inhibitors at lowering

TC and LDL-C clinically (both fluvastatin and dalvastatin are racemic, thus 50% of the administered dose is pharmacologically inactive).[141] Preliminary results with several chiral HMGR inhibitors have also been reported recently. Clinical studies with the pyrrole based inhibitor CI-981 (**54**, atorvastatin) have suggested that this compound may have enhanced potency and efficacy, producing reductions in TC and LDL-C of 45 and 58%, respectively, when dosed at 80 mg/day for two weeks in healthy volunteers. The minimum effective dose in this study was 2.5 mg, which produced significant reductions in TC and LDL-C of 13 and 22%, respectively.[142] CI-981 was well tolerated in this study. In a second study to assess the effect of AM vs. PM dosing, CI-981 dosed to healthy volunteers at 40 mg/day for 14 days was equally effective at lowering TC and LDL-C whether given in the AM or PM (48% vs. 49% reduction in LDL-C, respectively).[143] Preliminary clinical results have also been published recently for a second totally synthetic chiral HMGR inhibitor, rivastatin (**68**). In healthy volunteers, rivastatin produced reductions in TC and LDL-C of 20–25% and 30–36%, respectively, at the low doses of 100–400 µg/day.[144] This potency is consistent with the reports that rivastatin is 100 times more potent than lovastatin at inhibiting HMGR *in vitro*.[145] Rivastatin was well tolerated in this study up to and including the 300µg dose, however, one patient from the group dosed at 400 µg was dropped due to increased CPK levels. Several other HMGR inhibitors, including HR 780 (**69**),[118] BMY 21950 (**61**),[127] and SQ 33,600 (**62**),[135] have also been reported to be undergoing clinical evaluation, however, no human efficacy data has been reported for these compounds to date.

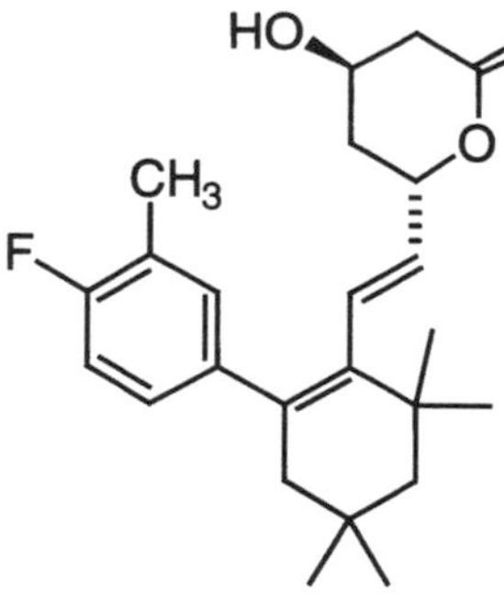

67 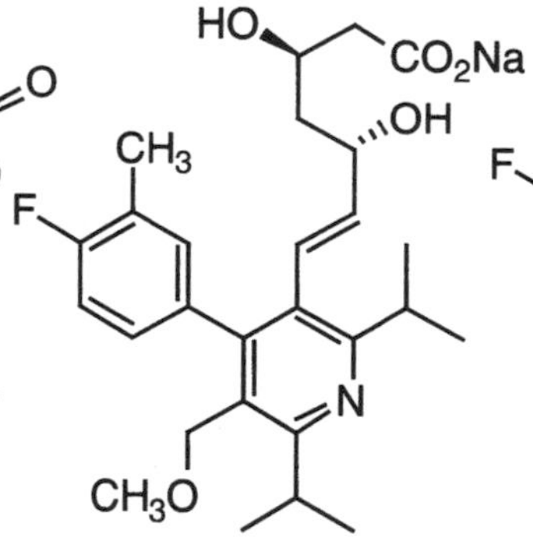   **68** 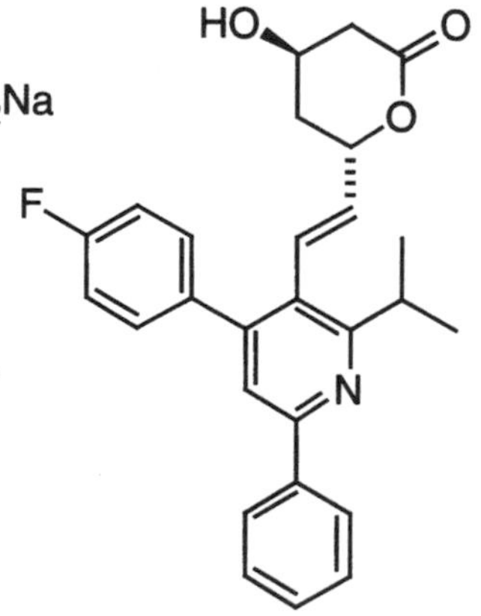   **69**

As an alternative to the inhibitors designed using the fungal metabolites as templates, attempts have been made to design inhibitors based on the reaction substrate HMG-CoA and co-factor NADPH. Thus, a series of CoA and pantetheine thioethers (**70, 71**)[146,147] and pantetheine hemithioacetals (**72, 73**)[148,149] have been reported, but these strategies produced compounds which inhibited HMGR only at the µM level.

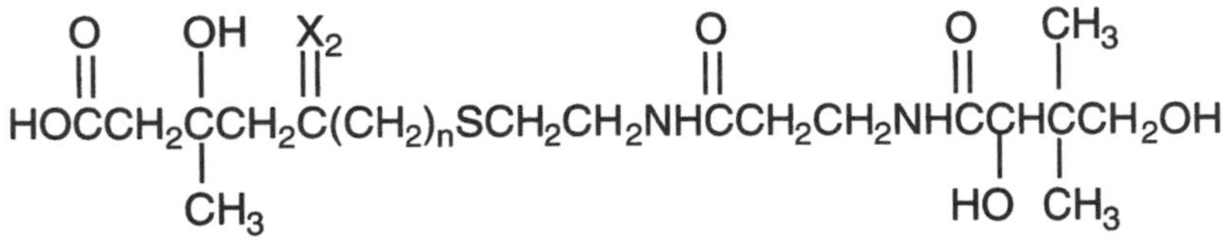

$$\underset{CH_3}{\overset{O \quad OH \quad X_2}{HOCCH_2CCH_2C(CH_2)_nSCH_2CH_2NHCCH_2CH_2NHCCHCCH_2OH}}$$

70, X=O, n=0
71, X=O, n=1
72, X=H$_2$, n=0
73, X=H$_2$, n=1

The strategy of developing inhibitors based on NADPH also produced weak (no inhibition at 10^{-4} M) inhibitors, regardless of the stereochemistry at C-4 of the mevalonolactone (**74**), suggesting that the nicotinamide moiety does not provide sufficient binding energy to the NADPH subsite.[150,151]

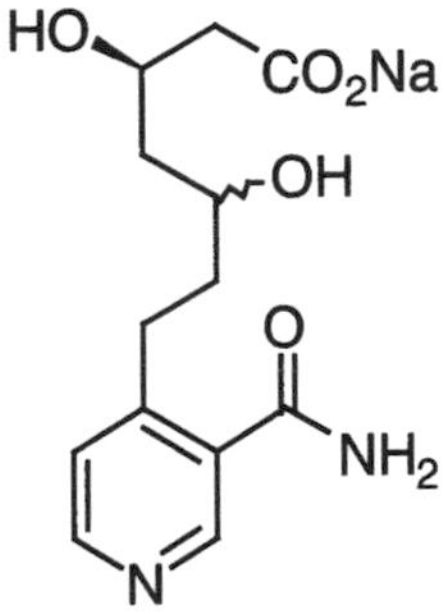

74

In addition to the competitive inhibitors, several other chemical classes have been identified which are reported to inhibit HMGR by less well defined mechanisms. Thus, a series of acetylenic acids, represented by **75**, have been isolated from a root bark which inhibit HMGR with IC$_{50}$s ranging from 0.5–10 μM. A kinetic analysis of their mode of inhibition did not fit any of the classical inhibitor models (competitive, non-competitive, uncompetitive), suggesting that their effects are nonspecific, possibly of a detergent type.[152] Crilvastatin (**76**) has been reported to induce non-competitive inhibition of rat liver HMGR and to lower plasma TC in hypercholesterolemic rats. However, the results of these studies suggested that inhibition of HMGR may not be the only mechanism by which crilvastatin affects cholesterol metabolism *in vivo*.[153]

2.3 Feedback Inhibitors

The role of oxygenated sterols in the regulation of cholesterol biosynthesis was discovered independently through the groundbreaking work of Brown and

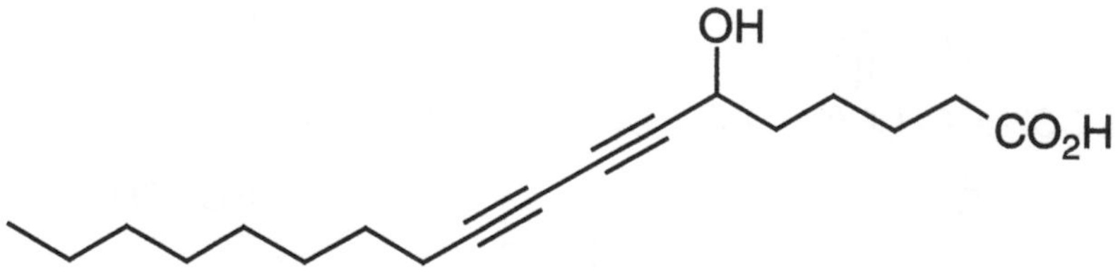

75

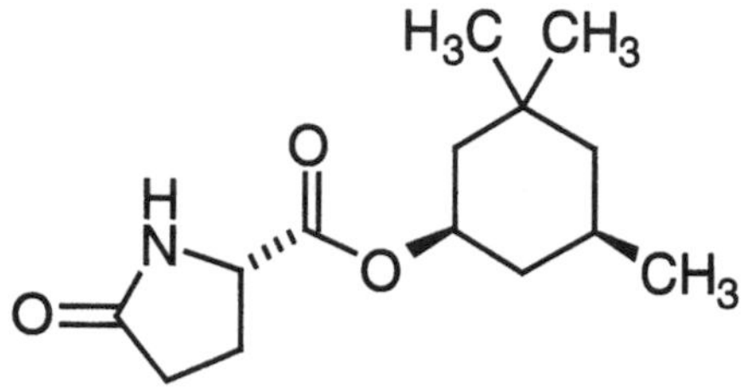

76

Goldstein[154] and Kandutsch and coworkers[155,156] on the effects of 7-keto (**77**) and 25-hydroxycholesterol (**78**) on sterol synthesis. Unlike the competitive inhibitors of this enzyme, which cause an increase in HMG-R and LDL-R activity and protein, oxygenated sterols lead to decreases in both.[44] This observation led to the postulate that HMG-R and the LDL-R were coordinately regulated and ultimately to the elucidation of the LDL-R pathway by Brown and Goldstein.[157] Further seminal work by Kandutsch and coworkers[158] and Brown and Goldstein[159] has led to the identification of a protein thought to be the putative oxysterol receptor through which these sterols act to down-regulate HMG-R activity. Several other hydroxysterols have also been identified which affect HMG-R by this mechanism.[160]

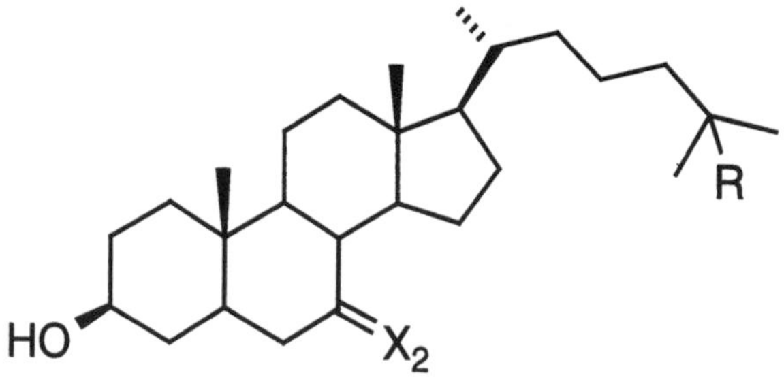

77, X_2 = O, R = H
78, X_2 = H_2, R = OH

Recently, it has been suggested that oxysterols may regulate cholesterol biosynthesis by both repression of the transcription of the genes for several enzymes involved in cholesterol genesis (including HMGR and HMGS) and acceleration of the degradation of HMGR.[161] This has been confirmed in a structure-activity study with a variety of oxysterols where it was demonstrated that both transcriptional regulation of HMGS and degradation of HMG-R were highly correlated with relative binding affinities to the oxysterol receptor.[162] Despite the mechanistic interest in these inhibitors, it has been felt that their therapeutic value would be limited, due to the undesirable down-regulation of the LDL-R observed with these compounds. Thus, these compounds have been studied to a much lesser degree than the competitive inhibitors.

In addition to the sterol regulators, there has been strong evidence for the existence of a non-sterol, mevalonate-derived regulator(s) of HMGR.[15,163] Recently, a family of compounds, the tocotrienols (79), were isolated from barley which suppressed HMGR activity *in vitro* and *in vivo* by a feedback mechanism.[164] Careful fractionation of an isolate from palm oil (a richer source of tocotrienols) revealed the presence of four major components (79a–d). Biological evaluation of these components in cultured rat hepatocytes indicated that the γ- and δ-tocotrienols (79c and 79d) were 5-fold more potent than α-tocotrienol (79a) at inhibiting cholesterol biosynthesis. γ-Tocotrienol ($IC_{50} = 3$ μM) was found to be 30 times more potent than 79a at inhibiting cholesterol biosynthesis in cultured Hep-G2 cells (a human hepatoma cell line more sensitive to cholesterol biosynthesis inhibitors than rat hepatocytes).[165] Mechanistic studies have demonstrated that unlike the sterol regulators of HMGR, these compounds only modestly diminished the rate of synthesis of HMGR, but the rate of HMGR degradation was increased 2.4 fold. Also, in contrast to the effects of 25-hydroxycholesterol, treatment of Hep-G2 cells with 79c led to an increase in LDL-R protein, thus suggesting that these compounds mimic the activity of the putative non-sterol regulator of HMGR.[166]

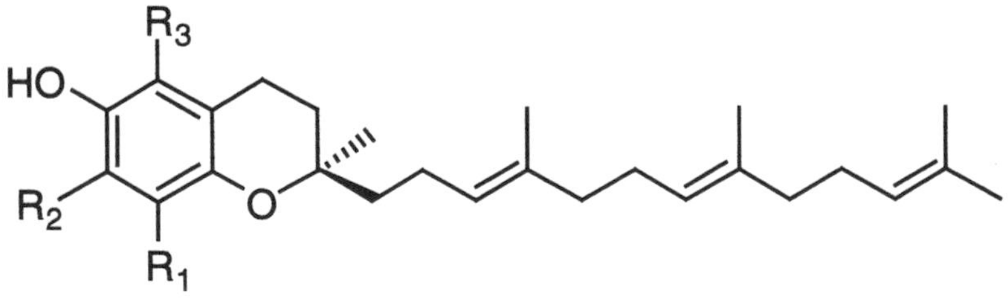

Compound	R_1	R_2	R_3
79a, α-tocotrienol	CH_3	CH_3	CH_3
79b, β-tocotrienol	CH_3	H	CH_3
79c, γ-tocotrienol	CH_3	CH_3	H
79d, δ-tocotrienol	CH_3	H	H

3. INHIBITORS OF STEPS AFTER MEVALONATE FORMATION

Mevalonate undergoes three separate phosphorylations to form 3-phospho-5-pyrophosphate which is subsequently decarboxylated (with concomitant hydrolysis of the phosphate group at the 3-position) to give 3-isopentyldiphosphate. These enzymes, as well as the rest that lead to the synthesis of farnesyl pyrophosphate, are soluble cytosolic proteins.[14a,167] The first enzyme of mevalonate metabolism is **mevalonate-5-phosphotransferase** (EC 2.7.1.36, mevalonate kinase). This enzyme catalyzes the ATP-dependent phosphorylation of mevalonic acid. The enzyme is specific for $R(+)$-mevalonate. No specific inhibitors are reported to date.

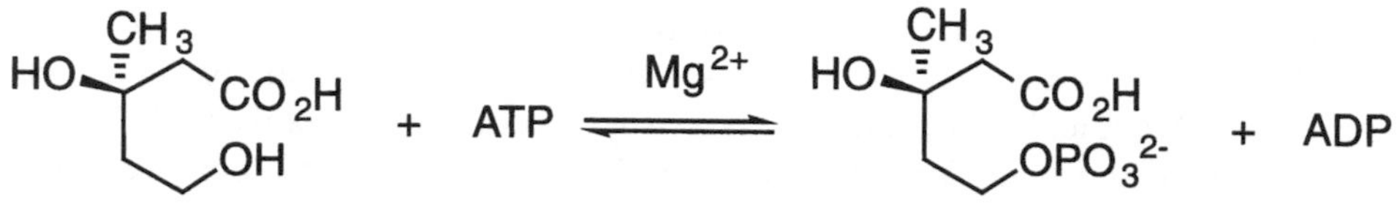

Mevalonate 5-pyrophosphate is formed by an ATP-dependent phosphorylation of 5-phosphomevalonic acid by the enzyme, **phosphomevalonate kinase** (EC 2.7.4.2). This is an unusual way to synthesize the pyrophosphate ester, since pyrophosphate esters are usually synthesized in one step by the direct transfer of pyrophosphate from a nucleotide triphosphate.[167]

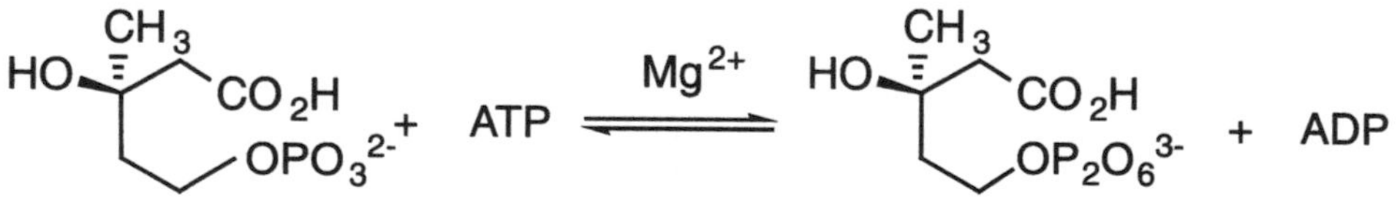

It has been shown that isosteric analogs of pyrophosphate esters can function as substrate analog type enzyme inhibitors in the biosynthetic pathway. The hypothesis behind the design of such analogs follows from observations that the actual pyrophosphate intermediates are effective inhibitors of cholesterol biosynthesis *in vitro*, but that their effectiveness *in vivo* would be doubtful due to the presence of phosphatases which would hydrolyze the compounds *in vivo*.[168] Replacement of the phosphate, or pyrophosphate, by methylenephosphonate or methylenepyrophosphate may provide more effective compounds *in vivo*, since there is no mammalian enzyme known to hydrolyze C-P bond. By utilizing this strategy, it has been shown that 3-hydroxy-3-methyl-6-phosphonohexanoic acid (**81**), an isosteric analog of 5-phosphomevalonate (**80**), specifically inhibits phosphomevalonate kinase with a K_i of 145 μM.[169] Although no *in vivo* activity is disclosed, the same authors describe cellular activity in rat hepatocytes for the closely related cholic acid analog (**82**).[170]

The next enzyme in mevalonate metabolism, **pyrophosphomevalonate decarboxylase** (EC 4.1.1.33) is unique, in that, it catalyzes the simultaneous decarboxylation and dehydration of mevalonate-5-pyrophosphate to Δ^3-isopentenyl

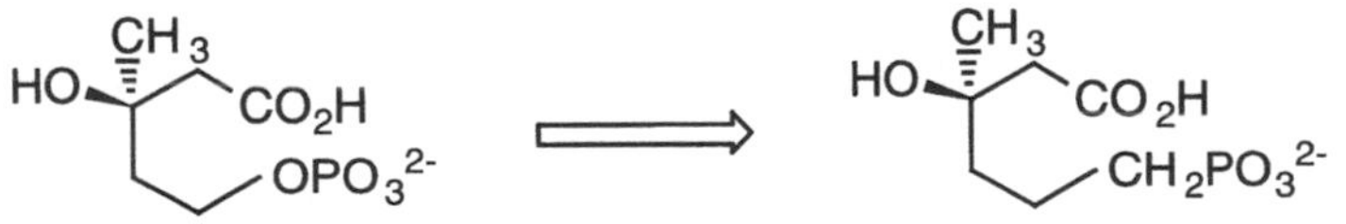

80 **81**

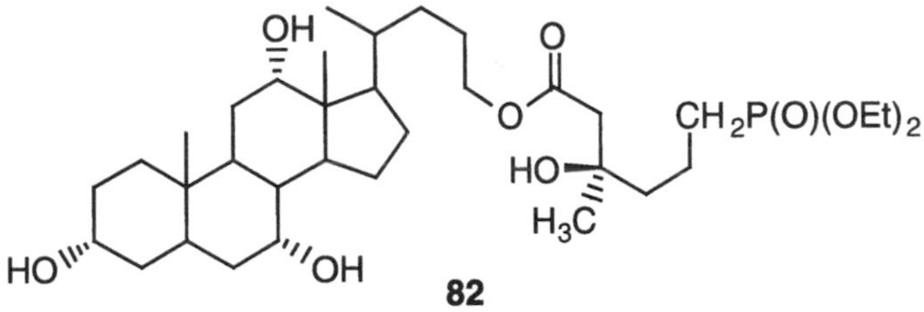

82

pyrophosphate. Evidence for an intermediate phosphorylation of the 3-OH group of mevalonate is suggestive of the existence of 3-phospho-5-pyrophosphomevalonate as a transient intermediate.[171]

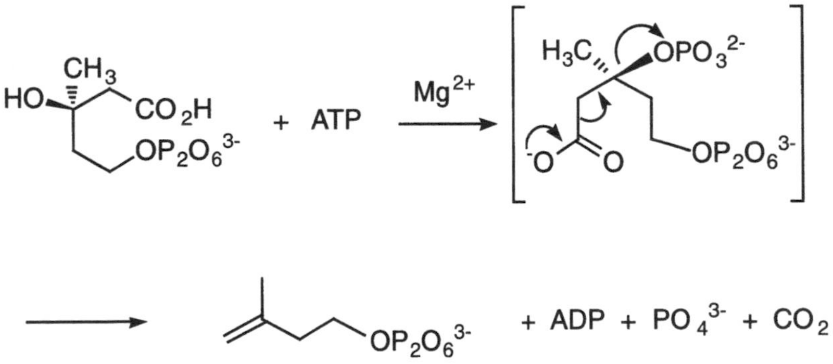

Nave has reported that 6-fluoromevalonate (**83**) is an inhibitor of this enzyme. It was shown in this study that **83** was phosphorylated by mevalonate-5-phosphokinase and phosphomevalonate kinase and that the pyrophosphorylated compound (**84**) inhibited the enzyme with a K_i of 10 nM.[172] Related compounds (**85, 86**) have also been shown to possess inhibitory activity.[173] The efficacy of these compounds *in vivo* is again probably limited due to the fact that the active pyrophosphate analogs would be hydrolyzed by phosphatases present. Isosteric replacement by methylene groups may yield compounds which are much more stable *in vivo*.

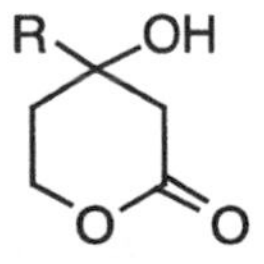

83, R=CH$_2$F
85, R=CHF$_2$
86, R=CF$_3$

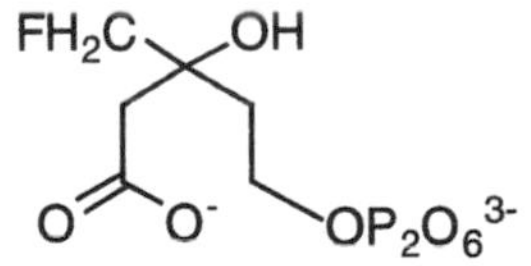

84

The first enzyme in the biosynthesis of the allylic pyrophosphates is the cytoplasmic **isopentenyl pyrophosphate isomerase** (EC 5.3.3.2). This enzyme catalyzes the interconversion of Δ^3-isopentenyl pyrophosphate and Δ^2-dimethylallyl pyrophosphate. At equlilibrium the Δ^2 allylic compound is favored nine to one.[167]

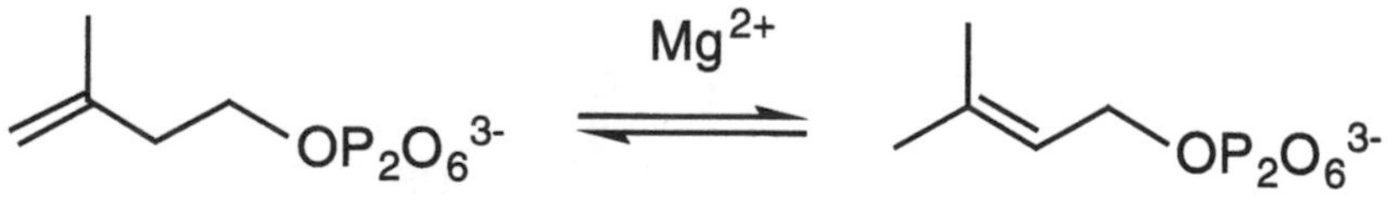

It has been shown that this enzyme can be inhibited by geranyl and farnesyl pyrophosphate ($K_i = 10$ μM). However, this enzyme has received little attention as a target for cholesterol biosynthesis inhibition.

Prenyltransferase (farnesyl pyrophosphate synthase, EC 2.5.1.1) catalyzes the head-to-tail condensation between an allylic pyrophosphate and isopentenyl pyrophosphate in the presence of either divalent magnesium or manganese. This is a polymerizing enzyme whose end products are various polyprenyl pyrophosphates. These compounds then serve as donors or intermediates in the synthesis of numerous products such as sterols, dolichols, ubiquinones, heme a,[149a,167] and certain proteins containing the carboxy-terminal sequence-CAAX (where C is cysteine, A is aliphatic, X is any amino acid). The prenylated proteins are a new class of post-translationally modified proteins which include p21[ras], nuclear lamins and the γ subunit of the heterotrimeric G proteins. Modification consists of the transfer of a 15-carbon farnesyl group or a 20-carbon geranylgeranyl (transfers specifically to proteins terminating in CAIL) group to the CAAX box of the protein.[174,175] In the sterol biosynthetic pathway, the initial reaction is a $1'-4$ condensation between Δ^3-isopentenyl pyrophosphate and Δ^2-dimethylallyl pyrophosphate to form the monoterpene, geranyl pyrophosphate.

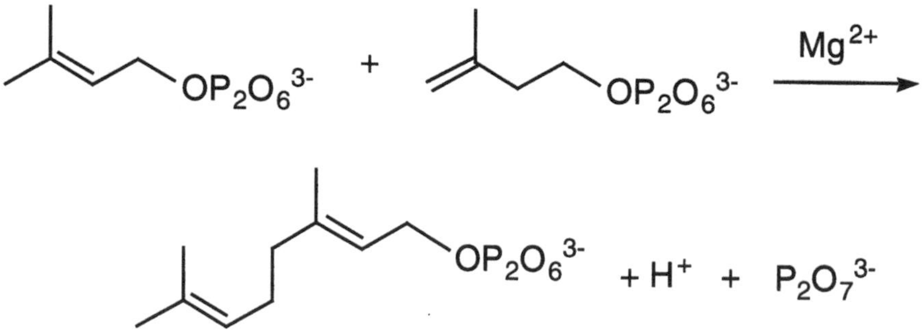

The enzyme then condenses isopentenyl pyrophosphate with geranyl pyrophosphate to form the sesquiterpene, farnesyl pyrophosphatase.

Many substrate analogs for this enzyme have been tested as inhibitors or substrates. One of the more novel approaches to the synthesis of substrate analogs is from the work of Poulter, who developed a strategy to design substrate analogs as inhibitors which are less susceptible to P-O-P hydrolysis by phosphatases. Replacement of the bridging oxygen in the diphosphate analogs of geraniol with carbon (methane-diphosphonates) or difluoromethyl (difluoromethane-diphosphonates) gave **87** and **88**. Comparative studies between compounds **87** and **88** have shown that all compounds underwent 1′–4 condensation with isopentenyl pyrophosphate when catalyzed by avian liver farnesyl pyrophosphate synthetase. In addition to the observation that these compounds could function as alternate substrates for this enzyme, it was also shown that they are resistant to attack by phosphatases.[176] This approach has been extended to the synthesis of compounds which contain the phosphonylphosphinyl system, a moiety in which the bridging oxygen moiety between the two phosphorous atoms of the pyrophosphate group and the bridging ester oxygen moiety are both replaced by methylene groups. Compounds **89** and **90**, the C-P-C-P analogs of isopentenyl and dimethylallyl pyrophosphate, respectively, were competitive inhibitors of avian liver farnesyl pyrophosphate synthetase. The K_i's for **89** and **90** were 19 and 71 µM, respectively.[177]

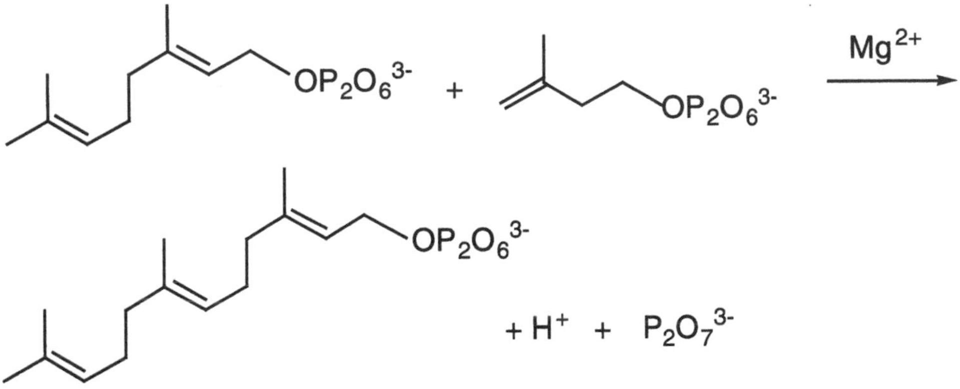

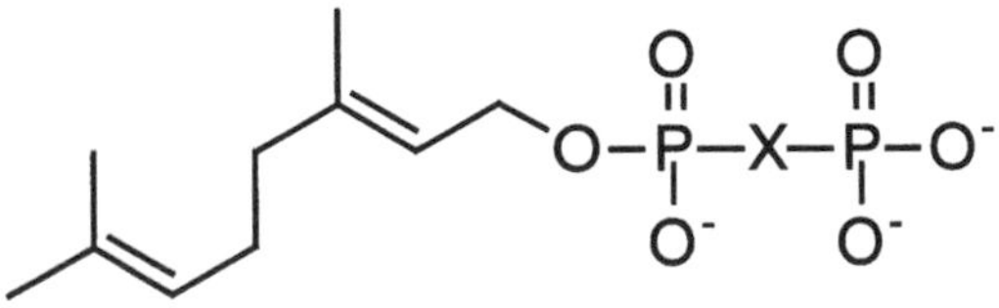

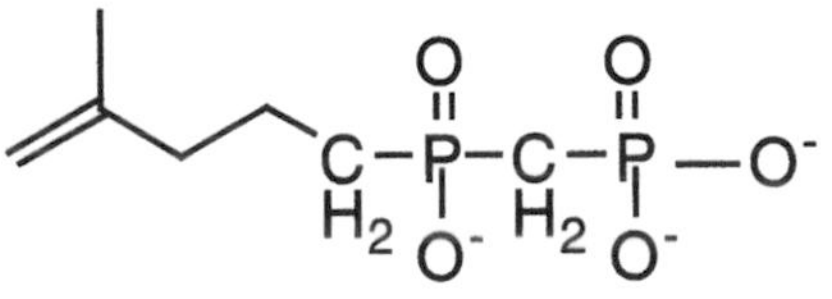

Although none have been identified as viable inhibitors of cholesterol biosynthesis, the compounds have elucidated some of the important structural features necessary for substrate binding.

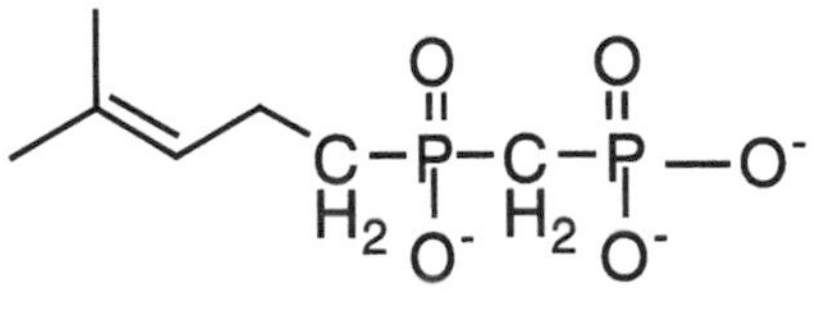

Currently, more resources are being invested in the search for inhibitors of farnesyl protein transferase (FPTase) as anticancer agents for Ras-dependent tumors (e.g. >50% of colon and pancreatic carcinoma).[178]

One of the most vigorous areas in the search for novel inhibitors of cholesterol biosynthesis is the development of inhibitors of **squalene synthase** (farnesyl-diphosphate:farnesyl diphosphate farnesyltransferase EC 2.5.1.21), a membrane-bound enzyme. As has been previously mentioned, farnesyl pyrophosphate is the branch point in the whole mammalian isoprenoid pathway. The pathway for the synthesis of the other isoprenoids such as dolichol, ubiquinone etc, diverge from the synthesis of sterols either at or before this important branch point. Thus, squalene synthase is the first enzyme in the sterol biosynthetic cascade which is solely committed to cholesterol biosynthesis. This observation has important ramifications, in that, inhibition at this site would not adversely affect the synthesis of other isoprenoids such as ubiquinone, which is essential for cell growth.[179] Thus, inhibition of this enzyme is theoretically more desirable, in terms of safety profile, than inhibition of HMGR, which has been shown to deplete other isoprenoids in man. Another advantage is the fact that farnesyl pyrophosphate, the substrate for the enzyme, is water soluble and thus may be readily metabolized, this is in marked contrast to the situation faced when cholesterol biosynthesis is inhibited after the sterol nucleus has formed. These intermediates are insoluble and cannot be readily metabolized, and, as was discussed previously with reference to triparanol, the accumulation of non-metabolizable sterols can have dire consequences.[24,25]

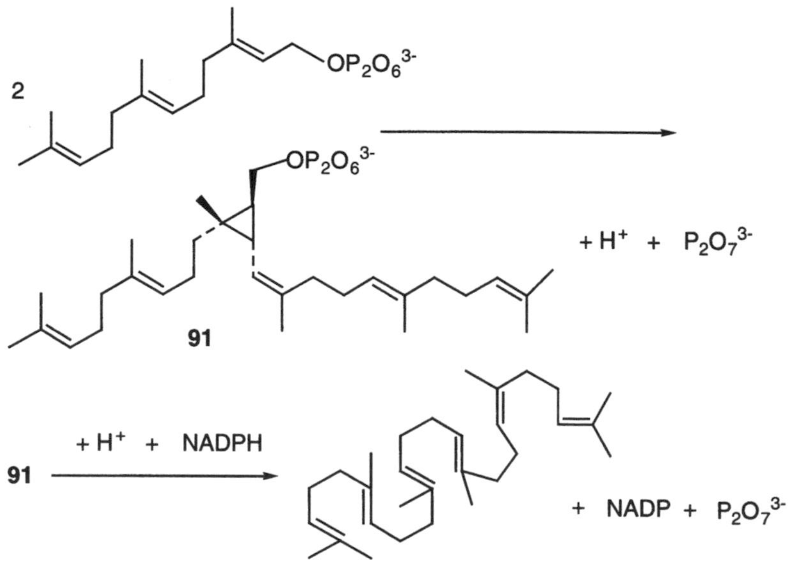

Squalene synthase catalyzes the formation of squalene, the first substrate in the sterol biosynthetic pathway that is insoluble in the aqueous environment of the cell, from farnesyl diphosphate in two distinct steps. In the first step, the enzyme catalyzes the reductive dimerization of two molecules of farnesyl pyrophosphate to form the intermediate chiral cyclopropane, presqualene pyrophosphate (**91**). This compound is then converted to squalene by a rearrangement that cleaves the two newly formed cyclopropane bonds and joins the C1 carbons of the two original farnesyl residues to generate the C_{30} triterpene, squalene.[167]

Squalene synthase has a divalent cation requirement (Mn or Mg) as well as an absolute requirement for the pyrophosphate of farnesyl pyrophosphate (the phosphate is inactive as a substrate). Requirements for the hydrocarbon moiety seem to be less stringent. Longer chain analogs are better substrates for the enzyme than shorter chain analogs[180] and the presence of the $\Delta^{6(7)}$-double bond seems crucial for binding.[181]

The first types of inhibitors of this enzyme reported were **substrate analogs** of farnesyl pyrophosphate. In one study, a number of farnesol analogs were prepared and it was shown that binding to this enzyme required the presence of the pyrophosphate moiety and relatively non-specific, lipophilic interactions. Thus, the free alcohol and the monophosphate analogs were shown to be inactive, however substitution at C-2, C-3 and C-4 (analogs **92–94**) was well tolerated. Saturation of the double bonds or shortening the chain greatly reduced binding. Kinetic analysis showed that these inhibitors were competitive or mixed inhibitors and that no irreversible inhibition was observed. The most potent compound of the series, **93**, had a K_i value of 0.5 μM. However, it was recognized that such pyrophosphates are of little therapeutic utility due to allylic C-O cleavage and attack by phosphatases.[182]

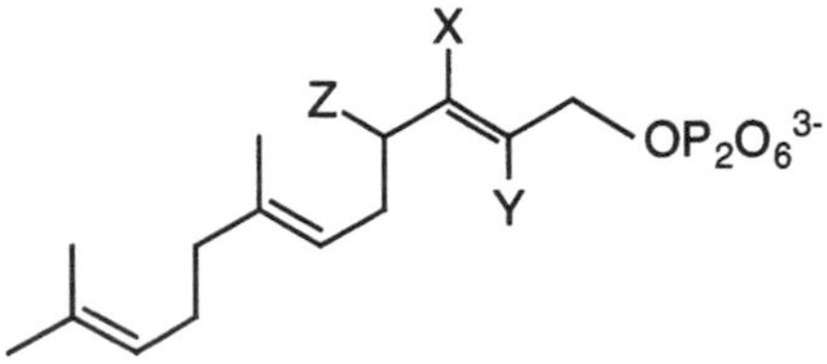

92, X=CH$_3$, Y=Z=H
93, X=Y=CH$_3$, Z=H
94, X=CH$_3$, Y=H, Z=CH$_3$S

Corey and Volante designed a series of inhibitors in which the allylic oxygen moiety was replaced by a methylene. Since the head-to-tail coupling of isoprene units involves intermolecular nucleophilic attack by a carbon-carbon double bond at a saturated carbon with elimination of a pyrosphate group, this modification to the substrates of the enzyme should prevent the C-C coupling from occuring. In fact, members of a series of C-substituted methylphosphonophosphates (**95**, **96**) were shown to be inhibitors of squalene synthetase in a crude assay system.[183]

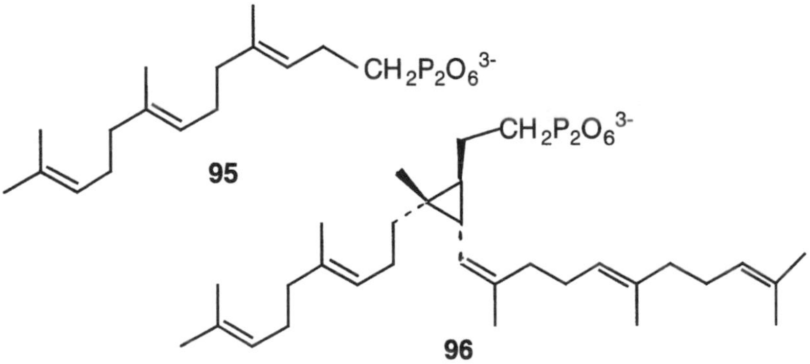

95

96

These compounds, however, are still susceptible to hydrolysis by phosphatases. This problem was overcome by Biller and co-workers, who replaced the pyrophosphate group in farnesyl pyrophosphate by the phosphonylphosphinyl group. In this series of compounds both the allylic and anhydride oxygen atoms of the pyrophosphate moiety were replaced by a methylene moiety. A number of analogs were synthesized and a number of SAR observations were made. Isosteres **97a**, **97b**, **98a** and **98b** are all effective inhibitors of squalene synthase with IC_{50} values of 31.5, 12.2, 29.9 and 15.5 µM respectively. The 1,3-dienyl analogs **97b** and **98b** are more potent than **97a** and **98a** indicating that the vinyl group may be a better substitute for the allylic C-O linkage. Homologation of **97a** to **97c** results in a twofold loss of potency ($IC_{50} = 67$ µM for **97c**) and the shorter chain geranyl derivative **97d** (40% inhibition at 600 µM) is much less potent than **97a**. It was also shown that **97a** is a competitive inhibitor with an apparent K_i of 10 µM, a figure comparable to that found for the natural substrate (14.8 µM).[184] Thus, it seems that neither the allylic or anhydride oxygen atoms are critical in substrate binding. This group was also employed by Poulter in the work on farnesyl pyrophosphate synthetase (vide supra).

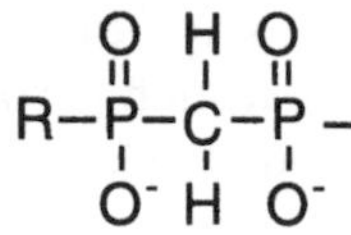

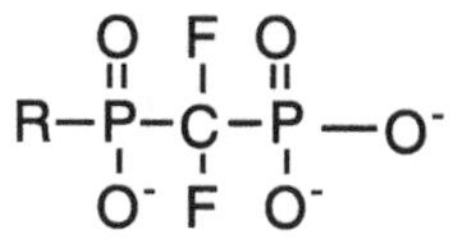

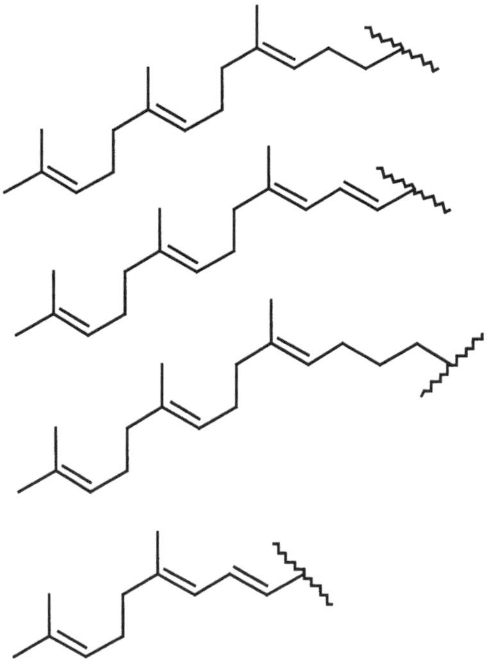

Although compounds such as **97a** do inhibit microsomal squalene synthase, they do not affect the incorporation of [^{14}C] acetate into cholesterol in freshly isolated rat hepatocytes.[185] It was hypothesized that the lack of activity in whole cells was due to the high overall charge of the phosphonylphosphinyl moiety and the effect upon penetration of cell membranes. This led to the development of a series of inhibitors whose structure was based upon phosphoformic acid (**99**, PFA), an antiviral agent.[186] In this series of inhibitors, one of the charged oxygen atoms of **99** was replaced with a C- or O-linked isoprenyl group.[185] The resulting analogs, **100a–c** and **101a,b**, respectively, possess only one ionizable acid function. It was found that **100b** was equipotent in the enzyme assay to **97a** despite its lower charge. It was also shown that **100b** inhibited cholesterol biosynthesis in whole cells, whereas **97a** was inactive. Introduction of an oxygen atom in the form of a phosphorous ester (**101a,b**) resulted in more active inhibitors than the corresponding C-linked isosteres (**100b,c**). The optimal overall chain length of both the C- and O-linked inhibitor series (**100b** and **101a**) corresponds to that which is isosteric to farnesyl pyrophosphate. Analog **101a** was shown to be a competitive inhibitor of squalene synthase with respect to farnesyl pyrophosphate ($IC_{50} = 8.7$ μM), it was also the most potent inhibitor in the rat hepatocyte assay ($IC_{50} = 6.0$ μM). Thus, it was demonstrated that the phosphinylformate moiety is a novel dianionic pyrophosphate surrogate capable of inhibiting cholesterol biosynthesis in whole cells.

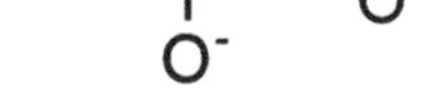

99

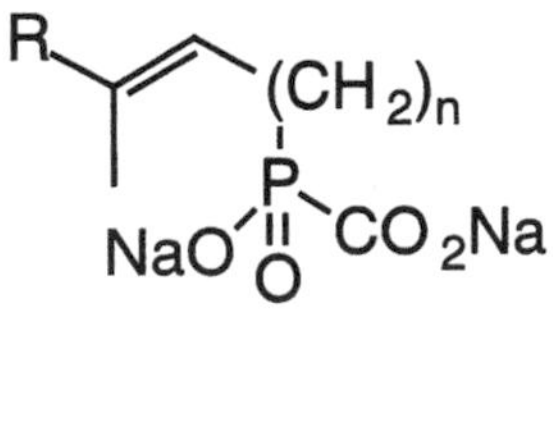

100a, n=2
100b, n=3
100c, n=4

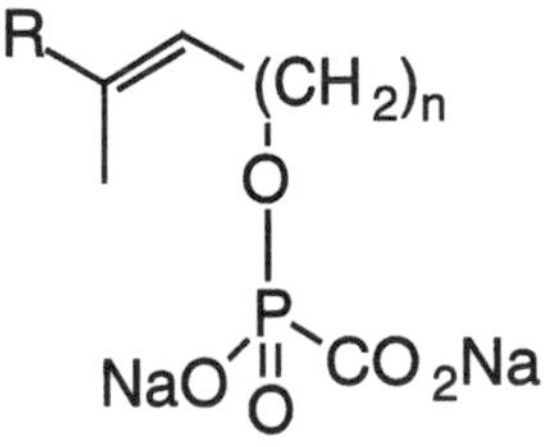

101a, n=2
101b, n=3

An even more potent inhibitor, **102** ($IC_{50} = 0.05$ μM), was rationally designed utilizing a proposal for the mechanism of the enzymatic reaction. It was also speculated that the tight binding of **102** to squalene synthase was due to H bonding of the ether oxygen with a key active-site catalyst.[187]

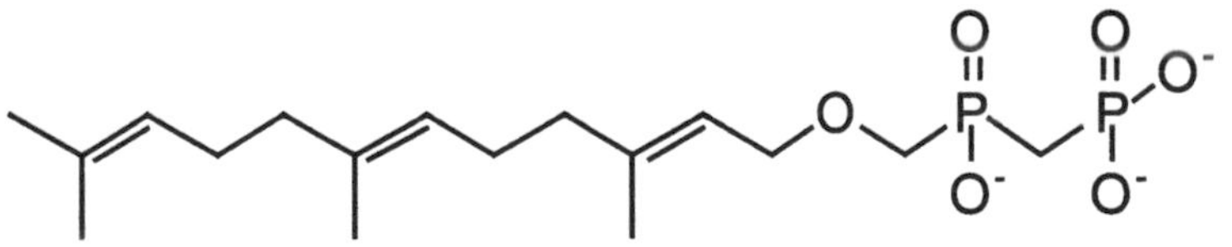

102

The bisphosphonates are well known inhibitors of bone resorption and are used in the treatment of several bone disorders (e.g., osteoporosis, Paget's disease, tumor osteolysis).[188] Some of these compounds have recently been shown to be inhibitors of squalene synthase.

The most potent compound was 103 ($IC_{50} = 64$ nm), which was shown to be a non-competitive inhibitor. Compounds 104 and 105 were less potent ($IC_{50} = 208$ and 311 nm, respectively). Compounds 106–109 were relatively inactive. Compounds 103–105 were also potent inhibitors of cholesterol and other sterol biosynthesis using radiolabelled mevalonate in a cell free system. The IC_{50} values for 103–105 were 17, 113 and 74 nm, respectively. Compounds 106 and 107 were poor inhibitors of squalene synthase but were potent inhibitors of sterol biosynthesis ($IC_{50} = 420$ and 168 nm, respectively). Compounds 108 and 109 were inactive.[189]

Recently, it was shown that compound 103 inhibited cholesterol biosynthesis in rat liver with an estimated ED_{50} of 30 mg/kg s.c. Since 103 is a non-competitive inhibitor of the enzyme, the bisphosphonates must inhibit the enzyme by a different mechanism than the competitive inhibitors (farnesyl pyrophosphate mimics) described by Biller.[184,185]

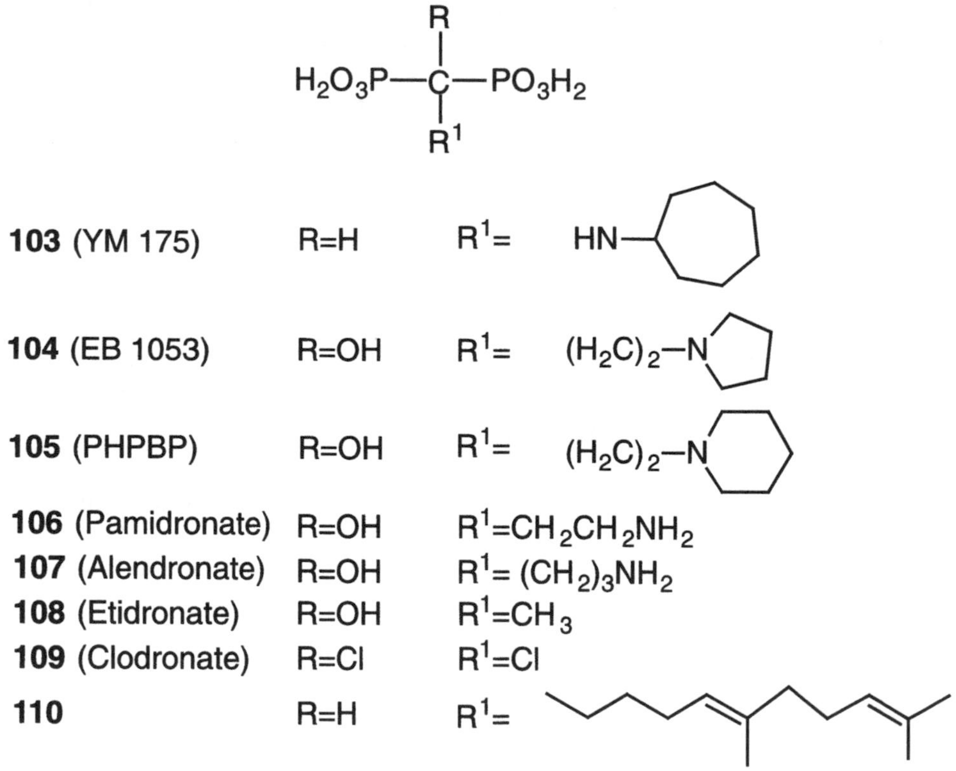

103 (YM 175)	R=H	R^1 = HN—(cycloheptyl)
104 (EB 1053)	R=OH	R^1 = $(H_2C)_2$—N(pyrrolidine)
105 (PHPBP)	R=OH	R^1 = $(H_2C)_2$—N(piperidine)
106 (Pamidronate)	R=OH	$R^1 = CH_2CH_2NH_2$
107 (Alendronate)	R=OH	$R^1 = (CH_2)_3NH_2$
108 (Etidronate)	R=OH	$R^1 = CH_3$
109 (Clodronate)	R=Cl	$R^1 = Cl$
110	R=H	R^1 =

Thus, it seems that 103 binds to a site on the enzyme which is different to the site of attachment of its substrate, farnesyl pyrophosphate. It is speculated that this compound mimics the carbocationic intermediates in the second step of the squalene synthase reaction. The design of inhibitors of these intermediates has been pioneered by Poulter. From the data shown, the presence of an aliphatic chain containing a positively charged nitrogen atom plays a role in enzyme inhibition. Drugs without an amine sidechain, e.g., **108** and **109**, were inactive. Also, primary amines (**106** and **107**) were less active than secondary (**103**) and tertiary (**104, 105**) amines.[189] However, bisphosphonate (**110**), which contains no amine function, was shown to be a very potent inhibitor ($IC_{50} = 0.95$ nM). It was also shown to lower plasma cholesterol in rats and hamsters. This compound had no effect on dolichol or coenzyme Q_{10} biosynthesis, even when cholesterol biosynthesis was >90% inhibited.[190]

The mechanism of the conversion of presqualene pyrophosphate (**91**) to squalene has attracted much attention. The mechanism as proposed by Rilling recognized that the reductive rearrangement of **91** to squalene could be rationalized in terms of the bond reorganizations typically observed for cyclopropylcarbinyl cations.[191] Poulter then demonstrated that the high degree of regiocontrol necessary for the biosynthesis of squalene was achieved through favorable interactions between pyrophosphate generated as a consequence of cleaving the C-O bond in **91** and carbocationic species en route to squalene.[192]

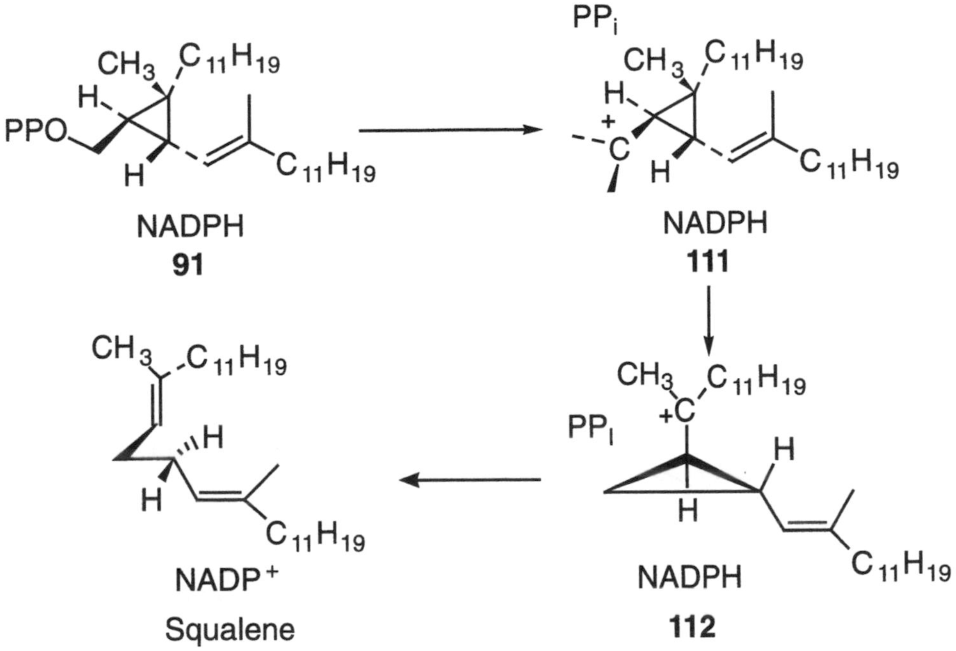

Poulter designed compounds **113–115** to mimic the electrostatic and topological properties of primary cation **111** and tertiary cation **112**, respectively. It was found that neither analog was an effective inhibitor of squalene synthase as measured in standard assays. However, in combination with pyrophosphate both became potent inhibitors ($IC_{50} = 10$ and 3 μM respectively with [pyrophosphate] = 1 mM). Inhibition studies with **113** and **114** demonstrated that they did not bind tightly to the active site of the enzyme, despite similarity to the carbocations **111** and **112**. However, it has been proposed that inhibition occurs through an ion pair mechanism in which unfavorable electrostatic interactions between the unshielded ammonium moieties and the enzyme are neutralized by the pyrophosphate anion.

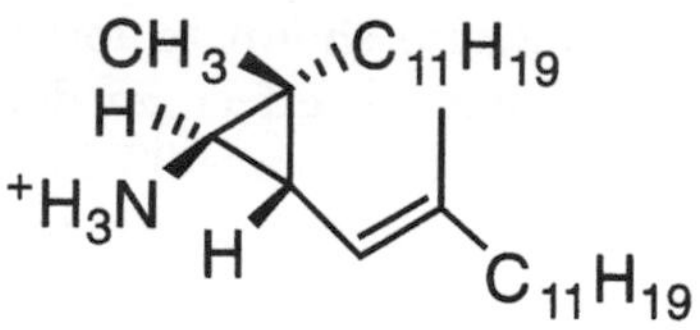

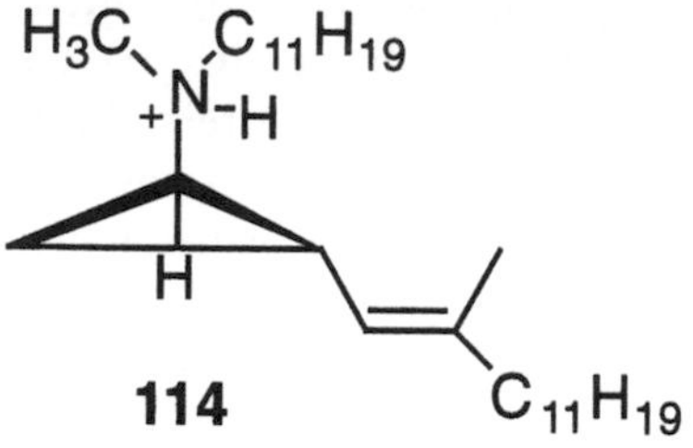

It was also shown that **115**, a tethered analog of the **113**-pyrophosphate ion pair, was a potent inhibitor in the absence of pyrophosphate. This was designed to increase the effective concentration of pyrophosphate in the presence of **113** by tethering a phosphonophosphate moiety to the ammonium bridge. This compound had a IC_{50} of 3–5 μM. This compares favorably to that found for **113** and **114** in the presence of pyrophosphate.[193]

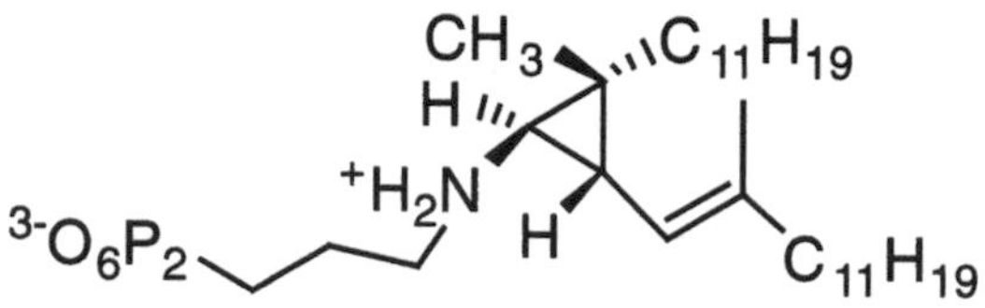

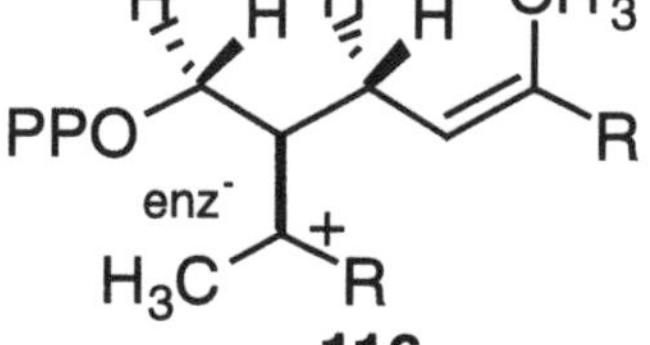

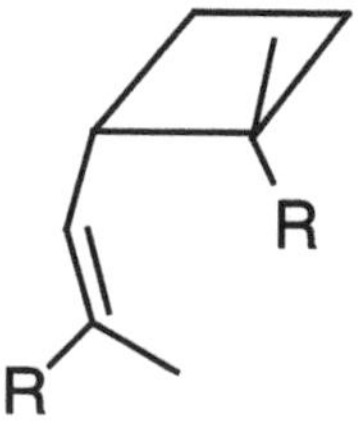

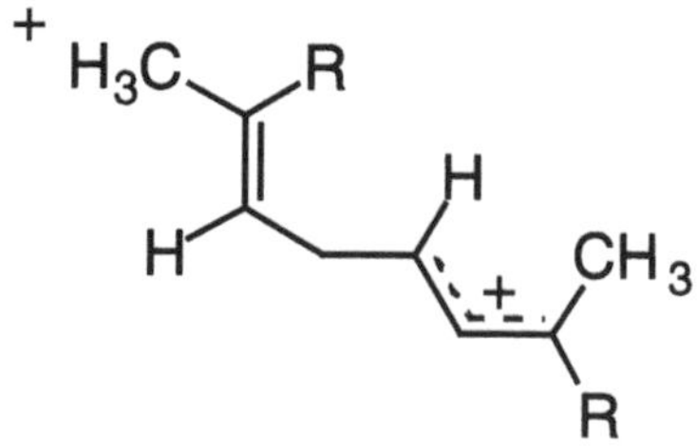

Other transition state analogs have been designed as inhibitors of squalene synthase using putative intermediates in the squalene synthase reaction. The farnesyl-enzyme cation **116** has been postulated to be involved in the formation of presqualene pyrophosphate. Cyclobutyl cation **117** is postulated to be a possible precursor to the squalene cation **118**, which is eventually reduced by NADPH to squalene.[194] Cyclobutanones **119** and **120** were designed as mimics of **117**, however they were shown to be poor inhibitors of yeast squalene synthase (10–15% inhibition at 5 mM).[195] Sulfonium ion mimics **121** and **122** were designed as mimics to the intermediates **116** and **118** and have been shown to be inhibitors of squalene synthase.[196]

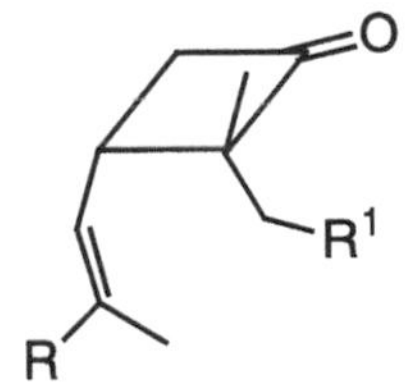

119, R^1=H
120, R^1=tetrahydrogeranyl

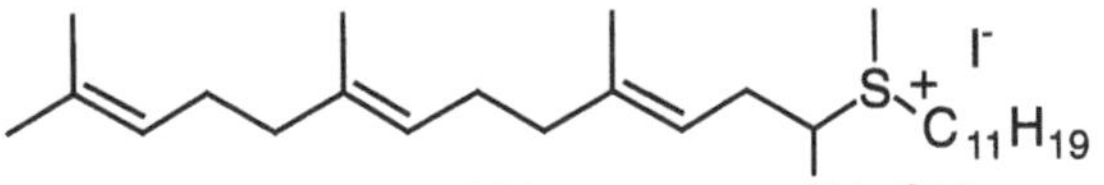

121

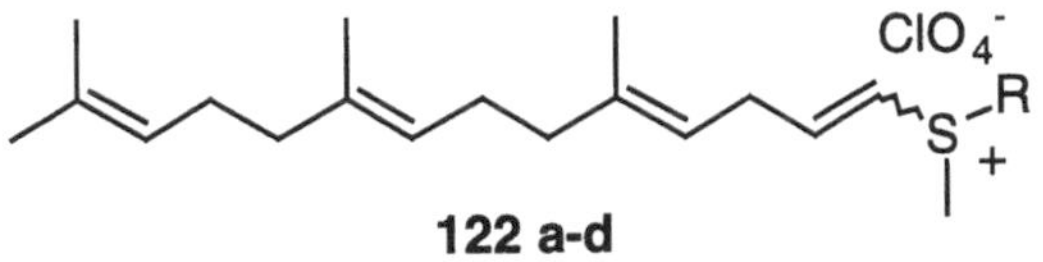

122 a-d
a=E-$C_{11}H_{19}$; b=Z-$C_{11}H_{19}$;
c=E-C_6H_{13};   d=Z-C_6H_{11}

A series of N-(arylalkyl)farnesylamine analogs were designed to mimic the carbocation **118**. It was postulated that such compounds would be protonated *in vivo* to yield an ammonium cation which may inhibit the enzyme. A number of compounds were shown to be potent inhibitors. Compound **123** had an IC_{50} value of 100 nM. When tested in the presence of pyrophosphate, this value decreased to 10 nM. Replacement of the phenyl ring in **123** by a 3-pyridyl group gave the most potent inhibitor (**124**) in this series with an IC_{50} of 4 nM in the presence of pyrophosphate. Any modifications of the farnesyl chain in this molecule led to a significant loss of activity.[197] Recent work from these same laboratories describes the synthesis of a tethered analog of **123** in which a pyrophosphate mimic is attached to **123** in order to mimic the **118**-pyrophosphate ion pair. The resulting compound, **125**, is a potent inhibitor of squalene synthase with an IC_{50} of 20 nM, five times more potent than **123**. There was no added enhancement upon addition of pyrophosphate. This suggests that the phosphinato(dimethylmethyl)phosphonate moiety in **125** binds the pyrophosphate binding region in the active site of squalene synthase.[198]

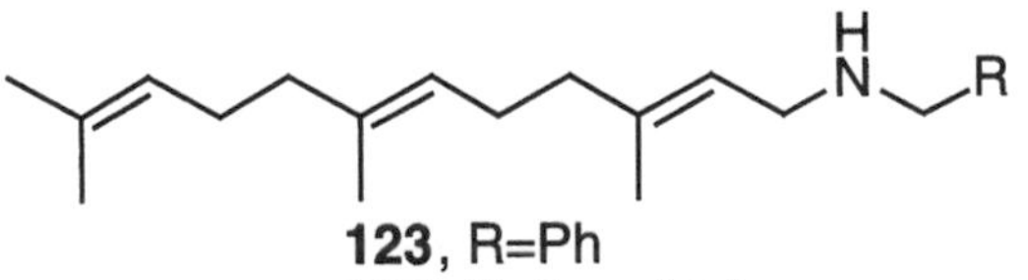

123, R=Ph
124, R=3-pyridyl

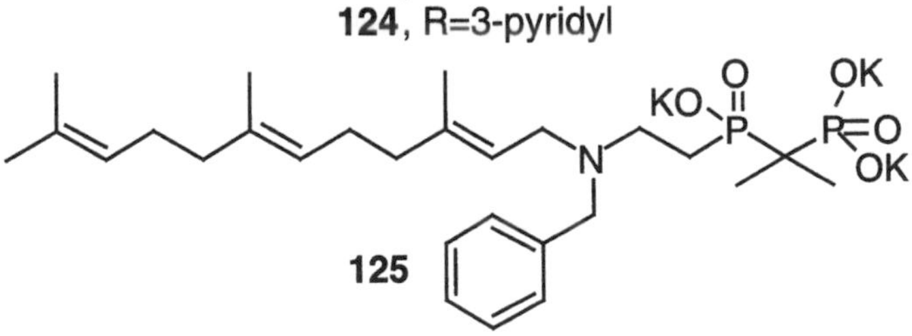

125

Altman *et al.* has shown that farnesylamine inhibits squalene synthase, but it is possible that this inhibition is non-specific and related to disruption of cellular membranes.[199] Workers at Merck have previously described compounds such as **126** as 5-HT$_3$ antagonists,[200] however, it has been shown using rat liver microsomes, that these compounds are potent inhibitors of squalene synthetase with IC$_{50}$ values ranging from 11–660 nM.[201] Similar compounds (such as **127**) have also been shown to be potent inhibitors (IC$_{50}$ values ranging from 1 nm to 25 µM). An *in vivo* rat screen to evaluate the ability of the compounds to inhibit the incorporation of [^{14}C]-acetate produced ED$_{50}$ values of 0.1 to 100 mg/kg.[202] Closely related compounds such as **128** have also been reported by workers at Rhone Poulenc Rorer.[203]

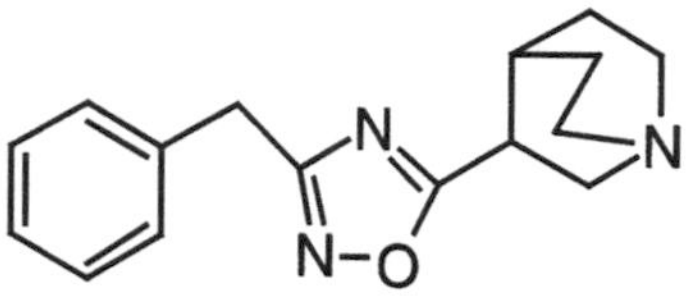

126 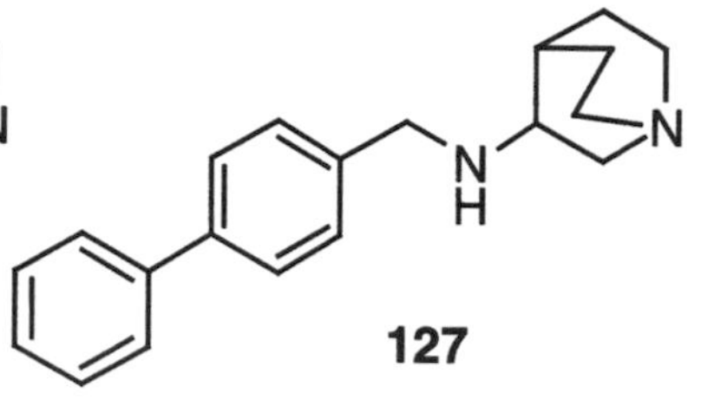   **127**

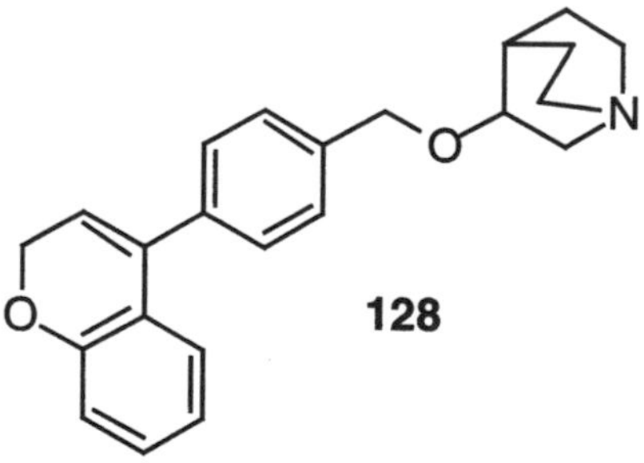

128

The revival of interest in inhibitors of squalene synthase is almost certainly due to the discovery of the squalestatins and zaragozic acids, two closely related series of compounds isolated from fungii and shown to be extremely potent inhibitors of squalene synthase.

The squalestatins were isolated from a fungal culture from the newly discovered *Phoma* sp. C2932, a soil fungus collected in Portugal.[204] The structures were elucidated from IR, MS and NMR spectral data and X-ray crystallographic analysis. The compounds have a highly functionalized bicyclic core, (1S-(1α, 3α, 4β, 5α, 6α, 7β)) 4,6,7-trihydroxy-2,8-dioxabicyclo[3.2.1] 3,4,5-tricarboxylic acid. The absolute stereochemistry of **129** and **130** was (1S (4S, 5R), 3S, 4S, 5R, 6R (2E, 4S, 6S), 7R), whereas **131** was (1S (4S, 5R), 3S, 4S, 5R, 6R, 7R).[205] Compounds **129–131** were shown to be potent inhibitors of squalene synthase using either [1-^{14}C]IPP or [1-^{14}C]FPP as substrates. The IC$_{50}$'s for **129–131** utilizing IPP as substrate were 21.5, 20.8 and 3.9 nM, respectively. Data has been presented for **129** (squalestatin 1) showing that it inhibits squalene synthase from rat and marmoset livers with an IC$_{50}$

of 12 nM, it has also been shown to inhibit cholesterol biosynthesis in rat hepatocytes with an IC_{50} of 39 nM. The compound was also shown to be an effective inhibitor of cholesterol biosynthesis when dosed to rats i.v. and followed immediately with an intraperitoneal injection of ^{14}C-acetate (50% inhibition at a dose of 0.1 mg/kg). Squalestatin 1 (**129**) also lowers plasma TC in adult marmosets, a primate with a lipoprotein metabolism similar to that of man. At a dose of 10 mg/kg a 51% reduction in plasma cholesterol was observed. A 100 mg/kg dose produced a 75% decrease. The effect was apparent in 24 hours and this effect could be maintained for as long as 8 weeks with prolonged dosing.[206]

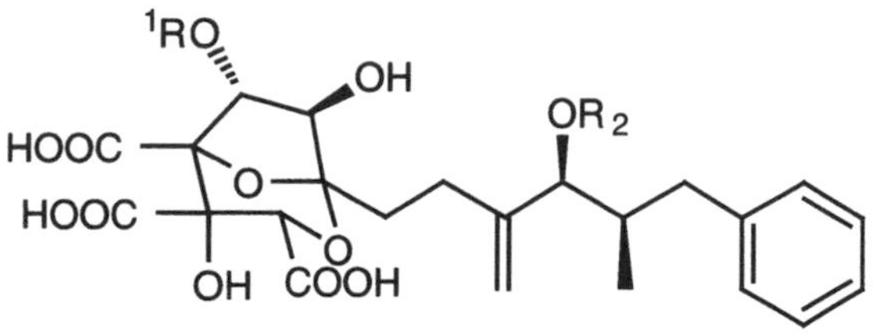

129 (Squalestatin 1), $R^1 =$ 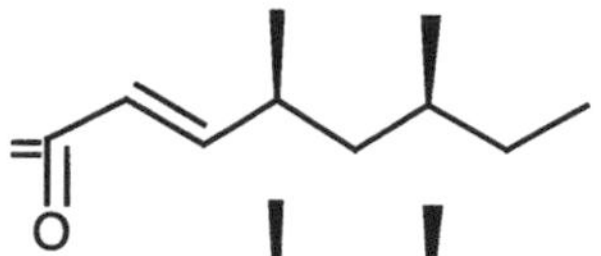 , $R^2 = COCH_3$

130 (Squalestatin 2), $R^1 =$, $R^2 = H$

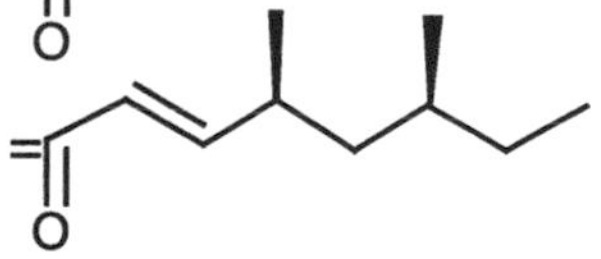

131 (Squalestatin 3), $R^1 = H$, $R^2 = COCH_3$

Three closely related compounds, zaragozic acids A (**132**), B (**133**), and C (**134**), were isolated from an unidentified sterile fungal culture (isolated from a water sample taken from the Jalon river in Zaragoza, Spain), *Sporormiella intermedia*, and *Leptodontium elatius*, respectively.[207–209]

Chemical studies led to the determination of the total absolute stereochemistry of zaragozic acid A (**132**). This was confirmed by single crystal X-ray crystallography on two crystalline derivatives. It was shown that zaragozic acid A is identical to squalestatin 1 (**129**). All three of these compounds were shown to be potent, competitive inhibitors of squalene synthase with apparent K_i values of 78 pM, 29 pM and 45 pM, respectively. These compounds were also shown to inhibit the incorporation of tritiated mevalonate into cholesterol in Hep G2 cells with IC_{50} values of 6 µM, 0.6 µM and 4 µM, respectively for compounds **132–134**. In addition, when dosed to mice subcutaneously, **132** exhibited an ED_{50} of 0.2 mg/kg for the inhibition of cholesterol biosynthesis *in vivo*.[207]

It has been postulated that these compounds inhibit squalene synthase by mimicking the binding of presqualene pyrophosphate to the enzyme. Interestingly, extracts from cells or animals after treatment with the zaragozic acids have been shown to contain farnesoic acid and a 15-carbon dicarboxylic acid derivative of farnesoic acid. Thus, the primary fate of mevalonate diverted from cholesterol biosynthesis by a zaragozic acid is excretion via the kidney as a water soluble metabolite.[207]

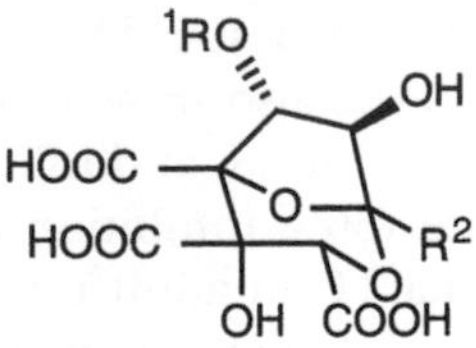

132 (Zaragozic acid A), R¹ = ...

R² = ...

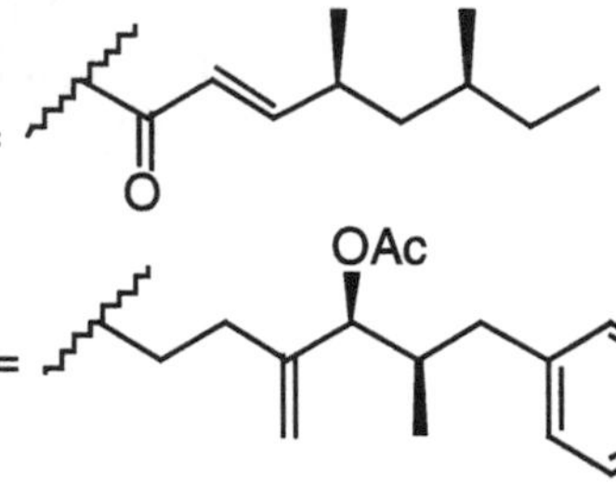

133 (Zaragozic acid B), R¹ = ...

R² = ...

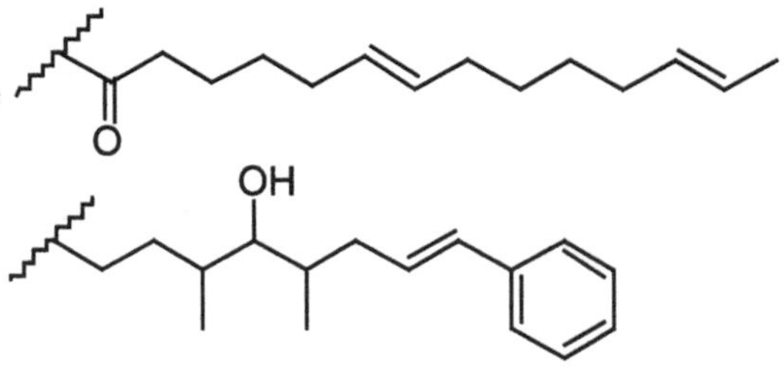

134 (Zaragozic acid C), R¹ = ...

R² = ...

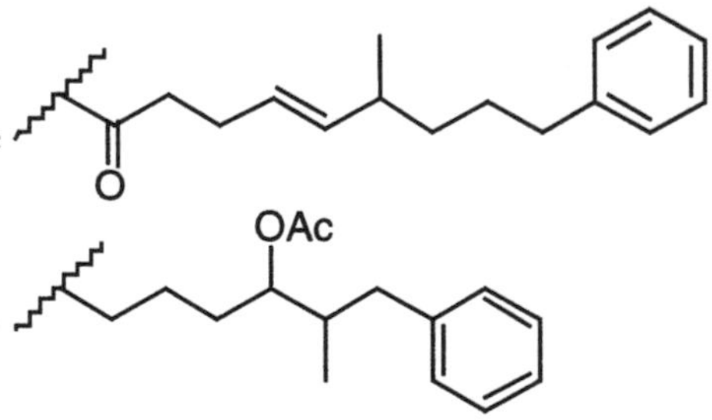

135 (Zaragozic acid D), R¹ = ...
n=1

136 (Zaragozic acid D₂) R² = ...
n=2

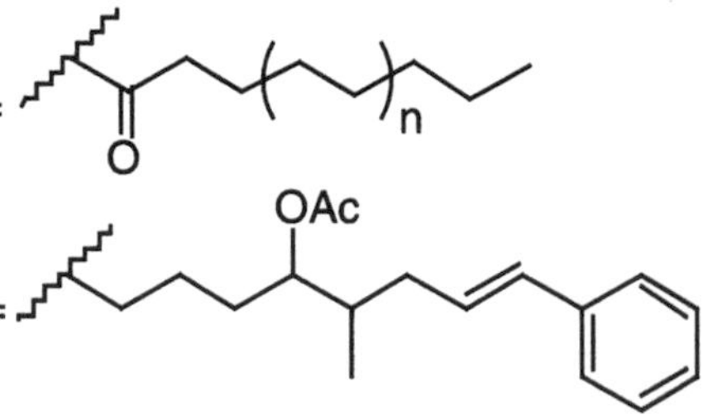

Recently, two new members of the zaragozic acids were isolated. Zaragozic acids D (**135**) and D_2 (**136**) were isolated from the keratinophilic fungus *Amauroascus niger*. These compounds were less potent than compounds **132–134**. The IC_{50} values for **135** and **136** were 6 nM and 2 nM, respectively (in this study the corresponding values for **132–134** were 0.5 nM, 0.2 nM and 0.4 nM, respectively). However, compounds **135** and **136** inhibit farnesyl transferase with IC_{50} values of 100 nM. Compounds **132** and **133** were less potent.

Zaragozic acid A (**132**) has been shown to be poorly bioavailable when dosed orally to mice. In an assay measuring the incorporation of tritiated mevalonolactone into cholesterol, the ED_{50} was 100 mg/kg. To improve the oral absorption of these agents, various diesters of the natural products were prepared and evaluated for their ability to inhibit rat liver squalene synthase and inhibit the incorporation of tritiated mevalonolactone into cholesterol when dosed orally to mice.

SAR studies revealed that esters at C3 were more potent than **132** in the mouse model (61% inhibition at 40 mg/kg for $R_1 = CH_2CH_2CH(CH_3)_2$, $R_2 = R_3 = H$). However, the greatest activity was observed when the acid moieties at positions 3 and 4 were esterified. The most potent compounds (**137**, $R_1 = CH_2CH_2CH(CH_3)_2$, $R_2 = CH_2OCOtBu$, $R_3 = H$ and **138**, $R_1 = CH_2CH_2CH(CH_3)_2$, $R_2 = CH_2OCOtBu$, $R_3 = H$) had ED_{50}'s of 9 and 6 mg/kg, respectively. The 3,5 diesters were not active at 24 mg/kg.[212]

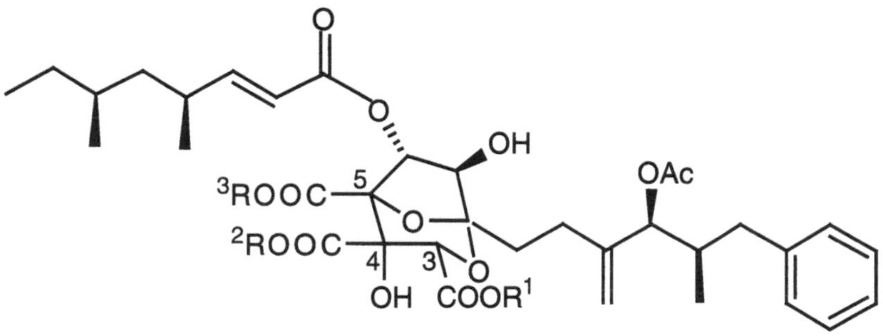

The conversion of the polyolefin, squalene (**139**), to the tetracyclic steroid, lanosterol (**141**), is one of the most remarkable biosynthetic transformations. There are two enzymes involved in this process, **squalene epoxidase** (SE) and **oxidosqualene cyclase** (OSC). The overall process is shown below.

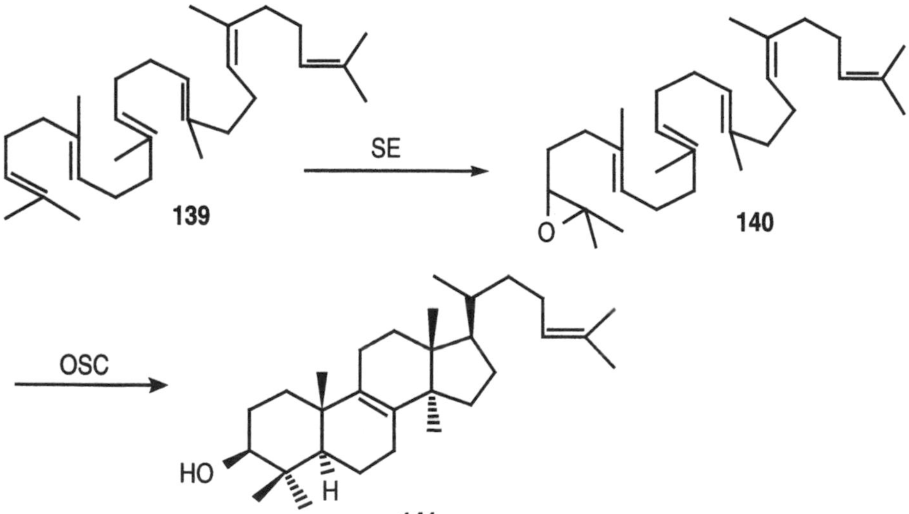

Squalene epoxidase (EC 1.14.99.7 abbreviated as SE) catalyses the conversion of squalene (**139**) to (3*S*)-2,3-oxidosqualene (**140**), in the first oxidative step of the biosynthetic pathway. The enzyme is present in the microsomes of eukaryotic cells, and requires the presence of cytoplasmic fraction to show activity. The reaction utilizes molecular oxygen, NADPH, FAD (flavin adenine dinucleotide), cytochrome P-450 reductase and a soluble protein factor (SPF), which is an intermembrane transporter for the substrate. It does not appear to be a heme-containing enzyme and no metals have been implicated in the active site.[149,167] Both enzymes, squalene epoxidase and oxidosqualene cyclase, have become targets for inhibition because of their position in the cholesterol biosynthetic pathway. Inhibition of either of these two enzymes should not interfere with the synthesis of other vital isoprenoids, such as dolichol and ubiquinone, and also should not lead to the accumulation of potentially harmful steroid precursors.

Research on this enzyme has concentrated on finding inhibitors of the fungal SE as a treatment for various fungal infections. Inhibition of fungal SE leads to a defiency of ergosterol (an essential component of the cell membrane) and thus arrested cell growth, and an accumulation of squalene, which is thought to be responsible for the fungicidal action.[213,214]

Naftifine (**142**), the first specific inhibitor identified, belongs to a class of allylamines, and has been used for the topical treatment of superficial mycoses. A related compound, terbinafine (**143**), has been shown to be orally active and more potent than **142**.

Using the microsomal SE system from *Candida albicans*, it has been shown that **142** and **143** are potent non-competitive inhibitors with K_i values of 1.1 and 0.3 µM, respectively. However, little inhibition was observed (K_i values of 144 and 77 µM) when tested against a mammalian SE (rat liver). Interestingly, **143** was much more potent when tested against a guinea pig SE ($IC_{50} = 5$ µM). In rat liver, this inhibition was shown to be competitive with respect to squalene and also with respect to the soluble cytoplasmic fraction.[214]

This selectivity was reversed with the disclosure of NB-598 (**144**) by Banyu. This compound contains the allylamine moiety found on the antimycotic SE inhibitors. However, this compound potently inhibited mammalian SE (microsomes from rat liver, dog liver, Hep-G2 cells) with IC_{50} values of 4.4, 2.0, 0.75 nM, respectively. Kinetic analysis showed that the inhibition was competitive. This compound also inhibited cholesterol biosynthesis from [^{14}C]-acetate in Hep-G2 cells with an IC_{50} of 3.4 nM. This compound also inhibited cholesterol biosynthesis from [^{14}C] acetate *in vivo*, the ED_{50} value in rats was 5.1 mg/kg.[215] At doses of 1, 3, and 10 mg/kg,

144 decreased plasma TC in dogs by 3%, 14% and 34%, respectively after 21 days of treatment. This lowering was comparable to that found with simvastatin (**34**) (32% lowering at 10 mg/kg after 21 days). The compound also causes an increase in serum squalene levels,[216] but it did not show any antifungal activity against a variety of fungii. Thus, it is a specific inhibitor of mammalian squalene epoxidase. The evolution of this compound was recently reviewed.[217]

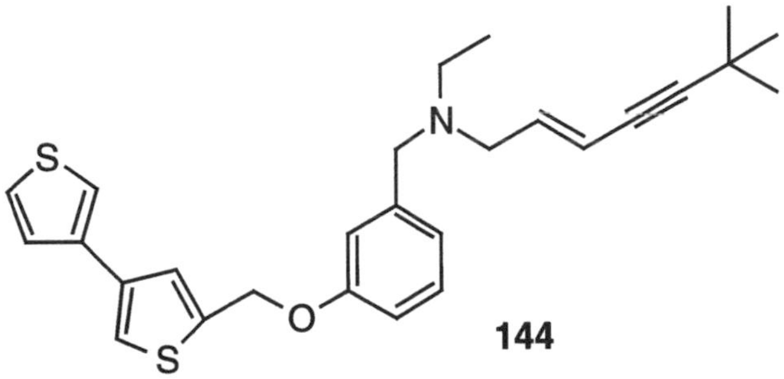

144

Mimics of the physiological substrate, squalene, have also been evaluated as inhibitors of SE. 2-Aza-2,3-dihydrosqualene (**145**) was one of the first squalene mimics disclosed to inhibit SE. The compound inhibited SE from both fungal and mammalian origins (IC$_{50}$ = 2.4 µM against rat liver SE).[218] Trinorsqualene alcohol (**146**) was shown to be a potent inhibitor of SE isolated from pig liver. This compound had an IC$_{50}$ of 4 µM and SAR studies showed that the primary alcohol functionality (the corresponding thiol was also active) and the trinorsqualenoid skeleton were required for activity. Kinetic studies showed that the inhibition was not competitive and that the mechanism of inhibition was clearly more complicated. It has been speculated that **146** functions as a bi-substrate analog, mimicking a reactive intermediate that incorporates both squalene and an activated form of squalene.[219] Later studies, using partially purified SE, confirmed that **146** was a noncompetitive inhibitor at pH 8.8 with a K_i value of 4 µM. It was also shown that 26-hydroxysqualene (**147**) was both a competitive inhibitor with a K_i value of 4 µM at pH 8.8 and a substrate for SE.[220] Analogs with a truncated and extended carbon skeleton have been prepared and shown to be poor inhibitors of pig liver SE.[221]

The rational design of inhibitors was attempted by Prestwick who designed molecules in which the isopropylidene moiety of squalene was replaced by a sterically or electronically perturbed π containing system at one or both ends of the poly-olefinic chain (e.g. analogs containing 1,1-dihaloalkene, acetylene, allene, diene and cyclopropane functionalities). These analogs were designed to function either as suicide substrates or prosuicidal substrates. However, most of these were poor inhibitors (IC$_{50}$ > 400 µM). The authors concluded, based on molecular mechanics calculations, that a good inhibitor should possess geometry and size similar to squalene, a hydrophobic moiety in the region of the isopropylidene methyl groups, an unpolarized, reactive double bond and a hydroxyl group in the isopropylidene region.[222] Interestingly, Cattel *et al.* showed that bis allene **148** inhibited rat liver SE with an IC$_{50}$ of 50 µM.[223]

It is known that a variety of cyclopropylamines are potent mechanism-based inactivators of cytochrome P-450 enzymes. Since SE is believed to be an external flavoprotein monoxygenase, a number of squalenoid cyclopropylamines were prepared and evaluated for their ability to inhibit mammalian SE. Secondary amine (**149**) was shown to be one of the most potent inhibitors of mammalian SE thus far identified ($IC_{50} = 2$ μM). The N-methyl analog (**150**) was a poor inhibitor ($IC_{50} = 100$ μM), while the analog bearing one less methylene moiety (**151**) was as potent as **149**.[224] Amine **149** does not appear to be a mechanism-based inhibitor, and it has been speculated that the interaction between the enzyme and inhibitor is electrostatic in nature. Both **149** and **150** inhibit cholesterol biosynthesis in Hep G2 cells with IC_{50}'s of 1.0 and 0.5 μM, respectively, however only **149** inhibited SE, **150** inhibited 2,3-oxidosqualene cyclase. Thus, it appears that cyclopropyl derivatives of squalene are effective inhibitors of cholesterol biosynthesis, and that nitrogen substitutions affect enzyme selectivity.

A series of difluoro olefins have been shown to be inhibitors of rat liver SE. Compounds **152** and **153** inhibit SE with IC_{50} values of 5.4 and 4.5 μM, respectively.[226] Hydroxylamine derivative (**154**) was shown to be a potent inhibitor of both SE and OSC ($IC_{50} = 13$ and 5 μM, respectively).[227]

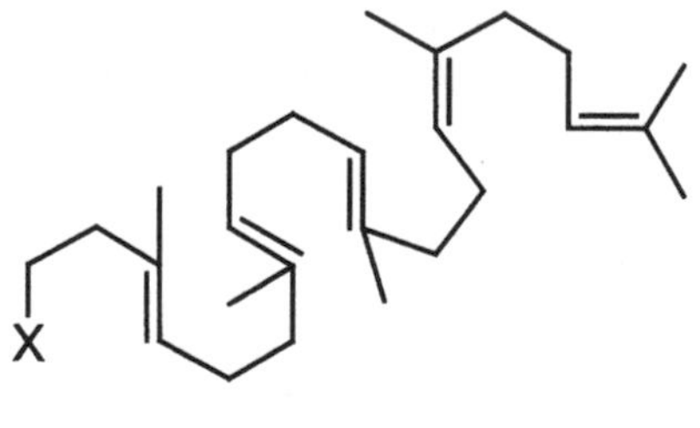

145, X=CH$_2$N(CH$_3$)$_2$
146, X=CH$_2$OH
149, X=CH$_2$NH-c-C$_3$H$_5$
150, X=CH$_2$N(CH$_3$)-c-C$_3$H$_5$
151, X=NH-c-C$_3$H$_5$
152, X=HC=CF$_2$
154, X=CH$_2$N(OH)CH$_3$

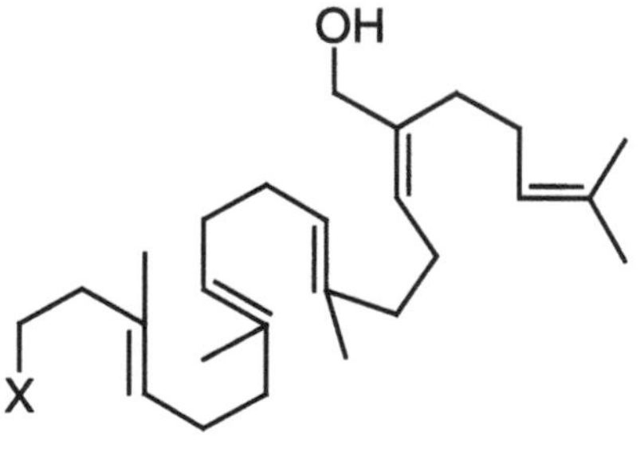

147

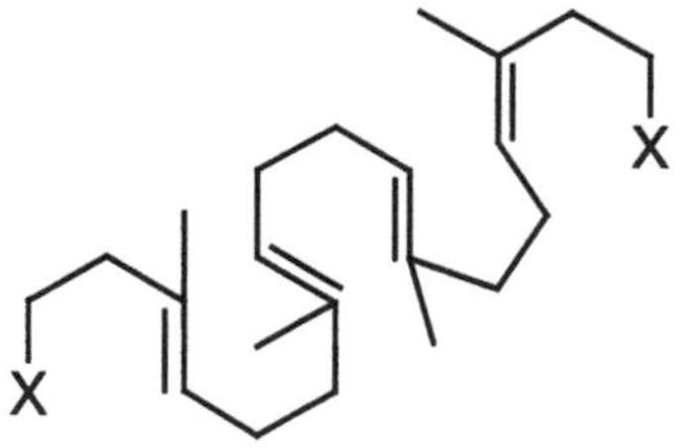

148, X=CH=C=CH-CH$_3$
153, X=HC=CF$_2$

2,3-Oxidosqualene-lanosterol cyclase (EC 5.4.99.7 abbreviated as OSC) catalyzes the most remarkable step in the cholesterol biosynthetic pathway, the conversion of (3S)-2,3-oxidosqualene (**140**) in a single step to lanosterol (**141**). The reaction proceeds in the absence of added cofactors and under anaerobic conditions to produce, in non-photosynthetic organisms, lanosterol, and in photosynthetic organisms, cycloartenol. The mechanism for the cyclization of (3S)-2,3-oxidosqualene to lanosterol has been the subject of intense research.[228-230] The initial step in the reaction involves the enzymic cyclization of 2,3-oxidosqualene through a series of suprafacial 1,2 shifts, with the formation of four new carbon-carbon bonds, to a cationic protosterol (recently studies have demonstrated that **155** is the likely structure).[231] Proton loss gives lanosterol (**141**).

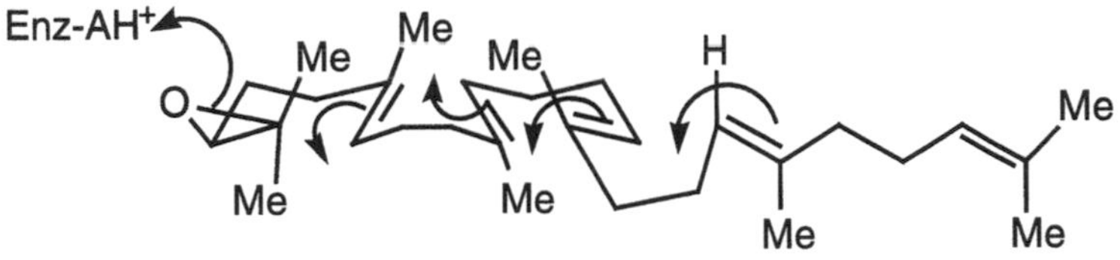

140

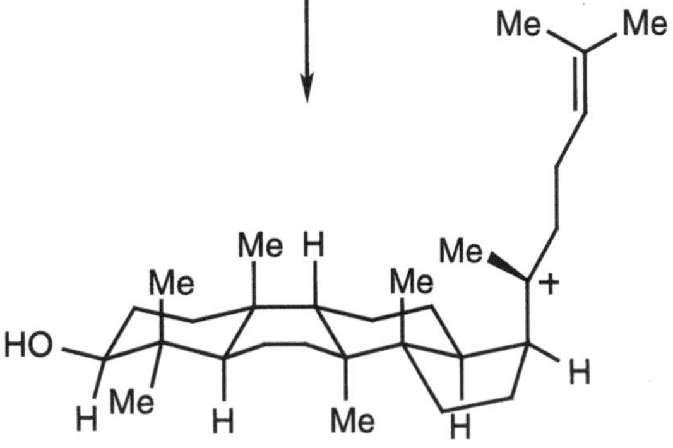

155

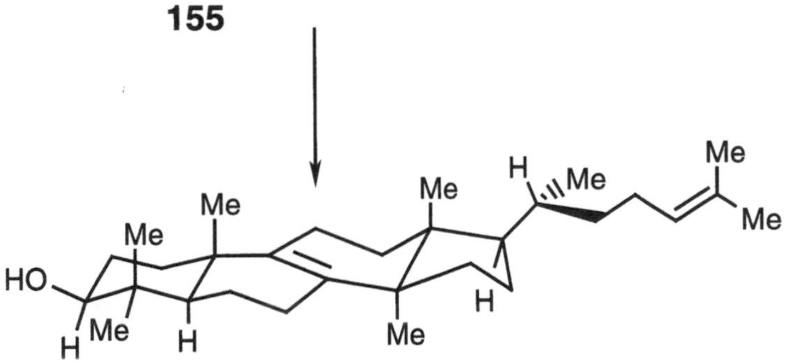

141

The search for inhibitors of this enzyme has intensified in recent years. The known OSC inhibitors fall into a number of different classes. One of the early OSC inhibitors was a substrate mimic, 2,3-iminosqualene (**156**), a close analog of

2,3-oxidosqualene (**140**), the physiological substrate for OSC which was shown to be an inhibitor of OSC isolated from pig liver microsomes. The corresponding 2,3-sulfidosqualene (**157**), in comparison, was a very weak inhibitor.[232]

It has also been recognized that the placement of a second epoxide in **140** could interfere with the cyclization to lanosterol. It has been shown that (3S,22S)-2,3;22,23-dioxidosqualene (**158**) is produced in animal cells and liver homogenates treated with OSC inhibitors and also that SE can convert **140** into **158** with one-half the efficiency of the epoxidation of squalene.[233] This compound, (**158**), is cyclized by OSC to 24,25-epoxylanosterol, which is then converted to 24S,25-epoxycholesterol, an important oxysterol regulator of HMG-R.[234] However, it is interesting to note that the accumulation of dioxidosqualenes can have negative physiological effects, e.g. angiotoxicity in rabbits.[235] It has been shown that **158** inhibits OSC with an IC_{50} of 16 µM. A more recent study, aimed at identifying more potent squalene dioxides as inhibitors of OSC, has identified 2,3;18,19-dioxidosqualene (**159**) as a potent inhibitor of rat liver OSC. This compound had an IC_{50} of 0.11 µM, compared to the IC_{50} of 142 µM obtained for **158**. Inhibition was shown to be non-competitive.[236] Interestingly, it is possible that **159** can be formed under physiological conditions in the organism itself.[237] Addition of a methylidene group, at the 29 position of compound **140**, produced **160**, which was a potent, irreversible, inhibitor of pig liver OSC ($IC_{50} = 0.5$ µM).[238]

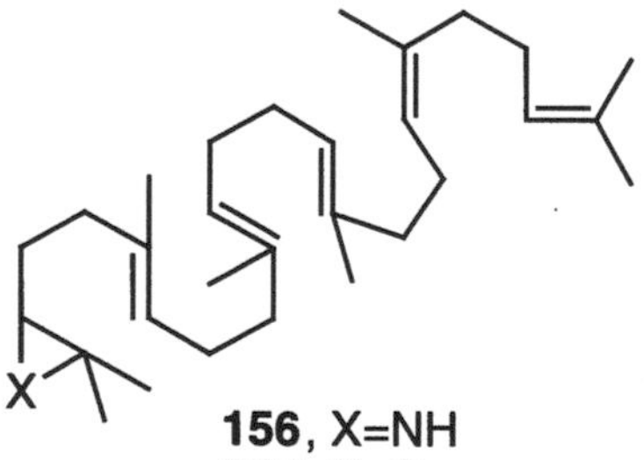

156, X=NH
157, X=S

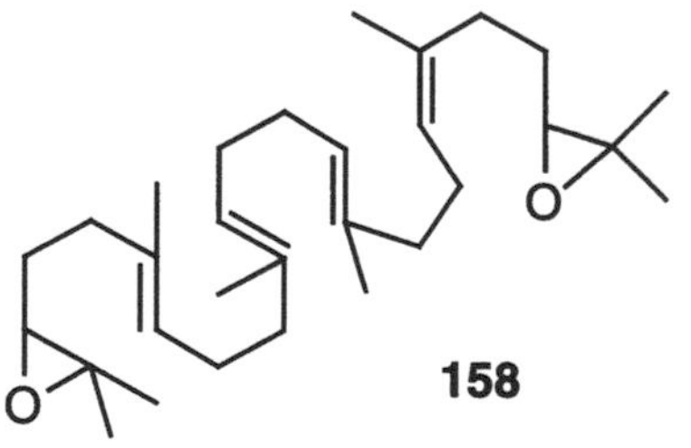

158

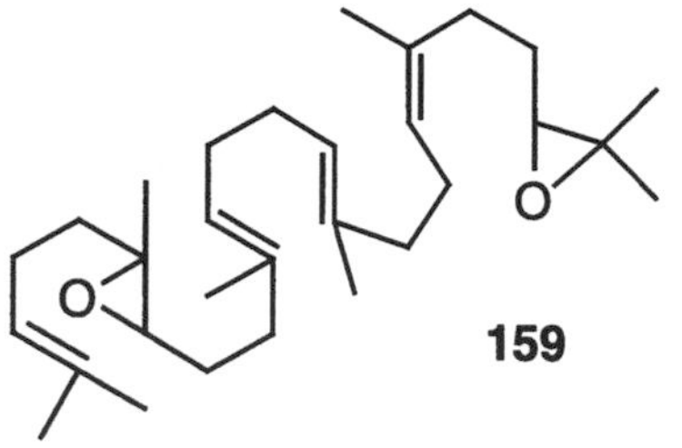

159

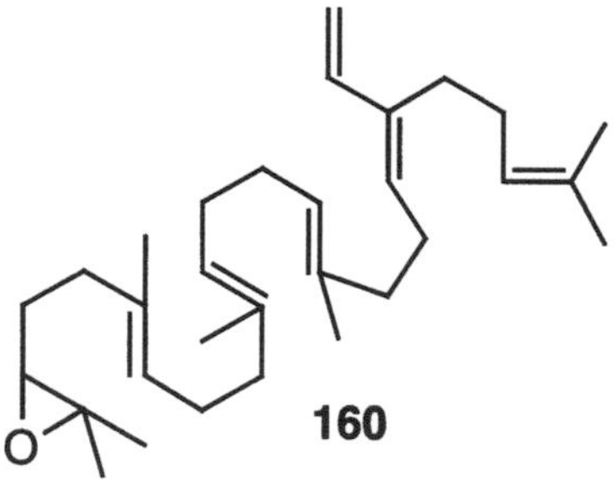

160

It has been hypothesized that the mechanism of action of some inhibitors may be product inhibition due to the fact that they resemble the product of the enzymatic reaction. The prototype of the product mimics is 4,4,10β-trimethyl-*trans*-decal-3β-ol (**161**, abbreviated as TMD). Intial experiments were undertaken using a standard S_{10} rat liver homogenate. At 2 mM concentration there was a 96% inhibition of cholesterol biosynthesis and significant accumulation of both squalene epoxide (**140**) and squalene dioxide (**158**). Similar results were obtained in CHO cells.[239,240] However, it has been reported that this compound is ineffective in primary rat hepatocytes and in whole animals. Evidence has accumulated which indicates that this lack of efficacy is due to rapid metabolism (hydroxylation at C7). To suppress this metabolism a small series of 7,7-disubstituted TMD derivatives were synthesized and evaluated for their ability to inhibit OSC in a rat liver microsomal assay. Preliminary data has shown that compounds **162–164** are as potent as **161**, however no data is available in an intact animal.[241]

161, $R^1 = R^2 = H$
162, $R^1, R^2 = CH_2CH_2$
163, $R^1 = R^2 = CH_3$
164, $R^1 = R^2 = F$

The most active area in the search for new OSC inhibitors has been the development of compounds which mimic certain transition states or high energy intermediates postulated in the cyclization of 2,3-oxidosqualene to lanosterol. There does exist some disagreement on the precise mechanism of the enzymic cyclization of squalene epoxide, the concertedness of the overall annulation is still debated and data exists which would suggest that the cyclization proceeds through a series of discrete conformationally rigid carbocationic intermediates.[229,231]

In an attempt to mimic the first transient carbocationic intermediate postulated to be involved in the epoxide ring opening, **145** was synthesized with the expectation that the resulting charged ammonium group would be capable of functioning as a stable isostere of a carbocation. This compound had an IC_{50} of 8.8 μM when tested against rat liver OSC, the corresponding diethyl compound was also active ($IC_{50} = 7.7$ μM), however, as the steric bulk increases beyond diethyl, the inhibition obtained decreases.[242] Compound **145** was also shown to inhibit cholesterol biosynthesis in 3T3 fibroblasts with an IC_{50} of 0.3 μM.[243] Initial experiments also showed that the enzyme, Δ^{24}-reductase, was another target of this compound. This compound and related analogs, such as the bis-azasqualene ammonium derivative (**165**, $IC_{50} = 1.5$ μM against OSC) were also shown to be effective anti-bacterials and anti-fungal agents.[223,244] N-oxides have often been incorporated into molecules to mimic polarized bonds involved in enzyme catalyzed reactions. The N-oxide derivatives, **166** and **167**, were shown to be twice as potent as the corresponding tertiary amines.[245] The 22,23-epoxy derivative of **166** was also shown to be a very potent inhibitor with an IC_{50} of 1.5 μM.[246] Interestingly, the N-oxide of compound **150** (an SE inhibitor) was shown to be inhibitory ($IC_{50} = 40$ μM).[224]

The related squalenoid oxaziridine (**168**) was shown to be a potent inhibitor of rat liver OSC ($IC_{50} = 1.5$ μM), but showed little activity in inhibiting cholesterol biosynthesis in 3T3 fibroblasts. This was probably due to the inherent chemical instability of the oxaziridine ring.[248]

There have been attempts to design "suicide" inhibitors of OSC. In theory this would form an oxenium ion after cyclization, which would then form a stable covalent bond with a postulated nucleophilic group present on the enzyme. Compound **169** was synthesized and shown to weakly inhibit OSC ($IC_{50} = 80$ μM) from rat liver. It was also shown to be a competitive inhibitor.

Compound **170** (*E*-isomer) was designed to mimic the C-8 carbonium ion formed during squalene epoxidase catalyzed cyclization. Both *E*- and *Z*-**171** were synthesized and shown to differ greatly in activity. Only the *E*-isomer, the carbocation analog with the same configuration as 2,3-oxidosqualene, possessed activity against rat and pig liver OSC's. The IC_{50}'s were approximately 5 μM. The *Z*-isomer was inactive even at the higher concentrations.[249]

165

166, $R^1 = R^2 = Me$
167, $R^1 = R^2 = Et$

168

169

170

171

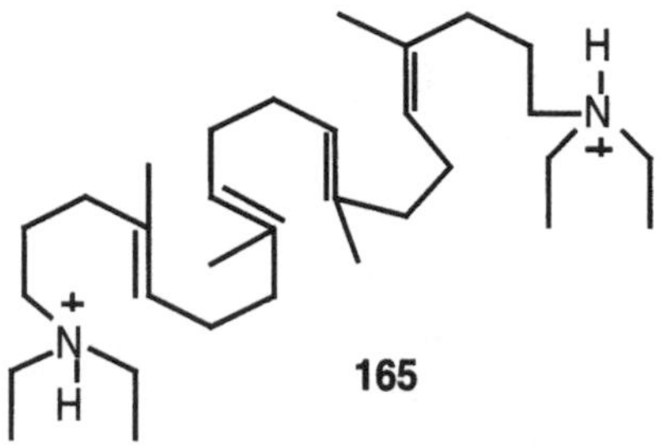

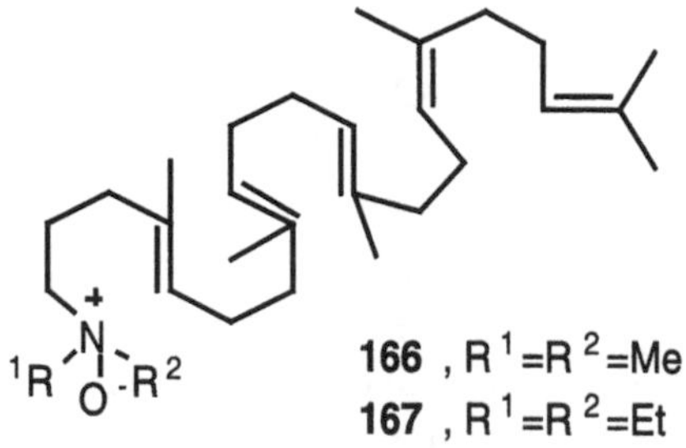

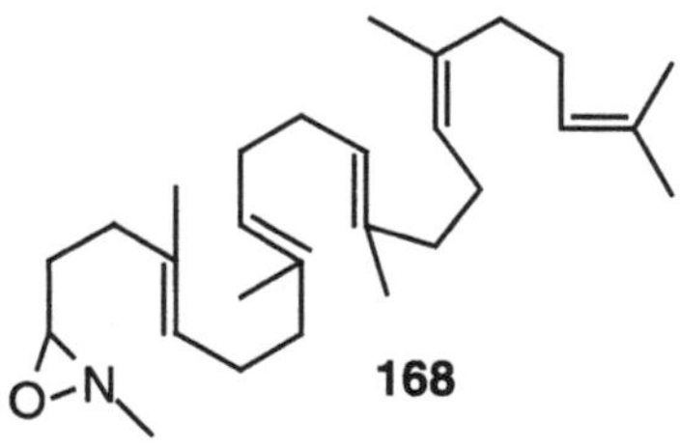

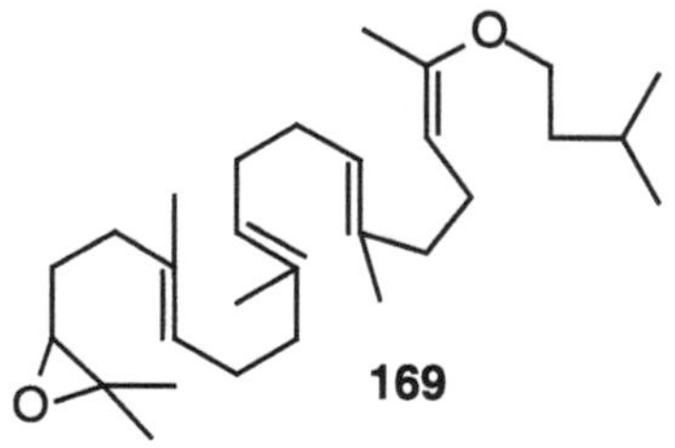

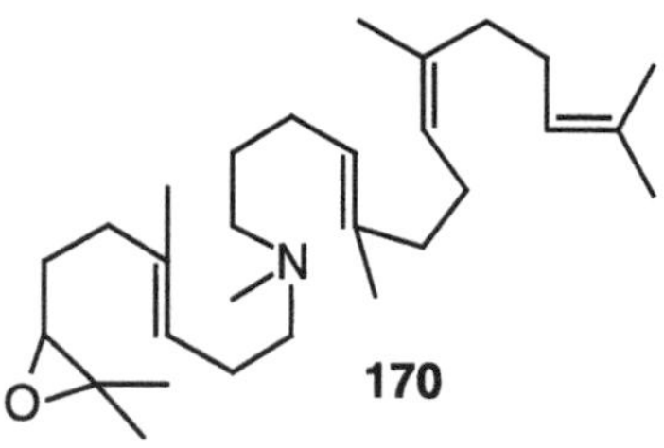

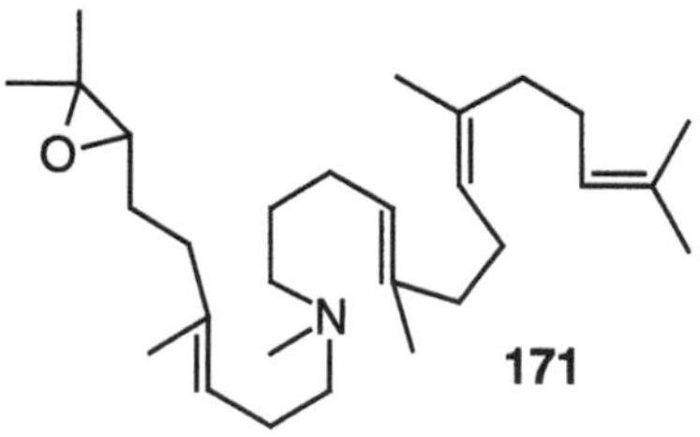

Another series of compounds shown to mimic the same C-8 carbonium ion are the azadecalins, as represented by **172**. This compound inhibits OSC from rat liver ($IC_{50} = 2.0$ μM). SAR studies showed that the presence of the charged nitrogen atom at the C-8 position of the decalin skeleton was not sufficient to mimic the C-8 carbocation, but the presence of an isoprenoid chain on the nitrogen, which mimics the C and D ring of the steroid nucleus is a necessity for potency.[250] In addition, **172** was shown to be a potent inhibitor of cholesterol biosynthesis in 3T3 fibroblasts with an IC_{50} of 20 nM. Interestingly, this inhibitor acts synergistically in inhibiting cholesterol biosynthesis, i.e., by decreasing the amount of lanosterol formed and repressing HMGR through the formation of oxysterols such as the 24,25-epoxysterols. It was found that treatment of the cells with increasing concentrations of **172** resulted in progressive reduction of the expression of HMGR activity. Unlike other ammonium containing OSC inhibitors, this compound did not inhibit Δ^{24}-reductase.[251]

Amide **173** was also shown to mimic the C-8 carbocation. It inhibited OSC from rat liver ($IC_{50} = 0.11$ μM) and cholesterol biosynthesis in Hep G2 cells ($IC_{50} = 0.7$ μM). It has a tight binding affinity to the enzyme ($K_i = 28$ nM). The less lipophilic N-acetyl derivative was much less potent ($IC_{50} = 50$ μM) and the more basic amine derivative had an IC_{50} of 0.55 μM and a K_i of 40 nM. The amine potently inhibited cholesterol biosynthesis in Hep G2-cells but predominantly by inhibiting Δ^{14}-reductase as evidenced by the accumulation of $\Delta^{8,14}$-cholestadienol in the culture medium.[252]

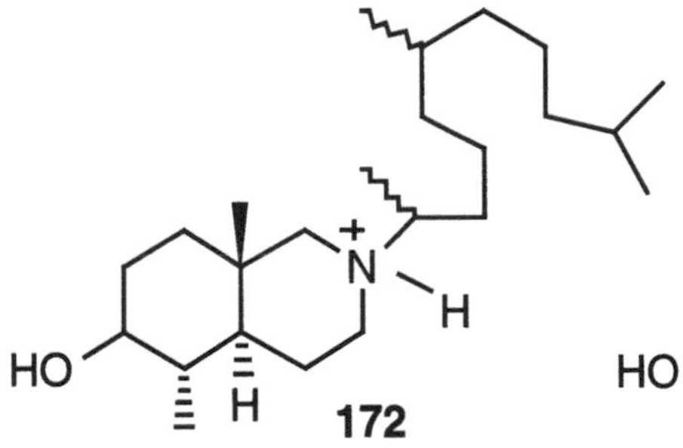

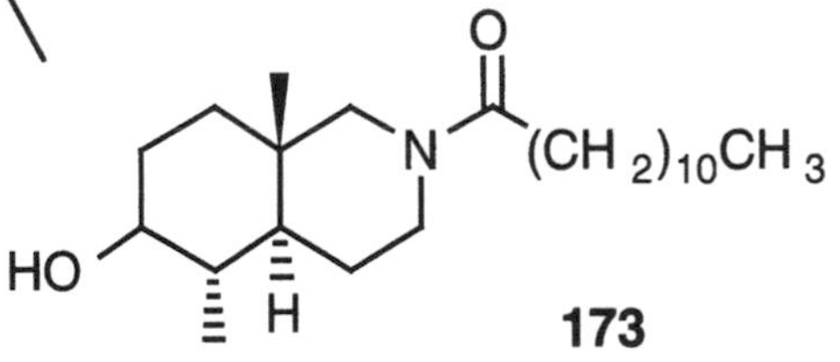

A series of 8-azadecalins bearing less lipophilic substituents at N-8 (such as Me, Bn) were synthesized and shown not to inhibit OSC, even though these compounds were shown to mimic similar carbocationic species observed in the reaction catalyzed by cycloeucalenol-obtusifoliol isomerase. Simple azadecalin analogs of TMD (**161**) were synthesized and evaluated for their ability to inhibit OSC from rat liver. Compound **174** was shown to be the most potent inhibitor ($IC_{50} = 165$ μM), compared to the IC_{50} of 65 μM obtained for **161** in this study. The dipolar canonical form of lactam **174** bears an obvious relationship to a carbocationic species thought to be formed during the cyclization of squalene epoxide.[253]

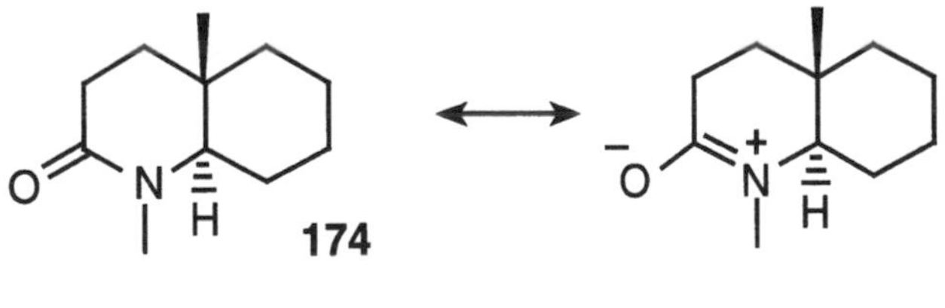

Piperidine **175** was shown to inhibit OSC from *Candida albicans* with an IC_{50} of 0.23 µM in a cell free system. This was 3 fold more potent than the compound lacking the gem-dimethyl groups. The methiodide salt of **175** was suprisingly less potent ($IC_{50} = 14$ µM) than the amine.[254] A closely related analog (**176**) was shown to be a potent inhibitor of rat liver OSC ($IC_{50} = 20$ µM). Saturation of the side chain eliminated all activity. It is postulated that this compound acts by mimicking carebenium intermediates possessing the positive charge at carbon pro-C-10.[255] However, piperidine analog (**177**) was shown to be a potent inhibitor of rat liver OSC ($IC_{50} = 0.42$ µM).[256] This compound was a significantly more potent inhibitor of cholesterol biosynthesis in Hep G2 cells ($IC_{50} = 40$ nM). The simple piperidine sulfone (**178**) was also shown to inhibit OSC ($IC_{50} = 5$ µM).[257]

Imidazole derivatives have long been known to inhibit cholesterol biosynhesis in rat liver. It was later shown that N-dodecylimidazole (**179**) actually potently inhibited rat liver OSC with an IC_{50} of 3.9 µM.[258] U18666A (**180**) has been shown to block the conversion 2,3-oxidosqualene into lanosterol in rat intestinal epithelial cultures, leading to the intracellular accumulation of 2,3;22,23-dioxidosqualene.[259] It was also shown that **180** inhibits rat liver OSC with an IC_{50} of 0.8 µM.[242]

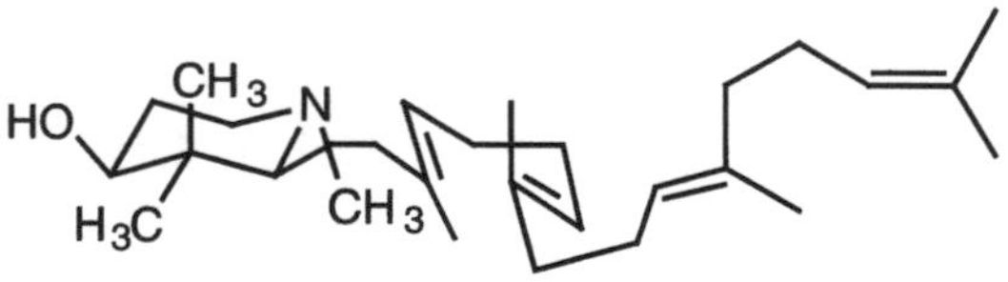

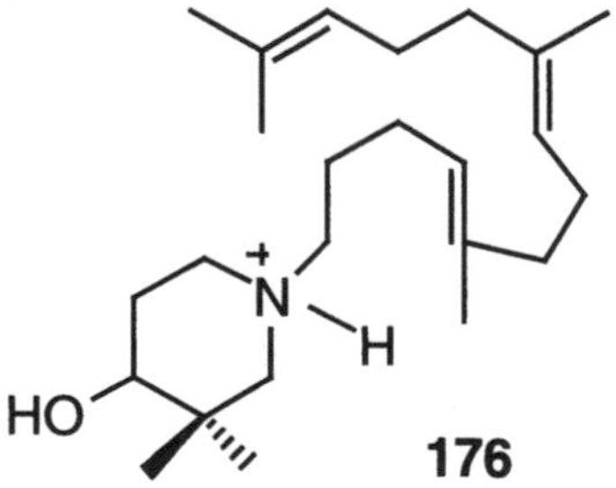

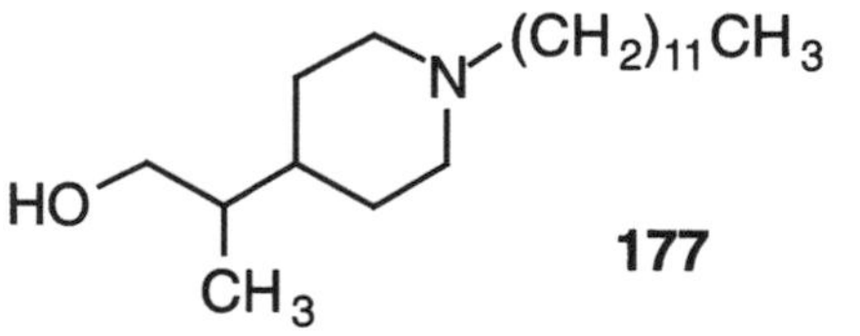

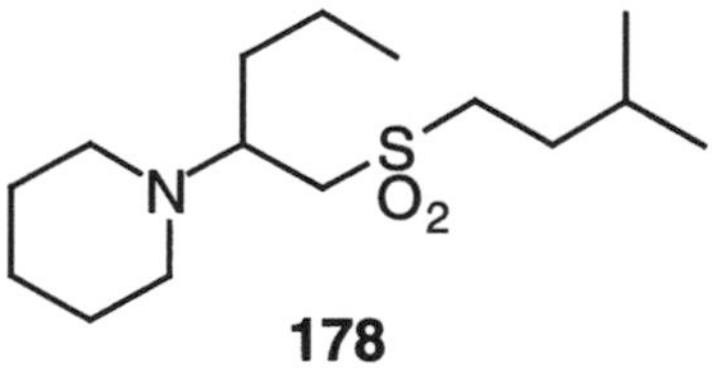

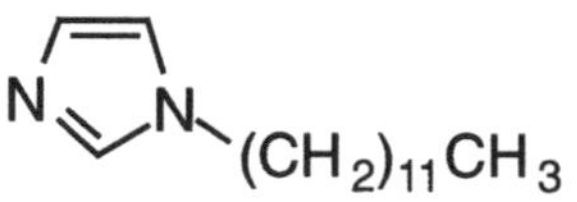

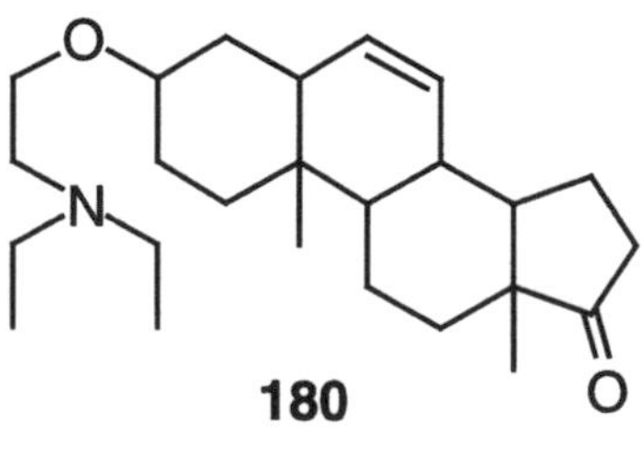

In contrast to the remarkable enzymatic cyclization of 2,3-oxidosqualene (**140**) to lanosterol (**141**), the conversion of lanosterol into cholesterol only requires the removal of three methyl groups (at positions 4,4,14α-), saturation of the double bond in the side chain and shift the double bond from the 8,9 to the 5,6 position in ring B, however 18 separate enzymatic reactions are required to effect these transformations.[14a]

Removal of the 14α-methyl group of lanosterol is a three step oxidative process requiring a cytochrome P-450 linked enzyme (**lanosterol 14α-demethylase**, P-450$_{14DM}$), NADPH-cytochrome P-450 reductase and molecular oxygen with subsequent formation of 32-hydroxy and 32-oxo intermediates. The 14α-methyl group is removed as formic acid with loss of the 15α-H and subsequent $\Delta^{8,14}$-diene sterol (**181**) formation. The enzymatic reduction of the $\Delta^{14(15)}$-double bond by **sterol-14-reductase** yields **182**, occurs under anaerobic conditions and requires NADPH.

The demethylation of the 4,4-gem-dimethyl groups is an oxidative process which results in the removal of the C30 and C31 methyl groups as carbon dioxide. The reaction requires three separate enzymes to demethylate a single 4α-methyl group from the starting substrate. The intial step is catalyzed by **4α-methyl sterol oxidase**, this is rate limiting for the overall demethylation process and requires NADPH and molecular oxygen during the conversion to the corresponding 4α-carboxylic acid. Decarboxylation of the acid is catalyzed by a NAD$^+$ dependent decarboxylase. This is accompanied by migration of the 4-β-methyl group to the 4α-position with loss of the 3α-H and subsequent 3-keto steroid formation. Reduction of this compound is accomplished with an NADPH specific 3-keto steroid reductase and results in the generation of the corresponding 3-β-hydroxy steroid. A second cycle of demethylation then occurs with the 4α-monomethyl sterol to yield **183**.

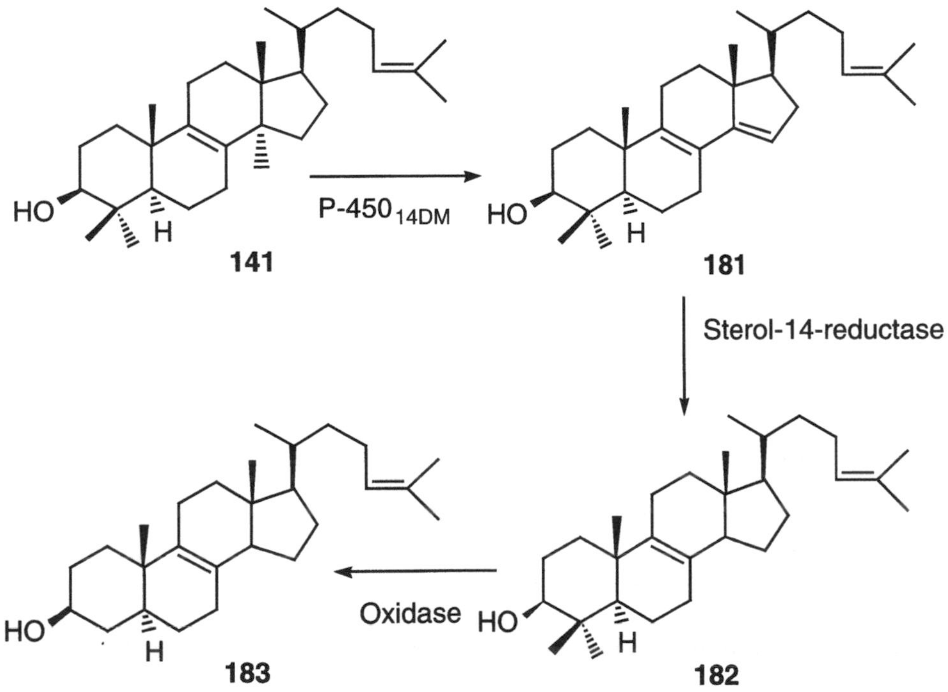

Introduction of the Δ^5-double bond found in cholesterol is intiated by isomerization of the $\Delta^{8,(9)}$-double bond to the $\Delta^{7,(8)}$ position by the enzyme, **steroid-8-ene isomerase**. No cofactors are required for catalytic activity and the reaction proceeds anaerobically to yield **184**. Formation of the Δ^5-double bond requires oxidative elimination of the 5α- and 6α-hydrogens. This is catalyzed by the enzyme, Δ^7**-sterol-5-desaturase** and requires NADPH and molecular oxygen for activity. Reduction of the Δ^7-double bond of the $\Delta^{5,7}$-diene sterol (**185**) is achieved by the enzyme, $\Delta^{5,7}$**-sterol-7-reductase**. The reaction requires NADPH and proceeds anaerobically to yield desmosterol (**186**). The final conversion to cholesterol (**1**) is achieved by side chain saturation of desmosterol and it is catalyzed by the enzyme, Δ^{24}**-sterol-24-reductase**.

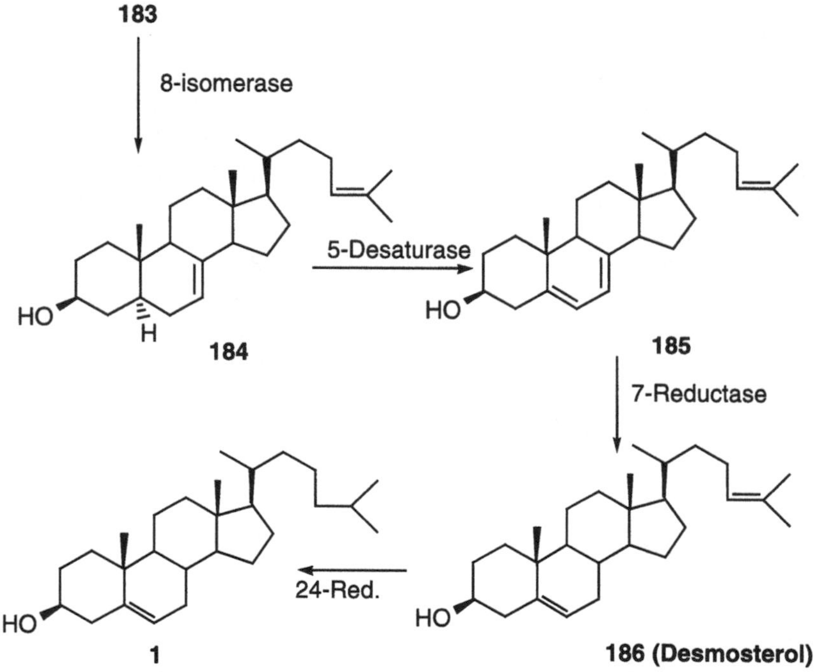

Most of the early research into the development of cholesterol biosynthesis inhibitors concentrated on inhibiting steps late in the biosynthetic pathway (after lanosterol formation). However, it has already been discussed that inhibition at these late steps can result in the accumulation of non-metabolizable sterol intermediates and subsequent toxicities. Since this time the efficacy of such inhibition has long been doubted and no serious attempts to develop such inhibitors as hypocholesterolemic agents have been recorded.

The vast majority of research into the search for late stage inhibitors of cholesterol biosynthesis has been in the development of novel anti-fungal agents.[260] The pre-eminent drug in this class is ketoconazole (**187**). This imidazole derivative inhibits ergosterol (a vital component of the cell wall of fungii) biosynthesis in fungii at low nanomolar concentrations with a broad spectrum of activity and low toxicity. Its mode of action is inhibition of the cytochrome P-450 mediated C14-demethylation of lanosterol, the first step in the conversion of lanosterol into cholesterol (or ergosterol). It appears to interact with P-450$_{14DM}$ at the heme iron site. In addition it also interferes with different P-450 enzyme systems in several organs, e.g., testis (C$_{17}$–C$_{20}$ lyase), adrenal gland (11β-hydroxylase and cholesterol side-chain cleavage enzyme), ovary (aromatase), kidney and liver.[261] Ketoconazole has been shown to inhibit the conversion of lanosterol into cholesterol in mammalian cells at micromolar concentrations. Inhibition results in the accumulation of dihydrolanosterol and lanosterol in the serum. When added to Hep-G2 cells in concentrations of 2–100 μM, cholesterol biosynthesis is inhibited. The maximal effect was seen at 100 μM. After 20 hours preincubation of the cells it was noted that the activity of HMGR was decreased, while receptor mediated binding, uptake and degradation of human LDL was increased. These observations suggest that an endogenous suppressor of HMGR can be formed and accumulate in the presence of the drug.[262]

It has been shown that P-450$_{14DM}$ is the site of action for the regulation of HMGR by ketoconazole in both CHO cells and primary rat hepatocytes by correlating the amount of HMGR protein with the production of 32-oxylanosterols (specifically 3β-hydroxylanost-8-ene-32-aldehyde) at varying concentrations. The partial inhibition of P-450$_{14DM}$ increases endogenous 32-oxylanosterols, resulting in suppressing the gene expression of HMGR.[263,264] The effects of ketoconazole on cholesterol metabolism in man has been assessed in patients with prostate cancer. At a dose of 1.2 grams per day in 5 patients the average level of TC fell by 27%, LDL by 41% and LDL apo B by 32%. The serum content of dihydrolanosterol and lanosterol was increased 250 fold. The findings of this study indicate that by inhibiting P-450$_{14DM}$, ketoconazole drains sterols from the body into the bile and faeces, mainly in the form of lanosterol, resulting in the reduction of serum TC.[265] These findings have inspired the search for specific inhibitors of P-450$_{14DM}$ without interfering with steroid hormone production. One of the more fruitful areas of research has been in the search for different oxygenated lanosterol derivatives which can serve as dual inhibitors of cholesterol biosynthesis by virtue of their ability to inhibit sterol P-450$_{14DM}$ and regulate HMGR gene expression.

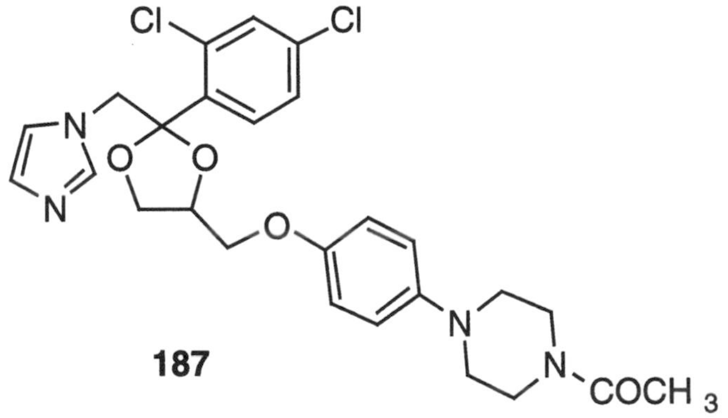

187

The fluoro analog (**188**) has been shown to be a competitive inhibitor of P450$_{14DM}$ ($K_i = 315\ \mu M$) as well as a substrate for the enzyme. When CHO cells are treated with **188** a decrease in HMGR activity and protein level was observed. It was also found that HMGR mRNA levels were not reduced as much as expected for a sterol regulator of HMGR and that the decrease in protein level is due to inhibition of enzyme synthesis, suggesting that this compound reduces the translational efficiency of HMGR mRNA. Collectively, these observations support the premise that oxysterols regulate HMGR gene expression through a post-translational process distinct from other previously described sterol regulatory mechanisms.[266]

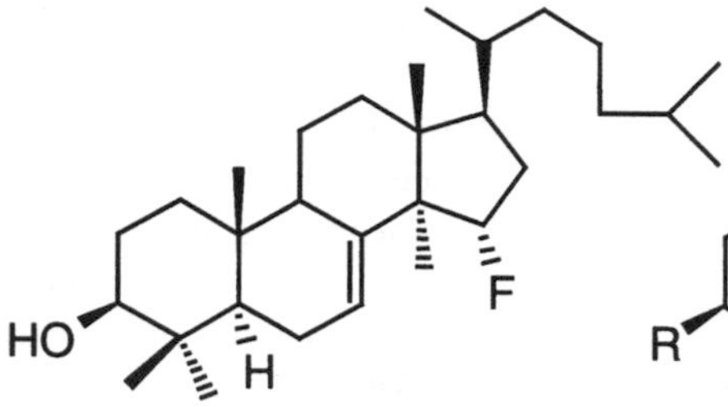

188

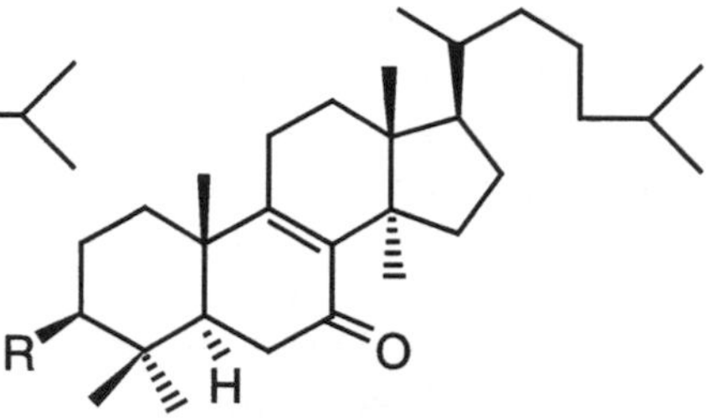

189 R=H
190 R=OAc

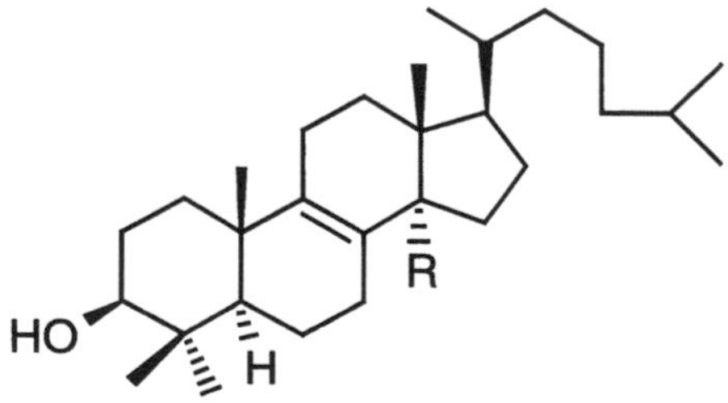

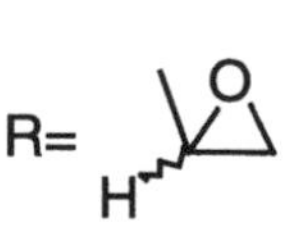

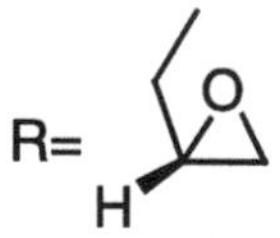

R= **191**   R= **192**

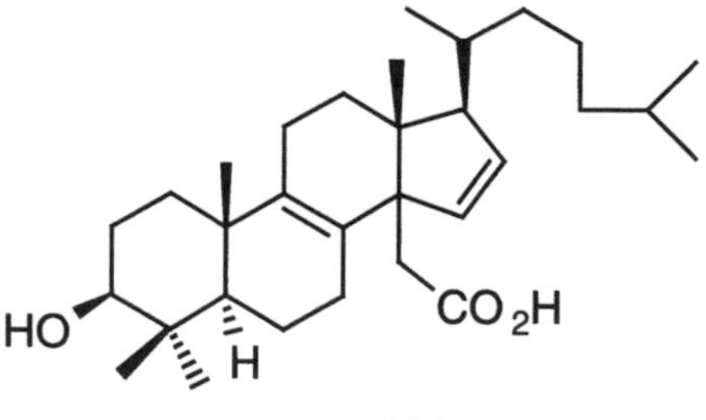

193

7-Oxo-dihydrolanosterol (DHL) (**189**) and 7-oxo-DHL-3-OAc (**190**) were shown to be potent inhibitors of cholesterol biosynthesis from 24,25-dihydrolanosterol in rat liver homogenates.[267] Sterols with functionalities at C14 were also shown to be potent inhibitors of cholesterol biosynthesis downstream from dihydrolanosterol. Oxiranes (**191**) and (**192**) inhibited cholesterol and lathosterol biosynthesis by >89% at 10 μM, comparable to that obtained with ketoconazole (90% at 10 μM).

Later studies with **192** showed that this compound was a potent, stereoselective, competitive inhibitor of P-450$_{14DM}$ with a K_i of 0.62 µM. Its diastereomer was less active with a K_i of 2 µM. The corresponding thiiranes were considerably less potent. SAR studies concluded that the active site of the enzyme is relatively insensitive to the size and degree of unsaturation of the C14α alkyl substituents up to and including propyl.[268]

SKF-104976 (**193**) was shown to be a highly potent inhibitor of P-450$_{14DM}$ with an IC$_{50}$ of 2 nM. It was also shown to inhibit the incorporation of ^{14}C-acetate into cholesterol in Hep-G2 cells and to decrease HMGR activity, however, it did not affect LDL uptake or LDL-R activity. This suggests that under these conditions HMGR and LDL-R are not coordinately regulated. These same authors also report data showing that either a mevalonate derived sterol or <u>non-sterol</u> product is essential for HMGR regulation. This suggests that oxylanosterols may not be the key mediators of HMGR in Hep-G2 cells upon P-450$_{14DM}$ inhibition.[269] Compounds **194–196** have been shown to be potent inhibitors of P-450$_{14DM}$ with IC$_{50}$'s of 0.33, 1.6 and 9.1 µM, respectively, in rat liver microsomes. They also inhibit HMGR activity by reducing the level of protein in CHO cells with IC$_{50}$'s of 3, 1.5 and 1 µM, respectively.[270]

Several series of steroidal acetylenes have been synthesized as mechanism based inhibitors of P-450$_{14DM}$. The terminal acetylene was chosen because compounds containing this moiety are among the most efficient and predictable inactivators of P-450 isozymes. Activation of the acetylene by P-450 can lead to the formation of a covalent adduct with the active site heme, rendering it catalytically inactive. Thus, lanosterol derivatives containing an appropriately positioned acetylene group could serve as mechanism based inhibitors.[271]

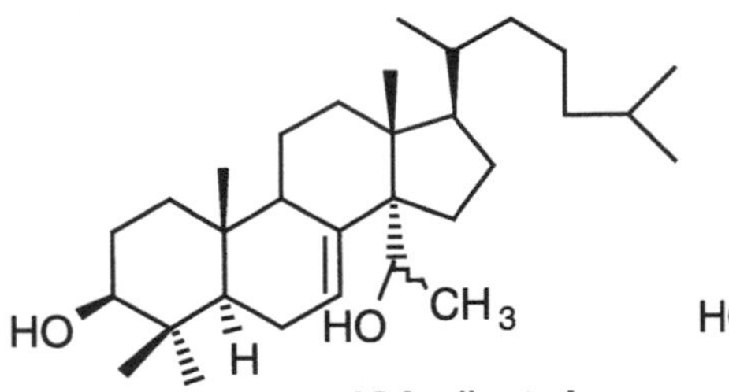

194, diast. A
195, diast. B

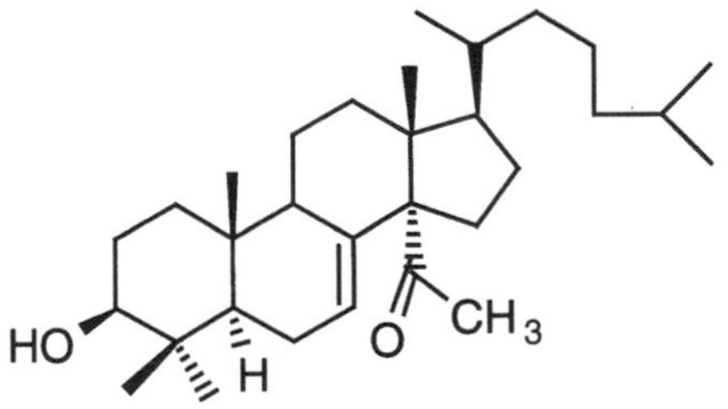

196

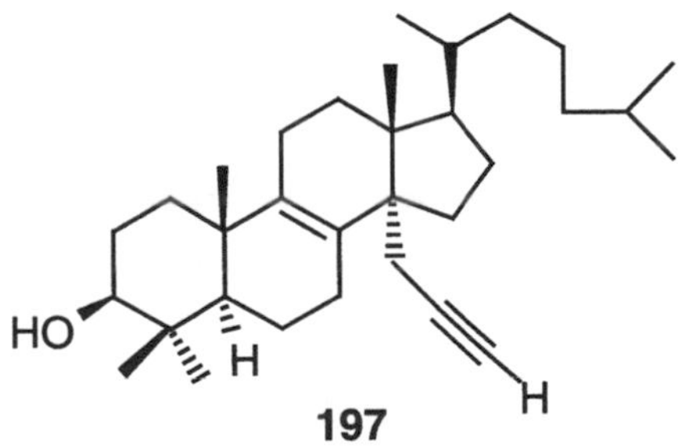

197

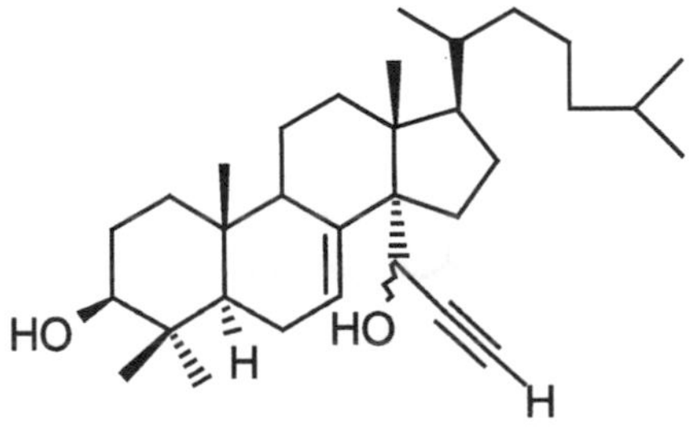

198, diast. A
199, diast. B

Compounds **197–199** were shown to be potent inhibitors of rat liver microsomal P-450$_{14DM}$ with K_i values of 1.2, 1.2 and 6.3 μM, respectively. Using a tritium release assay it was shown that **197** is a time dependent, irreversible inhibitor of P-450$_{14DM}$. Additional studies on other compounds indicate that this is a general characteristic of steroidal acetylenes.[271]

Oxygenated lanosterol derivatives (ganoderic acids and their derivatives) were isolated from *Ganoderma lucidum* and shown to be inhibitors of cholesterol biosynthesis from 24,25-dihydrolanosterol. At 40 μM, ganoderic acid B (**200**) showed little inhibition of cholesterol biosynthesis during incubation with the S-10 fraction of rat liver homogenates. However, **201**, a chemically modified analog had an IC$_{50}$ of 18 μM.[272]

Solacongestidine (**202**), a steroid alkaloid from *Solanum congestiflorum* was shown to inhibit cholesterol biosynthesis (59% inhibition at 40 μM) in a rat liver preparation. This compound was also shown to be a potent anti-fungal agent.[273]

More recently, the first non-steroidal, selective inhibitor (**203**, RS-21607) of mammalian P-450$_{14DM}$ has been reported. This compound competively inhibited P-450$_{14DM}$ in hepatic microsomes from rat, hamster and human with apparent K_i values of 2.5, 1.4 and 0.35 nM, respectively. Comparative values for the enantiomer (**203**, 2R,4R) and ketoconazole (**187**) were: (**203**, 2R,4R) 37, 40.4 and 17.7 nM and (**187**) 65, 24.5 and 63.5 nM, respectively. Selectivity, relative to other cytochrome P-450 enzymes involved in steroid biosynthesis, was determined using progesterone 17α,20-lyase, cholesterol-7α-hydroxylase, aromatase, corticoid 11β-hydroxylase and progesterone hepatic 6β-hydroxylase and the following K_i values were determined:

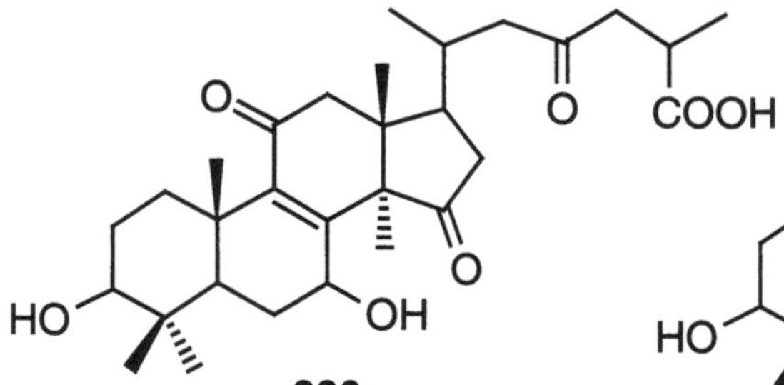

200

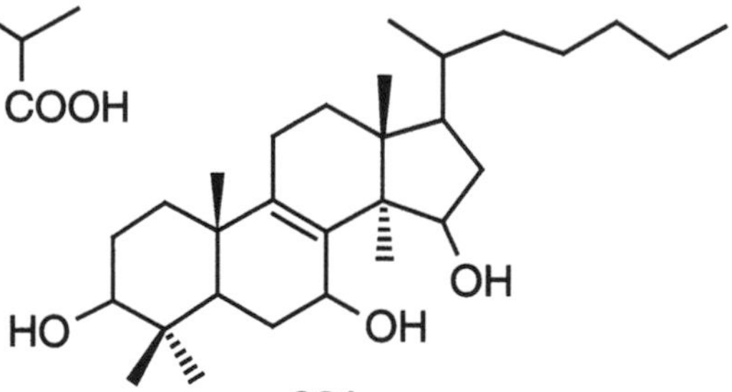

201

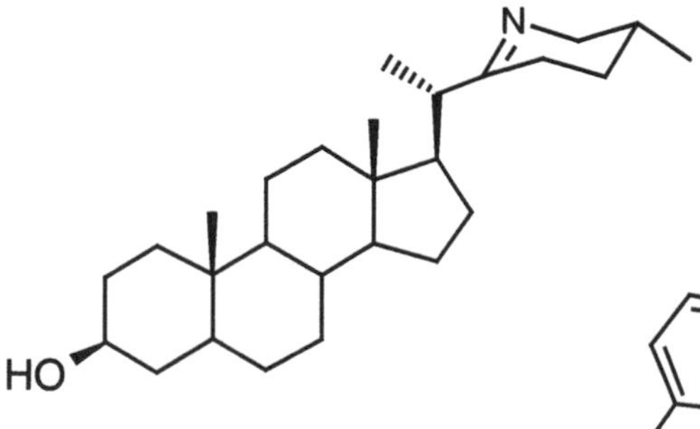

202

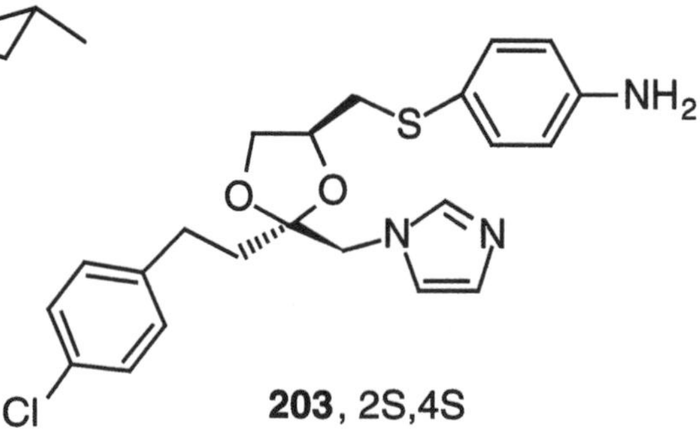

203, 2S,4S

(**203**, 2S,4S) 447, 1625 7.6, 35 and 28 nM, respectively and (**203**, 2R,4R) 54, 109, 13, 15 and 33 nM. Thus, **203** showed good selectivity for P-450$_{14DM}$ relative to other P-450 enzymes and a favorable separation of activity from its enantiomer. Inhibition of cellular cholesterol biosynthesis from ^{14}C-acetate was assessed using human fibroblasts. The IC$_{50}$'s for **203**, **203** (2R,4R) and ketoconazole were 0.09, 8 and 37 nM, respectively. Hypocholesterolemic effects were evaluated in male Syrian hamsters. At an oral dose of 100 mg/kg, TC was lowered by 62% with **203** (2S,4S) and 22% with **203** (2R,4R). Chronic dosing, at 50 mg/kg, over 14 days resulted in a 37% lowering of TC for **203** compared to 16% decrease for lovastatin (**18**) at the same dose. Both dihydrolanosterol and lanosterol levels increased in the liver. The elevations in the plasma were much less, thus, possibly negating the fears that sterol accumulation in peripheral tissues may be substantial and potentially toxic. It is reported that **203** is currently undergoing clinical evaluation as a hypo-cholesterolemic agent.[274]

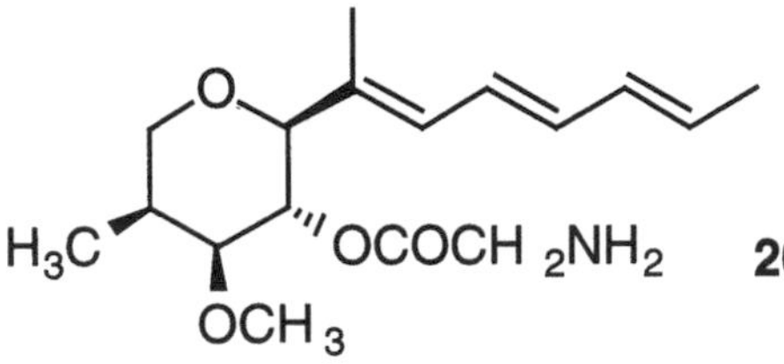

Lanomycin (**204**) a naturally occuring anti-fungal agent isolated from *Pycnidiophora dispersa* was shown to inhibit P-450$_{14DM}$.[275]

BM 15766 (**205**) has been shown to inhibit cholesterol biosynthesis at the $\Delta^{5,7}$-**sterol-7-reductase** step. This compound showed a dose dependent action on the incorporation of ^{14}C-acetate into non-saponifiable lipids. The inhibition was 10–12% at 100 μM. Simultaneously, 7-dehydrocholesterol levels rose.[276] In male rats, treated with 120 mg/kg for 28 days, plasma total sterol content was reduced to 25% of controls. Total sterol content of the liver decreased 50% after 7 days at 60 mg/kg.[277]

AY-9944 (**206**) was also shown to inhibit this enzyme. At 1 mM, **206** inhibits the incorporation of ^{14}C-mevalonate into cholesterol by liver homogenates of rat (81%), dog (21%) and monkey (59%). Additional sites of enzyme inhibition have been shown, e.g. $\Delta^{8,14}$-sterol-14-reductase.[278a]

Compound **207** was found to lower plasma TC levels in rats (by 85% at a dose of 56 mg/kg) with a concurrent accumulation of 3β-cholesta-5,7-dien-3-ol, suggesting inhibition of $\Delta^{5,7}$-sterol-7-reductase.[279]

Triparanol (**8**) is the prototypical Δ^{24}-sterol reductase inhibitor. The problems with its development have already been discussed. 20,25-diazacholesterol (**208**), which also inhibits Δ^{24}-sterol reductase, has also been shown to lead to an increase in desmosterol. Although **208** causes a net increase in sterol levels, a net increase in the activity of HMGR has been noted.[278b]

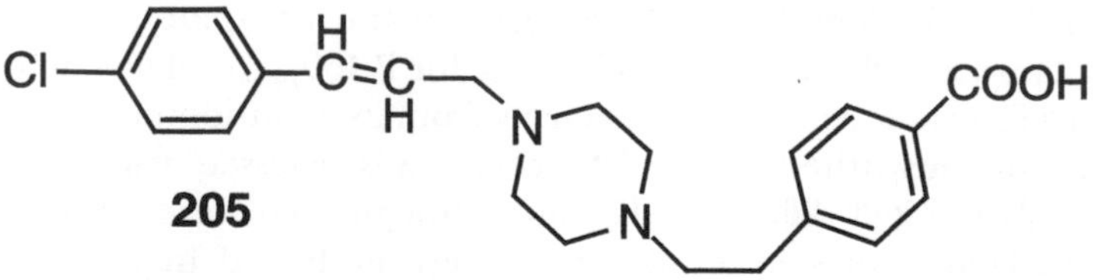

205

206

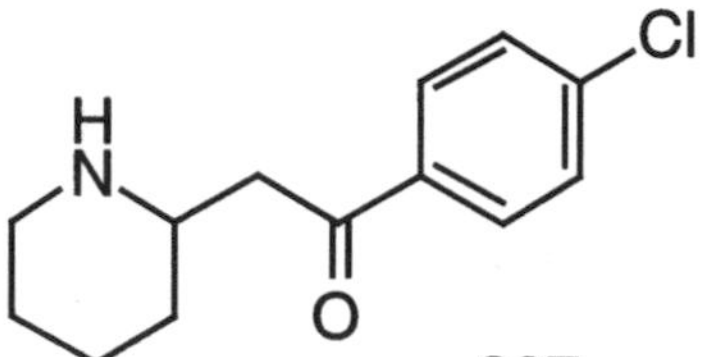

207

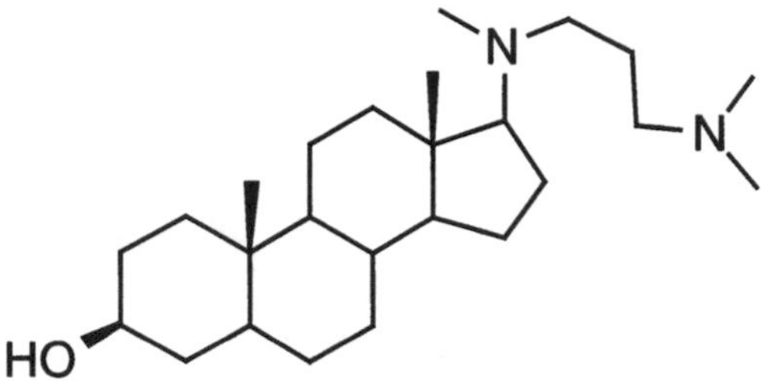

208

4. REFERENCES

1. Manson, J.E., Tosteson, P.H.H., Ridker, P.M., *et al.* (1992) The primary prevention of myocardial infarction. *N. Eng. J. Med.*, **21**, 1406–1416.

2. Virchow, R. (1856) *Phlogse and Thrombose in Gefassystem Gessamelete Abhandlungen zur Wissenschaftlichen Medicin.* p. 458. Frankfurt-am-Main: Meidinger Sohn.

3. von Rokitansky, C. (1852) *A manual of pathological anatomy,* **Vol. 4**, p. 261, Day, G.E. (trans.) London: Sydenham Society.

4. Anitschow, N.N. (1967) A history of experimentation on arterial atherosclerosis in animals. In *Cowdrys arteriosclerosis,* Blumenthal, H.T. (ed.), p. 21, Charles C. Thomas (pjb.).

5. Roberts, W.C. (1992) Atherosclerotic risk factors, are there ten or only one? *Atherosclerosis,* **97**, S5–S9.

6. Muldoon, M.F., Manuck, S.B. and Matthews, K.A. (1991) Mortality experience in cholesterol reduction trials. *N. Eng. J. Med.*, **324**, 922–923.

7. Lipid Research Clinics Program (1984) The Lipid Research Clinics Coronary Primary Prevention Trial results. *JAMA,* **251**, 365–373.

8. Frick, M.H., Elo, O., Haapa, K., *et al.* (1987) Helsinki Heart Study: primary prevention trial with Gemfibrozil in middle-aged men with dyslipidemias. *N. Eng. J. Med.*, **317**, 1237–1245.

9. Blankenhorn, D.H., Nessim, S.A., Johnson, R.L., Sanmarco, M.E., Azen, S.P. and Cashin-Hemphill, L. (1987) Beneficial effects of combined colestipol niacin therapy on coronary atherosclerosis and coronary venous bypass grafts. *JAMA,* **257**, 3233–3240.

10. Buchwald, H., Varco, R.L., Matts, J.D., *et al.* (1990) Effects of partial ileal bypass surgery on mortality and morbidity from coronary heart disease in patients with hypercholesterolemia: report of the Program On the Surgical Control of the Hyperlipidemias (POSCH). *N. Eng. J. Med.*, **323**, 946–955.

11. (a) McCarthy, P.A. (1993) New appproaches to atherosclerosis: an overview. *Med. Res. Rev.*, **13**, 139–159. (b) Gotto, A.M. (1993) Dyslipidemia and atherosclerosis. A forecast of pharmaceutical approaches. *Circulation*, **87 (Supp. III)**, III-54–III-59.

12. Turley, S.D. and Dietschy, J.M. (1982) Cholesterol metabolism and excretion. In *The Liver, Biology and Pathology*. Arias, I., Popper, H., Schacter, D., Shafritz, D.A. (eds.), p. 467. New York: Raven Press.

13. Dietschy, J.M. and Wilson, J.D. (1968) Cholesterol synthesis in the squirrel monkey: relative rates of synthesis in various tissues and mechanisms of control. *J. Clin. Invest.*, **47**, 166–174.

14. (a) Faust, J.R., Trzaskos, J.M. and Gaylor, J.L. (1988) Cholesterol biosynthesis. In *The Biology of Cholesterol*. Yeagle, P.L. (ed.), pp. 19–38. Boca Raton, FL: CRC Press. (b) Myant, N.B. (1981) The biology of cholesterol and related steroids. pp. 1–910. London: William Heinemann Medical Books.

15. Brown, M.S. and Goldstein, J.L. (1980) Multivalent feedback regulation of HMG-CoA reductase, a control mechanism coordinating isoprenoid synthesis and cell growth. *J. Lipid Res.*, **21**, 505–517.

16. Curran, G.L. and Azarnoff, D.L. (1958). Inhibition of cholesterol biosynthesis in man. *Arch. Intern. Med.*, **101**, 685–689.

17. Cottet, J., Mathirat, A. and Redel, J. (1954) Therapeutic study of a synthetic hypocholesterolemic phenylethylacetic acid. *Press Med.*, **62**, 939–941.

18. Bargeton, D., Krumm-Heller, C. and Tricaud, M.E. (1954) Influence de l'age et d'un acetate subsititue sur le cholesterol serique chez le rat. *C.R. Soc. Biol.*, **148**, 63–65.

19. Frederickson, D.S. and Steinberg, D. (1957) Failure of alpha-phenylbutyrate and beta-phenylvalerate in the treatment of hypercholesterolemia. *Circulation*, **15**, 391–396.

20. Barr, D.P. (1955) Influence of sex and sex hormones upon the development of atherosclerosis and upon the lipoproteins of the plasma. *J. Chronic Dis.*, **1**, 63–85.

21. Palpoli, F.P., Allen, R.E., Schumann, E.L. Day, W.L. and VanCampen, M.G. (1958) Estrogen antagonists substituted aminoalkoxy triarylethanols and their derivatives. In *Proceedings of the 134th National Meeting of the American Chemical Society*, p. 12.0 (abs.). Chicago, IL.

22. Blohm, T.R., Kariya, T., Laughlin, M.E. and Palopoli, F.P. (1959) Reduction of blood and tissue cholesterol by MER-29, a cholesterol biosynthesis inhibitor. *Fed. Proc.*, **18**, 369 (abs.).

23. Avigan, J., Steinberg, D. and Thompson, M.J. (1960) Studies of cholesterol biosynthesis I, the identification of desmosterol in serum and tissues of animals and man treated with MER-29. *J. Biol. Chem.*, **235**, 3123–3126.

24. Laughlin, R.C. and Carey, T.F. (1962) Cataracts in patients treated with triparanol. *JAMA*, **181**, 339–340.

25. Achor, R.W.P., Winkleman, R.K. and Perry, H.O. (1961) Cutaneous side effects from triparanol (MER-129), preliminary data on ichthyosis and loss of hair. *Proc. Staff Meet. Mayo Clin.*, **36**, 217–218.

26. Avigan, J., Steinberg, D., Thompson, M.J. and Mossettig, E. (1960) Mechanism of action of MER-129, an inhibitor of cholesterol biosynthesis. *Biochem. Biophys. Res. Commun.*, **2**, 63–65.

27. Gylling, H., Mantyla, E. and Miettinen, T.A. (1992) Tamoxifien decreases serum cholesterol by inhibiting cholesterol biosynthesis. *Atherosclerosis*, **96**, 245–247.

28. Kritchevsky, D.J. (1987) Inhibition of cholesterol biosynthesis. *J. Nutr.*, **117**, 1330–1334.

29. Holmes, W.L. and DiTullio, N.W. (1967) Current knowledge of drugs affecting lipid synthesis. *Prog. Biochem. Pharmacol.*, **2**, 1–20.

30. Middleton, B. (1973) The Oxoacyl-Coenzyme A Thiolases of Animal Tissues. *Biochem. J.*, **132**, 717–730.

31. Gehring, U. and Harris, J.I. (1970) The subunit structure of thiolase. *Eur. J. Biochem.*, **16**, 487–491.

32. Gehring, U. and Harris, J.I. (1970) The active site cysteines of thiolase. *Eur. J. Biochem.*, **16**, 492–498.

33. Palmer, M.A., Differding, E., Gamboni, R., Williams, S.F., Peoples, O.P., Walsh, C.T., Sinskey, A.J. and Masamune, S. (1991) Biosynthetic thiolase from *Zoogloea ramigera*. Evidence for a mechanism involving Cys-378 as the active site base. *J. Biol. Chem.*, **266**, 8369–8375.

34. Bloxham, D.P., Chalkley, R.A., Coghlin, S.J. and Salam, W. (1978) Synthesis of chloromethyl ketone derivatives of fatty acids. *Biochem. J.*, **175**, 999–1011.

35. Davis, J.T., Chen, H.H., Moore, R., Nishitane, Y., Masamune, S., Sinskey, A.J. and Walsh, C.T. (1987) Biosynthetic thiolase from *Zoogloea ramigera*. II. Inactivation with haloacetyl CoA analogs. *J. Biol. Chem.*, **262**, 90–96.

36. Salam, W. and Bloxham. D.P. (1987) Hypolipidemic effect of polymethylenemethane thiosulfonates: inhibitors of acetoacetyl coenzyme A thiolase. *J. Pharmacol. Exp. Ther.*, **241**, 1099–1105.

37. Greenspan, M.D., Yudkovitz, J.B., Chen, J.S., Hanf, D.P., Chang, M.N., Chaing, P.Y.C., Chabala, J.C. and Alberts, A.W. (1989) The inhibition of cytoplasmic acetoacetyl-CoA thiolase by a triyne carbonate (L-660,631) *Biochem. Biophys. Res. Commun.*, **163**, 548–553.

38. Lowe, D.M. and Tubbs, P.K. (1985) 3-Hydroxy-3-methylglutaryl-coenzyme A synthase from ox liver. Properties of its acetyl derivative. *Biochem. J.*, **227**, 601–607.

39. Ramachandran, C.K., Gray, S.L. and Melnykovych, G. (1980) Cytoplasmic 3-hydroxy-3-methylglutaryl coenzyme A synthase EC 4.1.3.5 and the regulation of sterol synthesis in tissue culture cells. *Biochim. Biophys. Acta*, **618**, 439–448.

40. Greenspan, M.D., Yudkovitz, J.B., Lo, C.-Y., Chen, J., Alberts, A.W., Chang, M.N., Yang, S.S., Thompson, K.L., Chaing, Y.-C., Chabala, J.C., Monaghan, R.L. and Schwartz, R.L. (1987) Inhibition of hydroxymethylglutaryl-coenzyme A synthase by L-659,669. *Proc. Nat. Acad. Sci. USA*, **84**, 7488–7492.

41. Omura, S., Tomoda, H., Kumagai, H., Greenspan, M.D., Yudkovitz, J.B., Chen, J.S., Alberts, A.W., Martin, I., Mochales, S., Monaghan, R.L., Chabala, J.C., Schwartz, R.L. and Patchett, A.A. (1987) Potent inhibitory effect of antibiotic 1233A on cholesterol biosynthesis which specifically blocks 3-hydroxy-3-methylglutaryl-coenzyme A synthase. *J. Antibiot.*, **40**, 1356–1357.

42. Mayer, R.J., Louis-Flamberg, P., Elliott, J.D., Fisher, M. and Leber, J. (1990) Inhibition of 3-hydroxy-3-methylglutaryl-coenzyme A synthase by antibiotic 1233A and other β-lactones. *Biochem. Biophys. Res. Commun.*, **169**, 610–616.

43. Greenspan, M.D., Bull, H.G., Yudkovitz, J.B., Hanf, D.P. and Alberts, A.W. (1993) Inhibition of 3-hydroxy-3-methylglutaryl-coenzyme A synthase and cholesterol biosynthesis by β-lactone inhibitors and binding of these inhibitors to the enzyme. *Biochem. J.*, **289**, 889–895.

44. Goldstein, J.L. and Brown, M.S. (1990) Regulation of the mevalonate pathway. *Nature*, **343**, 1450–1461.

45. Goldstein, J.L. and Brown, M.S. (1984) Progress in understanding the LDL receptor and HMG-CoA reductase, two membrane proteins that regulate the plasma cholesterol. *J. Lipid Res.*, **25**, 425–430.

46. Quenski, N. and Porter, J.W. (1981) Biosynthesis of Isoprenoid Compounds, **Vol. 1**, Porter, J.W. and Spurgen, J.L. (eds.) p. 47. New York: John Wiley and Sons.

47. Lupien, P.J., Moorjani, S., Brun, D. and Bielmann, P. (1979) Effects of 3-hydroxy-3-methylglutaric acid on plasma and low-density lipoprotein cholesterol levels in familial hypocholesterolemia. *J. Clin. Pharmacol.*, **19**, 120–126.

48. Endo, A., Kuroda, M. and Tsujita, Y. (1976) ML-236A, ML-236B and ML-236C, new inhibitors of cholesterogenesis produced by *Penicillium citrinum. J. Antibiot.*, **29**, 1346–1348.

49. Brown, A.G., Smale, T.C., King, T.J., Hassenkamp, R. and Thompson, R.H. (1976) Crystal and molecular structure of compactin, a new antifungal metabolite from *Penicillium brevicompactum. J. Chem. Soc. Perkin I.*, 1165–1170.

50. Endo, A. (1979) Monacolin K, A new hypocholesterolemic agent produced by a Monascus species. *J. Antibiot.*, **32**, 852–854.

51. Alberts, A.W., Chen, J., Kuron, G., *et al.* (1980) Mevinolin, a highly potent competitive inhibitor of hydroxymethylglutaryl coenzyme A reductase and a cholesterol lowering agent. *Proc. Natl. Acad. Sci. USA*, **77**, 3957–3961.

52. Lam, T.Y., Gullo, V.P., Goegelman, R.T., *et al.* (1981) Dihydrocompactin, a new potent inhibitor of 3-hydroxy-3-methylglutaryl coenzyme A reductase from *Penicillium citrinum. J. Antibiot.*, **34**, 614–616.

53. Albers-Schonberg, G., Joshua, H., Lopez, M.B., *et al.* (1981) Dihydromevinolin, a potent hypocholesterolemic metabolite produced by *Aspergillus terreus. J. Antibiot.*, **34**, 407–412.

54. Endo, A., Hasumi, K. and Negishi, S. (1985) Monacolin J and L, new inhibitors of cholesterol biosynthesis produced by *Monascus ruber. J. Antibiot.*, **38**, 420–422.

55. Endo, A., Hasumi, K., Nakamura, T. and Kunishima, M. (1985) Dihydromonacolin L and monocolin X, new metabolites that inhibit cholesterol biosynthesis. *J. Antibiot.*, **38**, 321–327.

56. Haruyama, H., Kuwano, M., Kinoshita, T., Terahara, A., Nishigaki, T. and Tamura, C. (1986) Structure elucidation of the bioactive metabolites of ML 236B (Mevastatin) isolated from dog urine. *Chem. Pharm. Bull.*, **54**, 1459–1467.

57. Serizawa, N., Serizawa, S., Nakagawa, K., Furuya, K., Okazaki, T. and Terahara, A. (1983) Microbial hydroxylation of ML-236B (compactin). *J. Antibiot.*, **36**, 887–891.

58. Tanzawa, K. and Endo, A. (1979) Kinetic analysis of reaction catalyzed by rat liver 3-hydroxy-3-methylglutaryl coenzyme A reductase using two specific inhibitors. *Eur. J. Biochem.*, **98**, 195–210.

59. Germershausen, J.I., Hunt, V.M., Bostedor, R.G., Baily, P.J., *et al.* (1989) Tissue selectivity of the cholesterol-lowering agents, lovastatin, simvastatin and pravastatin in rats *in vivo*. *Biochem. Biophys. Res. Commun.*, **158**, 667–675.

60. Slater, E.E., Alberts, A.W. and Smith, R.L. (1987) HMG-CoA reductase inhibitors. In *The Role of Cholesterol in Atherosclerosis, New Therapeutic Opportunities*. pp. 35–50. Grundy, S.M. and Bearn, A.G. (eds.) Philadelphia, PA: Hanley and Belfus, Inc.

61. Fears, R., Richards, D.H. and Ferres, H. (1980) The effect of compactin, a potent inhibitor of 3-hydroxy-3-methylglutaryl coenzyme A reductase activity on cholesterogenesis and serum cholesterol levels in rats and chicks. *Atherosclerosis*, **35**, 439–449.

62. Krause, B.R. and Newton, R.S. (1991) Animal models for the evaluation of inhibitors of HMG-CoA reductase. *Adv. Lipid Res.*, **1**, 57–72.

63. Kovanen, P.T., Bilheimer, D.W., Goldstein, J.L., Jaramillo, J.J. and Brown, M.S. (1981) Regulatory role for hepatic low-density lipoprotein receptors *in vivo* in the dog. *Proc. Nat. Acad. Sci. USA*, **78**, 1194–1198.

64. Alberts, A.W. (1988) Discovery, biochemistry and biology of lovastatin. *Am. J. Cardiol.*, **62**, 10J–15J.

65. Tsujita, Y., Kuroda, M., Tanzawa, K., Kitano, N. and Endo, A. (1979) Hypolipidemic effects in dogs of ML-236B, a competitive inhibitor of 3-hydroxy-3-methylglutaryl coenzyme A reductase. *Atherosclerosis*, **32**, 307–313.

66. Tsujita, Y., Kuroda, M., Shimada, Y., *et al.* (1986) CS-514, a competitive inhibitor of 3-hydroxy-3-methylglutaryl coenzyme A reductase: tissue selective inhibition of sterol synthesis and hypolipidemic effect on various animal species. *Biochem. Biophys. Acta*, **877**, 50–60.

67. Grundy, S.M. (1988) HMG-CoA reductase inhibitors for the treatment of hypercholesterolemia. *N. Eng. J. Med.*, **319**, 24–32.

68. MacDonald, J.S., Gerson, R.J., Kornbrust, D.J., *et al.* (1988) Preclinical evaluation of lovastatin. *Amer. J. Cardiol.*, **62**, 16J–27J.

69. Yamamoto, A., Sudo, H. and Endo, A. (1980) Therapeutic effects of ML-236B in primary hypercholesterolemia. *Atherosclerosis*, **35**, 259–266.

70. Yamamoto, A., Yamamura, T., Yokohama, S., Sudo, H. and Matsuzawa Y. (1984) Combined drug therapy-cholestyramine and compactin-for familial hypercholesterolemia. *Int. J. Clin. Pharmacol. Ther.*, **22**, 493–497.

71. Illingworth, D.R. and Sexton, G.J. (1984) Hypocholesterolemic effects of mevinolin in patients with heterozygous familial hypercholesterolemia. *J. Clin. Invest.*, **74**, 1972–1978.

72. Havel, R.J., Hunninghake, D.B., Illingworth, D.R., *et al.* (1987) Lovastatin (mevinolin) in the treatment of heterozygous familial hypercholesterolemia, a multi-center study. *Ann. Intern. Med.*, **107**, 609–615.

73. Carmena, R., DeOya, M., Franco, M., *et al.* (1989) Treatment of heterozygous familial hypercholesterolemia with pravastatin and/or cholestyramine, the Spanish multicentre Pravastatin study. *Atherosclerosis VIII*, pp. 757–760. Grepeldi, G., Gotto, A.M., Manzata, E. and Baggio, G. (eds.), London: Elsevier.

74. Blankenhorn, D.H., Azen, S.P., Kramsch, D.M., *et al.* (1993) Coronary angiographic changes with lovastatin therapy. *Ann. Intern. Med.*, **119**, 969–976.

75. Willis, R.A., Folkers, K., Tucker, J.L., Ye, C.-Q., Xia, L.-J. and Tamagawa, H. (1990) Lovastatin decreases coenzyme Q levels in rats. *Proc. Nat. Acad. Sci. USA*, **87**, 8928–8930.

76. Folkers, K., Langsjoan, P., Willis, R., *et al.* (1990) Lovastatin decreases coenzyme Q levels in humans. *Proc. Nat. Acad. Sci. USA*, **87**, 8931–8934.

77. Smith, R.L. (1991) HMG-CoA reductase inhibitors and lipid lipoprotein metabolism. In *Antilipidemic Drugs. Medicinal, Chemical and Biochemical Aspects*. pp. 121–158. Witiak, D.T., Newman, H.A. and Feller, D.R. (eds.), Amsterdam: Elsevier.

78. (a) Stokker, G.E., Rooney, C.S., Wiggins, J.M. and Hirshfield, J. (1986) Synthesis and X-ray characterization of 6(S)-mevinolin, a lactone epimer. *J. Org. Chem.*, **51**, 4931–4934. (b) Heathcock, C.H., Hadley, C.R., Rosen, T., Theisen, P.D. and Hecker, S.J. (1987) Total synthesis and biological evaluation of structural analogs of compactin and dihydromevinolin. *J. Med. Chem.*, **30**, 1858–1873.

79. Bartmann, W., Beck, G., Grunzer, E., Jendralla, H., Kerekjarto, B.V. and Wess, G. (1986) Convenient two-step stereospecific hydroxy-substitution with retention in β-hydroxy-δ-lactones. *Tetrahedron Lett.*, **27**, 4709–4712.

80. Lee, T.-J., Holtz, W.J. and Smith, R.L. (1982) Structural modifications of mevinolin. *J. Org. Chem.*, **47**, 4750–4757.

81. Endo, A. (1985) Drugs inhibiting HMG-CoA reductase. *J. Med. Chem.*, **28**, 401–405.

82. Lee, T.-J. (1987) Synthesis, SAR's and therapeutic potential of HMG-CoA reductase inhibitors. *TIPS*, **8**, 442–446.

83. Clive, D.J., Murthry, K.S.K., George, R. and Poznansky, M.J. (1990) Chemical synthesis of 3-ethylcompactin, an inhibitor of 3-hydroxy-3-methylglutaryl coenzyme A reductase. *J. Chem. Soc. Perkin I*, 2099–2108.

84. (a) Duggan, D.E. and Vickers, J. (1990) Physiological disposition of HMG-CoA reductase inhibitors. *Drug Metab. Rev.*, **22**, 333–362. (b) Stubbs, R.J., Schwartz, M.S., Gerson, R.J., Thornton, T.J. and Bayne, W.F. (1990) Comparison of plasma profiles of lovastatin (mevinolin), simvastatin (epistatin) and pravastatin (eptastatin) in the dog. *Drug Invest.*, **2** (**Supp. 2**), 18–28.

85. Heathcock, C.H., Davis, B.R. and Hadley, C.R. (1989) Synthesis and biological evaluation of a monocyclic fully functional analog of compactin. *J. Med. Chem.*, **32**, 197–202.

86. Karanewsky, D.S. (1991) Synthetic transformations of the mevinic acid nucleus: preparation of a monocylic analog of compactin. *Tetrahedron Lett.*, **32**, 3911–3914.

87. Willard, A.K. and Smith, R.L. (1982) Incorporation of 2(S)-methylbutanoic acid-1-^{14}C into the structure of mevinolin. *J. Labelled Compd. Radiopharm.*, **19**, 337–344.

88. Hoffman, W.F., Alberts, A.W., Anderson, P.S., Chen, J.S., Smith, R.L. and Willard, A.K. (1986) 3-hydroxy-3-methylglutaryl-coenzyme A reductase inhibitors. 4. Side chain ester derivatives of mevinolin. *J. Med. Chem.*, **29**, 849–852.

89. Lee, T.J., Holtz, W.J., Smith, R.L., Alberts, A.W. and Gilfillan, J.L. (1991) 3-hydroxy-3-methylglutaryl-coenzyme A reductase inhibitors. 8. Side chain ether analogs of lovastatin. *J. Med. Chem.*, **34**, 2474–2477.

90. Aggarwal, D., Saha, R.N., Gupta, J.K. and Gupta, S.P. (1988) A quantitative structure-activity relationship study of 3-hydroxy-3-methylglutaryl-coenzyme A reductase inhibitors. *J. Pharmacobio. Dyn.*, **11**, 591–599.

91. Chao, Y., Chen, Y.S., Hunt, V.M., *et al.* (1991) Lowering of plasma cholesterol levels in animals by lovastatin and simvastatin. *Eur. J. Clin. Pharmacol. Letters*, **40**, S311–S314.

92. Pietro, A.A. and Mantell, G. (1990) Simvastatin: a new HMG-CoA reductase inhibitor. *Cardiovasc. Drugs Rev.*, **8**, 220–228.

93. Duggan, M.E., Alberts, A.W., Bostedor, R.G., *et al.* (1991) 3-Hydroxy-3-methylglutaryl-coenzyme A reductase inhibitors. 7. Modification of the hexahydronapthalene moiety of simvastatin: 5-oxygenated and 5-oxa derivatives. *J. Med. Chem.*, **34**, 2489–2495.

94. Sliskovic, D.R., Roth, B.D. and Bocan, T.M.A. (1992) Tissue selectivity of HMG-CoA reductase inhibitors. *DN&P*, **5**, 517–533.

95. (a) Germershausen, J.I., Hunt, V.M., Bostedor, R.G., Bailey, P.J., Karkas, J.D. and Alberts, A.W. (1989) Tissue selectivity of cholesterol-lowering agents, lovastatin, simvastatin and pravastatin in rats *in vivo*. *Biochem. Biophys. Res. Commun.*, **158**, 667–675. (b) Duggan, D.E., Chen, I.-W., Bayne, W.F., *et al.* (1989) The physiological disposition of lovastatin. *Drug Metab. Dispos.*, **17**, 166–173.

96. Shaw, M.K., Newton, R.S., Sliskovic, D.R., Roth, B.D., Ferguson, E. and Krause, B.R. (1990) Hep-G2 cells and primary rat hepatocytes differ in their response to inhibitors of HMG-CoA reductase. *Biochem. Biophys. Res. Commun.*, **170**, 726–734.

97. Mencar, K.A., Patel, D., Clay, V., Howes, C. and Taylor, P.W. (1992) A novel approach to the site specific delivery of potential HMG-CoA reductase inhibitors *BioMed. Chem. Lett.*, **2**, 285–290.

98. Bone, E.A., Cunningham, E.M., Davidson, A.H., *et al.* (1992) The design and biological evaluation of a series of 3-hydroxy-3-methylglutaryl coenzyme A (HMG-CoA) reductase inhibitors related to dihydromevinolin. *BioMed. Chem. Lett.*, **2**, 223–228.

99. Bone, E.A., Davidson, A.H., Lewis, C.N. and Todd, R.S. (1992) Synthesis and biological evaluation of dihydroeptastatin, a novel inhibitor of 3-hydroxy-3-methylglutaryl coenzyme A reductase. *J. Med. Chem.*, **35**, 3388–3393.

100. Boots, M.R., Boots, S.G., Noble, C.M. and Guyer, K.E. (1973) Hypocholesterolemic agents II: Inhibition of β-hydroxy-β-methylglutaryl coenzyme A reductase by arylalkyl hydrogen succinates and glutarates. *J. Pharm. Sci.*, **62**, 952–957.

101. Guyer, K.E., Boots, S.G., Marecki, P.E. and Boots, M.R. (1976) Hypocholesterolemic agents III. Inhibition of β-hydroxy-β-methylglutaryl coenzyme A reductase by half acid esters of 1-(4-biphenylyl) pentanol. *J. Pharm. Sci.*, **65**, 548–552.

102. Boots, M.R., Yeh, Y.-M. and Boots, S.G. (1980) Hypocholesterolemic agents VII: Inhibition of β-hydroxy-β-methylglutaryl coenzyme A reductase by monoesters of substituted glutaric acids. *J. Pharm. Sci.*, **69**, 506–509.

103. Boots, S.G., Boots, M.R., Guyer, K.E. and Marecki, P.E. (1976) Hypocholesterolemic agents V: Inhibition of β-hydroxy-β-methylglutaryl coenzyme A reductase by substituted 4-biphenylalkyl carboxylic acids and methyl esters. *J. Pharm. Sci.*, **65**, 1374–1379.

104. DeBold, C.R. and Elwood, J.C. (1981) Mevalonic acid analogs as inhibitors of cholesterol biosynthesis. *J. Pharm. Sci.*, **70**, 1007–1010.

105. Sato, A., Ogiso, A., Noguchi, H., Mitsui, S., Kaneko, I. and Shimada, Y. (1980) Mevalonolactone derivatives as inhibitors of 3-hydroxy-3-methylglutaryl coenzyme A reductase. *Chem. Pharm. Bull.*, **28**, 1509–1525.

106. Stokker, G.E., Hoffman, W.F., Alberts, A.W., Cragoe, E.J., Jr., Deana, A.A., Gilfillan, J.L., Huff, J.W., Novello, F.C., Prugh, J.D., Smith, R.L. and Willard, A.K. (1985) 3-Hydroxy-3-methylglutaryl-coenzyme A reductase inhibitors. 1. Structural modification of 5-substituted 3,5-dihydroxypentanoic acids and their lactone derivatives. *J. Med. Chem.*, **28**, 347–358.

107. Stokker, G.E., Alberts, A.W., Anderson, P.S., Cragoe, E.J., Jr., Deana, A.A., Gilfillan, J.L., Hirshfield, J., Holtz, W.J., Hoffman, W.F., Huff, J.W., Lee, T.J., Novello, F.C., Prugh, J.D., Rooney, C.S., Smith, R.L. and Willard, A.K. (1986) 3-Hydroxy-3-methylglutaryl-coenzyme a reductase inhibitors. 3. 7-(3,5-Disubstituted [1,1'-biphenyl]-2-yl)-3,5-dihydroxy-6-heptenoic acids and their lactone derivatives. *J. Med. Chem.*, **29**, 170–181.

108. Hoffman, W.F., Alberts, A.W., Cragoe, E.J., Jr., Deana, A.A., Evans, B.E., Gilfillan, J.L., Gould, N.P., Huff, J.W., Novello, F.C., Prugh, J.D., Rittle, K.E., Smith, R.L., Stokker, G.E. and Willard, A.K. (1986) 3-Hydroxy-3-methylglutaryl-coenzyme a reductase inhibitors. 2. Structural modification of 7-(substituted aryl)-3,5-dihydroxy-6-heptenoic acids and their lactone derivatives. *J. Med. Chem.*, **29**, 150–169.

109. Stokker, G.E., Alberts, A.W., Gilfillan, J.L., Huff, J.W. and Smith, R.L. (1986) 3-Hydroxy-3-methylglutaryl-coenzyme A reductase inhibitors. 5. 6-(Fluorenyl-9-yl)- and 6-(fluoren-9-ylidenyl)-3,5-dihydroxy-6-hexanoic acids and their lactone derivatives. *J. Med. Chem.*, **29**, 852–855.

110. Prugh, J.D., Alberts, A.W., Deana, A.A., Gilfillan, J.L., Huff, J.W., Smith, R.L. and Wiggins, J.M. (1990) 3-Hydroxy-3-methylglutaryl-coenzyme A reductase inhibitors. 6. trans-6-[2-(substituted-1-naphthyl) ethyl(or ethenyl)]-3,4,5,6-tetrahydro-4-hydroxy-2H-pyran-2-ones. *J. Med. Chem.*, **33**, 758–765.

111. Roth, B.D., Ortwine, D.F., Hoefle, M.L., Stratton, C.D., Sliskovic, D.R., Wilson, M.W. and Newton, R.S. (1990) Inhibitors of cholesterol biosynthesis. 1. trans-6-(2-pyrrol-1-ylethyl)-4-hydroxypyran-2-ones, a novel series of HMG-CoA reductase inhibitors. 1. Effects of structural modifications at the 2- and 5-positions of the pyrrole nucleus. *J. Med. Chem.*, **33**, 21–31.

112. Kathawala, F.G. (1991) HMG-CoA reductase inhibitors: an exciting development in the treatment of hyperlipoproteinemia. *Med. Res. Rev.*, **11**, 121–146.

113. Roth, B.D., Blankley, C.J., Chucholowski, A.W., Ferguson, E., Hoefle, M.L., Ortwine, D.F., Newton, R.S., Sekerke, C.S., Sliskovic, D.R., Stratton, C.D. and Wilson, M.W. (1991) Inhibitors of cholesterol biosynthesis. 3. Tetrahydro-4-hydroxy-6-[2-(1H-pyrrol-1-yl) ethyl]-2H-pyran-2-one inhibitors of HMG-CoA reductase. 2. Effects of introducing substituents at positions four and five of the pyrrole nucleus. *J. Med. Chem.*, **34**, 357–366.

114. Jendralla, H., Baader, E., Bartmann, W., Beck, G., Bergmann, A., Granzer, E., Kerekjarto, B.v., Kesseler, K., Krause, R., Schubert, W. and Wess, G. (1990) Synthesis and biological activity of new HMG-CoA reductase inhibitors. 2. Derivatives of 7-(1H-pyrrol-3-yl)-substituted-3,5-dihydroxyhept-6(E)-enoic (-heptanoic) Acids. *J. Med. Chem.*, **33**, 61–70.

115. Chan, C., Bailey, E.J., Hartley, C.D., Hayman, D.F., Hutson, J.L., Inglis, G.G., Ross, B.C. and Watson, N.S. (1993) Inhibitors of cholesterol biosynthesis. 2. 3,5-Dihydroxy-7-(N-pyrrolyl)-6-heptenoates, a novel series of HMG-CoA reductase inhibitors. *J. Med. Chem.*, **36**, 3658–3662.

116. Procopiou, P.A., Draper, C.D., Hutson, J.L., Inglis, G.G.A., Jones, P.S., Keeling, S.E., Kirk, B.E., Lamont, R.B., Lester, M.G., Pritchard, J.M., Ross, B.C., Scicinski, J.J., Spooner, S.J., Smith, G., Steeples, I.P. and Watson, N.S. (1993) Inhibitors of cholesterol biosynthesis. 1. 3,5-Dihydroxy-7-(N-imidazolyl)-6-heptenoates and heptanoates, a novel series of HMG-CoA reductase inhibitors. *J. Med. Chem.*, **36**, 3646–3657.

117. Sliskovic, D.R., Roth, B.D., Wilson, M.W., Hoefle, M.L. and Newton, R.S. (1990) Inhibitors of cholesterol biosynthesis. 2. 1,3,5-Trisubstituted [2-(tetrahydro-4-hydroxy-2-oxopyran-6-yl)ethyl]pyrazoles. *J. Med. Chem.*, **33**, 31–38.

118. Beck, G., Kesseler, K., Baader, E., Bartmann, W., Bergmann, A., Granzer, E., Jendralla, H., Kerekjarto, B.V., Krause, R., Paulus, E., Schubert, W. and Wess, G. (1990) Synthesis and biological activity of new HMG-CoA reductase inhibitors. 1. Lactones of pyridine- and pyrimidine-substituted 3,5-dihydroxy-6-heptenoic (-heptanoic) Acids. *J. Med. Chem.*, **33**, 52–60.

119. Sliskovic, D.R., Picard, J.A., Roark, W.H., Roth, B.D., Ferguson, E., Krause, B.R., Newton, R.S., Sekerke, C. and Shaw, M.K. (1991) Inhibitors of cholesterol biosynthesis. 4. trans-6-[2-(substituted-quinolinyl)ethenyl/ethyl] tetrahydro-4-hydroxy-2H-pyran-2-ones, a novel series of HMG-CoA reductase inhibitors. *J. Med. Chem.*, **34**, 367–373.

120. Amin, D., Gustafson, S.K., Weinacht, J.M., Cornell, S.A., Neuenschwander, K., Kosimider, B., Scotese, A.C., Regan, J.R. and Perrone, M.H. (1993) RG 12561 (Dalvastatin): A novel synthetic inhibitor of HMG-CoA reductase and cholesterol-lowering agent. *Pharmacology*, **46**, 13–22.

121. Jendralla, H., Granzer, E., Kerekjarto, B.V., Krause, R.U., Schacht, U., Baader, E., Bartmann, W., Beck, G., Bergmann, A., Kesseler, K., Wess, G., Chen, L.-J., Granata, S., Herchen, J., Kleine, H., Schussler, H. and Wagner, K. (1991) Synthesis and biological activity of new HMG-CoA reductase inhibitors. 3. Lactones of 6-phenoxy-3,5-dihydroxyhexanoic acids. *J. Med. Chem.*, **34**, 2962–2983.

122. Connolly, P.J., Westin, C.D., Loughney, D.A. and Minor, L.K. (1993) HMG-CoA reductase inhibitors: design, synthesis, and biological activity of tetrahydroindazole-substituted 3,5-dihydroxy-6-heptenoic acid sodium salts. *J. Med. Chem.*, **36**, 3674–3685.

123. Stokker, G.E. and Pitzenberger, S.M. (1987) Synthesis and characterization of a novel 6-heteroaryl-3,6-dihydro-2H-pyran-2-acetic acid. *Heterocycles*, **26**, 157–162.

124. Louis-Flamberg, P., Peishoff, C.E., Bryan, D.L., Elliott, J.D., Metcalf, B.W. and Mayer, R.J. (1990) Slow binding inhibitors of 3-hydroxy-3-methylglutaryl-coenzyme A reductase. *Biochemistry*, **29**, 4115–4120.

125. Baader, E., Bartmann, W., Beck, G., Below, P., Bergmann, A., Jendrella, H., Kesseler, K. and Wess, G. (1989) Synthesis of a novel HMG-CoA reductase inhibitor. *Tetrahedron Lett.*, **30**, 5115–5118.

126. Baader, E., Bartmann, W., Beck, G., Bergmann, A., Jendrella, H., Kesseler, K., Wess, G., Schubert, W., Granzer, E., Kerekjarto, B.V. and Krause, R. (1988) Enantioselective synthesis of a new fluoro-substituted HMG-CoA reductase inhibitor. *Tetrahedron Lett.*, **29**, 929–930.

127. Balasubramanian, N., Brown, P.J., Catt, J.D., Han, W.T., Parker, R.A., Sit, S.Y. and Wright, J.J. (1989) A potent, tissue-selective inhibitor of HMG-CoA reductase. *J. Med. Chem.*, **32**, 2038–2041.

128. Sit, S.Y., Parker, R.A., Motoc, I., Han, W., Balasubramanian, N., Catt, J.D., Brown, P.J., Harte, W.E., Thompson, M.D. and Wright, J.J. (1990) Synthesis, biological profile, and quantitative structure-activity relationship of a series of novel 3-hydroxy-3-methylglutaryl coenzyme A reductase inhibitors. *J. Med. Chem.*, **33**, 2982–2999.

129. Harwood, H.J., Jr., Silva, M., Chandler, D.E., Mikolay, L., Pellarin, L.D., Barbacci-Tobin, E., Wint, L.T. and McCarthy, P.A. (1990) Efficacy, tissue distribution and biliary excretion of methyl (3R*, 5R*)-(E)-3,5-dihydroxy-9,9-diphenyl-6,8-nonadienoate (CP-83101), a hepatoselective inhibitor of HMG-CoA reductase activity in the rat. *Biochem. Pharmacol.*, **40**, 1281–1293.

130. Balasubramanian, N., Brown, P.J., Parker, R.A., Sit, S.Y. and Wright, J.J. (1992) HMG-CoA reductase inhibitors 4. Tetrazole series: conformational constraints and structural requirements at the hydrophobic domain. *BioMed. Chem. Lett.*, **2**, 99–104.

131. Parker, R.A., Clark, R.W., Sit, S.Y., Lanier, T.L., Grosso, R.A. and Wright, J.J.K. (1990) Selective inhibition of cholesterol synthesis in liver versus extrahepatic tissues by HMG-CoA reductase inhibitors. *J. Lipid Res.*, **31**, 1271–1282.

132. Roth, B.D., Bocan, T.M.A., Blankley, C.J., Chucholowski, A.W., Creger, P.L., Creswell, M.W., Ferguson, E., Newton, R.S., O'Brien, P., Picard, J.A., Roark, W.H., Sekerke, C.S., Sliskovic, D.R. and Wilson, M.W. (1991) Relationship between tissue selectivity and lipophilicity for inhibitors of HMG-CoA reductase. *J. Med. Chem.*, **34**, 463–466.

133. Bocan, T.M.A., Ferguson, E., McNally, W., Uhlendorf, P.D., Bak Mueller, S., DeHart, P., Sliskovic, D.R., Roth, B.D., Krause, B.R. and Newton, R.S. (1992) Hepatic and nonhepatic sterol synthesis and tissue distribution following administration of a liver selective HMG-CoA reductase inhibitor, CI-981: comparison with selected HMG-CoA reducase inhibitors. *Biochim. Biophys. Acta*, **1123**, 133–144.

134. Ziegler, K. and Stunkel, W. (1992) Tissue-selective action of pravastatin due to hepatocellular uptake by a sodium-dependent bile acid transporter. *Biochim. Biophys. Acta*, **1139**, 203–209.

135. Karanewsky, D.S., Badia, M.C., Ciosek, C.P., Jr., Robl, J.A., Sofia, M.J., Simpkins, L.M., DeLange, B., Harrity, T.W., Biller, S.A. and Gordon, E.M. (1990) Phosphorus-containing inhibitors of HMG-CoA reductase. 1. 4-[(2-Arylethyl)hydroxyphosphinyl]-3-hydroxy-butanoic acids: a new class of cell-selective inhibitors of cholesterol biosynthesis. *J. Med. Chem.*, **33**, 2952–2956.

136. Prugh, J.D., Alberts, A.W., Gilfillan, J.L. Huff, J.W. and Wiggins, J.M. (1987) The preparation and HMG-Co-A reductase inhibitory activities of methyl 7-aryl-2-hydroxy-5-oxo-5 thiaheptanone and related 5-thiaheptanoates. *Abstr. Pap. Am. Chem. Soc.*, MEDI 120, Denver, CO.

137. Ashton, M.J., Hills, S.J., Newton, C.G., Taylor, J.B. and Tondu, S.C.D. (1989) Synthesis of 6-aryl-4-hydroxypiperidin-2-ones and a possible application to the synthesis of a novel HMG-CoA reductase inhibitor. *Heterocycles*, **28**, 1015–1035.

138. Baran, J.S., Laos, I., Langford, D.D., Miller, J.E., Jett, C., Taite, B. and Rohrbacher, E. (1985) 3-Alkyl-3-hydroxyglutaric acids: a new class of hypocholesterolemic HMG-CoA reductase inhibitors. *J. Med. Chem.*, **28**, 597–601.

139. Nakamura, C.E. and Abeles, R.H. (1985) Mode of interaction of β-hydroxy-β-methylglutaryl coenzyme A reductase with strong binding inhibitors: compactin and related compounds. *Biochemistry*, **24**, 1364–1376.

140. Yuan, J., Tsai, M.Y., Hegland, J. and Hunninghake, D.B. (1991) Effects of fluvastatin (XU 62-320), an HMG-CoA reductase inhibitor, on the distribution and compostion of low density lipoprotein subspecies in humans. *Atherosclerosis*, **87**, 147–157.

141. Levy, D., Eff, J., Ferguson, D., Rom, D. and Saunders, M. (1991) Dalvastatin for two weeks provides dose-related reductions in cholesterol. *Abs. 9th Int. Symp. on Atherosclerosis*, Rosemont, IL, Oct. 6–11, 1991, p. 36.

142. Cilla, D.D., Posvar, E.L. and Sedman, A.J. (1992) Multiple-dose tolerance and pharmacologic effect of CI-981. *J. Clin. Pharm.*, **32**, 749.

143. Cilla, D.D., Posvar, E.L. and Sedman, A.J. (1992) CI-981 lipid response as a function of dosing time. *J. Clin. Pharm.*, **32**, 749.

144. Mazzu, A.L., Lettieri, J., Kaiser, L. and Frey, R. (1993) Ascending multiple dose safety, tolerability and pharmacodynamics of rivastatin. *Clin. Pharm. Ther.*, **54**, 230.

145. Ritter, W., Frey, R., Krol, G. and Kuhlmann, J. (1993) Rivastatin single dose pharmacokinetics. *Clin. Pharm. Ther.*, **54**, 210.

146. Nguyen, T.-G., Aigner, H. and Eggerer, H. (1981) (3R,S)-3-Hydroxy-3-methyl-4-carboxybutyl-CoA, a specific inhibitor of 3-hydroxy-3-methylglutaryl-CoA reductase. *FEBS Lett.*, **128**, 145–148.

147. Nguyen, T.-G., Gerbing, K. and Eggerer, H. (1984) New substrates and inhibitors of 3-hydroxy-3-methylglutaryl-CoA reductase. *Z. Physiol. Chem.*, **365**, 1–8.

148. Fischer, G.C., Turakhia, R.H. and Morrow, C.J. (1985) Irreducible analogues of mevaldic acid coenzyme a hemithioacetal as potential inhibitors of HMG-CoA reductase. Synthesis of a carbon-sulfur interchanged analogue of mevaldic acid pantetheine hemithioacetal. *J. Org. Chem.*, **50**, 2011–2019.

149. Turakhia, R.H., Fischer, G.C., Morrow, C.J., Maschhoff, B.L., Toubbeh, M.I. and Zbur-Wilson, J.L. (1986) Irreducible analogues of mevaldic acid coenzyme a hemithioacetal as potential inhibitors of HMG-CoA reductase. 2. Synthesis of a secondary amide analogue of mevaldic acid pantetheine hemithioacetal and an amide analogue of 3-hydroxy-3-methylglutaryl-S-pantetheine. *J. Org. Chem.*, **51**, 1955–1960.

150. Barth, M., Bellamy, F.D., Renaut, P., Samreth, S. and Schuber, F. (1990) Towards a new type of HMG-CoA reductase inhibitor. *Tetrahedron*, **46**, 6731–6740.
151. Boquel, P., Taillefumier, C., Chapleur, Y., Renaut, P., Samreth, S. and Bellamy, F.D. (1993) Towards a new type of HMG-CoA reductase inhibitors: part II: dramatic substituents effects in the C-5 epimerization of carbohydrate derivatives. *Tetrahedron*, **49**, 83–96.
152. Path, A.D., Chan, J.A., Lois-Flamberg, P., Mayer, R.J. and Westley, J.W. (1989) Novel acetylenic acids from the root bark of *Paramacrolobium Caeruleum*: inhibitors of 3-hydroxy-3-methylglutaryl coenzyme A reductase. *J. Nat. Prod.*, **52**, 153–161.
153. Clerc, T., Jomier, M., Chautan, M., Portugal, H., Senft, M., Pauli, A.-M., Laruelle, C., Morel, O., LaFont, H. and Chanussot, F. (1993) Mechanisms of action in the liver of crilvastatin, a new hydroxymethylglutaryl-coenzyme A reductase inhibitor. *Eur. J. Pharmacol.*, **235**, 59–68.
154. Brown, M.S. and Goldstein, J.L. (1974) Suppression of 3-hydroxy-3-methylglutaryl coenzyme A reductase activity and inhibition of growth of human fibroblasts by 7-ketocholesterol. *J. Biol. Chem.*, **249**, 7306–7314.
155. Kandutsch, A.A., Heiniger, H.-J. and Chen, H.W. (1977) Effects of 25-hydroxycholesterol and 7-ketocholesterol, inhibitors of sterol synthesis, administration orally to mice. *Biochim. Biophys. Acta*, **486**, 260–272.
156. Kandutsch, A.A., Chen, H.W. and Heiniger, H.-J. (1978) Biological activity of some oxygenated sterols. *Science*, **201**, 498–501.
157. Brown, M.S. and Goldstein, J.L. (1988) The LDL receptor concept: clinical and therapeutic implications. *Atherosclerosis Rev.*, **18**, 85–121.
158. Taylor, F.R., Saucier, S.E., Shown, E.P., Parish, E.J. and Kandutsch, A.A. (1984) Correlation between oxysterol binding to a cytosolic binding protein and potency in the repression of hydroxymethylglutaryl coenzyme A reductase. *J. Biol. Chem.*, **259**, 12382–12387.
159. Dawson, P.A., Ridgeway, N.D., Slaughter, C.A., Brown, M.S. and Goldstein, J.L. (1989) cDNA cloning and expression of oxysterol-binding protein, an oligomer with a potential leucine zipper. *J. Biol. Chem.*, **264**, 16798–16803.
160. (a) Schroepfer, G.J., Jr., Parrish, E.J., Chen, H.W. and Kandutsch, A.A. (1977) Inhibition of sterol synthesis in L cells and mouse liver cells by 15-oxygenated sterols. *J. Biol. Chem.*, **252**, 8975–8980. (b) Erickson, K.A. and Nes, W.R. (1982) Inhibition of hepatic cholesterol synthesis in mice by sterols with shortened and stereochemically varied side chains. *Proc. Nat. Acad. Sci. USA*, **79**, 4873–4877. (c) Kim, H.-S., Wilson, W.K., Needleman, D.H., Pinkerton, F.D., Wilson, D.K., Quiocho, F.A. and Schroepfer, G.J., Jr. (1989) Inhibitors of sterol synthesis. Chemical synthesis, structure, and biological activities of (25R)-3β, 26-dihydroxy-5α-cholest-8(14)en-15-one, a metabolite of 3β-hydroxy-5α-cholest-8(14)en-15-one. *J. Lipid Res.*, **30**, 247–261.
161. Faust, J.R., Luskey, K.L., Chin, D.J., Goldstein, J.L. and Brown, M.S. (1982) Regulation of synthesis and degradation of 3-hydroxy-3-methylglutaryl-coenzyme A reductase by low density lipoprotein and 25-hydroxycholesterol in UT-1 cells. *Proc. Nat. Acad. Sci. USA*, **79**, 5205–5209.
162. Taylor, F.R. (1992) Correlation among oxysterol potencies in the regulation of the degradation of 3-hydroxy-3-methylglutaryl CoA reductase, the repression of 3-hydroxy-3-methylglutaryl CoA synthase and affinities for the oxysterol receptor. *Biochem. Biophys. Res. Comm.*, **186**, 182–189.
163. Boogaard, A., Griffioen, M. and Cohen, L.H. (1987) Regulation of 3-hydroxy-3-methylglutaryl-coenzyme A reductase in human hepatoma cell line, Hep G2. *Biochem. J.*, **241**, 345–351.
164. Qureshi, A.A., Burger, W.C., Peterson, D.M. and Elson, C.E. (1986) The structure of an inhibitor of cholesterol biosynthesis isolated from barley. *J. Biol. Chem.*, **261**, 10544–10550.
165. Pearce, B.C., Parker, R.A., Deason, M.E., Qureshi, A.A. and Wright, J.J.K. (1992) Hypocholesterolemic activity of synthetic and natural tocotrienols. *J. Med. Chem.*, **35**, 3595–3606.
166. Parker, R.A., Pearce, B.C., Clark, R.W., Gordon, D.A. and Wright, J.J.K. (1993) Tocotrienols regulate cholesterol production in mammalian cells by post-translational suppression of 3-hydroxy-3-methylglutaryl-coenzyme A reductase. *J. Biol. Chem.*, **268**, 11230–11238.
167. Rilling, H.C. and Chayet, L.T. (1985) Biosynthesis of cholesterol. In *Sterols and Bile Acids*. Danielson, H. and Sjovall, J. (eds.), pp. 1–72. Amsterdam: Elsevier.
168. Popjak, G., Holloway, P.W., Bond, R.P.M. and Roberts, M. (1969) Analogues of geranyl pyrophosphates as inhibitors of prenyl transferase. *Biochem. J.*, **111**, 333–343.

169. Popjak, G., Parker, T.S., Sarin, V., Tropp, B.E. and Engel, R. (1978) Inhibition of 5-phosphomevalonate kinase by an isosteric anlogue of 5-phosphomevalonate. *J. Amer. Chem. Soc.*, **100**, 8014–8016.

170. Engel, R.R., Sarin, V.K., Gotlinsky, B., Tropp, B.E. and Parker, T.S. (1981) *In vivo* inhibitors of cholesterol biosynthesis. *US Patent 4,279,898.*

171. Lindberg, M., Yvan, C., deWaard, H. and Bloch, K. (1962) On the mechanism of formation of isopentenyl pyrophosphates. *Biochemistry*, **1**, 182–188.

172. Nave, J.-F., d'Orchymont, H., Ducep, J.-B., Piriou, F. and Jung, M.J. (1985) Mechanism of the inhibition of cholesterol biosynthesis by 6-fluoromevalonate. *Biochem. J.*, **227**, 247–254.

173. Reardon, J.E. and Abeles, R.H. (1987) Inhibition of cholesterol biosynthesis by fluorinated mevalonate analogs. *Biochemistry*, **26**, 9717–9722.

174. Sinensky, M. and Lutz, R.J. (1992) The prenylation of proteins. *Bioessays*, **14**, 25–31.

175. Glomset, J., Gelb, M. and Farnsworth, C. (1991) The prenylation of proteins. *Curr. Op. Lipid.*, **2**, 118–124.

176. Stremler, K.E. and Poulter, C.D. (1987) Methane- and difluoromethanediphosphonate analogs of geranyl diphosphate: Hydrolysis inert alternate substrates. *J. Amer. Chem. Soc.*, **109**, 5542–5544.

177. McClard, R.W., Fujita, T.S., Stremler, K.E. and Poulter, C.D. (1987) Novel phophonylphophinyl (P-C-P-C-) analogs of biochemically interesting diphosphates. Synthesis and properties of P-C-P-C-analogs of isopentenyl diphosphate and dimethylallyl diphosphate. *J. Amer. Chem. Soc.*, **109**, 5544–5545.

178. Kohl, N.E., Mosser, S.D., deSolms, S.J., *et al.* (1993) Selective inhibition of *ras*-dependent transformation by a farnesyl transferase inhibitor. *Science*, **260**, 1934–1942.

179. Popjak, G. (1960) Inhibition of cholesterol biosynthesis by farnesoic acid and its analogues. *Lancet*, **1**, 1270–1273.

180. Ogura, K., Koyama, T. and Seto, S. (1972) Enzymic formation of squalene homologs from farnesyl pyrophosphate analogues. *J. Amer. Chem. Soc.*, **94**, 307–309.

181. Poulter, C.D. and Rilling, H.C. (1981) *Biosynthesis of isoprenoid Compounds*, **Vol. 1**, pp. 413–441, Porter, J.W. and Spurgeon, S.L. (eds.) New York: John Wiley.

182. Ortiz de Montellano, P.R., Wei, J.S., Castillo, R., Hsu, C.K. and Boparai, A. (1977) Inhibition of squalene synthase by farnesyl pyrophosphate analogues. *J. Med. Chem.*, **20**, 243–249.

183. Corey, E.J. and Volante, R.P. (1976) Application of unreactive analogs of terpenoid pyrophosphates to studies of multistep biosynthesis. Demonstration that "presqualene pyrophosphate" is an essential intermediate on the path to squalene. *J. Amer. Chem. Soc.*, **98**, 1291–1293.

184. Biller, S.A., Forster, C., Gordon, E.M., Harrity, T., Scott, W.A. and Ciosek, C.P. (1988) Isoprenoid (phosphinylmethyl) phosphonates as inhibitors of squalene synthase. *J. Med. Chem.*, **31**, 1869–1871.

185. Biller, S.A., Forster, C., Gordon, E.M., *et al.* (1991) Isoprenyl phosphinylformates: New inhibitors of squalene synthase. *J. Med. Chem.*, **34**, 1912–1914.

186. Oberg, B. (1989) Antiviral effects of phosphonoformate (PFA, Foscarnet sodium). *Pharmacol. Ther.*, **40**, 213–285.

187. Biller, S.A., Sofia, M.J., DeLange, B., *et al.* (1991) The first potent inhibitor of squalene synthase: a profound contribution of an ether oxygen to inhibitor-enzyme interaction. *J. Amer. Chem. Soc.*, **113**, 8522–8524.

188. Fleisch, H. (1983) Bisphosphonates: Mechanisms of action and clinical applications. *Bone Miner. Res.*, **1**, 319–357.

189. Amin, D., Cornell, S.A. and Gustafson, J.K. (1992) Bisphosphonates used for the treatment of bone disorders inhibited squalene synthase and cholesterol biosynthesis. *J. Lipid. Res.*, **33**, 1657–1663.

190. Ciosek, C.P., Magnin, D.R., Harrity, T.W., *et al.* (1993) Lipophilic 1,1-bisphosphonates are squalene synthase inhibitors and orally active cholesterol lowering agents. *J. Biol. Chem.*, **268**, 24832–24837.

191. Rilling, H.C., Poulter, C.D., Epstein, W.W. and Larsen, B.R. (1971) Studies on the mechanism of squalene biosynthesis, presqualene pyrophosphate sterochemistry and a mechanism for its conversion to squalene. *J. Amer. Chem. Soc.*, **93**, 1783–1785.

192. Poulter, C.D., Marsh, L.L., Hughes, J.M., *et al.* (1977) Model studies of the biosynthesis of non-head-to-tail terpenes. Rearrangements of the chrysanthemyl system. *J. Amer. Chem. Soc.*, **99**, 3816–3823.

193. Poulter, C.D., Capson, T.L., Thompson, M.D. and Bard, R.S. (1989) Squalene synthetase. Inhibition by ammonium analogues of carbocationic intermediates in the conversion of presqualene diphosphate to squalene. *J. Amer. Chem. Soc.*, **111**, 3734–3739.

194. (a) Beytia, E., Qureshi, A.A. and Porter, J.W. (1973) Squalene synthetase. *J. Biol. Chem.*, **248**, 1856–1867. (b) Dugan, R.E. and Porter, J.W. (1972) Hog liver squalene synthetase. The partial purification of the particulate enzyme and kinetic analysis of the reaction. *Arch. Biochem. Biophys.*, **152**, 28–35.

195. Ortiz de Montellano, P.R. and Castillo, R. (1976). Prenyl substituted cyclobutanene as squalene synthetase inhibitors. *Tetrahedron Lett.*, **17**, 4115–4118.

196. Ochlschlager, A.C., Singh, S.M. and Sharma, S. (1991) Squalene synthetase inhibitors: Synthesis of sulfonium ion mimics of the carbocationic intermediates. *J. Org. Chem.*, **56**, 3856–3861.

197. Prashad, M., Kathawala, F.G. and Scallen, T. (1993) N-(arylalkyl)farnesylamines: New potent squalene synthetase inhibitors. *J. Med. Chem.*, **36**, 1501–1504.

198. Prashad, M. (1993) Synthesis and squalene synthetase inhibitory activity of tripotassium 1-methyl-1-[(N-benzyl-N-farnesyl) aminoethyl phosphinate] ethyl phosphanate as a tethered analog of N-benzyl-N-farnesylamine-inorganic pyrophosphate ion pair. *Bioorg. Med. Chem. Lett.*, **3**, 2051–2054.

199. Bertolino, A., Altman, L.-J., Vasak, J. and Rilling, H.C. (1978) Polyisoprenoid amphiphilic, compounds as inhibitors of squalene synthesis and other microsomal enzymes. *Biochem. Biophys. Acta*, **530**, 17–23.

200. Swain, C.J., Baker, R., Kneen, C., *et al.* (1991) Novel 5-HT$_3$ antagonists. Indole oxadiazoles. *J. Med. Chem.*, **34**, 140–151.

201. Alberts, A.W., Berger, G.D. and Bergstrom, J.D. (1992) Squalene synthetase inhibitors. *US Patent 5,135,935.*

202. Brown, G.R. and Mallion, K.B. (1993) Quinuclidine derivatives as squalene synthetase inhibitors. *WO 9313096.*

203. Neuenschwander, K., Amin, D., Scotese, A.C. and Morris, R.L. (1992) Multicyclic teritiary amine polyaromatic squalene synthetase inhibitors. *WO 9215579.*

204. Dawson, M.J., Farthing, J.E., Marshall, P.S., *et al.* (1992) The squalestatins, novel inhibitors of squalene synthase produced by a species of *Phoma*. I. Taxonomy, fermentation, isolation, physico-chemical properties and biological activity. *J. Antibiot.*, **45**, 639–647.

205. Sidebottom, P.J., Highcock, R.M., Lane, S.J., Procopiou, P.A. and Watson, N.S. (1992) The squalestatins, novel inhibitors of squalene synthase produced by a spieces of *Phoma*. II. Structure elucidation. *J. Antibiot.*, **45**, 648–658.

206. Baxter, A., Fitzgerald, B.J., Hutson, J.L., *et al.* (1992) Squalestatin 1, a potent inhibitor of squalene synthase which lowers serum cholesterol *in vivo. J. Biol. Chem.*, **267**, 11705–11708.

207. Bergstrom, J.D., Kurtz, M.M., Rew, D.J., *et al.* (1993) Zaragozic acids: A family of fungal metabolites that are picomolar competitive inhibitors of squalene synthase. *Proc. Natl. Acad. Sci.*, **90**, 80–84.

208. Hensens, O.D., Dufresne, C., Liesch, J.M., Zink, D.L., Reamer, R.A. and VanMiddlesworth, I. (1993) The zaragozic acids: Structure elucidation of a new class of squalene synthase inhibitors. *Tetrahedron Lett.*, **34**, 399–402.

209. Dufresne, C., Wilson, K.E., Zink, D., *et al.* (1992) The isolation and structure elucidation of zaragozic acid, a novel potent squalene synthase inhibitor. *Tetrahedron*, **48**, 10221–10226.

210. Wilson, K.E., Back, R.M., Biftu, T., Ball, R.G. and Hoogsteen, K. (1992) Zaragozic acid, a potent inhibitor of squalene synthase: Initial chemistry and absolute stereochemistry. *J. Org. Chem.*, **57**, 7151–7158.

211. Dufrense, C., Wilson, K.E., Sing, S.B., *et al.* (1993) Zaragozic acids D and D$_2$: Potent inhibitors of squalene synthase and of *ras* farnesyl-protein transferase. *J. Natl. Prod.*, **56**, 1923–1929.

212. Chiang, Y.-C.P., Biftu, T., Doss, G.A., *et al.* (1993) Diesters of zaragozic acid A: Synthesis and biological activity. *Bioorg. Med. Chem. Lett.*, **3**, 2029–2034.

213. Petranyi, G, Ryder, N.S. and Stutz, A. (1984) Allylamine derivatives: New class of synthetic antifungal agents inhibiting fungal squalene epoxidase. *Science*, **224**, 1239–1241.

214. Ryder, N.S. (1991) Squalene epoxidase as a target for the allylamines. *Biochem. Soc. Trans.*, **19**, 774–777.

215. Horie, M., Tsuchiya, Y., Hayashi, M., *et al.* (1990) NB-598: A potent competitive inhibitor of squalene epoxidase. *J. Biol. Chem.*, **265**, 18075–18078.

216. Horie, M., Sawasaki, Y., Fukuzumi, H., *et al.* (1991) Hypolipidemic effects of NB-598 in dogs. *Atherosclerosis*, **88**, 183–192.

217. Iwasawa, Y. and Horie, M. (1993) Mammalian squalene epoxidase inhibitors and structure-activity relationships. *Drugs of the Future*, **18**, 911–918.

218. Ryder, N.S., Dupont, M.-C. and Frank, I. (1986) Inhibition of fungal and mammalian sterol biosynthesis by 2-aza-213-dihydrosqualene. *FEBS Lett.*, **204**, 239–242.

219. Sen, S.E. and Prestwich, G.D. (1989) Trinorsqualene alcohol, a potent inhibitor of vertebrate squalene epoxidase. *J. Amer. Chem. Soc.*, **111**, 1508–1510.

220. Bai, M. and Preswtich, G.D. (1992) Inhibition and activation of porcine squalene epoxidase. *Arch. Biochem. Biophys.*, **293**, 305–313.

221. Sen, S.E., Wawrzenczyk, C. and Prestwich, D. (1990) Inhibition of vertebrate squalene epoxidase by extended and truncated analogs of trinorsqualene alcohol. *J. Med. Chem.*, **33**, 1698–1701.

222. Sen, S.E. and Prestwich, G.D. (1984) Squalene analogs containing isopropylidene mimics as potential inhibitors of pig liver squalene epoxidase and oxidosqualene cyclase. *J. Med. Chem.*, **32**, 2152–2158.

223. Cattel, L., Ceruti, M., Balliano, G., Viola, F., Grosa, G. and Schuber, F. (1989) Drug design based on biosynthetic studies. Synthesis, biological activity, and kinetics of new inhibitors of 2,3-oxidosqualene-cyclase and squalene epoxidase. *Steroids*, **53**, 363–391.

224. Sen, S.E. and Prestwich, G.D. (1989) Trinorsqualene cyclopropylamine: A reversible, tight-binding inhibitor of squalene epoxidase. *J. Amer. Chem. Soc.*, **111**, 8761–8762.

225. VanSickle, W.A., Angelastro, M.R., Wilson, P., Cooper, J.R., Marquart, A. and Flanagan, M.A. (1992) Inhibition of cholesterol synthesis by cyclopropylamine derivatives of squalene in human heptoblastoma cells in culture. *Lipids*, **27**, 157–160.

226. Moore, W.R., Schatzman, G.L., Jarvi, E.T., Gross, R.S. and McCarthy, J.R. (1992) Terminal difluoro olefin analogues of squalene are time-dependent inhibitors of squalene epoxidase. *J. Amer. Chem. Soc.*, **114**, 360–361.

227. Anstead, G.M., Lin, H.-K. and Prestwich, G.D. (1993) Trinorsqualene methylhydroxylamine: A potent dual inhibitor of mammalian squalene epoxidase and oxidosqualene cyclase. *Bioorg. Med. Chem. Lett.*, **3**, 1319–1322.

228. (a) Corey, E.J., Russey, W. and Ortiz de Montellano, P.R. (1966) 2,3-oxidosqualene. An intermediate in the biological synthesis of sterols from squalene. *J. Amer. Chem. Soc.*, **88**, 4750–4751. (b) Corey, E.J. and Russey, W.E. (1966) Metabolic fate of 10,11-dihydrosqualene in sterol producing rat liver homogenate. *J. Amer. Chem. Soc.*, **88**, 4751–4752. (c) Van Tamelen, E.E., Willet, J.D., Clayton, R.B. and Lord, K.E. (1966) Enzymic conversion of squalene 2,3-oxide to lanosterol and cholesterol. *J. Amer. Chem. Soc.*, **88**, 4752–4754. (d) Eschenmoser, A., Rizicka, L., Jeger, O. and Arigoni, D. (1955) Eine stereochemische interpretation der biogentischen isoprenregel bei den triterpenen. *Helv. Chim. Acta*, **38**, 1890–1904.

229. Corey, E.J., Virgil, S.C. and Sarsher, S. (1991) New mechanistic and stereochemical insights on the biosynthesis of sterols from 2,3-oxidosqualene. *J. Amer. Chem. Soc.*, **113**, 8171–8172.

230. Corey, E.J., Virgil, S.C., Liu, D.R. and Sarshar, S. (1992) The methyl group at C(10) of 2,3-oxido-squalene is crucial to the correct folding of this substrate in the cyclization rearrangement step of sterol biosynthesis. *J. Amer. Chem. Soc.*, **114**, 1524–1525.

231. Corey, E.J. and Virgil, S.C. (1991) An experimental demonstration of the stereochemistry of enzymic cyclization of the protosterol system, forerunner of lanosterol and cholesterol. *J. Amer. Chem. Soc.*, **113**, 4025–4026.

232. Corey, E.J., Ortiz de Montellano, P.R., Lin, K. and Dean, P.D.G. (1967) 2,3-Iminosqualene, a potent inhibitor of the enzymatic cyclization of 2,3-oxidosqualene to sterols. *J. Amer. Chem. Soc.*, **89**, 2797–2798.

233. Bai, M., Xiao, X.-Y. and Prestwich, G.D. (1992) Epoxidation of 2,3-oxidosqualene to 2,3:22,23-squalene dioxide by squalene epoxidase. *Biochem. Biophys. Res. Commun.*, **185**, 323–329.

234. Taylor, F.R., Kandutsch, A.A., Gayen, A.K., *et al.* (1986) 24,25-Epoxysterol metabolism in cultured mammalian cells and repression of 3-hydroxy-3-methylglutaryl-CoA reductase. *J. Biol. Chem.*, **260**, 15039–15044.

235. Imai, H., Werthessen, N.T., Subramanyam, V., LeQuesne, P.W., Soloway, A.H. and Kanisawa, M. (1980) Angiotoxicity of oxygenated sterols and possible precursors. *Science*, **207**, 651–653.

236. Abad, J.-L., Casas, J., Sanchez-Baeza, F. and Messeguer, A. (1992) 2,3:18,19-Dioxidosqualene: Synthesis and activity as a potent inhibitor of 2,3-oxidosqualene-lanosterol cyclase in rat liver microsomes. *Bioorg. Med. Chem. Lett.*, **2**, 1239–1242.

237. Abad, J.-L., Casa, J., Sanchez-Baeza, F. and Messeguer, A. (1993) Dioxidosqualenes: Characterization and activity as inhibitors of 2,3-oxidosqualene-lanosterol cyclase. *J. Org. Chem.*, **58**, 3991–3997.

238. Xiao, X.-Y. and Prestwich, G.D. (1991) 29-Methylidene-2,3-oxidosqualene: a potent mechanism-based inactivator of oxidosqualene cyclase. *J. Amer. Chem. Soc.*, **113**, 9673–9674.

239. Nelson, J.A., Czarny, M.R., Spencer, T.A., Limanek, J.S., McCrue, K.R. and Chang, T.-Y. (1978) A novel inhibitor of steroid biosynthesis. *J. Amer. Chem. Soc.*, **100**, 4900–4902.

240. Chang, T.-Y., Schiavoni, E.S., McCrae, K.R., Nelson, J.A. and Spencer, T.A. (1979) Inhibition of cholesterol biosynthesis in Chinese hamster ovary cells by 4,4,10-β-trimethyl-trans-decal-3β-ol. *J. Biol. Chem.*, **254**, 11258–11263.

241. Raveendranath, P.C., Newcomb, L.F., Ray, N.C., Clark, D.S. and Spencer, T.A. (1990) 7,7-Disubstituted derivatives of 4,4,10ß-trimethyl-trans-decal-3β-ol(TMD). *Syn. Commun.*, **20**, 2723–2731.

242. Duriatti, A., Bouvier-Nave, P., Benveniste, P., *et al.* (1985) *In vitro* inhibition of animal and higher plants 2,3,-oxidosqualene-sterol cyclases by 2-aza-2,3-dihydrosqualene and derivatives, and by other ammonium-containing molecules. *Biochem. Pharmacol.*, **34**, 2765–2777.

243. Gerst, N., Schuber, F., Viola, F. and Cattel, L. (1986) Inhibition of cholesterol biosynthesis in 3T3 fibroblasts by 2-aza-dihydrosqualene, a rationally designed 2,3-oxidosqualene cyclase inhibitor. *Biochem. Pharmacol.*, **35**, 4243–4250.

244. Ceruti, M., Balliano, G., Viola, F., Cattel, L., Gerst, N. and Schuber, F. (1987) Synthesis and biological activity of azasqualenes, bis-azalqualenes and derivatives. *J. Eur. Med. Chem.*, **22**, 199–208.

245. Ceruti, M., Delprino, L., Cattel, L., *et al.* (1985) N-oxide as a potential function in the design of enzyme inhibitors. Application to 2,3-epoxysqualene-sterol cyclases. *J. Chem. Soc. Chem. Commun.*, 1054–1055.

246. Viola, F., Ceruti, M., Balliano, G., Caputo, O. and Cattell, L. (1990) 22,23-Epoxy-2-aza-2,3-dihydro-squalene derivatives: Potent new inhibitors of squalene 2,3,-oxide-lanosterol cyclase. *Il Farmaco*, **45**, 965–978.

247. Ceruti, M., Viola, F., Balliano, G., *et al.* (1988) Synthesis of a squalenoid oxaziridine and other new classes of squalene derivatives as inhibitors of cholesterol biosynthesis. *Eur. J. Med. Chem.*, **23**, 533–537.

248. Ceruti, M., Viola, F., Dosio, F., *et al.* (1988) Stereospecific synthesis of squalenoid epoxide vinyl ethers as inhibitors of 2,3-oxidosqualene cyclase. *J. Chem. Soc. Chem. Perkin Trans. I*, 461–469.

249. Ceruti, M., Balliano, G., Viola, F., Gross, G., Rocco, F. and Cattel, L. (1992) 2,3-Epoxy-10-aza-10,11-dihydrosqualene, a high energy intermediate analogue inhibitor of 2,3-oxidosqualene cyclase. *J. Med. Chem.*, **35**, 3050–3058.

250. Taton, M., Benveniste, P. and Rahier, A. (1986) N-[(1,5,9)-trimethyldecyl]-4α,10-dimethyl-8-aza-trans-decal-3β-ol, a novel potent inhibitor of 2,3-oxidosqualene cycloartenol and lanosterol cyclases. *Biochem. Biophys. Res. Commun.*, **138**, 764–770.   ,

251. Gerst, N., Duriatti, A., Schuber, F., Taton, M., Benveniste, P. and Rahier, A. (1988) Potent inhibition of cholesterol biosynthesis in 3T3 fibroblasts by N-[(1,5,9)-trimethyldecyl]-4α,10-dimethyl-8-aza-trans-decal-3β-ol, a new 2,3-oxydosqualene cycloartenol and lanosterol cyclase inhibtor. *Biochem. Pharmacol.*, **37**, 1955–1964.

252. Wannamaker, M.W., Waid, P.P., VanSickle, W.A., *et al.* (1992) N-(1-oxododecyl)-4α,10-dimethyl-8-aza-trans-decal-3β-ol: A potent competitive inhibitor of 2,3-oxidosqualene cyclase. *J. Med. Chem.*, **135**, 3581–3583.

253. Ruhl, K.K., Anzalone, L., Arguropoulos, E.D., Gayen, A.K. and Spencer, T.A. (1989) Azadecalin analogs of 4,4,10β-trimethyl-trans-decal-3β-ol: Synthesis and assay as inhibitors of oxidosqualene cyclase. *Bioorg. Chem.*, **17**, 108–120.

254. Dodd, D.S., Oehlschlager, A.C., Geragopapadakou, N.H., Polak, A.-M. and Hartman, P.G. (1992) Synthesis of inhibitors of 2,3-oxidosqualene-lanosterol cyclase. 2. Cyclocondensation of γ,δ-unsaturated β-keto esters with imines. *J. Org. Chem.*, **57**, 7226–7234.

255. Taton, M., Benveniste, P., Rahier, A., Johnson, W.S., Liu, H.-T. and Sudhaker, A.R. (1992) Inhibition of 2,3-oxidosqualene cyclases. *Biochemistry*, **31**, 7892–7898.

256. Wannamaker, M.W., Waid, P.P., Moore, W.R., Schatzman, G.L., VanSickle, W.A. and Wilson, P.K. (1993) Inhibition of 2,3-oxidosqualene cyclase by N-alkyl-piperdines. *BioMed. Chem. Lett.*, **3**, 1175–1178.

257. Barney, C.L., McCarthy, J.R. and Wannamaker, M.W. (1972) Novel piperidylethers and thioethers as inhibitors of cholesterol biosynthesis. *EP 468,434-A.*

258. Atkin, S.D., Morgan, B., Baggaley, K.H. and Green, J. (1972) The isolation of 2,3-oxidosqualene from the liver of rats treated with 1-dodecylimidazole, a novel hypocholesterolemic agent. *Biochem. J.*, **130**, 153–157.

259. Panini, S.R., Sexton, R.C., Gupta, A.K., Parish, E.J. Chitrakorn, S. and Rudney, H. (1986) Regulation of 3-hydroxy-3-methylglutaryl coenzyme A reductase activity and cholesterol biosynthesis by oxylanosterols. *J. Lipid. Res.*, **27**, 1190–1204.

260. Yoshida, Y. and Aoyama, Y. (1991) Sterol 14α-demethylase and its inhibition of yeast. Structural considerations on the interaction of azole antifungal agents with lanosterol 14α-demethylase (P-450$_{14DM}$). *Biochem. Soc. Trans.*, **19**, 778–787.

261. Sonino, N. (1987) The use of ketoconazole as an inhibitor of steroid production. *N. Eng. J. Med.*, **317**, 812–818.

262. Kempen, H.J., VanSon, K., Cohen, L.H., Griffioen, M., Verboom, H. and Havekes, L. (1987) Effect of ketoconzaole on cholesterol synthesis and on HMG-CoA reductase and LDL-receptor activities in Hep-G2 cells. *Biochem. Pharmacol.*, **36**, 1245–1249.

263. Favata, M.F., Trzaskos, J.M., Chen, H.W., Fischer, R.T. and Greenberg, R.S. (1987) Modulation of 3-hydroxy-3-methylglutaryl-coenzyme A reductase by azole antimycotics requires lanosterol demethylation, but not 24,25-epoxy lanosterol formation. *J. Biol. Chem.*, **262**, 12254–12260.

264. Trzaskos, J.M., Favata, M.F., Fisher, R.T. and Stam, S.H. (1987) *In situ* accumulation of 3β-hydroxylanost-8-en-32-aldehyde in hepatocyte cultures. *J. Biol. Chem.*, **262**, 12261–12268.

265. Miettinen, T.A. (1988) Cholesterol metabolism during ketoconzaole treatment in man. *J. Lipid. Res.*, **29**, 43–51.

266. Trzasko, J.M., Magolda, R.L., Favata, M.F., *et al.* (1993) Modulation of 3-hydroxy-3-methylglutaryl-CoA reductase by 15α-flourolanost-7-en-3β-ol. *J. Biol. Chem.*, **268**, 22591–22599.

267. Sonoda, Y., Sekigawa, Y. and Sato, Y. (1988) *In vitro* effects of oxygenated lanosterol derivatives on cholesterol biosynthesis from 24,25-dihydrolanosterol. *Chem. Pharm. Bull.*, **36**, 966–973.

268. Tuck, S.F., Robinson, C.H. and Silverton, J.V. (1991) Assessment of the active site requirements of lanosterol 14α-demethylase: Evaluation of novel substrate analogues as competitive inhibitors. *J. Org. Chem.*, **56**, 1260–1266.

269. Mayer, R.J., Adams, J.L., Bossard, M.J. and Berkhout, T.A. (1991) Effects of a novel lanosterol 14α-demethylase inhibitor on the regulation of 3-hydroxy-3-methylglutaryl-coenzyme A reductase in Hep G2 cells. *J. Biol. Chem.*, **266**, 20070–20078.

270. Frye, L.L., Cusack, K.P. and Leonard, D.A. (1993) 32-Methyl-32-oxylanosterols: Dual action inhibitors of cholesterol biosynthesis. *J. Med. Chem.*, **36**, 410–416.

271. Bossard, M.J., Tomaszek, T.A., Gallagher, T.F., Metcalf, B.W. and Adams, J.L. (1991) Steroid acetylenes: Mechanism-based inactivators of lanosterol 14α-demethylase. *Bioorg. Chem.*, **19**, 418–432.

272. Komoda, Y., Shimizu, M., Sonoda, Y. and Sato, Y. (1989) Ganoderic acid and its deriviatives as cholesterol synthesis inhibitors. *Chem. Pharm. Bull.*, **37**, 531–533.

273. Kusano, G., Takahishi, A., Nozoe, S., Sonoda, Y. and Sato, Y. (1987) Solanum alkaloids as inhibitors of enzymatic conversion of dihydrolanosterol into cholesterol. *Chem. Pharm. Bull.*, **35**, 4321–4323.

274. Walker, K.A.M., Kertesz, D.J., Rotstein, D.M., *et al.* (1993) Selective inhibition of mammalian lanosterol 14α-demethylase: a possible strategy for cholesterol lowering. *J. Med. Chem.*, **36**, 2235–2237.

275. Phillipson, D., Remsburg, B., Kirsch, D. Fisher, S. and Lai, M. (1992) Lanomycin, a natural product inhibitor of lanosterol demethylase. In *Abstr. Pap. Am. Chem. Soc.*, MEDI 155, San Francisco, CA.

276. Aufenanger, J., Pill, J., Schmidt, F.H. and Stegmeier, K. (1986) The effects of BM 15766, an inhibitor of 7-dehydrocholesterol-Δ^7-reductase, on cholesterol biosynthesis in primary rat hepatocytes. *Biochem. Pharmacol.*, **35**, 911–916.

277. Baumgart, E., Stegmeier, K.H., Schmidt, F.H. and Fahimi, H.D. (1987) Proliferation of peroxisomes in pericentral hepatocytes of rat liver after administration of a new hypocholesterolemic agent (BM 15766). *Lab. Invest.*, **56**, 554–564.

278. (a) Dvornik, D., Kraml, M., Dubuc, J., Givner, M. and Gaudry, R. (1963) A novel mode of inhibition of cholesterol biosynthesis. *J. Amer. Chem. Soc.*, **85**, 3309. (b) Niemiro, R. and Fumagalli, R. (1965) Studies on the inhibitory mechanism of some hypocholesterolemic agents on 7-dehydrocholesterol-Δ^7-bond reductase activity. *Biochem. Biophys. Acta*, **98**, 624–631.

279. Grisar, J.M., Claxton, G.P., Stewart, K.T., MacKenzie, R.D. and Kariya, T. (1976) (2-Piperidine)-and (2-pyrrolidine) ethanones and -ethanols as inhibitors of blood platelet aggregation. *J. Med. Chem.*, **19**, 1195–1201.

4. THE DESIGN AND SYNTHESIS OF INHIBITORS OF HIV PROTEINASE

GARETH J. THOMAS

Roche Research Centre, Roche Products Ltd., Broadwater Road, Welwyn Garden City, Hertfordshire AL7 3AY, United Kingdom

The rapid spread and serious consequences of AIDS have led to an unprecedented effort to define the replicative processes of HIV, the causative agent of AIDS, and to identify suitable targets for chemotherapy. One attractive target is an essential virally encoded proteinase. A brief account of the characterisation of HIV proteinase and of various strategies which have been employed in the design of inhibitors is followed by a more detailed account of work carried out in the Roche laboratories. This includes a systematic study of structure-activity relationships among a series of peptide mimetics which incorporate a hydroxyethylamine transition-state isostere, culminating in the synthesis of Ro 31-8959, a highly potent and selective inhibitor of HIV proteinase which exhibits potent anti-HIV activity. As part of its further development towards clinical trials, new synthetic routes which are suitable for large scale synthesis of Ro 31-8959 are described, as well as a more detailed evaluation of its biological properties. Finally, inhibitors of HIV proteinase which have been studied in other laboratories are briefly reviewed.

HIV AND AIDS

In the Spring of 1981 five young men in Los Angeles were found to have *Pneumocystis carinii* pneumonia,[1] and later that year twenty-six individuals in New York and California developed Kaposi's sarcoma, an uncommon type of skin cancer.[2] Before this both conditions had been extremely rare in the USA. The extraordinary occurrence of these diseases in previously healthy young men prompted the Centers for Disease Control to set up a task force to determine the extent of this phenomenon and to identify those at risk. It became evident that the early cases all involved homosexual men, and that another common feature was a severely impaired immune response. This led to the recognition of a new clinical entity known as the acquired immune deficiency syndrome or AIDS. More than five hundred cases of AIDS were reported over the next fifteen months and it became clear that a serious public health problem existed.[3] New cases of AIDS were found not only among homosexual men, but to a lesser extent in intravenous drug abusers, haemophiliacs, and blood transfusion recipients. These observations suggested that AIDS was caused by an infectious agent, and the consistent finding of depleted levels of $CD4^+$ lymphocytes in patients with AIDS suggested that this cell was the target for infection.[4] The known tropism of a retrovirus, human T-cell leukaemia virus type 1 (HTLV-I) for $CD4^+$ cells[5] suggested that the etiological agent of AIDS might also be a retrovirus, and less than three years after AIDS was first described its cause was conclusively shown to be a retrovirus of the *Lentiviridae* family.[6-8] This virus was variously termed lymphadenopathy-associated virus (LAV), human T-cell lymphotropic virus type III (HTLV-III), and AIDS-associated retrovirus (ARV), but is now universally referred to as the human immunodeficiency virus (HIV).[9] A second, genetically distinct virus originating from different geographical areas was later identified,[10–12] and the two types are now known as HIV-1 and HIV-2. Continued surveillance of the AIDS epidemic has led to the identification of increasing numbers of cases in most areas of the world. The World Health Organisation has recently estimated that as of mid-1993, over thirteen million people have become infected with HIV, and that this figure will increase to forty million by the year 2000.[13] To date, over two million HIV-infected adults have developed AIDS, and most of them have died. The rapid spread of AIDS and its extremely serious consequences have, however, led to an unprecedented effort to understand the disease and to develop suitable therapies. Rapid progress in the elucidation of the replicative processes of HIV has facilitated the identification of a number of molecular targets for potential chemotherapeutic intervention in the treatment of AIDS.

REPLICATIVE PROCESSES OF HIV AND POTENTIAL THERAPEUTIC TARGETS

One of the principal challenges in the design of chemotherapeutic agents is that of achieving selectivity. This is particularly problematical in the design of antiviral agents because many processes involved in viral replication are catalysed by enzymes of the host cell. Any agent designed to prevent viral replication by inhibiting these processes would therefore interfere with host cell metabolism, giving rise to potential toxicity. However, it has become clear that many viruses make use of virally encoded proteins as well as cellular proteins in their replication, and these provide an opportunity for the design of inhibitors which are selective for virus-specific targets.

A considerable body of knowledge has accumulated over recent years on the molecular biology of HIV.[14,15] Several virus-specific processes have been identified, each presenting a potential target for chemotherapy.[16–18] Retroviruses are characterised by the fact that following entry of the virus into the host cell and removal of outer viral proteins, the viral genetic material, single stranded RNA, is utilised as a template for assembly of a corresponding molecule of DNA. This is then integrated into the host's chromosomes, and provides the basis for subsequent viral replication. Transcription of viral RNA into DNA is catalysed by a virally encoded enzyme, reverse transcriptase (RT), and it is not therefore surprising that the earliest approaches to anti-HIV agents were based on inhibition of this enzyme. Indeed, RT has been termed "a pivotal target for antiretroviral therapy".[18] 3'-Deoxy-3'-azidothymidine (AZT) which, after conversion by cellular kinases to its triphosphate, is a potent inhibitor of RT, was shown to inhibit replication of HIV,[19] and became the first agent licensed for treatment of HIV infections. Studies of other nucleoside analogues followed,[20–22] and two additional compounds, 2',3'-dideoxycytidine (ddC) and 2',3'-dideoxyinosine (ddI) have recently been licensed as anti-HIV agents. Non-nucleoside derivatives, which act as allosteric inhibitors of RT, have also been extensively studied.[23–25] However, increasing knowledge of the processes involved in the replication of HIV has indicated other targets for therapy. These include blocking of virus binding to the cell membrane,[26–28] inhibition of viral regulatory proteins such as tat,[29] and, in particular, inhibition of a virally encoded proteinase.[30–35]

HIV PROTEINASE AS A TARGET

The genome of HIV is more complex than those of other retroviruses studied so far. It contains an elaborate set of regulatory genes as well as three genes which are common to all retroviruses, *gag*, *pol*, and *env*.[36,37] These three genes are expressed in host cells as polyproteins which are subsequently cleaved to give smaller functional proteins. The *gag* gene is translated as a 55 kD precursor protein which is processed at a late stage in viral replication to give structural proteins of the virus core. These comprise the matrix protein MA[38] (p17), which connects the viral core structure to the outer envelope,[39] the capsid protein CA (p24), which is the major structural protein of the viral core,[40] the nucleocapsid protein NC (p7), which is an RNA binding protein,[41] and a proline rich protein (p6). The *pol* gene is translated only as a 160 kD gag-pol fusion protein resulting from a frame shift between the overlapping *gag* and *pol* reading frames.[42] Processing of gag-pol gives viral enzymes (proteinase, reverse transcriptase, RNase H, and integrase) as well as the gag proteins.[43] The *env* gene encodes for a 160 kD protein which is cleaved to give the viral envelope glycoproteins gp120 and gp41. Processing of the env protein appears to be catalysed by a host cell proteinase, but the gag and gag-pol proteins are totally processed by the proteinase encoded within the *pol* gene.[44] It follows that inhibition of HIV proteinase should profoundly compromise the ability of the virus to replicate. Indeed, even before this enzyme was isolated it was suggested as a potential therapeutic target when its essential role in processing the gag polyprotein was demonstrated.[45] When recombinant HIV gag-pol protein was expressed in yeast cells processing of the gag protein was observed. When, however, a frame shift mutation was made in the proteinase region of *pol*, gag processing was lost, indicating the inability, at least in yeast, of cellular proteinases to take over the role of the viral enzyme. In addition, site-directed mutagenesis which rendered HIV proteinase inactive led to the production of immature, non-infectious virions.[46] These, and similar experiments,[47,48] provided ample evidence of the crucial role of the proteinase in HIV replication, and made selective inhibition of this enzyme an attractive strategy for chemotherapy of AIDS.

ISOLATION AND PURIFICATION OF HIV PROTEINASE

Because isolation of HIV proteinase directly from virus particles gave only small amounts of the pure enzyme,[49] larger quantities for further study have been obtained either by total synthesis, or through recombinant techniques. Syntheses of both HIV-1[50–52] and HIV-2[52,53] proteinases have been reported. Recombinant technology has been used to obtain constructs expressing HIV-1[54–56] or HIV-2[57] proteinase sequences which have been cloned and expressed in yeast and in insect cell cultures.[58] Constructs which contain *gag-pol* fusion sequences have also been produced. Expression of these extended constructs resulted in autoprocessing to give the 11 kD form of HIV proteinase.[55,59–66] A variety of methods have been employed for purification of the enzyme, among which ligand-affinity chromatography appears to be particularly successful.[57,66,67]

CHARACTERISATION OF HIV PROTEINASE

Comparison of the amino acid sequences of a number of retroviral proteinases revealed a highly conserved Asp-Thr-Gly triad which was reminiscent of the catalytic centre of a known group of proteinases, the aspartic proteinases, and it was suggested that HIV proteinase might belong to this class.[68] Site directed mutagenesis of the putative active site Asp to Ala,[44] Asn,[54] or Thr[69] abolished catalytic activity, confirming the mechanistic class of this enzyme. Moreover, HIV proteinase was inhibited by pepstatin, a known inhibitor of aspartic proteinases.[50,54,70–72] Nevertheless, significant differences were noted between HIV proteinase and classical mammalian and fungal aspartic proteinases. These comprise more than 200 amino acids and consist of two homologous domains, each containing one Asp-Thr-Gly motif.[73] The two Asp residues come together to form the catalytic centre. In contrast, HIV proteinase consists of only 99 amino acids, and contains only one Asp-Thr-Gly triad. This led to the suggestion that HIV proteinase functions as a homodimer, each monomer contributing one aspartic acid residue to the active site.[74] Both the dimeric nature of HIV proteinase and its mechanistic class were subsequently confirmed through determination of crystal structures of both the native enzyme and enzyme-inhibitor complexes.

SUBSTRATE SPECIFICITY

HIV proteinase has been shown to cleave the gag and gag-pol polyproteins at nine specific sites.[49,61,75–81] (Figure 1). However, these cleavage sites appear to have little sequence homology. While the cleavage sites are generally hydrophobic, polar and charged residues also occur. Several classification schemes have been proposed in an attempt to understand the specificity of retroviral proteinases.[30,82–84] Henderson *et al.* divided the cleavage sequences from HIV-1, HIV-2, and the simian immuno-deficiency virus (SIV) into three classes.[82] Those assigned to Class 1 contain the rather unusual consensus sequence Ser/Thr-X-Y-Tyr/Phe-Pro in positions P4 to P1' (sites 1, 6, and 7 in Figure 1). Cleavage of amide bonds N-terminal to proline is unusual in mammalian biochemistry, but has been observed for a number of retroviral proteinases.[85] It is not entirely clear how HIV proteinase is able to effect highly specific cleavages in such a diverse range of sequences, but it has been suggested that conformational factors, as well as sequence might play an important role.[53,62]

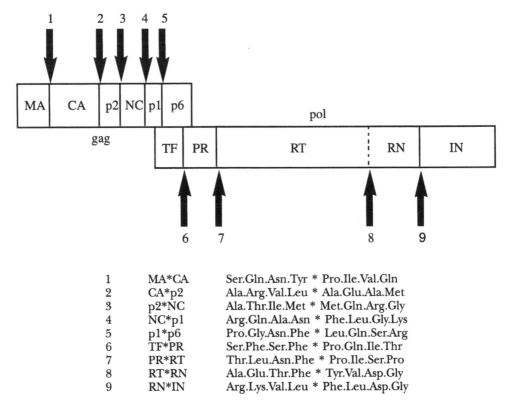

1	MA*CA	Ser.Gln.Asn.Tyr * Pro.Ile.Val.Gln
2	CA*p2	Ala.Arg.Val.Leu * Ala.Glu.Ala.Met
3	p2*NC	Ala.Thr.Ile.Met * Met.Gln.Arg.Gly
4	NC*p1	Arg.Gln.Ala.Asn * Phe.Leu.Gly.Lys
5	p1*p6	Pro.Gly.Asn.Phe * Leu.Gln.Ser.Arg
6	TF*PR	Ser.Phe.Ser.Phe * Pro.Gln.Ile.Thr
7	PR*RT	Thr.Leu.Asn.Phe * Pro.Ile.Ser.Pro
8	RT*RN	Ala.Glu.Thr.Phe * Tyr.Val.Asp.Gly
9	RN*IN	Arg.Lys.Val.Leu * Phe.Leu.Asp.Gly

Figure 1 Cleavage of gag and gag-pol polyproteins by HIV-1 proteinase.[81]
Abbreviations:[38] MA, matrix protein; CA, capsid protein; NC, nucleocapsid protein;
TF, transframe protein; PR, proteinase; RT, reverse transcriptase; RN, RNase H;
IN, integrase.

In an attempt to further understand the substrate specificity of HIV proteinase, and to facilitate the design of inhibitors, numerous synthetic oligopeptide substrates have been studied. Oligopeptides corresponding to most of the cleavage sites found in the *gag* and *gag-pol* gene products are cleaved efficiently by HIV proteinase, but the efficiency of cleavage varies greatly among different sequences.[60,79,80,86,87] For efficient and specific cleavage of the P1-P1' scissile bond a heptapeptide spanning the S4-S3' subsites is required. As expected, the preferred residues in peptide substrates correspond to those found at the polyprotein cleavage sites.

ENZYME ASSAYS

An essential requirement in any programme of inhibitor design is a suitable assay of enzyme activity. Both continuous[87–90] and discontinuous assay systems have been developed for HIV proteinase. In general, discontinuous systems are more sensitive, since assay conditions can be optimised independently of conditions required for enzyme activity. Sensitive, but qualitative methods using gel electrophoresis[91] or immunoblot assays[29,51,55,61,62,92] have been developed for following cleavage of polyprotein substrates. These methods are not, however, ideal for the high throughput required in a drug discovery programme, and assay systems based on the use of synthetic oligopeptide substrates and HPLC analysis of cleavage products have been developed.[66,71,79,93,94]

At Roche we developed an assay based on cleavage of a synthetic peptide substrate containing a Phe-Pro linkage. Because N-terminal proline residues are not often encountered in standard peptides and proteins, such an assay should be suitable for crude enzyme preparations, since extraneous proteins should not interfere. The heptapeptide 1, based on the P5-P2' sequence of the Tyr-Pro cleavage site in the gag polyprotein (site 1 in Figure 1), but in which Tyr is replaced by Phe for synthetic convenience, was chosen as a potential substrate. The N- and C-termini were protected in order to prevent cleavage of the peptide by any exopeptidases present in crude enzyme preparations, the N-terminal succinyl group also enhancing aqueous solubility. This heptapeptide proved to be a satisfactory substrate for HIV-1 proteinase, with a K_m of 0.79 mM.[95] Determination of the N-terminal proline containing cleavage product 3 was based on the formation of a deep blue colour on reaction with isatin. The quantitative reaction of free proline and other cyclic secondary amines with isatin to give resonance structures such as 4, 5 had been known since the late nineteenth century,[96,97] but reaction with peptides containing N-terminal proline requires harsher reaction conditions, and had previously only been applied qualitatively.[98] However, it was found that reaction of 3 with isatin in the presence of an acid catalyst occurs quantitatively and reproducibly under mild conditions to give a deep blue solution which may be assayed spectrophotometrically at a wavelength of 599 nm.[95] This provided a simple and specific assay system which is suitable for a high throughput screening programme, and which has been used routinely in the Roche laboratories to evaluate more than five hundred potential inhibitors of HIV proteinase.[99]

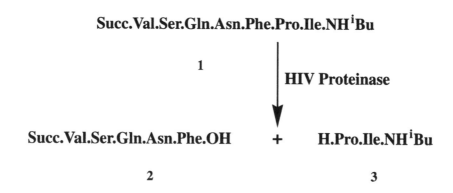

Succ.Val.Ser.Gln.Asn.Phe.Pro.Ile.NH^{i}Bu

1

HIV Proteinase

Succ.Val.Ser.Gln.Asn.Phe.OH   +   H.Pro.Ile.NH^{i}Bu

2   3

Figure 2   Cleavage of protected heptapeptide substrate 1 by HIV proteinase.

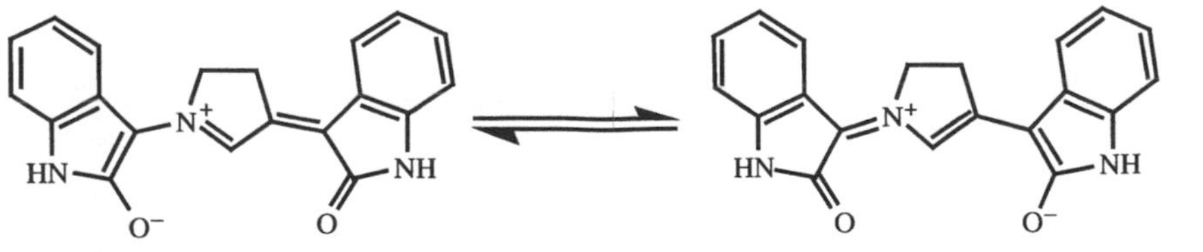

Figure 3   Proposed structure of the blue product formed on reaction of proline with isatin.[97]

STRUCTURE OF HIV PROTEINASE

The crystal structures of both synthetic[100] and recombinant[59,101] HIV proteinases were solved almost simultaneously in a number of laboratories, confirming predictions made previously on the basis of modelling studies using known structures of eukaryotic[74,102] and Rous sarcoma virus[103] proteinases as templates. The enzyme exists as a homodimer whose structure shares many common features with those of the two-domain fungal and mammalian aspartic proteinases. It contains four interdigitated short strands of β-sheet, rather than the six long strands typical of non-viral aspartic proteinases. A glycine rich β-hairpin loop, or flap, projecting from each monomer encloses the catalytic cleft. In the native HIV proteinase these flaps are about 7 Å away from the active site, resulting in a very open conformation. The active site triad Asp^{25}-Thr^{26}-Gly^{27} is located within a loop whose structure is stabilised by a network of hydrogen bonds with the corresponding loop of the other monomer, forming the "fireman's grip" arrangement which is characteristic of aspartic proteinases.[104]

At least 170 crystal structures of HIV proteinase-inhibitor complexes have been determined, although most of these remain unpublished.[105] The published structures include complexes with acetyl pepstatin[106] and inhibitors which incorporate reduced amide,[107] hydroxyethylamine,[108,109] and hydroxyethylene transition state mimetics,[110–114] as well as C_2 symmetric inhibitors designed to take advantage of the symmetry of the dimeric enzyme.[115–117] Despite the diverse nature of these inhibitors, the enzyme structure is well conserved in each complex. The core structure of the proteinase is similar, in each case, to that of the native enzyme, but the two flaps are considerably displaced, their tips moving by as much as 7 Å (Figure 4). The flaps thus form one side of a pocketed hydrophobic tunnel 23 Å long in which the pockets serve as subsites.[115] With the exception of the active site aspartic acid residues, subsites S3 to S3′ comprise mainly hydrophobic residues, and make extensive van der Waal's contacts with the side chains of inhibitors. Inhibitors bind in an extended conformation, and an extensive network of hydrogen bonds is formed with both the floor and flap regions of the enzyme. For inhibitors other than reduced amides, the hydroxyl group of the non-scissile moiety is located between the carboxyl groups of the two catalytic Asp residues, making hydrogen bonds with at least one carboxyl oxygen of each. These features are illustrated for a hydroxyethylamine- HIV-1 proteinase inhibitor complex in Figure 5.

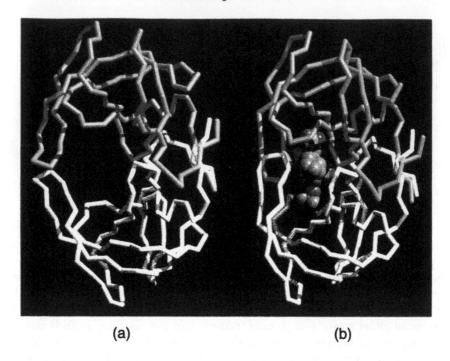

(a) (b)

Figure 4 (a), Native HIV-1 proteinase and (b), HIV-1 proteinase complexed with the inhibitor Ro 31-8959.

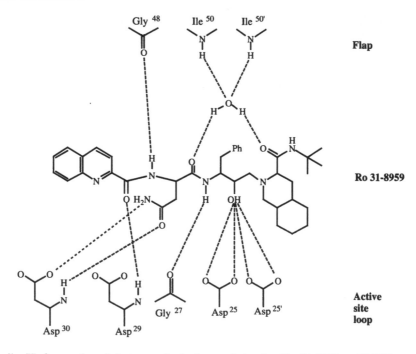

Figure 5 Hydrogen bonds between the hydroxyethylamine Ro 31-8959 and HIV-1 proteinase.

One feature common to almost all the HIV proteinase-inhibitor complexes which is not found in archetypal aspartic acid proteinase-inhibitor structures is a tightly bound water molecule buried deep within the binding site. This forms hydrogen bonds with the P2 and P1′ carboxyl groups of the inhibitor and with the amide hydrogens of Ile[50] and Ile[50′] of the flaps. The presence of such a water molecule in an enzyme-substrate complex would, by bridging the substrate and enzyme flap, exert extra strain on the scissile amide bond and render it more susceptible to hydrolysis.[108]

THE TRANSITION-STATE MIMETIC APPROACH TO INHIBITORS OF HIV PROTEINASE

As a result of a variety of kinetic[54,87,93,94,118,119] and structural[106–110,115] studies the mechanism of action of HIV proteinase is thought to resemble the general acid-general base mechanism which has been proposed for the monomeric aspartic proteinases.[120–123] The pH dependence of kinetic parameters for a number of peptide substrates indicated that both active site aspartic acid residues of HIV proteinase are involved in catalysis, and that one is protonated during the reaction, while the other is not.[119] A mechanism which is consistent with these, and other[118] observations is shown in Figure 6. In the proposed mechanism the oxygen of the scissile amide bond is protonated by Asp[25], while Asp[25′] deprotonates the lytic water molecule. The resulting hydroxide ion attacks the amide to form the tetrahedral intermediate **7**. Collapse of **7** yields the cleavage products, and the aspartic acid residues are restored to their initial protonation states. Based on the premise, originally proposed by Pauling,[124] that an enzyme has a higher affinity for the transition state than for either substrates or products,[125] peptide analogues in which

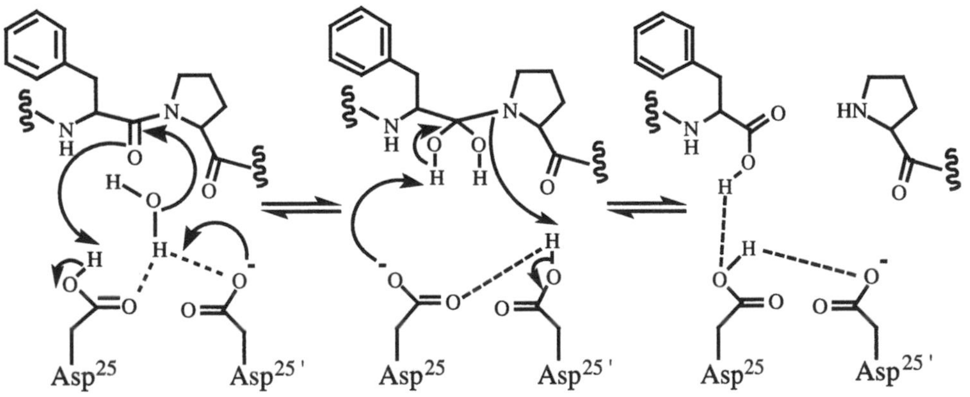

Figure 6 Proposed mechanism for cleavage of a Phe-Pro amide bond by HIV proteinase.

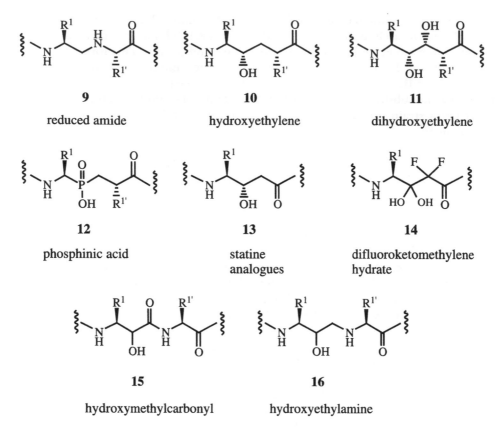

Figure 7 Transition-state mimetics which have been incorporated into inhibitors of aspartic proteinases.

the P1-P1′ residues are replaced by a non-hydrolysable dipeptide isostere which mimics the transition state should provide potent inhibitors. This concept, which had previously been applied successfully to the design of inhibitors of aspartic proteinases such as the mammalian enzyme renin, has, more recently, formed the basis for the design of numerous inhibitors of HIV proteinase.

Various transition-state mimetics have been incorporated into peptide analogues as potential inhibitors of aspartic proteinases (Figure 7). (Some of these may be more accurately regarded as collected substrate analogues.[126]) Each mimetic moiety incorporates some, but not all of the structural features of the presumed transition-state **7**. The tetrahedral geometry at the scissile bond carbon atom in **7** is reproduced in all these isosteres, but other aspects are satisfied to different degrees in various mimetics. Dreyer *et al.* have carried out a systematic study of a number of transition-state mimetics inserted into a common amino acid sequence (Table 1).[127] The reduced amide isostere **17**, which lacks a hydroxyl group to interact with the active site aspartic acid residues, is a poor inhibitor of HIV proteinase, and other reduced amides are also relatively weak inhibitors.[86,93,107] In contrast, hydroxyethylene **18** is

Table 1   Transition-state mimetics incorporated into a common peptide template

Compound	Structure	K_i (nM)
17	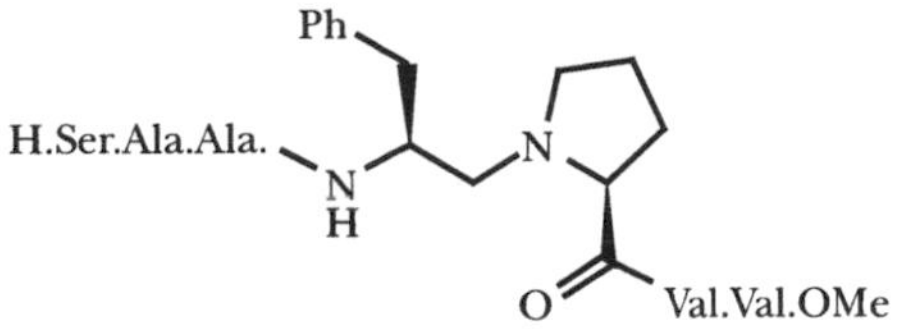	19,000
18	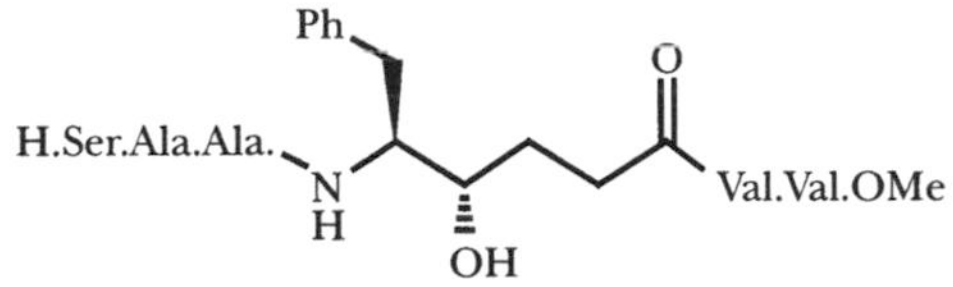	62
19	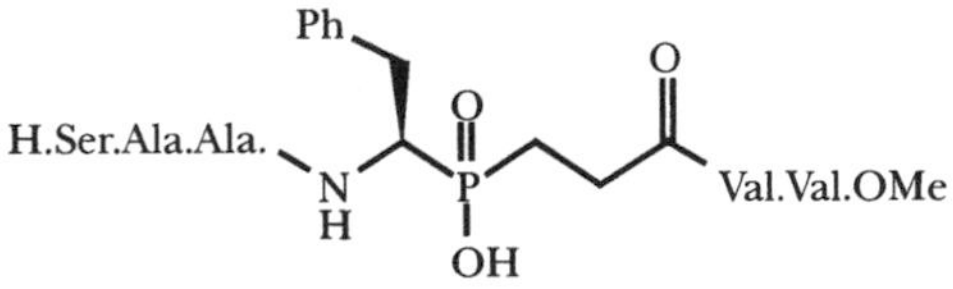	4,400
20	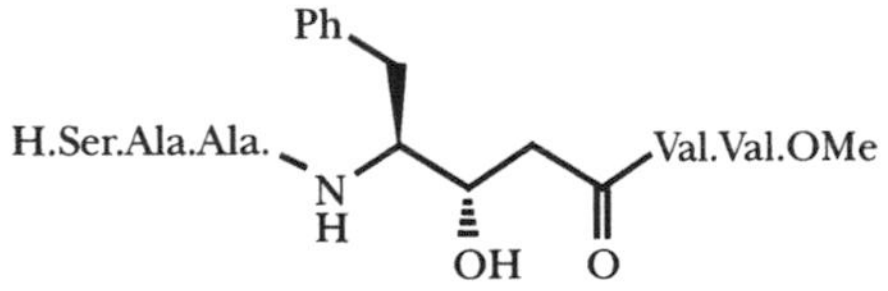	810
21	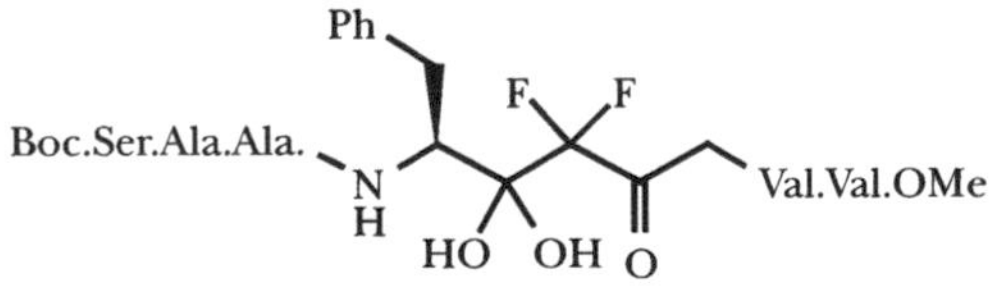	160

the most potent inhibitor in the series, and other hydroxyethylenes,[128–132] as well as compounds based on the dihydroxyethylene isostere **11**[133] are also very potent inhibitors. The hydroxyethylene isostere, in particular, has become one of the most extensively utilised transition-state mimetics, much of the methodology arising from earlier work on hydroxyethylene inhibitors of renin.[134]

Inhibitors which incorporate phosphinic acid transition-state mimetics **12** show, in general, only modest potency, possibly due to repulsion between the anionic phosphinate group and the carboxylate group of the active site aspartate. The binding affinity of these inhibitors is strongly pH dependent, affinity increasing at lower pH values, when the phosphinic acid group may be in the neutral, protonated form. Peptide mimetics which incorporate phosphonamidate moieties are also relatively modest inhibitors.[135] Recently some more potent phosphinic acid isosteres have been described.[136]

The foregoing transition-state mimetics maintain the same spacing between P1 and P1' side chains as in the substrates. In contrast, the statine type mimetic **13** is shortened by one atom, and lacks a P1' side chain. Although inhibition by the natural product pepstatin A is a characteristic property of aspartic proteinases, this compound is a relatively poor inhibitor of HIV-1 proteinase.[50,54,71,137] The *N*-acetyl analogue is, however, substantially more potent against HIV-1 proteinase,[128] and also inhibits HIV-2 proteinase with a K_i of 5 nM.[138] Other peptide mimetics which contain statine,[139] or the corresponding cyclohexyl[140] or phenyl[140,141] analogues, have shown modest potency.

The difluoroketomethylene isostere **14**, which is presumed to exist as the hydrate, and therefore uniquely mimics the *gem* diol functionality of the transition-state **7**, gives a reasonably potent inhibitor **21**. More potent derivatives have been obtained through incorporation of the difluoroketomethylene moiety into shorter analogues which contain a phenyl residue which may bind at the S1' subsite.[142,143]

Peptide analogues in which the spacing between P1 and P1' side chains is lengthened in comparison with the substrates have also provided potent inhibitors. The hydroxymethylcarbonyl transition-state mimetic **15** has been incorporated into a number of very active inhibitors.[144–146] The hydroxyethylamine moiety **16** contains both an alcohol group to interact with the active site aspartic acid residues and an amine function which might become protonated and undergo an ionic interaction with the enzyme. This transition state mimetic has provided some very potent inhibitors of HIV proteinase, and has been a major focus of the drug discovery programme carried out in the Roche laboratories.

THE ROCHE HIV PROTEINASE INHIBITOR PROGRAMME

Inhibitor Design

As noted earlier, one of the most challenging aspects of the design of chemotherapeutic agents, and, in particular, antiviral agents, is that of achieving selectivity. Although we, and others, had identified HIV proteinase as an attractive therapeutic target, the problem remained of how to design inhibitors which are selective for the viral enzyme, as opposed to the mammalian aspartic proteinases renin, pepsin, gastricsin, cathepsin D, and cathepsin E. Any compound which inhibited these enzymes might cause undesirable effects. In addition, pepsin, gastricsin, and cathepsin E are present in high concentrations in the gastrointestinal tract. Binding to these enzymes might thus reduce oral absorption, and it was already known from studies of renin inhibitors that peptide mimetics of the kind proposed tend to have low oral bioavailability.

The design of enzyme inhibitors based on the transition-state mimetic concept involves replacement of the scissile amide bond of a minimum peptide substrate by a non-hydrolysable transition-state isostere. The amino acid sequence of such a peptide substrate would be based on the cleavage sites found in the polyprotein substrates. Among the diverse cleavage sequences identified for HIV proteinase, we were particularly attracted to the Tyr-Pro and Phe-Pro cleavage sites (1, 6, and 7 in

Figure 1). Amide bonds N-terminal to proline are rarely cleaved by mammalian endopeptidases, and we reasoned that inhibitors based on these amino acid sequences might therefore be highly selective for the viral enzyme. Of the various transition-state mimetics which were considered, it was felt that the reduced amide **9** and hydroxyethylamine **16** isosteres would most readily accommodate the imino acid moiety of the Tyr/Phe-Pro cleavage sites. We also considered hydroxyethylene **10** and phosphinic acid **12** mimetics. Although the former isostere has provided potent inhibitors of aspartic proteinases, the considerable body of earlier work on renin inhibitors made the patent situation more difficult for such inhibitors. Our initial studies with phosphinic acid derivatives yielded some potent inhibitors of HIV proteinase, but these had poor selectivity for the viral enzyme, and were also inactive in cell-based assays of antiviral activity. A number of reduced amide derivatives were synthesised, but these were weak inhibitors, as was later confirmed in other laboratories.[86,93,107,127] In contrast, our early studies with hydroxyethylamine containing analogues gave us much encouragement, and we therefore carried out a programme to define structure-activity relationships and to optimise potency among a series of hydroxyethylamine inhibitors of HIV proteinase.[99,147]

Synthesis of Hydroxyethylamines

Because we could not predict the stereochemical requirement for the secondary alcohol function, we wished, in the first instance, to prepare both *R* and *S* hydroxy-ethylamines as potential inhibitors of HIV proteinase. Reduction of an α-aminoketone derived from *L*-phenylalanine could, in principle, be carried out under chelation or non-chelation control to give the desired alcohols. The known chloromethyl ketone **24**[148,149] provided a suitable precursor. Condensation with various proline derivatives followed by reduction and separation of diastereomers gave amino-alcohols which were further elaborated to yield a series of peptide mimetics with amino acid sequences based on the pol fragment Leu[165]-Ile[169] (cleavage site 7 in Figure 1) (Table 2). A more convenient procedure, however, involved reduction of the ketone **24** and conversion of the resulting alcohols to epoxides which could subsequently be opened with proline derivatives (Figure 8). Reduction of **24** with sodium borohydride gave a 3:1 mixture of alcohols **25** and **26** which were readily separated by crystallisation. These were converted to the corresponding epoxides **27** and **28** which were treated with a variety of imino acid derivatives to give hydroxyethylamines which were further elaborated to yield a series of HIV proteinase inhibitors.

Assignment of the stereochemistry of epoxides **27** and **28** was based on the ^{1}H NMR spectra of the 2-oxazolidinones **30** and **31** obtained on treatment with sodium thiophenoxide. Signals due to H-5 for **30** and **31** appeared at $\delta4.76$ and $\delta4.37$ respectively, compared with values of $\delta4.7$–4.8 and $\delta4.2$–4.3 reported for analogous *cis* and *trans* oxazolidinones derived from N-Boc.Leucine.[150] Similarly, coupling constants $J_{4,5}$ for **30** and **31** were 7.1 Hz and 4.4 Hz, compared with values of 7–8 Hz and 4–5 Hz for the leucine derived analogues. In addition, a strong n.O.e. was observed between the 4- and 5-protons for **30**, while no such effect was seen for **31**.

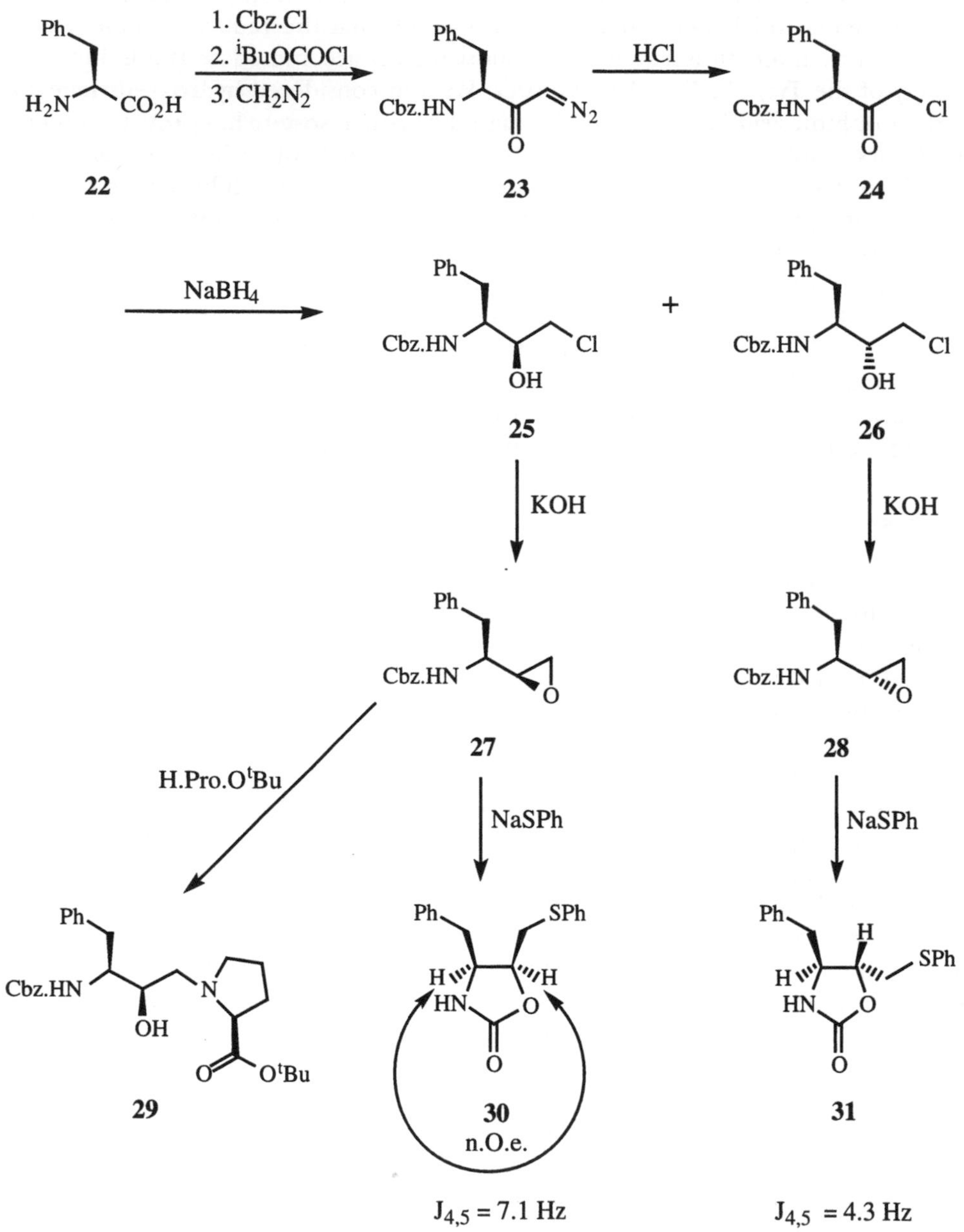

Figure 8   Synthesis and stereochemical assignment of hydroxyethylamines.

Table 2   Determination of optimal inhibitor length[*]

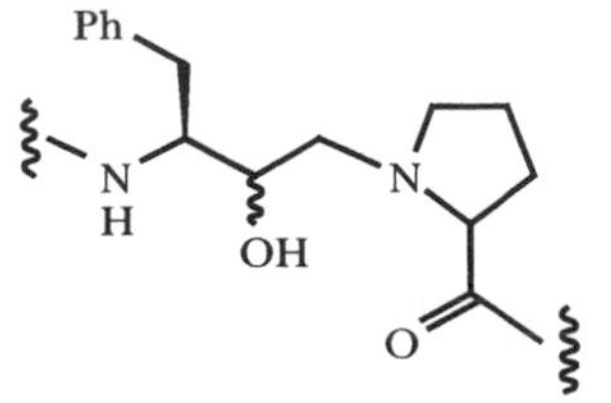

Pheψ[CH(OH)CH₂N].Pro

Compound	Structure	IC_{50} (nM)
32, 33	Cbz.Pheψ[CH(OH)CH₂N].Pro.O^{t}Bu	6,500 and 30,000
34, 35	Cbz.Asn.Pheψ[CH(OH)CH₂N].Pro.O^{t}Bu	140 and 300
36, 37	Cbz.Leu.Asn.Pheψ[CH(OH)CH₂N].Pro.O^{t}Bu	600 and 1,100
38, 39	Cbz.Asn.Pheψ[CH(OH)CH₂N].Pro.Ile.NH^{i}Bu	130 and 2,400
40, 41	Cbz.Leu.Asn.Pheψ[CH(OH)CH₂N].Pro.Ile.NH^{i}Bu	750 and 10,000

[*] The more potent diastereomer of **34**, **35** has the *R* configuration at the secondary alcohol position. Stereochemical assignments were not made for other pairs of isomers.

Structure-Activity Relationships

Peptide mimetics **32–41** were prepared in order to determine the minimum sequence required for potent inhibition of HIV proteinase (Table 2). Clearly, smaller molecules would offer an advantage with respect to the ease and cost of synthesis, as well as the probable pharmacokinetic properties of any compound which might eventually be considered for clinical development. The protected dipeptide mimetics **32** and **33** showed only weak inhibitory activity, but extension of the N-terminus by one amino acid (**34** and **35**) gave a fifty to one hundred fold increase in potency. Further extension towards the N-terminus (**36** and **37**), extension towards the C-terminus (**38** and **39**), or extension in both directions to the pentapeptide mimetics **40** and **41** did not result in any additional increase in potency. The more active of the tripeptide mimetics **34** and **35** was shown to have the *R* configuration at the secondary alcohol function, and this compound was considered a suitable candidate for investigation of structure-activity relationships in order to optimise inhibitory activity.

At the time **34** was identified as a promising lead there was no X-ray crystallographic data which might aid the design of more potent analogues. We therefore set out to elucidate structure-activity relationships and to identify the optimal side-chain and terminal residues through systematic modification of each residue of **34** in turn. Replacement of the N-terminal Cbz group of **34** by smaller groups such as acetyl, or even Boc, caused a considerable loss in potency (Table 3). In contrast, the dihydrocinnamoyl derivative **44**, which is isosteric with **34**, is a potent inhibitor, as is the more conformationally restricted cinnamoyl analogue **45**. Introduction of

Table 3 Replacement of the Cbz group of **34**[*]

R.Asn.Pheψ[CH(OH)CH$_2$N].Pro.O^{t}Bu

Compound	R	IC$_{50}$ (nM)
34	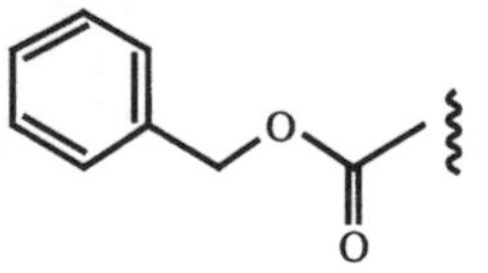	140
42	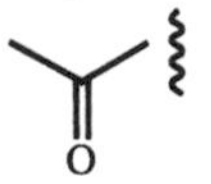	8,600
43	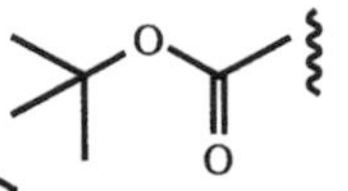	8,000
44	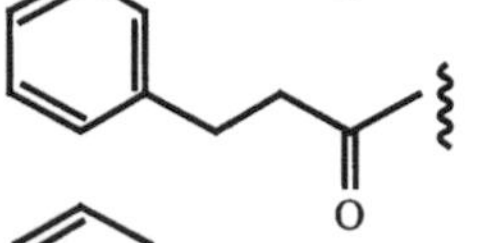	240
45	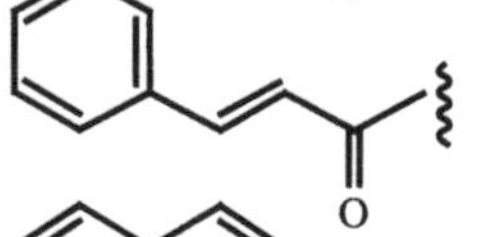	240
46	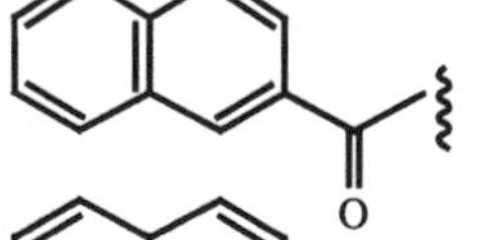	46
47	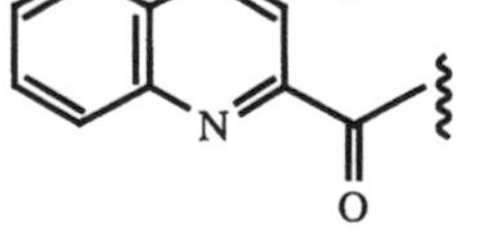	23

[*] All compounds have the R configuration at the secondary alcohol function.

larger lipophilic groups such as β-naphthoyl (**46**) gave increased potency, while the quinoline carbonyl analogue **47** is the most potent inhibitor in this series.

Replacement of the asparagine residue at the P2 position of **34** by alanine or glutamine reduced activity (Table 4). On the other hand, the β-cyanoalanine and S-methylcysteine analogues were comparable to the parent compound. However, no improvement over asparagine was found for this position. (Following our disclosure of the optimised inhibitor Ro 31-8959[99,109] Thompson *et al.* synthesised an analogue in which the asparagine residue at P2 is replaced by tetrahydrofuranylglycine on the basis that the conformationally constrained heterocyclic oxygen atom might replace the amide carbonyl oxygen of asparagine as a hydrogen bond acceptor. This resulted in a four-fold increase in potency *vs.* HIV-1 proteinase.[151])

No significant improvement was found over the benzyl side-chain of phenylalanine at the P1 position. The non-aromatic leucine **54** and cyclohexylalanine **55** analogues

Table 4 Replacement of the Asn residue of **34**[*]

	Cbz.**Asn**.Pheψ[CH(OH)CH$_2$N].Pro.X^{t}Bu		
Compound	**Asn** Replacement	X	IC$_{50}$ (nM)
34	Asn	O	140
48	Ala	O	1,900
49	Gln	O	1,200
50	Asp	NH	270
51	Asp(OMe)	NH	500
52	β-cyanoalanine	NH	160
53	S-methylcysteine	NH	240

[*] All compounds have the R configuration at the secondary alcohol function.

were poor inhibitors, whereas the substituted phenylalanine derivatives **56** and **57** were comparable to **34**. Rather surprisingly in view of the ability of HIV proteinase to cleave the gag polyprotein substrate between tyrosine and proline residues, the tyrosine analogue **59** was a weak inhibitor, causing only 40% inhibition at a concentration of 1000 nM.

The most marked improvements in potency were achieved through modification of the imino acid residue at the P1′ position. Early modelling studies had suggested that imino acids other than proline might be accommodated in this position, a suggestion which was later confirmed when X-ray crystallographic studies of HIV proteinase showed the S1 and S1′ subsites to be identical as a consequence of the C$_2$ symmetry of the native enzyme. Replacement of the proline residue by acyclic amino acids such as valine (**61**) caused a loss of potency, although activity was

Table 5 Replacement of the phenylalanine mimetic moiety of **34**[*]

	Cbz.Asn.**Phe**ψ[**CH(OH)CH$_2$N**].Pro.X^{t}Bu		
Compound	**Phe mimetic** Replacement	X	IC$_{50}$ (nM)
34	Phe	O	140
54	Leu	O	>10,000
55	cyclohexylalanine	O	1,900
56	p-methoxyphenylalanine	O	140
57	p-fluorophenylalanine	O	77
58	β-naphthylalanine	NH	440
59	Tyr	NH	>1,000

[*] All compounds have the R configuration at the secondary alcohol function.

Table 6   Replacement of the Pro residue of **34**[*]

Cbz.Asn.Pheψ[CH(OH)CH$_2$N].**Pro**.NH^{t}Bu

Compound	Pro Replacement	IC$_{50}$ (nM)
60	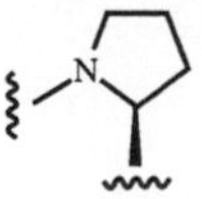	210
61	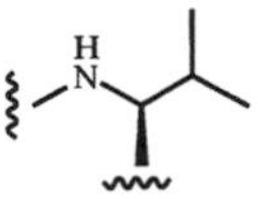	780
62	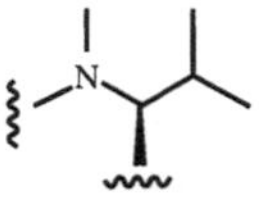	430
63	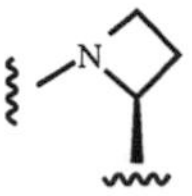	>1,000
64	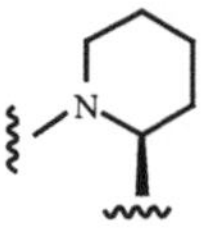	18
65	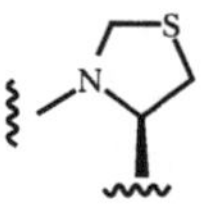	8.4
66	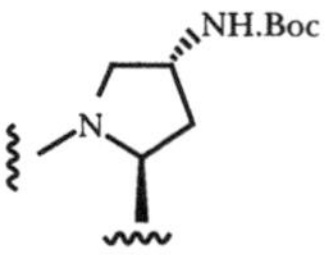	16
67	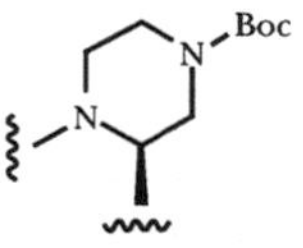	5.3
68	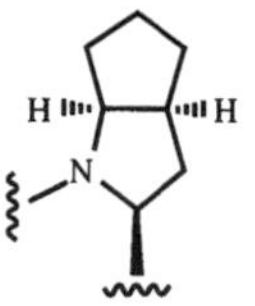	5.6
69	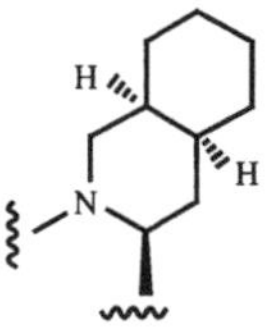	2.7

[*] All compounds have the *R* configuration at the secondary alcohol function. Compound **61** was tested as the tert.-butyl ester.

Table 7   Replacement of the tert.butyl ester group of **34**[*]

Cbz.Asn.Pheψ[CH(OH)CH$_2$N].Pro.**R**		
Compound	R	IC$_{50}$ (nM)
34	O^{t}Bu	140
60	NH^{t}Bu	210
70	NHMe	670
71	NHCH$_2$Ph	160
72	N(Me)^{t}Bu	>1,000

[*] All compounds have the *R* configuration at the secondary alcohol function.

partially restored on N-methylation (**62**) (Table 6). (More recently some reasonably potent inhibitors which contain phenylalanine at P1′ have been described.[152]) Variation of ring size in imino acid residues showed the azetidine **63** to be inactive, while the piperidine derivative **64** was a very potent inhibitor. The thiazole **65** was also a potent inhibitor, possibly as a result of the steric similarity of the thiazole ring to the 6-membered ring in **64**. That large lipophilic residues could be accommodated in the S1′ subsite was shown by the very potent inhibition exhibited by the imino acid derivatives **66** and **67**. This subsite could also be filled, and in a more conformationally constrained manner, through introduction of bicyclic residues. These gave extremely potent inhibitors, and the 4a*S*,8a*S*-decahydroisoquinoline moiety of **69** became the best residue which we have so far identified for the P1′ position. The *S,S,S* stereochemistry of these bicyclic residues was optimal for activity, other isomers being much less potent.

The C-terminal tert.-butyl ester function of **34** could be replaced by other lipophilic groups such as tert.-butylamide or benzylamide groups without significant change in potency, but no better replacement was found (Table 7). On grounds of chemical and probable metabolic stability the tert.-butylamide function of **60** was selected as the C-terminal group of choice for additional inhibitors.

Configuration of Hydroxyethylamines

One feature which is common to most potent inhibitors of aspartic proteinases is a critical hydroxyl group which forms hydrogen bonds to the catalytic aspartic acid residues. Inhibitors which contain hydroxyethylene **10** or statine type **13** transition-state isosteres show a marked preference for the *S* configuration at the secondary alcohol function. It was therefore interesting to note a slight preference for the *R* configuration in the hydroxyethylamine tripeptide mimetics **34** and **35**, although similar observations had previously been noted for hydroxyethylamine inhibitors of renin.[153,154] When the proline residue at the P1′ position was replaced by a piperidine or 4a*S*,8a*S*-decahydroisoquinoline residue, however, preference for the *R*

Table 8 Effect of configuration and chain extension on activity

Structure	Configuration of alcohol function			
	R		S	
	Compound	IC_{50} (nM)	Compound	IC_{50} (nM)
Cbz.Asn.Pheψ[CH(OH)CH₂N].Pro.OᵗBu	34	140	35	300
Cbz.Asn.Pheψ[CH(OH)CH₂N].PIP.NHᵗBu	64	18	73	960
Cbz.Asn.Pheψ[CH(OH)CH₂N].DIQ.NHᵗBu	69	2.7	74	>1,000
Cbz.Asn.Pheψ[CH(OH)CH₂N].Pro.Ile.Val.OMe	75	>>100	76	13
Ac.Ser.Leu.Asn.Pheψ[CH(OH)CH₂N].Pro.Ile.Val.OMe	77	65	78	3.4
Cbz.Asn.Pheψ[CH(OH)CH₂N].DIQ.Ile.Val.OMe	79	>>100	80	>>100

*Abbreviations: PIP, piperidine-2(S)-carbonyl; DIQ, 4a(S),8a(S)-decahydro-3(S)-isoquinolinecarbonyl. Figures for **77** and **78** are taken from reference 155.

configuration became much more marked (**64** vs. **73** and **69** vs. **74** in Table 8). Rather surprisingly, when inhibitors having proline at P1′ were extended as far as the P3′ position the stereochemical preference was reversed, the S diastereomer **76** being much more potent than the R isomer **75**. Similar findings have been published by Rich et al. for the heptapeptide mimetics **77** and **78**.[155] In contrast to proline containing inhibitors, extension of decahydroisoquinoline derivatives abolished activity, irrespective of the configuration of the secondary alcohol. Modelling studies carried out by ourselves,[156] and, independently, by Rich et al.[155] suggested that both R and S hydroxyethylamine derivatives can bind to HIV proteinase in a way that

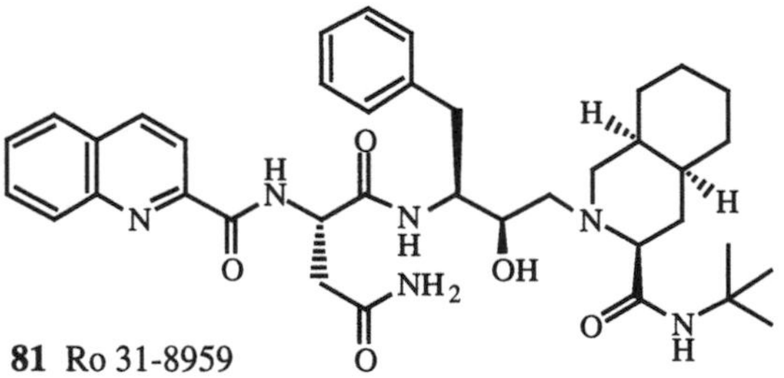

81 Ro 31-8959

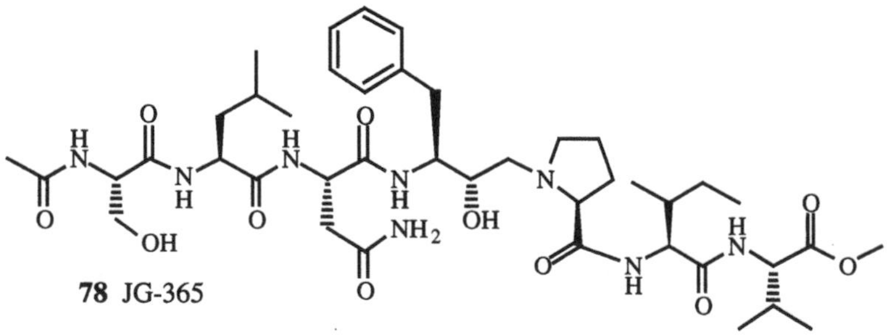

78 JG-365

Figure 9 (R)- and (S)-hydroxyethylamine containing inhibitors Ro 31-8959 and (S)-JG-365.

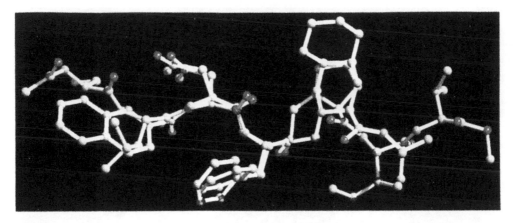

Figure 10 X-ray crystallographic structures of Ro 31-8959 and (*S*)-JG-365 in inhibitor HIV-1 proteinase complexes.

enables hydrogen bonding of the hydroxyl group to the catalytic aspartic acid residues, and this was later confirmed through X-ray crystallography of enzyme-inhibitor complexes of both classes.

Binding Modes of *R*- and *S*-Hydroxyethylamines Ro 31-8959 and JG-365

The X-ray crystal structure of HIV-1 proteinase complexed with the *R*-hydroxy-ethylamine Ro 31-8959 **81**, one of our optimised inhibitors (*vide infra*), has recently been determined.[30,109] This is compared with the enzyme-bound structure of the *S*-hydroxyethylamine JG-365 **78**[108] in Figure 10. Both inhibitors bind in an extended conformation, the Asn and Phe side chains making similar contacts with the enzyme in each case, while the N-terminal quinoline residue of **81** fits tightly into the S3 subsite. Despite having different configurations, the hydroxyl group of each inhibitor is located between the catalytic aspartic acid residues of the proteinase. Consequently, the backbones of these inhibitors diverge after the hydroxymethylene carbon atom. The decahydroisoquinoline residue of **81** is in the preferred chair-chair conformation, and occupies almost the entire S1' subsite, making hydrophobic contact with both the flap and floor regions of the enzyme (Figure 11). The ring nitrogen atom of the decahydroisoquinoline system of **81** has the opposite configuration to that of the proline residue in **78**, the bond between the methylene group and nitrogen being equatorial (Figure 12). The carbonyl groups of both decahydroisoquinoline and proline carboxamides are suitably orientated to hydrogen bond to the water molecule which bridges the inhibitor and the flaps of the enzyme, but the adjacent nitrogen atoms are considerably displaced. The tert.-butylamide group of **81** does not continue along the peptide backbone, but rather occupies the S2' subsite which, in the case of **78** is occupied by the isoleucine side chain. Extension of decahydroisoquinoline derivatives into the S3' subsite is therefore no longer possible, and this explains the loss of activity in the extended analogue **79**. The crystal

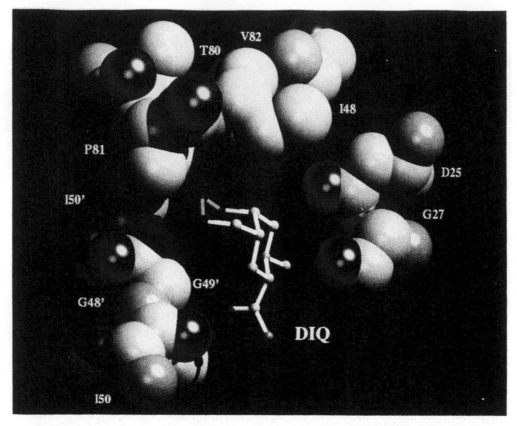

Figure 11 Crystal structure of HIV-1 proteinase-compound **81** complex showing the S1′ subsite (van der Waal's spheres) and the decahydroisoquinoline residue (ball and stick representation).

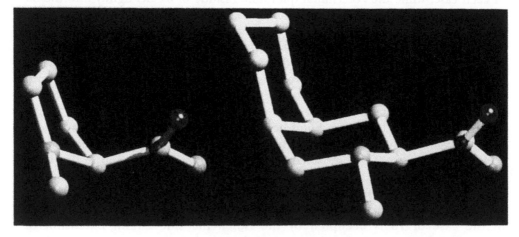

Figure 12 Crystal conformation of the proline and decahydroisoquinoline residues in enzyme-bound structures **78** and **81**.

Table 9 Inhibition of HIV-1 and HIV-2 proteinases by optimised inhibitors[*]

Compound	Structure	Proteinase IC_{50} (nM)	
		HIV-1	HIV-2
34	Cbz.Asn.Pheψ[CH(OH)CH$_2$N].Pro.O^{t}Bu	140	330
46	BN.Asn.Pheψ[CH(OH)CH$_2$N].Pro.O^{t}Bu	46	50
82	QC.SMC.Pheψ[CH(OH)CH$_2$N].PIP.NH^{t}Bu	12	15
83	QC.Asn.Pheψ[CH(OH)CH$_2$N].PIP.NH^{t}Bu	2.0	9.5
81	QC.Asn.Pheψ[CH(OH)CH$_2$N].DIQ.NH^{t}Bu	<0.4 (K_i 0.12)	<0.8 (K_i < 0.10)

[*] Abbreviations: BN, β-naphthoyl; QC, quinoline-2-carbonyl; SMC, S-methylcysteinyl; PIP, piperidine-2(S)-carbonyl; DIQ, 4a(S),8a(S)-decahydro-3(S)-isoquinolinecarbonyl. IC_{50} values for **81** limited by mutual depletion.

structure of the much less potent S diastereomer of **81** complexed with HIV proteinase[157] shows the conformation of the hydroxyethylamine isostere portion of the inhibitor to be very similar to that of the S-alcohol JG-365 **78**. To allow the decahydroisoquinoline residue to occupy the S1' subsite the bond between the methylene group and nitrogen is forced into a high energy axial conformation, and this may explain the 1000-fold difference in binding constants between the two diastereomers.

It has thus become clear that hydroxyethylamine containing inhibitors of HIV proteinase can adopt different modes of binding, dependent on both sequence and chain length. Consequently, caution must be exercised in extrapolating structure-activity relationships from one series of inhibitors of HIV proteinase to another.

Optimised Inhibitors

Having identified optimal residues at each position of the lead structure **34**, a number of analogues which contain combinations of the preferred side chains were synthesised (Table 9). Several very potent inhibitors of HIV-1 proteinase were obtained, showing the effect of individual residues to be additive. These compounds were also potent inhibitors of HIV-2 proteinase. It was found that structure-activity relationships were similar for both enzymes, but that these compounds were, in general, somewhat more potent vs. HIV-1 proteinase than the type 2 enzyme.

The original premise, that inhibition of HIV proteinase should prevent viral replication, was vindicated when the optimised inhibitors were shown to possess in vitro anti-HIV activity, their IC_{50} values correlating well with their potencies as proteinase inhibitors (Table 10). Cellular assays of cytotoxicity were also carried out. No significant cytotoxicity was seen at considerably higher concentrations than those required for antiviral activity. The most potent antiviral agent, Ro 31-8959, showed a greater than 5000-fold selectivity index. On the basis of its potent and selective (vide infra) inhibition of both HIV-1 and HIV-2 proteinase, its

Table 10 HIV-proteinase inhibition, antiviral activity, and cytotoxicity of optimised inhibitors

Compound	Inhibition of HIV-1 proteinase IC_{50} (nM)	Antiviral activity IC_{50} (nM)	Cytotoxicity TD_{50} (nM)
60	210	400	>100,000
34	140	300	>100,000
46	46	130	>10,000
82	12	13	>10,000
83	2.0	17	>100,000
81	<0.4	2.0	>10,000

anti-HIV activity, and its lack of cellular toxicity, this compound was selected for further development, leading eventually to clinical evaluation as an anti-HIV agent.

THE DEVELOPMENT OF RO 31-8959

New Synthetic Routes to Ro 31-8959

As a consequence of the decision to progress towards clinical development of Ro 31-8959 it became necessary to reevaluate its synthesis. While mg quantities of each analogue had been sufficient for primary biological evaluation, it was now necessary to synthesise Kg amounts of Ro 31-8959, and to start planning in terms of tonnes! The original route (Figure 8), which had been very satisfactory for the synthesis of relatively small quantities of a wide range of hydroxyethylamines, could not be scaled up to the required extent due to the use of diazomethane. We therefore required an alternative route to **81** which was suitable for large scale synthesis. However, since the reaction of the epoxide **27** with various imino acid derivatives had proved to be very successful in the preparation of hydroxyethylamines, we did not wish to diverge more than was necessary from the existing route. We therefore required a new synthesis of a protected amino epoxide such as **27**, or a synthetic equivalent.

Published syntheses of α-amino epoxides were unsuitable for our purposes. Epoxidation of olefins obtained through Wittig olefination of N-protected α-amino aldehydes gives predominantly the undesired diastereomer.[150] Reaction of sulphonium ylides with N-Boc protected α-amino aldehydes gives an approximately 1:1 mixture of diastereomers,[158] although analogous reaction of doubly protected α-amino aldehydes such as N,N-dibenzyl derivatives does occur with a high degree of non-chelation control to give the desired diastereomer.[159] However, the relatively harsh conditions required to remove N-benzyl groups meant that this approach was also unsuitable.

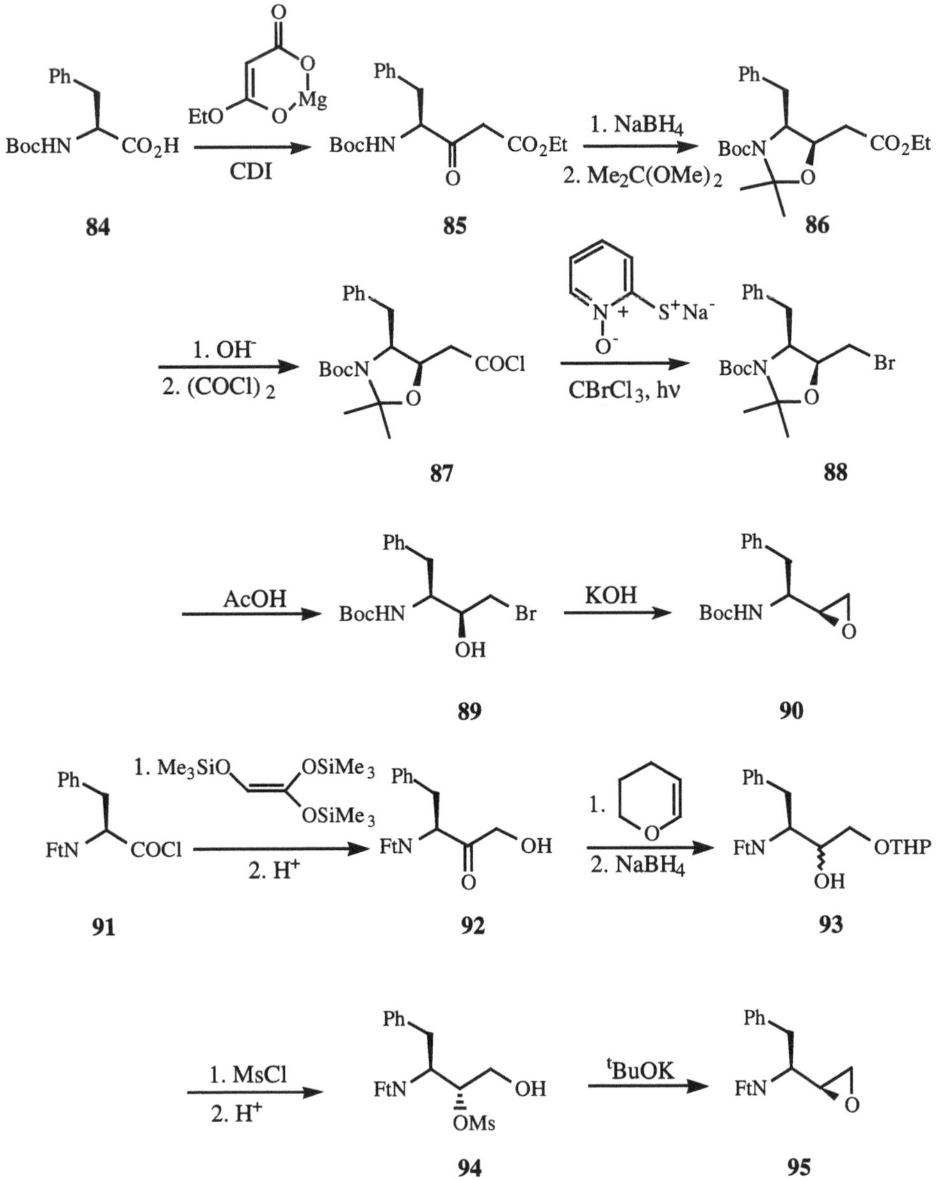

Figure 13 Synthesis of protected α-amino epoxides from L-phenylalanine.

In considering novel approaches to protected α-amino epoxides we explored strategies which, like our original route, and all the published syntheses, start with an L-amino acid derivative. We also considered approaches based on asymmetric epoxidation of achiral starting materials, and routes starting from precursors having both chiral centres already present.

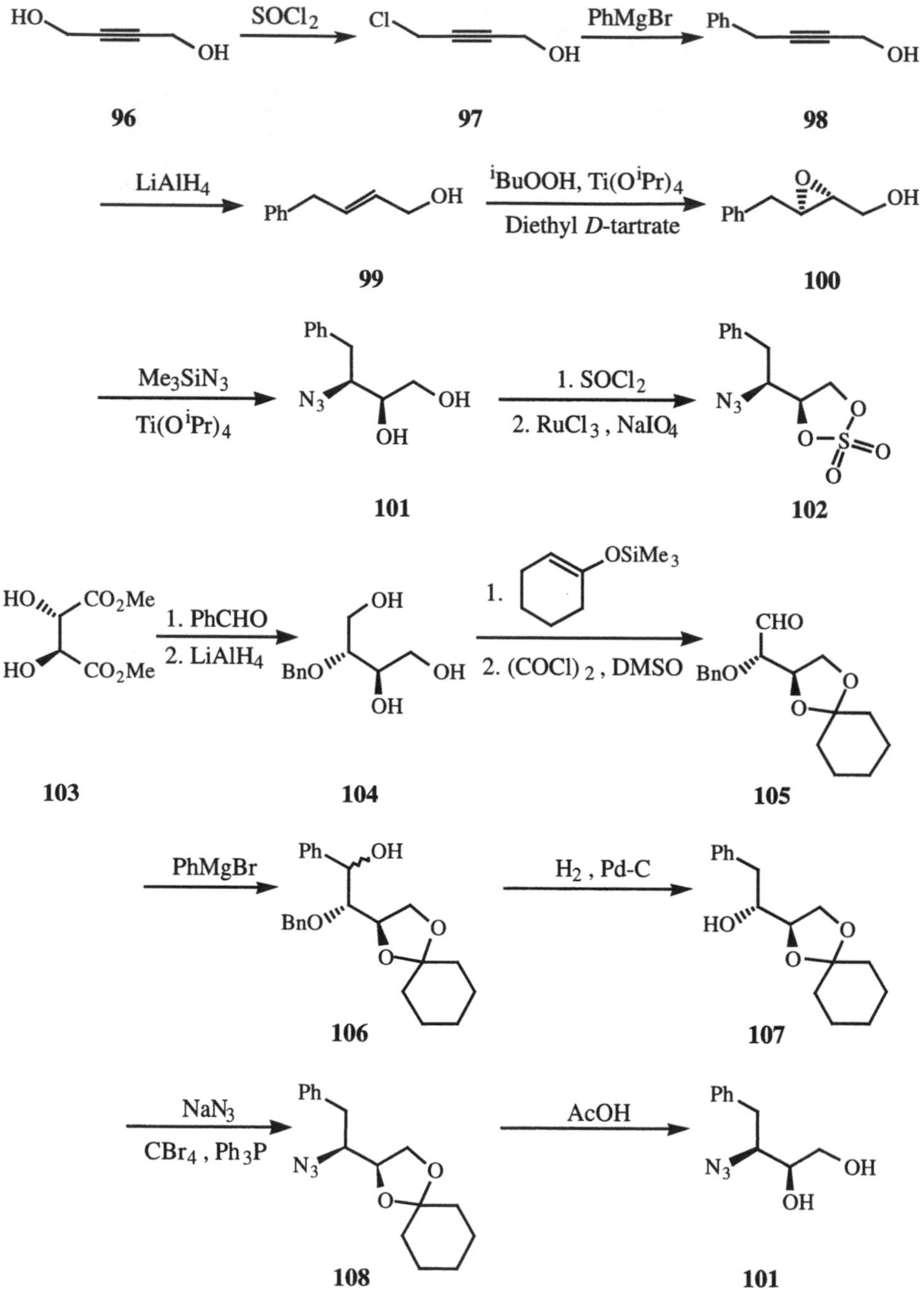

Figure 14 Synthesis of azidosulphates *via* Sharpless epoxidation and from dimethyl *D*-tartrate.

In the first of our new routes from *L*-phenylalanine the magnesium enolate of ethyl malonate provided an alternative to diazomethane for introduction of an additional carbon atom (Figure 13). Reaction with *N*-Boc phenylalanine gave the ketoester **85**.[160] Reduction of **85** yielded a mixture of alcohols from which the major product, the desired 3*R*,4*S* diastereomer, could easily be separated by recrystallisation. Conversion to the protected acid chloride **87** was followed by decarboxylative bromination[161] to give the bromide **88**. Deprotection and treatment with base then gave the epoxide **90**.

A second alternative to diazomethane for introduction of an additional carbon atom involved reaction of a silylated ketene acetal[162] with the *N*-phthaloyl acid chloride **91**. Reduction of the resulting ketone **92** gave predominantly the un-desired *S*-alcohol arising from non-chelation control. It was therefore necessary to invert the configuration at this centre, and this was conveniently achieved through protection of the primary alcohol group prior to reduction of the ketone. The secondary alcohol was then converted to the mesylate and the tetrahydropyranyl protecting group was removed to give the homochiral product **94** which crystallised out from the reaction mixture in a pure state. Treatment with base gave the desired epoxide **95**.

A recent report of the regiospecific opening of 2,3-epoxy alcohols, including **100** (Figure 14), using the [Ti(O^iPr)$_2$(N$_3$)$_2$] reagent[163] prompted us to investigate the synthesis of α-amino epoxides from the readily available 2-butyne-1,4-diol **96**. This was converted to the *trans* allylic alcohol **99** which gave the epoxide **100** with greater than 92% enantiomeric excess on Sharpless epoxidation in the presence of molecular sieves.[164] Although we succeeded in converting the azide **101** to the protected amino epoxide **27**, we felt that the azide function could be considered a protected form of an amine, allowing us to avoid protection and deprotection steps. The azido diol could be converted to the corresponding azido epoxide, but it was found more convenient to prepare the cyclic sulphate **102** using the Sharpless procedure.[165] This derivative is a synthetic equivalent of a protected α-amino epoxide, and displayed somewhat higher reactivity than epoxides in reaction with imino acid derivatives.

Finally we investigated a synthetic approach in which both chiral centres of the product are already present in a starting material derived from the chiral pool. The starting material in this case was dimethyl *D*-tartrate **103**. (After completion of this work an alternative synthesis of the azide **101** from **103** was published.[166]) Dimethyl *D*-tartrate was converted to the known triol **104**.[167,168] Protection of the 1.2-diol function and oxidation gave the aldehyde **105** which, on treatment with phenyl magnesium bromide, gave the diastereomeric alcohols **106**. Catalytic hydrogenation of **106** resulted in rapid removal of the benzyl ether protecting group, but hydrogenolysis of the benzylic secondary alcohol function was considerably slower. Treatment of the resulting alcohol **107** with sodium azide in the presence of carbon tetrabromide and triphenylphosphine[169] gave the azide **108** with inversion of configuration. Removal of the cyclohexylidene protecting group then gave the same azido diol **101** as was obtained by the earlier route.

Each of these new routes to α-amino epoxides or their equivalent avoids the use of hazardous reagents and, for the most part, chromatographic separations. Each

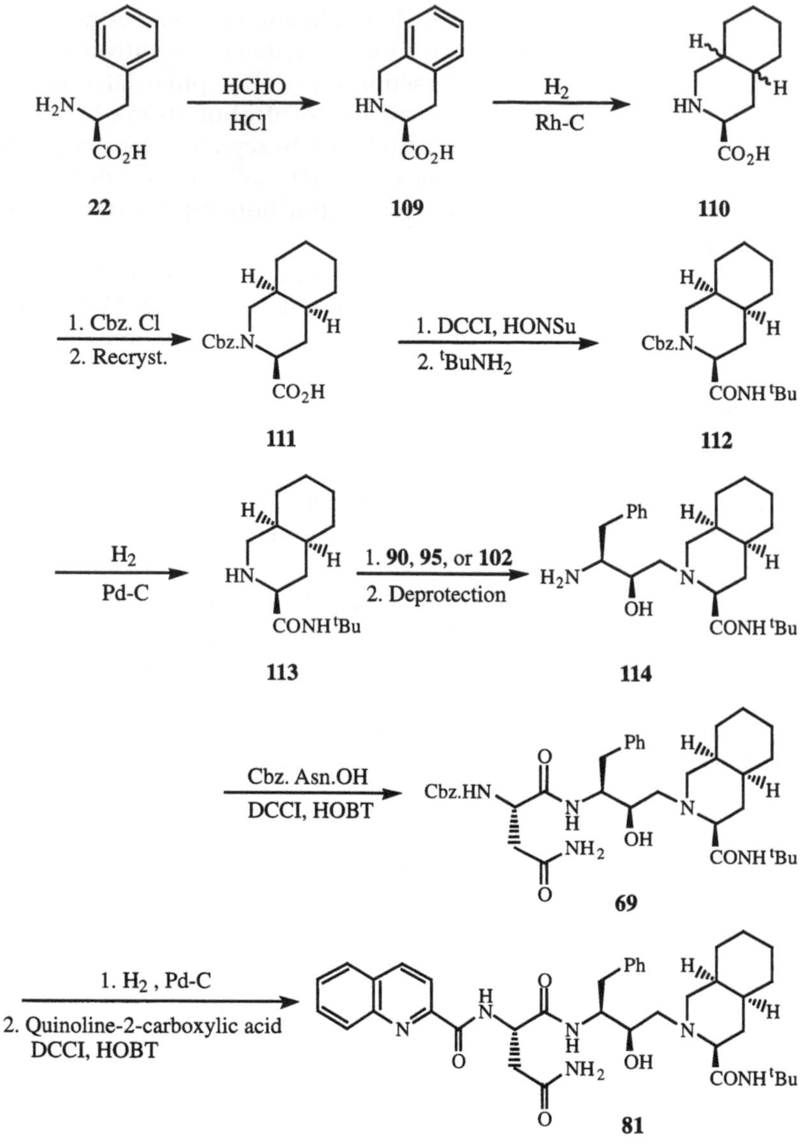

Figure 15 Synthesis of Ro 31-8959.

is potentially applicable to large scale synthesis of Ro 31-8959. Treatment of either of the epoxides **90, 95,** or of the cyclic sulphate **102** with imino acid derivative gave hydroxyethylamines in good yield. The imino acid amide **113** required for the synthesis of Ro 31-8959 was derived from *L*-phenylalanine through Pictet-Spengler reaction to the tetrahydroisoquinoline **109** followed by catalytic hydrogenation (Figure 15). The desired 4a*S*,8a*S* diastereomer was obtained by recrystallisation of the *N*-Cbz derivative **111**. Conversion to the tert.-butyl amide and deprotection gave **113**. An alternative route to **113** from a homochiral cyclohexene carboxylate obtained either through classical resolution, or through enzymic hydrolysis of a *meso* diester, has recently been published.[170] The amine **113** could be coupled with either **90, 95** or **102** to give, after deprotection, the hydroxyethylamine **114** which was further elaborated to Ro 31-8959 **81** by standard procedures.

Biological Evaluation of Ro 31-8959

Following the selection of Ro 31-8959 as a candidate for further development, it was necessary to conduct a more detailed evaluation of its biological properties. This would include determination of its antiviral activity *vs.* clinical isolates of HIV, as well as the laboratory strains which had been studied hitherto, and also studies of the possible emergence of resistance, and of synergy with other antiviral agents.

It was important to establish that a compound which had been optimised for inhibition of cleavage of a small synthetic substrate was also effective in blocking processing of the natural polyprotein substrates. This was demonstrated for **81** when complete inhibition of processing of HIV-1 gag polyprotein by both HIV-1 and HIV-2 proteinases was observed at a concentration of 1 μM in a mixed bacterial lysate assay assessed by Western blot analysis.[99]

As noted earlier, an attractive feature of inhibitors based on Phe/Tyr-Pro cleavage sequences is that these might be selective for viral aspartic proteinases. That the desired level of selectivity had indeed been achieved was shown by the complete lack of inhibition by Ro 31-8959 **81** of a series of mammalian aspartic proteinases, as well as representatives of other classes of proteinases (Table 11).

Table 11 Inhibition of proteinases by Ro 31-8959 **81**

Proteinase	Class	IC_{50} (nM)
HIV-1 Proteinase	Aspartic	<0.40
HIV-2 Proteinase	Aspartic	<0.80
Human Renin	Aspartic	>>10,000
Human Pepsin	Aspartic	>>10,000
Human Gastricsin	Aspartic	>10,000
Human Cathepsin D	Aspartic	>10,000
Human Cathepsin E	Aspartic	>>10,000
Human Leucocyte Elastase	Serine	>>10,000
Bovine Cathepsin B	Cysteine	>>10,000
Human Synovial Fibroblast Collagenase	Metallo	>>10,000

IC_{50} values for HIV-1 and HIV-2 proteinase limited by mutual depletion. K_i values for HIV-1 and HIV-2 proteinase 0.12 nM and <0.10 nM respectively.

Table 12 Antiviral activity of Ro 31-8959 compared with AZT and ddC[171]

			Ro 31-8959		Standards	
Virus Strain	Cell Line	Assay Method	Antiviral IC_{50} (nM)	Cytotoxicity TD_{50} (nM)	Antiviral AZT	IC_{50} (nM) ddC
HTLV-III$_{RF}$	C8166	P$_{24}$ ELISA	3.0	45,000	3.7	190
HTLV-III$_{RF}$	C8166	P$_{24}$ ELISA	<1.5	75,000	19	24
HIV-1$_{GB8}$	JM	P$_{24}$ ELISA	<1.5	75,000	1,900,000	<380
HIV-1$_{NDK}$	Molt4	CPE (Cell no.)	<1500	400,000	<37	<47
HTLV-III$_{B}$	MT4	CPE (MTT)	<0.75	12,500	1.1	230
HIV-2$_{ROD}$	MT4	CPE (MTT)	<1.2	17,500	1.3	270

Potent *in vitro* antiviral activity of Ro 31-8959 has been demonstrated *vs.* a variety of strains of HIV, including AZT resistant strains, and also HIV-2, in test systems which use different cell lines and different assay methods (Table 12).[171,172] This compound is also active *vs.* SIV,[173] but not against murine retroviruses.[174] Typical antiviral IC_{50} values are in the range 1–10 nM, several orders of magnitude lower than the cytotoxic concentrations on uninfected host cells. Ro 31-8959 is effective whether added early or late to acutely infected cell cultures, and is also effective in chronic infection.[172] The fact that addition of the inhibitor can be delayed post infection without loss of antiviral activity shows that it acts at a late stage in the viral life cycle, consistent with inhibition of virus maturation. This contrasts with reverse transcriptase inhibitors, which act at an early stage in viral replication, and are ineffective if addition is delayed by more than two or three hours post infection. Electron microscopy of virions produced in chronically infected cells confirmed that Ro 31-8959 arrests maturation of the virus, blocking conversion of the immature "doughnut" form into infectious virus particles having a fully formed capsid (Figure 16).[172] Viral titre of immature virus particles produced in the presence of 100 nM of Ro 31-8959 showed their infectivity to be reduced 1000 fold, and these particles did not regain infectivity when the inhibitor was removed.

A major problem in the treatment of viral infections is the emergence of resistance. The clinical use of the reverse transcriptase inhibitor AZT has led to resistant strains of HIV, and resistance has been shown to arise particularly rapidly in *in vitro* studies with non-nucleoside reverse transcriptase inhibitors. It was thought that development of resistance might be less likely with a HIV proteinase inhibitor which closely mimics the natural substrate. Repeated *in vitro* passage of HIV-1 with increasing concentrations of Ro 31-8959 did eventually result in decreased sensitivity, but this occurred much more slowly than in parallel studies with both nucleoside and non-nucleoside inhibitors of reverse transcriptase. The emergence and significance of resistance to Ro 31-8959 in a clinical situation remains to be determined. It is likely, however, that combination therapy will provide significant advantages with respect to the emergence of resistance, as well as reduced drug toxicity. *In vitro* studies of Ro 31-8959 in combination with other antiviral agents, including AZT and ddC, have showed clear synergy of antiviral action.[174]

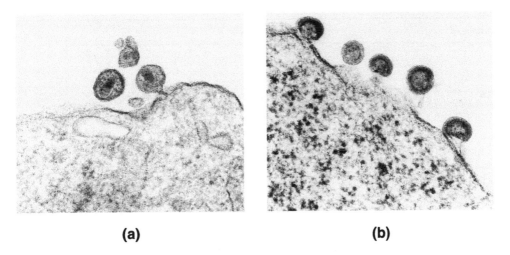

(a)   **(b)**

Figure 16 (a), Untreated chronically infected CEM cell showing mature virus particles which have a central condensed core; and (b), Chronically infected CEM cells treated with 100 nM Ro 31-8959 for 24 h showing immature virus particles in the "doughnut" form.

The lack of a suitable animal model of HIV infection meant that progression to clinical trials would have to occur in the absence of any *in vivo* efficacy data. It was difficult to predict, with confidence, how the *in vitro* antiviral activity of Ro 31-8959 would relate to the clinical situation. In addition, our ultimate goal was an oral anti-HIV agent. It was therefore necessary to establish, at an early stage, that reasonable plasma concentrations could be achieved. Pharmacokinetic studies in the rat showed that a single oral dose 10 mg/kg of Ro 31-8959 gave peak plasma levels of about 150 nM and that plasma levels remained above 15 nM for over six hours.[174] Animal toxicity studies showed the compound to be very well tolerated, and this allowed phase 1 clinical studies to be carried out in healthy volunteers. In these studies multiple dosing of 600 mg orally every eight hours maintained a plasma concentration of at least 70 nM of Ro 31-8959 over the dosing period, a level which may be compared with typical *in vitro* antiviral IC_{50} and IC_{90} values of 1–10 nM and 5–50 nM respectively.[174] Data on tolerability and drug exposure justified progression to phase 2. Clinical trials in HIV infected patients have now begun in several countries, making Ro 31-8959 the first HIV proteinase inhibitor to be evaluated clinically for the treatment of HIV infection.

RECENT DEVELOPMENTS AMONG HIV PROTEINASE INHIBITORS OF OTHER CLASSES

Whereas hydroxyethylamine containing isosteres constituted the main part of the Roche HIV proteinase inhibitor programme, other groups have concentrated on different transition state mimetics. The hydroxyethylene isostere **10** has received

Table 13   Inhibition of HIV-1 proteinase by hydroxyethylenes

Compound	Structure	IC_{50} (nM)
115	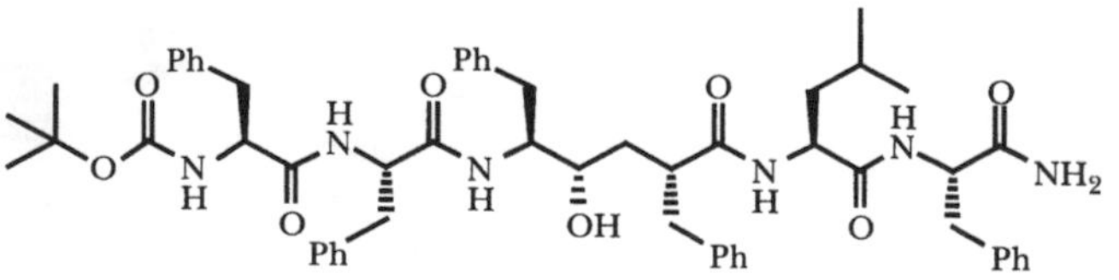	1.0
116	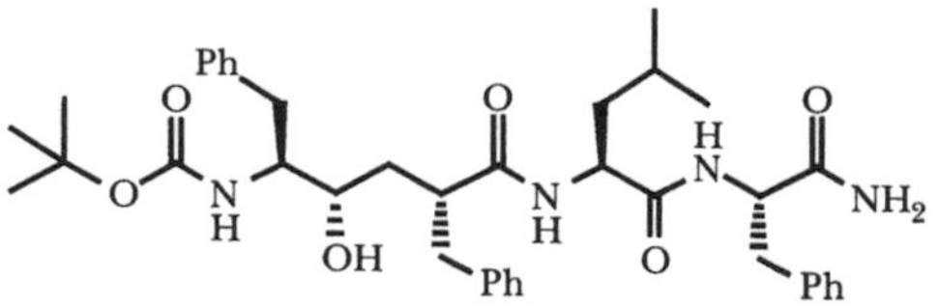	0.6
117	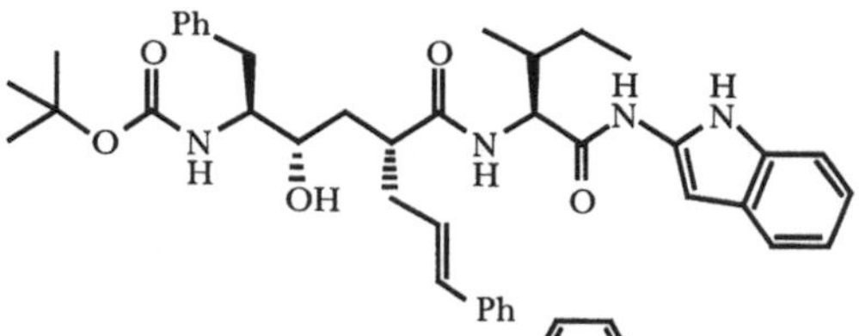	0.03
118	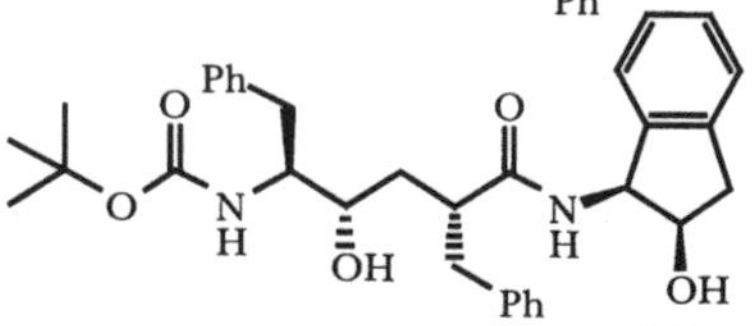	0.35
119	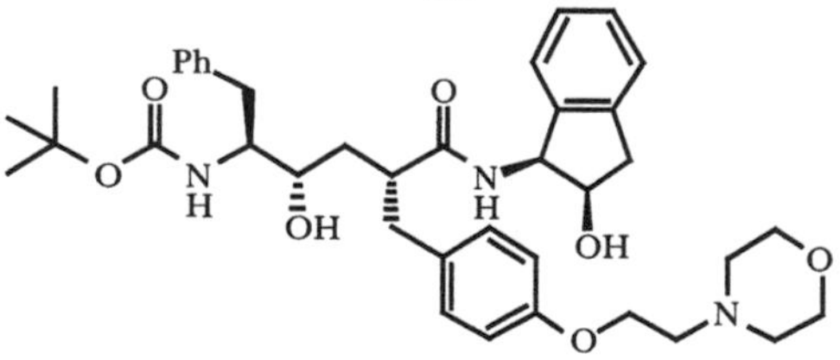	0.45
120	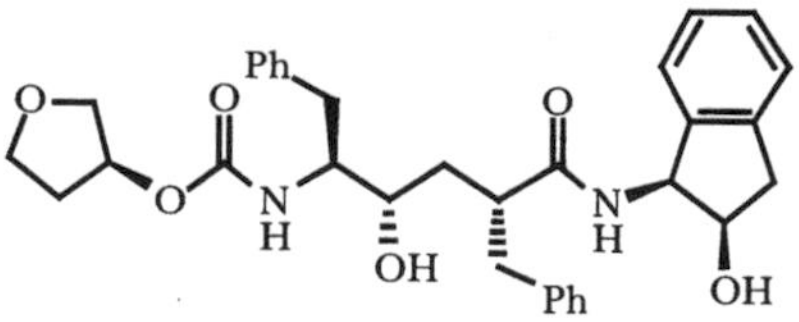	<0.03

most attention, largely as a result of the availability of inhibitors and synthetic intermediates derived from earlier renin inhibitor programmes. Renin inhibitors have been screened for inhibition of HIV proteinase in order to generate lead structures which might then be modified to optimise inhibition of HIV proteinase, and to abolish renin inhibitory activity. This strategy was employed by research workers at Merck who identified the protected hexapeptide mimetic **115** as a potent inhibitor of HIV-1 proteinase (Table 13). This compound was also a reasonably potent renin inhibitor (IC_{50} values of 1.0 nM and 73 nM respectively). Deletion of two amino acid residues from the N-terminus to give **116** resulted in increased affinity for HIV proteinase, while inhibition of renin was lost (IC_{50} values of 0.6 nM and >10,000 nM respectively).[175] Optimisation of residues at P1′ and P3′ gave a very potent inhibitor **117** which completely blocked HIV replication *in vitro* at a concentration of 12 nM. Smaller, less peptidic inhibitors were obtained through replacement of the C-terminal isoleucine amide portion of these derivatives by an aminoindanol moiety, as in **118**,[176] the indanol hydroxyl group forming a hydrogen bond with the NH of Asp[29] [177] in an analogous manner to the P2′ carbonyl oxygen in other HIV proteinase-inhibitor complexes. Further modifications, in an attempt to improve aqueous solubility, led to the morpholine derivative **119** which had similar HIV proteinase inhibiting activity to the unsubstituted analogue **118**, but was considerably more active in an *in vitro* antiviral assay (IC_{95} values of 12 nM and 400 nM respectively).[177] Very recently these inhibitors have been further modified through replacement of the N-terminal Boc group by 3-tetrahydrofuranyl and 3-tetrahydropyranyl urethane moieties.[178] The 3(S)-tetrahydrofuranyl urethane **120**, in particular, was more than ten times as active as the Boc derivative **118** *vs.* HIV proteinase, and was much more active in an antiviral assay (IC_{95} values of 3 nM and 400 nM respectively).

Dreyer *et al.* have carried out a systematic study of the effect of substitution at P1′ in a series of hydroxyethylene isosteres (Table 14).[179] The Phe-Gly isostere **121** was a poor inhibitor, but activity increased dramatically with increasing size of the P1′ side-chain. Introduction of a P3′ valine residue also increased potency by a factor of 20–30 (**126** *vs.* **122**), whereas extension towards the N-terminus had less effect.

In the Roche laboratories renin inhibitors containing a Leu-Val hydroxyethylene isostere, which we reasoned might resemble the Leu-Ala cleavage site for HIV proteinase (site 2 in Figure 1), were screened for inhibition of HIV proteinase. The pentapeptide mimetic **127** (Table 15) was identified as a lead structure, and optimisation led to the very potent inhibitor **128**.[30] Interestingly, research workers at the Upjohn company independently synthesised a very close analogue **129**.[130]

Although caution must be exercised in comparing results obtained under different assay conditions, it is clear that **129** is considerably less potent than **128**. The crystal structure of **128** complexed with HIV-1 proteinase[30,114] suggested a possible explanation for this difference in potency. A weak hydrogen bond is formed between the backbone oxygen atom of the Boc group and a water molecule, which, in turn, hydrogen bonds to an arginine side-chain. The N-terminal tert.-butylacetyl group precludes such interactions for **129**. In addition, the pyridine ring at the C-terminus of **128** forms a stacking interaction with Arg[8]. The shorter aminomethylpyridine residue in **129** may be unable to participate in a similar interaction.

An interesting variant of the hydroxyethylene transition state mimetic in which the P1′ α-carbon atom is replaced by trigonal nitrogen has recently been reported.[180] Many of these hydroxyethyl ureas are potent and selective inhibitors of HIV proteinase, and also show *in vitro* antiviral activity. Rather surprisingly in view of their similarity to the hydroxyethylene transition state mimetics, these inhibitors have the opposite

Table 14 Effect of P1′ side-chain on inhibition of HIV-1 proteinase by hydroxyethylene transition state mimetics

Compound	Structure	K_i (nM)
121	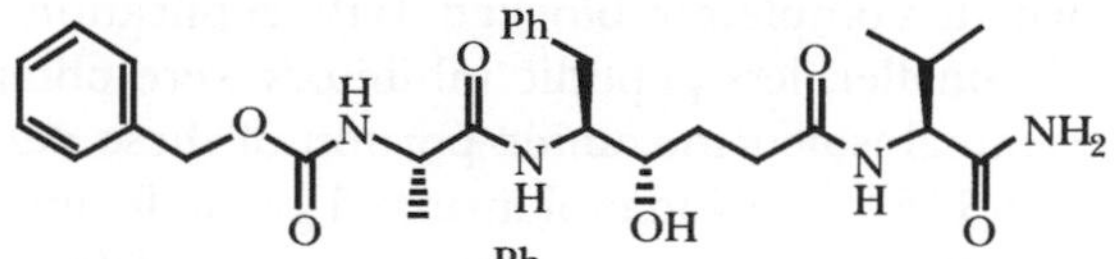	6,500
122	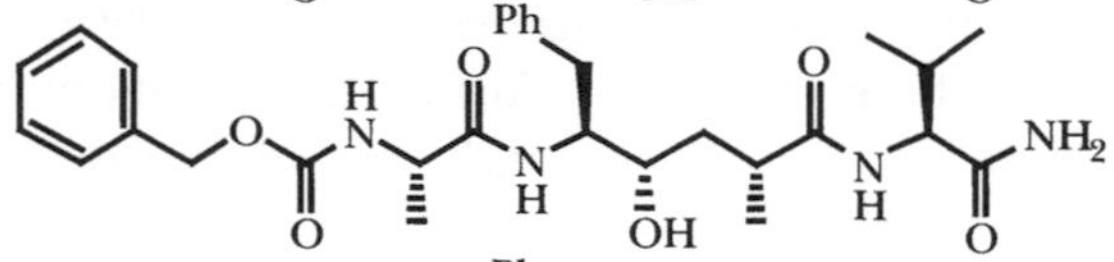	50
123	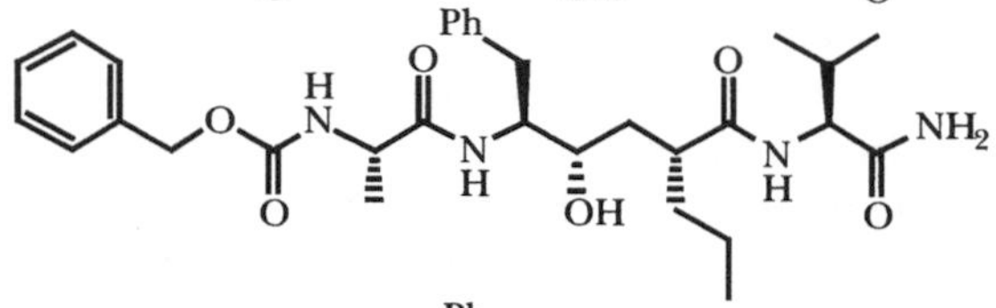	20
124	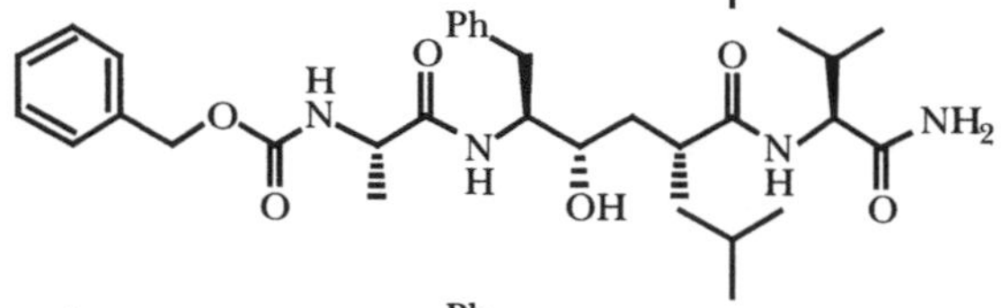	3.9
125	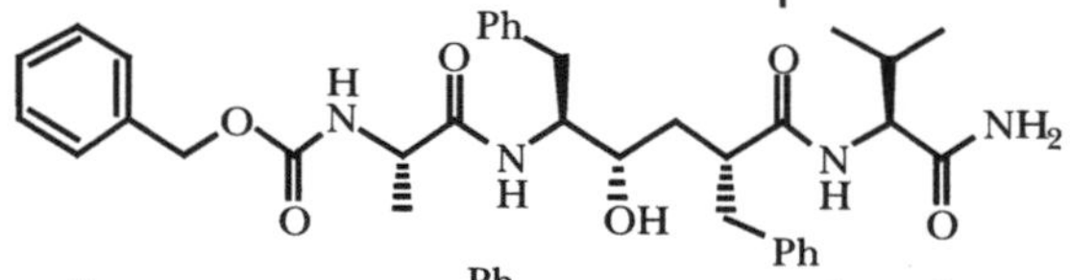	0.6
126	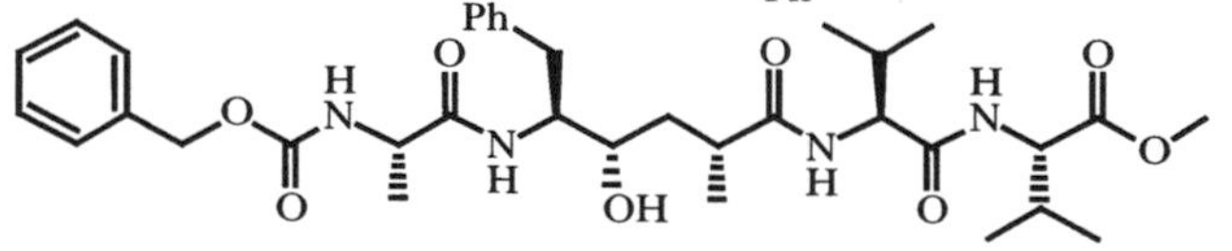	1.6

stereochemical requirement for the secondary alcohol function. The *R* alcohol **130** (Table 16) is 1700 times as potent as the *S* diastereomer. X-ray crystallography showed **130** and other analogues to have an unexpected mode of binding to HIV proteinase. The isobutyl group does not bind in the S1′ subsite, but rather resides in the S2′ subsite. Similarly, the N-terminal alkyl group binds in the S1′ subsite. This juxtapositioning of side-chains provides another example of the way in which subtle changes in inhibitor structure can influence the mode of binding.

Starting with a potent renin inhibitor which also inhibited HIV proteinase,[130] Thaisrivongs *et al.* synthesised a series of dihydroxyethylene isosteres such as **131**.[181] This very potent inhibitor was active in an *in vitro* antiviral assay with an IC_{50} value in the range 1–10 nM. Replacement of the P2 histidine residue by an *ortho* substituted benzoyl group which might follow the N-terminal backbone of **131** gave the less peptidic analogue **132**. Although somewhat reduced in potency *vs.* HIV proteinase, this compound had considerably greater antiviral activity (IC_{50} < 1 nM).[182]

Table 15 Hydroxyethylene isosteres based on Leu-Val and Cyclohexylalanine-Val residues

Compound	Structure	IC_{50} or K_i (nM)
127	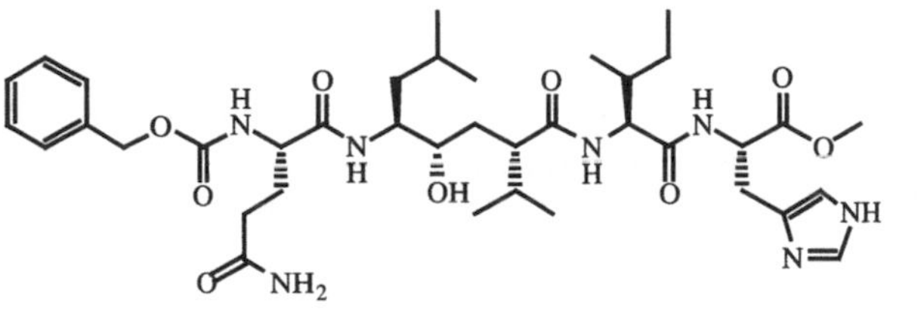	IC_{50} 27
128	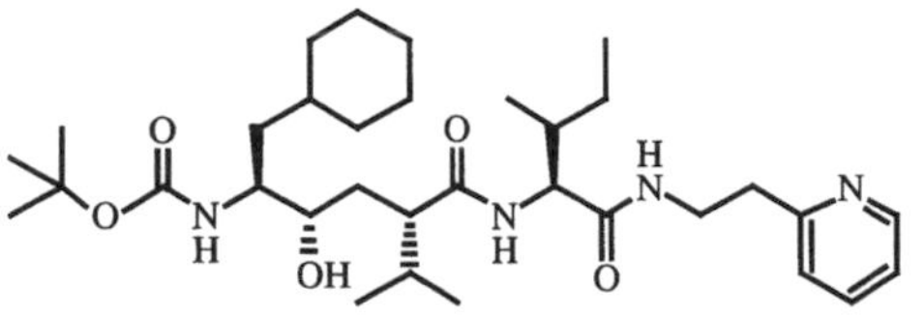	IC_{50} < 2.5 K_i 0.3
129	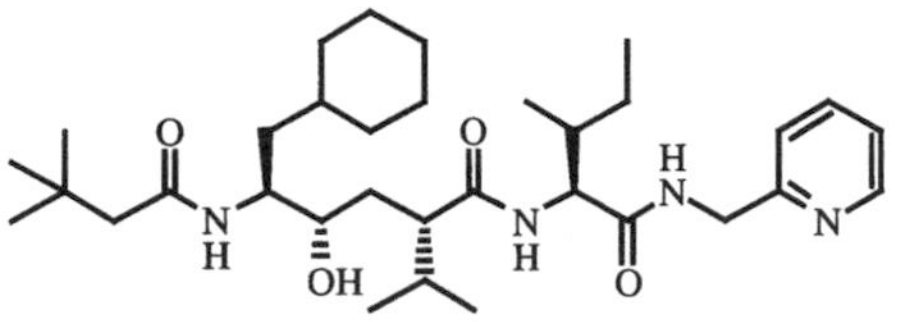	K_i 80

Table 16 (Hydroxyethyl)urea and dihydroxyethylene inhibitors of HIV-1 proteinase

Compound	Structure	IC_{50} or K_i (nM)
130	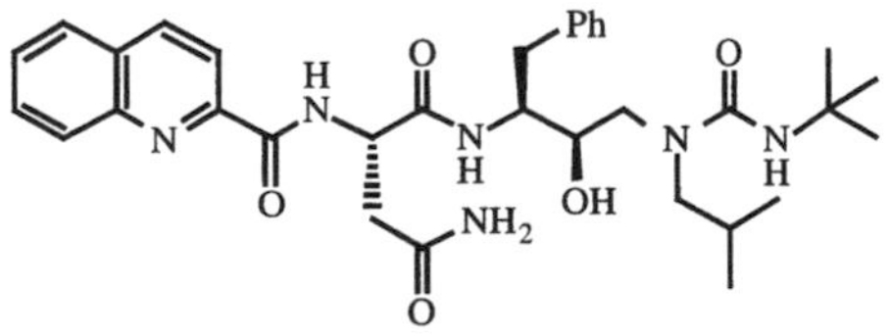	IC_{50} 6.3
131	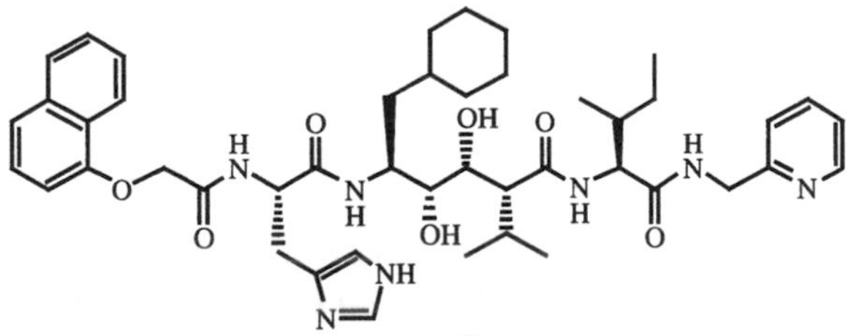	K_i < 1.0
132	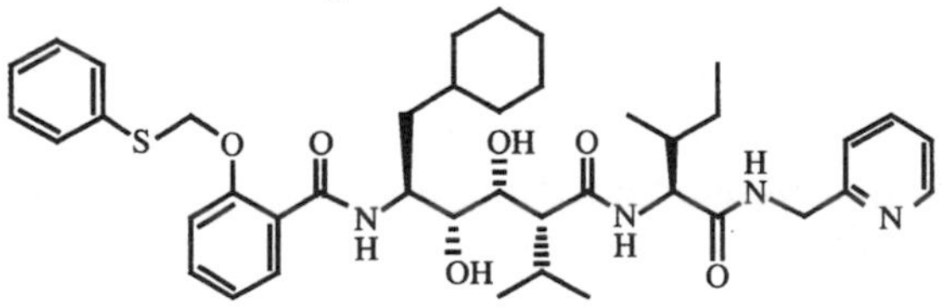	K_i 5

A novel approach to inhibition of HIV proteinase which involves the design of C_2-symmetric compounds which match the unique C_2-symmetry of retroviral proteinases has been described by Kempf et al.[115,183] Conceptually this involves definition of a C_2 axis in the tetrahedral transition state 7 of peptide bond cleavage, arbitrary deletion of one "half" of the substrate (the choice of the C-terminal portion for deletion was based on the greater importance observed for the N-terminal region in previously reported inhibitors of aspartic proteinases), and two fold rotation of the remaining portion of the substrate to give a symmetric structure. Location of the C_2 axis on the tetrahedral carbon atom of the transition state 7 leads to a pseudosymmetric diaminoalcohol core structure, exemplified in 133 (Table 17), while dissection of the scissile carbon-nitrogen bond in 7 generates three stereochemically distinct diols, two symmetric and one pseudosymmetric, as shown in 134–136. The pseudosymmetric alcohol 133 proved to be a potent inhibitor of HIV-1 proteinase, and, in an analogous approach, Bone et al. deleted the N-terminal portion of the hydroxyethylene inhibitor 118 to give compound 137 having a more extended diamidoalcohol core structure.[116] The diols 134–136, which maintain the same spacing between the P1 and P1' side chains as the substrates, are an order of magnitude more potent than the mono-hydroxy derivative 133. As a result of the potent enzyme inhibition and promising in vitro antiviral activity of the diols 134–136 (anti-HIV IC_{50} values of 20 nM, 60 nM, and 20 nM respectively), a considerable number of analogues having modified terminal groups were synthesised in an attempt to improve the physicochemical and pharmacokinetic profiles of these derivatives.[184] Among these was the pyridylmethyl urea 138 (A77003), which has considerably enhanced aqueous solubility, but, unfortunately, low oral bioavailability.[185] This inhibitor has recently entered clinical trials as an intravenous formulation. Interestingly, the configuration of the alcohol functions in these diols has little effect on potency, possibly indicating a high degree of flexibility of the central core structure, and in dramatic contrast to more traditional hydroxyl containing inhibitors. X-ray crystallographic analysis of the pseudosymmetric inhibitor 138 complexed with HIV proteinase shows it to bind unsymmetrically.[186] The R-hydroxyl group is situated close to the C_2-symmetry axis of the enzyme, and forms hydrogen bonds with both catalytic aspartic acid residues, while the S-hydroxyl is orientated away from the active site, within hydrogen bonding distance of only one aspartic acid residue. The asymmetry of binding cannot, however, be attributed to the pseudosymmetric nature of the inhibitor, since the pseudosymmetric mono-hydroxy derivative 133 binds in a symmetric fashion.[115] Furthermore, recent crystal structures of symmetric R,R-diols 134[186] and 139[117] show these inhibitors to bind asymmetrically. It would appear that in order to optimise hydrogen bonding between one hydroxyl group and the catalytic aspartic acid residues the inhibitor is translated along the active site, its midpoint being displaced by approximately 0.9 Å from the C_2-axis of the enzyme.[157] In contrast, the C_2-symmetric inhibitor 142 (Table 18), which lacks hydroxyl groups to interact with the aspartic acid residues, binds symmetrically to HIV proteinase.[157] This compound was derived from a high throughput screening programme. Structural modification of the initial lead compound gave the optimised inhibitor 142.[187,188] Other symmetry-based inhibitors

Table 17 Symmetric and pseudosymmetric inhibitors of HIV-1 proteinase[*]

Compound	Structure	IC_{50} (nM)
133	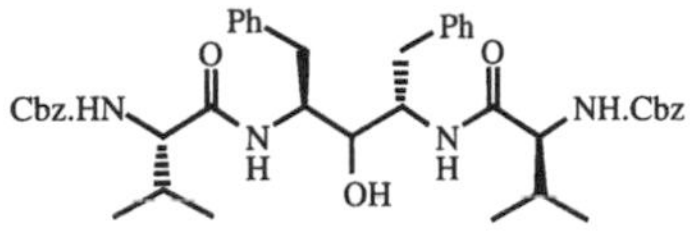	3.0
134	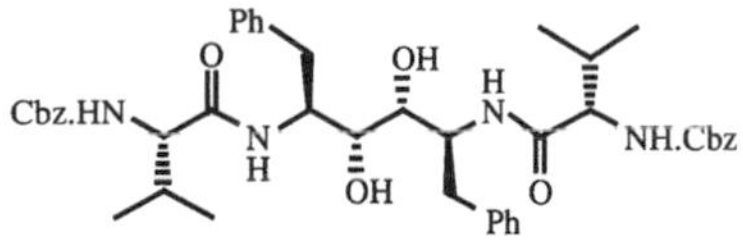	0.22
135	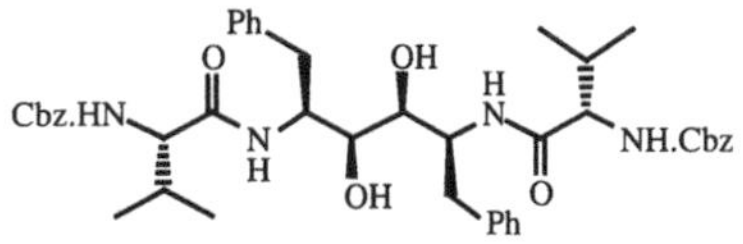	0.38
136	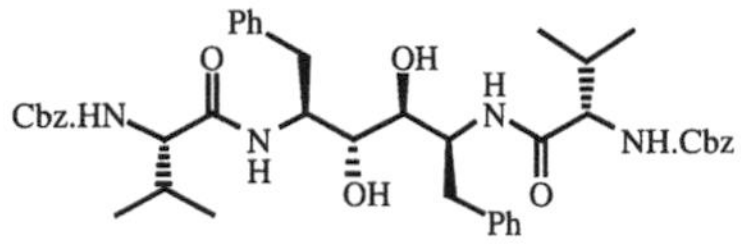	0.22
137	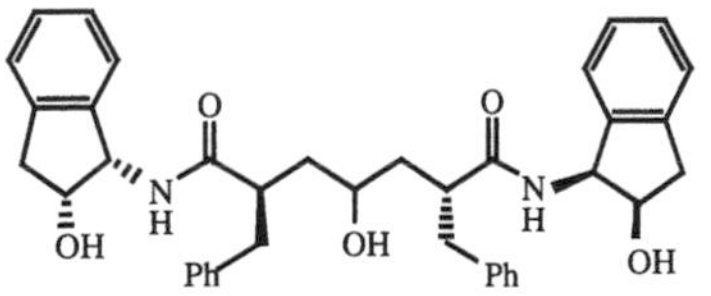	0.67
138	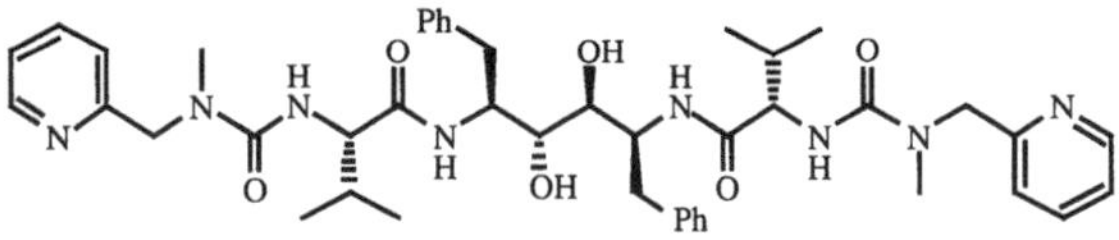	0.15
139	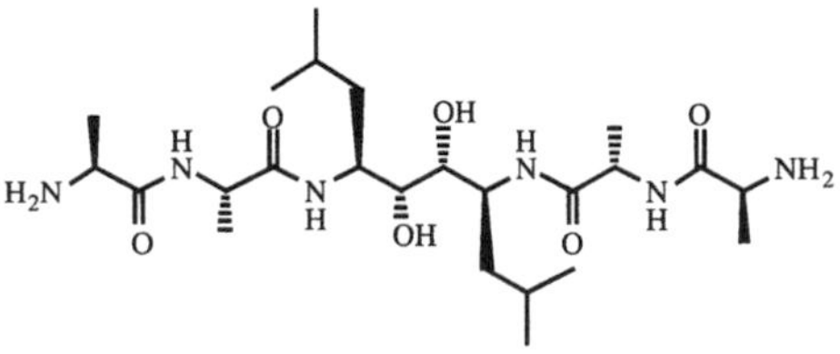	800[*]
140	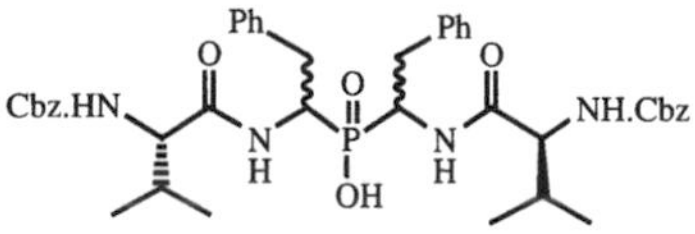	36
141	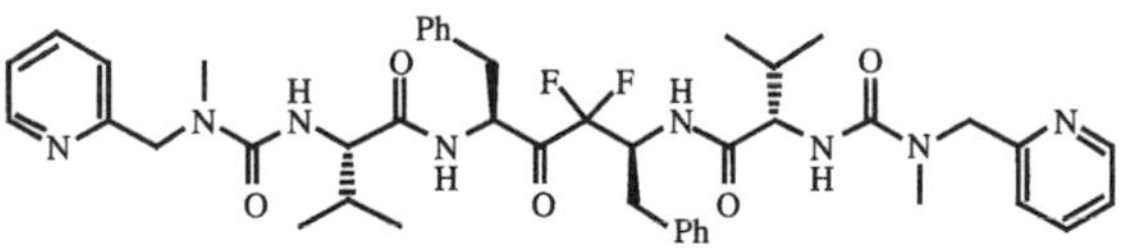	0.02

[*] K_i shown for **139**.

Table 18 Non-peptidic inhibitors of HIV-1 proteinase

Compound	Structure	IC_{50} or K_i (nM)
142	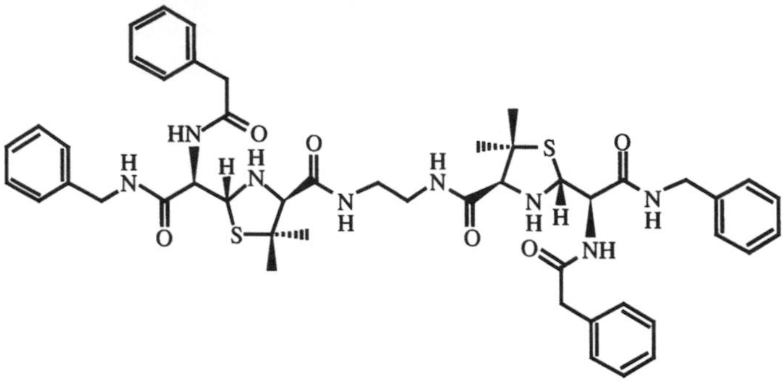	IC_{50} 0.9
143	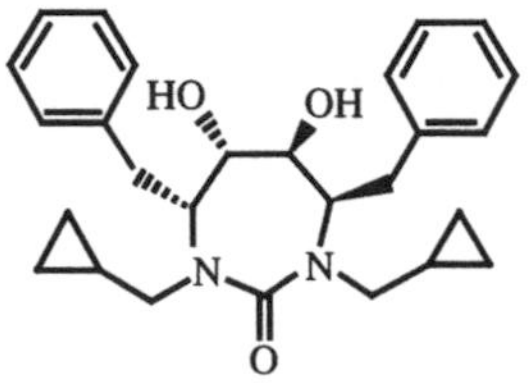	K_i 2.1
144	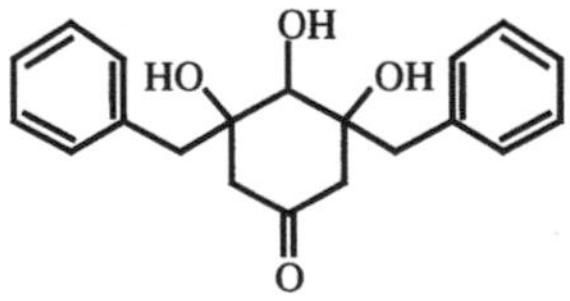	K_i <25,000
145	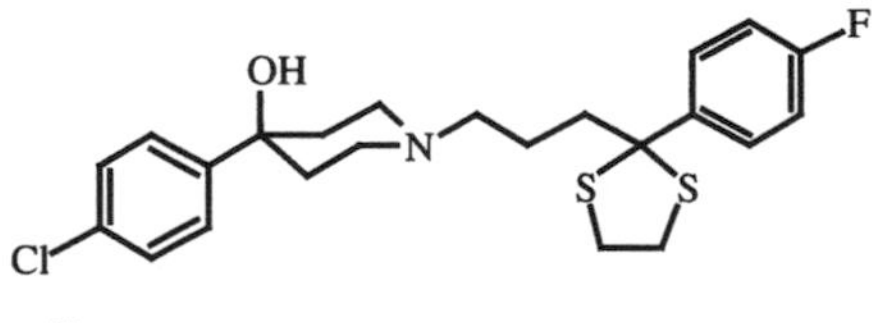	K_i 15,000
146		K_i 5,300

of HIV proteinase include the phosphinic acid **140**, tested as a mixture of diastereomers,[189] and the difluoroketone **141**, which binds in the *gem*-diol hydrated form, and is one of the most potent symmetry-based inhibitors reported to date.[190]

Whereas many HIV proteinase inhibitor research programmes were initiated before any crystal structures were available, and employed, at least as a starting point, peptide mimetics based on substrate sequences, the availability of numerous X-ray crystal structures now allows the *de novo* design of inhibitors. In one such approach structure-based modelling combined with a three-dimensional pharmacophore search strategy led to the very potent cyclic urea **143**.[191] This C_2-symmetric inhibitor binds to HIV proteinase in a symmetrical fashion, and, uniquely, displaces the tightly bound water molecule found in other HIV proteinase-inhibitor complexes.[186] Instead, the carbonyl oxygen of the urea function acts as a hydrogen bond acceptor for hydrogen bonds with Ile^{50} and $Ile^{50'}$ on the flap of the enzyme. In an analogous approach the SKB group have recently disclosed a series of cyclic triols such as **144** which inhibit HIV proteinase and also exhibit antiviral activity, but full details of their biological activities have not been published.[192] It remains to be seen whether these compounds or analogues will prove to have useful antiviral activity in a clinical setting.

In an alternative approach to the identification of new inhibitor structures DesJarlais *et al.* employed the program DOCK to search the Cambridge Crystallographic Database of 10,000 molecules for steric, and to some extent chemical complementarity with the active site of HIV-1 proteinase. Bromoperidol was identified as an interesting candidate, and the closely related antipsychotic agent haloperidol was tested and found to inhibit both HIV-1 and HIV-2 proteinases with K_i values in the region of 100 mM.[193] Conversion of the ketone to the corresponding thioketal **145** gave a more potent inhibitor, and two crystal structures of **145** complexed with HIV proteinase have recently been determined.[194] While it is unlikely that **145** itself will prove clinically useful, this approach has generated a new structural lead which has potential for further development to give more potent analogues.

One of the most extraordinary inhibitors of HIV proteinase reported to date is the fullerene derivative **146**.[195] Modelling studies suggested that an icosahedral C_{60} fullerene molecule has approximately the same radius as the cylindrical active site of HIV proteinase in its open, uncomplexed form. Since the active site of HIV proteinase is lined mainly by hydrophobic residues, it was suggested that a C_{60} fullerene should undergo a strong hydrophobic interaction with the active site of the enzyme. This hypothesis was vindicated when the water soluble fullerene derivative **146** was shown to inhibit HIV proteinase, and also to possess *in vitro* antiviral activity (anti-HIV IC_{50} 7,000 nM). While this inhibitor is considerably less potent than some peptidic inhibitors, it was suggested that potency could be increased by several orders of magnitude through introduction of amino groups to interact with the catalytic aspartic acid residues. Modelling of a 1,4-diamino fullerene derivative suggested that this should undergo hydrophobic interaction with HIV proteinase in the same way as **146**, but that it should also form salt bridges with the catalytic aspartates. This has yet to be verified experimentally.

OUTLOOK

Tremendous progress has been made in the design of inhibitors of HIV proteinase since this enzyme was first identified as a potential target for antiviral chemotherapy. Various transition state mimetic moieties have been incorporated into highly potent and selective inhibitors, many of which exhibit *in vitro* anti-HIV activity at nanomolar concentrations. The majority of inhibitors reported to date were designed using the classical approach of incorporating stable dipeptide isosteres into peptide mimetics based on substrate cleavage sequences, much of the methodology arising from earlier work on inhibitors of renin. However, the availability of high resolution crystal structures has, more recently, facilitated structure-based approaches, in which compounds are designed on the basis of a detailed understanding of the precise interactions between the enzyme and inhibitors. Such approaches may lead to structures which are less peptide-like in nature, and which, consequently, have potential for improved pharmacokinetic properties. Earlier studies of peptide mimetic renin inhibitors showed such derivatives to have, in general, rather poor oral bioavailability. Nevertheless, the Roche hydroxyethylamine containing HIV proteinase inhibitor Ro 31-8959 **81** has adequate oral bioavailability, and this, together with its very high potency, has allowed clinical trials using the oral route of administration. Initial results from Phase 2 clinical trials carried out in the United Kingdom, France, and Italy show Ro 31-8959 to be well tolerated, and to achieve immunological and antiviral effects based on laboratory markers of disease. While this inhibitor is the most advanced in terms of development, other compounds have already entered clinical trials. It is too early to predict with certainty the eventual outcome of treatment of AIDS with inhibitors of HIV proteinase, but the very high potency and selectivity, the low potential for toxicity and for development of viral resistance, and the clear synergy of action with other classes of anti-HIV agents all lend considerable encouragement to the hope that HIV proteinase inhibitors such as Ro 31-8959 will play an important role in the clinical treatment of AIDS.

ACKNOWLEDGEMENTS

I am grateful to Drs. A. Kröhn, B. Sherborne, and N. Borkakoti for computer graphics, to Drs. C. Craig and D. Hockley (National Institute for Biological Standards and Control) for electron micrographs, and to Mr. G. Towerzey for photography.

REFERENCES

1. Centers for Disease Control (1981) *Pneumocystis* pneumonia — Los Angeles. *Morbidity and Mortality Weekly Report,* **30**, 250–252.
2. Centers for Disease Control (1981) Kaposi's sarcoma and *Pneumocystis* pneumonia among homosexual men — New York City and California. *Morbidity and Mortality Weekly Report,* **30**, 305–308.
3. Centers for Disease Control (1982) Update on acquired immune deficiency syndrome (AIDS) — United States. *Morbidity and Mortality Weekly Report,* **31**, 507–514.

4. Klatzmann, D., Barré-Sinoussi, F., Nugeyre, M.T., Dauguet, C., Vilmer, E., Griscelli, C., Brun-Vézinet, F., Rouzioux, C., Gluckman, J.C., Chermann, J.C. and Montagnier, L. (1984) Selective tropism of lymphadenopathy associated virus (LAV) for helper-inducer T lymphocytes. *Science*, **225**, 59–63.

5. Poiesz, B.J., Ruscetti, F.W., Gazdar, A.F., Bunn, P.A., Minna, J.D. and Gallo, R.C. (1980) Detection and isolation of type C retrovirus particles from fresh and cultured lymphocytes of a patient with cutaneous T-cell lymphoma. *Proceedings of the National Academy of Sciences of the USA*, **77**, 7415–7419.

6. Barré-Sinoussi, F., Chermann, J.C., Rey, F., Nugeyre, M.T., Chamaret, S., Gruest, J., Dauguet, C., Axler-Blin, C., Vézinet-Brun, F., Rouzioux, C., Rozenbaum, W. and Montagnier, L. (1983) Isolation of a T-lymphotropic retrovirus from a patient at risk for acquired immune deficiency syndrome (AIDS). *Science*, **220**, 868–871.

7. Popovic, M., Sarngadharan, M.G., Read, E. and Gallo, R.C. (1984) Detection and continuous production of cytopathic retroviruses (HTLV-III) from patients with AIDS and pre-AIDS. *Science*, **224**, 497–500.

8. Gallo, R.C., Salahuddin, S.Z., Popovic, M., Shearer, G.M., Kaplan, M., Haynes, B.F., Palker, T.J., Redfield, R., Oleske, J., Safai, B., White, G., Foster, P. and Markham, P.D. (1984) Frequent detection and isolation of cytopathic retroviruses (HTLV-III) from patients with AIDS and at risk for AIDS. *Science*, **224**, 500–503.

9. Gallo, R.C. and Montagnier, L. (1988) AIDS in 1988. *Scientific American*, **259**, 25–32.

10. Clavel, F., Guyader, M., Guétard, D., Sallé, M., Montagnier, L. and Alizon, M. (1986) Molecular cloning and polymorphism of the human immune deficiency virus type 2. *Nature*, **324**, 691–695.

11. Guyader, M., Emerman, M., Sonigo, P., Clavel, F., Montagnier, L. and Alizon, M. (1987) Genome organisation and transactivation of the human immunodeficiency virus type 2. *Nature*, **326**, 662–669.

12. Levy, J.A. (1989) Human immunodeficiency viruses and the pathogenesis of AIDS. *Journal of the American Medical Association*, **261**, 2997–3006.

13. Merson, M.H. (1993) Slowing the Spread of HIV: Agenda for the 1990s. *Science*, **260**, 1266–1268.

14. Wong-Staal, F. (1989) Molecular Biology of Human Immunodeficiency Viruses. In *Current Topics in AIDS*, **Vol. 2**, pp. 81–102. John Wiley and Sons Ltd.

15. Haseltine, W.A. (1990) Molecular Biology of HIV-1. In *AIDS and the New Viruses*, pp. 11–40. Academic Press Ltd.

16. Sandstrom, E. (1989) Antiviral therapy in human immunodeficiency virus infection. *Drugs*, **38**, 417–450.

17. Kuehl, P.G. and Pau, A.K. (1990) The AIDS frontline: New drugs in research. *Drug Topics*, 7 May, 52–61.

18. Mitsuya, H., Yarchoan, R. and Broder, S. (1990) Molecular targets for AIDS therapy. *Science*, **249**, 1533–1543.

19. Mitsuya, H., Weinhold, K.J., Furman, P.A., St. Clair, M.H., Lehrman, S.N., Gallo, R.C., Bolognesi, D., Barry, D.W. and Broder, S. (1985) 3′-Azido-3′-deoxythymidine (BW A509U): An antiviral agent that inhibits the infectivity and cytopathic effect of human T-lymphotropic virus type III/lymphadenopathy-associated virus *in vitro*. *Proceedings of the National Academy of Sciences of the USA*, **82**, 7096–7100.

20. Mitsuya, H. and Broder, S. (1987) Strategies for antiviral therapy in AIDS. *Nature*, **325**, 773–778.

21. Mitsuya, H., Matsukura, M. and Broder, S. (1987) Rapid *in vitro* systems for assessing activity of agents against HTLV-III/LAV. In *AIDS: Modern concepts and therapeutic challenges*, pp. 303–333, Marcel-Dekker, New York.

22. De Clercq, E. (1989) Potential drugs for the treatment of AIDS. *Journal of Antimicrobial Chemotherapy*, **23 Suppl. A**, 35–46.

23. Pauwels, R., Andries, K., Desmyter, J., Schols, D., Kukla, M.J., Breslin, H.J., Raeymaeckers, A., Van Gelder, J., Woestenborghs, R., Heykants, S., Scellekens, K., Janssen, M.A.C., De Clercq, E. and Jansses, P.A.J. (1990) Potent and selective inhibition of HIV-1 replication *in vitro* by a novel series of TIBO derivatives. *Nature*, **343**, 470–474.

24. Saari, W.S., Hoffman, J.M., Wai, J.S., Fisher, T.E., Rooney, C.S., Smith, A.M., Thomas, C.M., Goldman, M.E., O'Brien, J.A., Nunberg, J.H., Quintero, J.C., Schlief, W.A., Emini, E.A. and Anderson, P.S. (1991) 2-Pyridone derivatives: A new class of non-nucleoside, HIV-1-specific reverse transcriptase inhibitors. *Journal of Medicinal Chemistry*, **34**, 2922–2925.

25. Saari, W.S., Wai, J.S., Fisher, T.E., Thomas, C.M., Hoffman, J.M., Rooney, C.S., Smith, A.M., Jones, J.H., Bamberger, D.L., Goldman, M.E., O'Brien, J.A., Nunberg, J.H., Quintero, J.C., Schlief, W.A., Emini, E.A. and Anderson, P.S. (1992) Synthesis and evaluation of 2-pyridone derivatives as HIV-1-specific reverse transcriptase inhibitors. 2. Analogues of 3-aminopyridin-2(H)-one. *Journal of Medicinal Chemistry*, **35**, 3792–3802.

26. Smith, D.H., Byrn, R.A., Marsters, S.A., Gregory, T., Groopman, J.E. and Capon, D.J. (1987) Blocking of HIV infectivity by a soluble, secreted form of CD4 antigen. *Science*, **238**, 1704–1707.

27. Deen, K.C., McDougal, J.S., Inacker, R., Folena-Wasserman, G., Arthos, J., Rosenberg, J., Maddon, P.J., Axel, R. and Sweet, R.W. (1988) A soluble form of CD4 (T4) protein inhibits AIDS virus infection. *Nature*, **331**, 82–84.

28. Fisher, R.A., Bertonis, J.M., Meier, W., Johnson, V.A., Costopoulos, D.S., Liu, T., Tizard, R., Walker, B.D., Hirsch, M.S., Schooley, R.T. and Flavell, R.A. (1988) HIV infection blocked *in vitro* by recombinant soluble CD4. *Nature*, **331**, 76–78.

29. Hsu, M.-C., Schutt, A.D., Holly, M., Slice, L.W., Sherman, M.I., Richman, D.D., Potash, M.J. and Volsky, D.J. (1991) Inhibition of HIV replication in acute and chronic infections *in vitro* by a tat antagonist. *Science*, **254**, 1799–1802.

30. Martin, J.A. (1992) Recent advances in the design on HIV proteinase inhibitors. *Antiviral Research*, **17**, 265–278.

31. Huff, J.R. (1991) HIV Protease: A novel chemotherapeutic target for AIDS. *Journal of Medicinal Chemistry*, **34**, 2305–2314.

32. Norbeck, D.W. and Kempf, D.J. (1991) HIV Protease inhibitors. *Annual Reports in Medicinal Chemistry*, **26**, 141–150.

33. Tomasselli, A.G., Howe, W.J., Sawyer, T.K., Wlodawer, A. and Heinrikson, R.L. (1991) The complexities of AIDS: An assessment of the HIV protease as a therapeutic target. *Chimica Oggi*, **May**, 6–27.

34. Meek, T.D. (1992) Inhibitors of HIV-1 protease. *Journal of Enzyme Inhibition*, **6**, 65–98.

35. Debouck, C. (1992) The HIV-1 protease as a therapeutic target for AIDS. *AIDS Research and Human Retroviruses*, **8**, 153–164.

36. Ratner, L., Haseltine, W., Patarca, R., Livak, K.J., Starcich, B., Josephs, S.F., Doran, E.R., Rafalski, J.A., Whitehorn, E.A., Baumeister, K., Ivanoff, L., Petteway, S.R., Jr., Pearson, M.L., Lautenberger, J.A., Papas, T.S., Ghrayeb, J., Chang, N.T., Gallo, R.C. and Wong-Staal, F. (1985) Complete nucleotide sequence of the AIDS virus, HTLV-III. *Nature*, **313**, 277–284.

37. Hellen, C.U.T., Krausslich, H.-G. and Wimmer, E. (1989) Proteolytic processing of polyproteins in the replication of RNA viruses. *Biochemistry*, **28**, 9881–9890.

38. Leis, J., Baltimore, D., Bishop, J.M., Coffin, J., Fleissner, E., Goff, S.P., Oroszlan, S., Robinson, H., Skalka, A.M., Temin, H.M. and Vogt, V. (1988) Standardized and simplified nomenclature for proteins common to all retroviruses. *Journal of Virology*, **62**, 1808–1809.

39. Yu, X., Yuan, X., Matsuda, Z., Lee, T.-H. and Essex, M.J. (1992) The matrix protein of human immunodeficiency virus type 1 is required for incorporation of viral envelope protein into mature virions. *Journal of Virology*, **66**, 4966–4971.

40. Sarngadharan, M., Popovic, M., Bruch, L., Schupbach, J. and Gallo, R.C. (1984) Antibodies reactive with human T-lymphotropic retrovirus (HTLV-III) in the serum of patients with AIDS. *Science*, **224**, 506–508.

41. De Rocquigny, H., Gabus, C., Vincent, A., Fournie-Zaluski, M.-C., Roques, B. and Darlix, J.-L. (1992) Viral RNA annealing activities of human immunodeficiency virus type 1 nucleocapsid protein require only peptide domains outside the zinc fingers. *Proceedings of the National Academy of Sciences of the USA*, **89**, 6472–6476.

42. Jacks, T., Power, M.D., Masiarz, F.R., Luciw, P.A., Barr, P.J. and Varmus, H.E. (1988) Characterization of ribosomal frameshifting in HIV-1 *gag-pol* expression. *Nature*, **331**, 280–283.

43. Farmerie, W.G., Loeb, D.D., Casavant, N.C., Hutchinson, C.A., III, Edgell, M.H. and Swanstrom, R. (1987) Expression and processing of the AIDS virus reverse transcriptase in Escherichia coli. *Science*, **236**, 305–308.

44. Le Grice, S.F.J., Mills, J. and Mous, J. (1988) Active site mutagenesis of the AIDS virus protease and its alleviation by *trans* complementation. *EMBO Journal*, **7**, 2547–2553.

45. Kramer, R.A., Schaber, M.D., Skalka, A.M., Ganguly, K., Wong-Staal, F. and Reddy, E.P. (1986) HTLV-III *gag* protein is processed in yeast cells by the virus *pol*-protease. *Science*, **231**, 1580–1584.

46. Kohl, N.E., Emini, E.A., Schlief, W.A., Davis, L.J., Heimbach, J.C., Dixon, R.A., Scolnick, E.M. and Sigal, I.S. (1988) Active human immunodeficiency virus protease is required for viral infectivity. *Proceedings of the National Academy of Sciences of the USA*, **85**, 4686–4690.

47. Göttlinger, H.G., Sodroski, J.G. and Haseltine, W.A. (1989) Role of capsid precursor processing and myristoylation in morphogenesis and infectivity of human immunodeficiency virus type 1. *Proceedings of the National Academy of Sciences of the USA*, **86**, 5781–5785.

48. Peng, C., Ho, B.K., Chang, T.W. and Chang, N.T. (1989) Role of human immunodeficiency virus type 1-specific protease in core protein maturation and viral infectivity. *Journal of Virology*, **63**, 2550–2556.

49. Lillehoj, E.P., Salazar, F.H.R., Mervis, R.J., Raum, M.G., Chan, H.W., Ahmad, N. and Venkatesan, S. (1988) Purification and structural characterisation of the putative gag-pol protease of human immunodeficiency virus. *Journal of Virology*, **62**, 3053–3058.

50. Nutt, R.F., Brady, S.F., Darke, P.L., Ciccarone, T.M., Colyon, C.D., Nutt, E.M., Rodkey, J.A., Bennett, C.D., Waxman, L.H., Sigal, I.S., Anderson, P.S. and Veber, D.F. (1988) Chemical synthesis and enzymatic activity of a 99-residue peptide with a sequence proposed for the human immunodeficiency virus protease. *Proceedings of the National Academy of Sciences of the USA*, **85**, 7129–7133.

51. Schneider, J. and Kent, S.B. (1988) Enzymatic activity of a synthetic 99 residue protein corresponding to the putative HIV-1 protease. *Cell*, **54**, 363–368.

52. Copeland, T.D. and Oroszlan, S. (1988) Genetic locus, primary structure, and chemical synthesis of human immunodeficiency virus protease. *Gene Analysis Techniques*, **5**, 109–115.

53. Wu, J.C., Carr, S.F., Jarnagin, K., Kirsher, S., Barnett, J., Chow, J., Chan, H.W., Chen, M.S., Medzihradszky, D., Yamashiro, D. and Santi, D.V. (1990) Synthetic HIV-2 protease cleaves the GAG precursor of HIV-1 with the same specificity as HIV-1 protease. *Archives of Biochemistry and Biophysics*, **277**, 306–311.

54. Darke, P.L., Leu, C.-T., Davis, L.J., Heimbach, J.C., Diehl, R.E., Hill, W.S., Dixon, R.A.F. and Sigal, I.S. (1989) Human immunodeficiency virus protease. Bacterial expression and characterisation of the purified aspartic protease. *Journal of Biological Chemistry*, **264**, 2307–2312.

55. Graves, M.C., Lim, J.J., Heimer, E.P. and Kramer, R.A. (1988) An 11-kDa form of human immuno-deficiency virus protease expressed in *Escherichia coli* is sufficient for enzymatic activity. *Proceedings of the National Academy of Sciences of the USA*, **85**, 2449–2453.

56. Cheng, Y.S., McGowan, M.H., Kettner, C.A., Schloss, J.V., Erickson, V.S. and Yin, F.H. (1990) High level synthesis of recombinant HIV-1 protease and the recovery of active enzyme from inclusion bodies. *Gene*, **87**, 243–248.

57. Rittenhouse, J., Turon, M.C., Helfrich, R.J., Albrect, K.S., Weigl, D., Simmer, R.L., Mordini, F., Erickson, J. and Kohlbrenner, W.E. (1990) Affinity purification of HIV-1 and HIV-2 proteases from recombinant E. coli strains using pepstatin-agarose. *Biochemical and Biophysical Research Communications*, **171**, 60–66.

58. Overton, H.A., Fujii, Y., Price, I.R. and Jones, I.M. (1989) The protease and *gag* gene products of the human immunodeficiency virus: Authentic cleavage and post-translational modification in an insect cell expression system. *Virology*, **170**, 107–116.

59. Lapatto, R., Blundell, T., Hemmings, A., Overington, J., Wilderspin, A., Wood, S., Merson, J.R., Whittle, P.J., Danley, D.E., Geoghegan, K.F., Hawrylik, S.J., Lee, S.E., Scheld, K.G. and Hobart, P.M. (1989) X-ray analysis of HIV-1 proteinase at 2.7 Å resolution confirms structural homology among retroviral enzymes. *Nature*, **342**, 299–302.

60. Billich, S., Knoop, M.T., Hansen, J., Strop, P., Sedlacek, J., Mertz, R. and Moelling, K. (1988) Synthetic peptides as substrates and inhibitors of human immunodeficiency virus-1 protease. *Journal of Biological Chemistry*, **263**, 17905–17908.

61. Debouck, C., Gorniak, J.G., Strickler, J.E., Meek, T.D., Metcalf, B.W. and Rosenberg, M. (1987) Human immunodeficiency virus protease expressed in *Escherichia coli* exhibits autoprocessing and specific maturation of the gag precursor. *Proceedings of the National Academy of Sciences of the USA*, **84**, 8903–8906.

62. Hansen, J., Billich, S., Schulze, T., Sukrow, S. and Moelling, K. (1988) Partial purification and substrate analysis of bacterially expressed HIV protease by means of monoclonal antibody. *EMBO Journal*, **7**, 1785–1791.

63. Hostomsky, Z., Appelt, K. and Ogden, R.C. (1989) High level expression of self-processed HIV-1 protease in *Escherichia coli* using a synthetic gene. *Biochemical and Biophysical Research Communications,* **161**, 1056–1063.

64. Korant, B.D. and Rizzo, C.J. (1990) Expression in *Escherichia coli* of the AIDS virus aspartic protease through a protein fusion. *Biol. Chem. Hoppe-Seyler,* **371**, 271–275.

65. Hirel, Ph.-H., Parker, F., Boiziau, J., Jung, G., Outerovitch, D., Dugue, A., Peltiers, C., Giuliacci, C., Boulay, R., Lelievre, Y., Cambou, B., Mayaux, J.-F. and Cartwright, T. (1990) HIV-1 aspartic proteinase: high level production and automated fluormetric screening assay of inhibitors. *Antiviral Chemistry and Chemotherapy,* **1**, 9–15.

66. Danley, D.E., Geoghegan, K.F., Scheld, K.G., Lee, S.E., Merson, J.R., Hawrylik, S.J., Rickett, G.A., Ammirati, M.J. and Hobart, P.M. (1989) Crystallizable HIV-1 protease derived from expression of the viral *pol* gene in *Escherichia coli. Biochemical and Biophysical Research Communications,* **165**, 1043–1050.

67. Heimbach, J.C., Garsky, V.M., Michelson, S.R., Dixon, R.A.F., Sigal, I.S. and Darke, P.L. (1989) Affinity purification of the HIV-1 protease. *Biochemical and Biophysical Research Communications,* **164**, 955–960.

68. Toh, H., Ono, M., Saigo, K. and Miyata, T. (1985) Retroviral protease-like sequence in the yeast transposon Ty 1. *Nature,* **315**, 691–692.

69. Seelmeier, S., Schmidt, H., Turk, V. and von der Helm, K. (1988) Human immunodeficiency virus has an aspartic-type protease that can be inhibited by pepstatin A. *Proceedings of the National Academy of Sciences of the USA,* **85**, 6612–6616.

70. Kräusslich, H.-G., Scneider, H., Zybarth, G., Carter, C.A. and Wimmer, E. (1988) Processing of *in vitro* synthesised *gag* precursor proteins of human immunodeficiency virus (HIV) type 1 by HIV proteinase generated in *Escherichia coli. Journal of Virology,* **62**, 4393–4397.

71. Kräusslich, H.G., Ingraham, R.H., Skoog, M.T., Wimmer, E., Pallai, P.V. and Carter, C.A. (1989) Activity of purified biosynthetic proteinase of human immunodeficiency virus on natural substrates and synthetic peptides. *Proceedings of the National Academy of Sciences of the USA,* **86**, 807–811.

72. Richards, A.D., Roberts, R., Dunn, B.M., Graves, M.C. and Kay, J. (1989) Effective blocking of HIV-1 proteinase activity by characteristic inhibitors of aspartic proteinases. *FEBS Letters,* **247**, 113–117.

73. Tang, J., James, M.N.G., Hsu, I.N., Jenkins, J.A. and Blundell, T.L. (1977) Structural evidence for gene duplication in the evolution of the acid proteases. *Nature,* **271**, 618–621.

74. Pearl, L.H. and Taylor, W.R. (1987) A structural model for the retroviral proteases. *Nature,* **329**, 351–354.

75. Casey, J.M., Kim, Y., Andersen, P.R., Watson, K.F., Fox, J.L. and Devare, S.G. (1985) Human T-cell lymphotropic virus type III: Immunologic characterization and primary structure analysis of the major internal protein, p24. *Journal of Virology,* **55**, 417–423.

76. Lightfoote, M.M., Coligan, J.E., Folks, T.M., Fauci, A.S., Martin, M.S. and Venkatesan, S. (1986) Structural characterization of reverse transcriptase and endonuclease polypeptides of the acquired immune deficiency syndrome retrovirus. *Journal of Virology,* **60**, 771–775.

77. Veronese, F.D., Copeland, De Vico, A.L., Rahman, R., Oroszlan, S., Gallo, R.C. and Sarngadharan, M.G. (1986) Characterization of highly immunogenic p66/p51 as the reverse transcriptase of HTLV-III/LAV. *Science,* **231**, 1289–1291.

78. Veronese, F.D., Copeland, T.D., Oroszian, S., Gallo, R.C. and Sarngadharan, M.G. (1988) Biochemical and immunological analysis of human immunodeficiency virus gag gene products p17 and p24. *Journal of Virology,* **62**, 795–801.

79. Darke, P.L., Nutt, R.F., Brady, S.F., Garsky, V.M., Ciccarone, T.M., Leu, C.-T., Lumma, P.K., Freidinger, R.M., Veber, D.F. and Sigal, I.S. (1988) HIV-1 protease specificity of peptide cleavage is sufficient for processing of gag and pol polyproteins. *Biochemical and Biophysical Research Communications,* **156**, 297–303.

80. Mizrahi, V., Lazarus, G.M., Miles, L.M., Meyers, C.A. and Debouck, C. (1989) Recombinant HIV-1 reverse transcriptase: Purification, primary structure, and polymerase/ribonuclease H activities. *Archives of Biochemistry and Biophysics,* **273**, 347–348.

81. Pettit, S.C., Michael, S.F. and Swanstrom, R. (1993) The Specificity of HIV-1 protease. *Perspectives in Drug Discovery and Design,* **1**, 69–83.

82. Henderson, L.E., Benveniste, R.E., Sowder, R., Copeland, T.D., Schultz, A.M. and Oroszlan, S. (1988) Molecular characterization of *gag* proteins from simian immunodeficiency virus (SIV$_{Mne}$). *Journal of Virology*, **62**, 2587–2595.

83. Pettit, S.C., Simsic, J., Loeb, D.D., Everitt, L., Hutchinson, C.A., III and Swanstrom, R. (1991) Analysis of retroviral protease cleavage sites reveals two types of cleavage sites and the structural requirements of the P1 amino acid. *Journal of Biological Chemistry*, **266**, 14539–14547.

84. Poorman, R.A., Tomasselli, A.G., Heinrikson, R.L. and Kézdy, F.J. (1991) A cumulative specificity model for proteases from human immunodeficiency virus types 1 and 2, inferred from statistical analysis of an extended substrate data base. *Journal of Biological Chemistry*, **266**, 14554–14561.

85. Pearl, L.H. and Taylor, W.R. (1987) Sequence specificity of retroviral proteases. *Nature*, **328**, 482.

86. Moore, M.L., Bryan, W.M., Fakhoury, S.A., Magaard, V.W., Huffman, W.F., Dayton, B.D., Meek, T.D., Hyland, L., Dreyer, G.B., Metcalf, B.W., Strickler, J.E., Gorniak, J. and Debouck, C. (1989) Peptide substrates and inhibitors of HIV-1 protease. *Biochemical and Biophysical Research Communications*, **159**, 420–425.

87. Richards, A.D., Phylip, L.H., Farmerie, W.G., Scarborough, P.E., Alvarez, A., Dunn, B., Hirel, P.H., Konvalinka, J., Strop, P., Pavlickova, L., Kostika, V. and Kay, J. (1990) Sensitive soluble chromogenic substrates for HIV-1 proteinase. *Journal of Biological Chemistry*, **265**, 7733–7736.

88. Nashed, N.T., Louis, J.M., Sayer, J.M., Wondrak, E.M., Mora, P.T., Oroszlan, S. and Jerina, D.M. (1989) Continuous spectrophotometric assay for retroviral proteases of HIV-1 and AMV. *Biochemical and Biophysical Research Communications*, **163**, 1079–1085.

89. Matayoshi, E.D., Wang, G.T., Krafft, G.A. and Erickson, J. (1990) Novel flourogenic substrates for assaying retroviral proteases by resonance energy transfer. *Science*, **247**, 954–958.

90. Geohegan, K.F., Spencer, R.W., Danley, D.E., Contillo, L.G., Jr. and Andrews, G.C. (1990) Fluorescence-based continuous assay for the aspartyl protease of human immunodeficiency virus-1. *FEBS Letters*, **262**, 119–122.

91. Tomasselli, A.G., Hui, J.O., Sawyer, T.K., Staples, D.J., Fitzgerald, D.J., Chaudhary, V.K., Pastan, I. and Heinrikson, R.L. (1990) Interdomain hydrolysis of a truncated *Pseudomonas* exotoxin by the human immunodeficiency virus-1 protease. *Journal of Biological Chemistry*, **265**, 408–413.

92. Giam, C.Z. and Boros, I. (1988) *In vivo* and *in vitro* auto processing of human immunodeficiency virus protease expressed in *Escherichia coli*. *Journal of Biological Chemistry*, **263**, 14617–14620.

93. Tomasselli, A.G., Olsen, M.K., Hui, J.O., Staples, D.J., Sawyer, T.K., Henrikson, R.L. and Tomich, C.-S.C. (1990) Substrate analogue inhibition and active site titration of purified recombinant HIV-1 protease. *Biochemistry*, **29**, 264–269.

94. Meek, T.D., Dayton, B.D., Metcalf, B.W., Dreyer, G.B., Strickler, J.E., Gorniak, J.G., Rosenberg, M., Moore, M.L., Magaard, V.W. and Debouck, C. (1989) Human immunodeficiency virus 1 protease expressed in Escherichia coli behaves as a dimeric aspartic protease. *Proceedings of the National Academy of Sciences of the USA*, **86**, 1841–1845.

95. Broadhurst, A.V., Roberts, N.A., Ritchie, A.J., Handa, B.K. and Kay, C. (1991) Assay of HIV-1 proteinase: A colorimetric method using small peptide substrates. *Analytical Biochemistry*, **193**, 280–286.

96. Schotten C. (1891) Ueber isatinblau, einen aus der verbindung von isatin und piperidin entstehenden farbstoff. *Berichte der Deutschen Chemischen Gesellschaft*, **24**, 1366–1373.

97. Johnson, A.W. and McCaldin, D.J. (1957) The structure of isatin blue. *Journal of the Chemical Society*, 3470–3477.

98. Grassmann, W. and von Arnim, K. (1935) Uber neue farbreaktionen des pyrrolidins und prolins. *Justus Liebigs Annalen fur Chemie*, **519**, 192–208.

99. Roberts, N.A., Martin, J.A., Kinchington, D., Broadhurst, A.V., Craig, J.C., Duncan, I.B., Galpin, S.A., Handa, B.K., Kay, J., Krohn, A., Lambert, R.W., Merrett, J.M., Mills, J.S., Parkes, K.E.B., Redshaw, S., Ritchie, A.J., Taylor, D.L., Thomas, G.J. and Machin, P.J. (1990) Inhibitors of HIV proteinase. *Science*, **248**, 358–361.

100. Wlodawer, A., Miller, M., Jaskolski, M., Sathyanarayana, B.K., Baldwin, E., Weber, I.T., Selk, L.M., Clawson, L., Schneider, J. and Kent, S.B.H. (1989) Conserved folding in retroviral proteases: crystal structure of a synthetic HIV-1 protease. *Science*, **245**, 616–621.

101. Navia, M.A., Fitzgerald, P.M.D., McKeever, B.M., Leu, C.-T., Heimbach, J.C., Herber, W.K., Sigal, I.S., Darke, P.L. and Springer, J.P. (1989) Three-dimensional structure of aspartyl protease from human immunodeficiency virus HIV-1. *Nature*, **337**, 615–620.

102. Pechik, I.V., Gustchina, A.E., Andreeva, N.S. and Fedorov, A.A. (1989) Possible role of some groups in the structure and function of HIV-1 protease as revealed by molecular modeling studies. *FEBS Letters*, **247**, 118–122.

103. Weber, I.T., Miller, M., Jaskolski, M., Leis, J., Skalka, A.M. and Wlodawer, A. (1989) Molecular modeling of the HIV-1 protease and its substrate binding site. *Science*, **243**, 928–931.

104. Pearl, L. and Blundell, T. (1984) The active site of aspartic proteinases. *FEBS Letters*, **174**, 96–101.

105. Wlodawer, A. and Erickson, J.W. (1993) Structure-based inhibitors of HIV-1 protease. *Annual Review of Biochemistry*, **62**, 543–585.

106. Fitzgerald, P.M.D., McKeever, B.M., VanMiddlesworth, J.F., Springer, J.P., Heimbach, J.C., Leu, C.-T., Herber, W.K., Dixon, R.A.F. and Darke, P.L. (1990) Crystallographic analysis of a complex between human immunodeficiency virus type 1 protease and acetyl-pepstatin at 2.0 Å resolution. *Journal of Biological Chemistry*, **265**, 14209–14219.

107. Miller, M., Schneider, J., Sathyanarayana, B.K., Toth, M.V., Marshall, G.R., Clawson, L., Selk, L., Kent, S.B.H. and Wlodawer, A. (1989) Structure of complex of synthetic HIV-1 protease with a substrate-based inhibitor at 2.3 Å resolution. *Science*, **246**, 1149–1150.

108. Swain, A.L., Miller, M.M., Green, J., Rich, D.H., Schneider, J., Kent, S.B.H. and Wlodawer, A. (1990) X-ray crystallographic structure of a complex between a synthetic protease of human immuno-deficiency virus 1 and a substrate-based hydroxyethylamine inhibitor. *Proceedings of the National Academy of Sciences of the USA*, **87**, 8805–8809.

109. Kröhn, A., Redshaw, S., Ritchie, J.C., Graves, B.J. and Hatada, M. (1991) Novel binding mode of highly potent HIV-1 proteinase inhibitors incorporating the (*R*)-hydroxyethylamine isostere. *Journal of Medicinal Chemistry*, **34**, 3340–3342.

110. Jaskolski, M., Tomasselli, A.G., Sawyer, T.K., Staples, D.G., Heinrikson, R.L., Schneider, J., Kent, S.B. and Wlodawer, A. (1991) Structure at 2.5 Å resolution of chemically synthesized human immuno-deficiency virus type 1 protease complexed with a hydroxyethylene-based inhibitor. *Biochemistry*, **30**, 1600–1609.

111. Murthy, K.H., Winborne, E.L., Minnich, M.D., Culp, J.S. and Debouck, C. (1992) The crystal structures at 2.2 Å resolution of hydroxyethylene-based inhibitors bound to human immunodeficiency virus type 1 protease show that inhibitors are present in two distinct orientations. *Journal of Biological Chemistry*, **267**, 22770–22778.

112. Dreyer, G.B., Lambert, D.M., Meek, T.D., Carr, T.J., Tomaszek, T.A., Jr., Fernandez, A.V., Bartus, H., Cacciavillani, E., Hassell, A.M., Minnich, M., Petteway, S.R., Jr. and Metcalf, B.W. (1992) Hydroxyethylene isostere inhibitors of human immunodeficiency virus 1 protease: structure activity analysis using enzyme kinetics, X-ray crystallography, and infected T-cell assays. *Biochemistry*, **31**, 6646–6659.

113. Thompson, W.J., Fitzgerald, P.M.D., Holloway, M.K., Emini, E.A., Darke, P.L., McKeever, B.M., Schlief, W.A., Quintero, J.C., Zugay, J.A., Tucker, T.J., Schwering, J.E., Homnick, C.F., Nunberg, J., Springer, J.P. and Huff, J. (1992) Synthesis and antiviral activity of a series of HIV-1 protease inhibitors with functionality tethered to the P1 or P1′ phenyl substituents: X-ray crystal structure assisted design. *Journal of Medicinal Chemistry*, **35**, 1685–1701.

114. Graves, B.J., Hatada, M.H., Miller, J.K., Graves, M.C., Roy, S., Kröhn, A., Martin, J.A. and Roberts, N.A. (1992) The three-dimensional X-ray crystal structure of HIV-1 protease complexed with a hydroxyethylene inhibitor. In *Structure and Function of the Aspartic Proteinases*, (B. Bunn, ed.) pp. 455–460. Plenum Press, New York.

115. Erickson, J., Niedhart, D.J., VanDrie, J., Kempf, D.J., Wang, X.C., Norbeck, D.W., Plattner, J.J., Rittenbouse, J.W., Turon, M., Wideburg, N., Kohlbrenner, W.E., Simmer, R., Helfrich, R., Paul, D.A. and Knigge, M. (1990) Design, activity, and 2.8 Å crystal structure of a C_2 symmetric inhibitor complexed to HIV-1 protease. *Science*, **249**, 527–533.

116. Bone, R., Vacca, J.P., Anderson, P.S. and Holloway, M.K. (1991) X-ray crystal structure of the HIV protease complex with L-700,417, an inhibitor with pseudo C_2 symmetry. *Journal of the American Chemical Society*, **113**, 9382–9384.

117. Dreyer, G.B., Boehm, J.C., Chenera, B., DesJarlais, R.L., Hassell, A.M., Meek, T.D., Tomaszek, T.A., Jr. and Lewis, M. (1993) A symmetric inhibitor binds HIV-1 protease asymmetrically. *Biochemistry*, **32**, 937–947.

118. Hyland, L.J., Tomaszek, T.A., Jr., Roberts, G.D., Carr, S.A., Magaard, V.W., Bryan, H.L., Fakhoury, S.A., Moore, M.L., Minnich, M.D., Culp, J.S., DesJarlais, R.L. and Meek, T.D. (1991) Human immunodeficiency virus-1 protease. 1. Initial velocity studies and kinetic characterization of reaction intermediates by ^{18}O isotope exchange. *Biochemistry*, **30**, 8441–8453.

119. Hyland, L.J., Tomaszek, T.A., Jr. and Meek, T.D. (1991) Human immunodeficiency virus-1 protease. 2. Use of pH rate studies and solvent kinetic isotope effects to elucidate details of chemical mechanism. *Biochemistry*, **30**, 8454–8463.

120. James, M.G.N., Hsu, I.-N. and Delbaere, L.J.J (1977) Mechanism of acid protease catalysis based on the crystal structure of penicillopepsin. *Nature*, **267**, 808–813.

121. Suguna, K., Padlan, E.A., Smith, C.W., Carlson, W.D. and Davies, D.R. (1987) Binding of a reduced peptide inhibitor to the aspartic proteinase from *Rhizopus chinensis*: Implications for a mechanism of action. *Proceedings of the National Academy of Sciences of the USA*, **84**, 7009–7013.

122. James, M.N.G. and Sielecki, A.R. (1985) Stereochemical analysis of peptide bond hydrolysis catalyzed by the aspartic proteinase penicillopepsin. *Biochemistry*, **24**, 3701–3713.

123. Bott, R., Subramanian, E. and Davies, D.R. (1982) Three-dimensional structure of a complex of the *Rhizopus chinensis* carboxyl proteinase and pepstatin at 2.5 Å. *Biochemistry*, **21**, 6956–6962.

124. Pauling, L. (1946) Molecular architecture and biological reactions. *Chemical and Engineering News*, **24**, 1375–1377.

125. Wolfenden, R. (1972) Analog approaches to the structure of the transition state in enzyme reactions. *Accounts of Chemical Research*, **5**, 10–18.

126. Rich, D.H. (1985) Pepstatin-derived inhibitors of aspartic proteinases. A close look at an apparent transition-state analogue inhibitor. *Journal of Medicinal Chemistry*, **28**, 263–273.

127. Dreyer, G.B., Metcalf, B.W., Tomaszek, T.A., Jr., Carr, T.A., Chandler, A.C., III, Hyland, L., Fakhoury, S.A., Magaard, V.W., Moore, M.L., Strickler, J.E., Debouck, C. and Meek, T.D. (1989) Inhibition of human immunodeficiency virus 1 protease *in vitro*: Rational design of substrate analogue inhibitors. *Proceedings of the National Academy of Sciences of the USA*, **86**, 9752–9756.

128. Richards, A.D., Roberts, R., Dunn, B.M., Graves, M.C. and Kay, J. (1989) Effective blocking of HIV-1 proteinase activity by characteristic inhibitors of aspartic proteinases. *FEBS Letters*, **247**, 113–117.

129. Meek, T.D., Lambert, D.M., Dreyer, G.B., Carr, T.J., Tomaszek, T.A., Moore, M.L., Strickler, J.E., Debouck, C., Hyland, L.J., Matthews, T.J., Metcalf, B.W. and Petteway, S.R. (1990) Inhibition of HIV-1 protease in infected T-lymphocytes by synthetic peptide analogues. *Nature*, **343**, 90–92.

130. Tomasselli, A.G., Hui, J.O., Sawyer, T.K., Staples, D.J., Bannow, C., Reardon, I.M., Howe, W.J., DeCamp, D.L., Craik, C.S. and Heinrikson, R.L. (1990) Specificity and inhibition of proteases from human immunodeficiency viruses 1 and 2. *Journal of Biological Chemistry*, **265**, 14675–14683.

131. Vacca, J.P., Guare, J.P., deSolms, S.J., Sanders, W.M., Giuliani, E.A., Young, S.D., Darke, P.L., Zugay, J., Sigal, I.S., Schlief, W.A., Quintero, J.C., Emini, E.A., Anderson, P.S. and Huff, J.R. (1991) L-687,908, a potent hydroxyethylene-containing HIV protease inhibitor. *Journal of Medicinal Chemistry*, **34**, 1225–1228.

132. Lyle, T.A., Wiscount, C.M., Guare, J.P., Thompson, W.J., Anderson, P.S., Darke, P.L., Zugay, J.A., Emini, E.A., Schlief, W.A., Quintero, J.C., Dixon, R.A.F., Sigal, I.S. and Huff, J.R. (1991) Benzocycloalkyl amines as novel X-termini for HIV protease inhibitors. *Journal of Medicinal Chemistry*, **34**, 1228–1230.

133. Ashorn, P., McQuade, T.J., Thaisrivongs, S., Tomasselli, A.G., Tarpley, W.G. and Moss, B. (1990) An inhibitor of the protease blocks maturation of human and simian immunodeficiency viruses and spread of infection. *Proceedings of the National Academy of Sciences of the USA*, **87**, 7472–7476.

134. Szelke, M. (1985) Chemistry of renin inhibitors. In *Aspartic Proteinases and Their Inhibitors*, (V. Kostka, ed.), pp. 412–441. de Gruyter, Berlin.

135. Camp, N.P., Hawkins, P.C.D., Hitchcock, P.B. and Gani, D. (1992) Synthesis of stereochemically defined phosphonamidate-containing peptides: Inhibitors for the HIV-1 proteinase. *Bioorganic and Medicinal Chemistry Letters*, **2**, 1047–1052.

136. Grobelny, D., Wondrak, E.M., Galardy, R.E. and Oroszlan, S. (1990) Selective phosphinate transition-state analogue inhibitors of the protease of human immunodeficiency virus. *Biochemical and Biophysical Research Communications*, **169**, 1111–1116.

137. Tamburini, P.P., Dreyer, R.N., Hansen, J., Letsinger, J., Elting, J., Gore-Willse, A., Dally, R., Hanko, R., Osterman, D., Kamarck, M.E. and Yoo-Warren, H. (1990) A fluorometric assay for HIV-protease activity using high-performance liquid chromatography. *Analytical Biochemistry*, **186**, 363–368.

138. Richards, A.D., Broadhurst, A.V., Ritchie, A.J., Dunn, B.M. and Kay, J. (1989) Inhibition of aspartic proteinase from HIV-2. *FEBS Letters*, **253**, 214–216.

139. Hui, K.Y., Manetta, J.V., Gygi, T., Bowdon, B.J., Kieth, K.A., Shannon, W.M. and Lai, M.-H.T. (1991) A rational approach in the search for potent inhibitors against HIV proteinase. *Federation of American Societies for Experimental Biology Journal*, **5**, 2606–2610.

140. Kahn, M., Nakanishi, H., Chrusciel, R.A., Fitzpatrick, D. and Johnson, M.E. (1991) Examination of HIV-1 protease secondary structure specificity using conformationally constrained inhibitors. *Journal of Medicinal Chemistry*, **34**, 3395–3399.

141. Raju, B., Deshpande, M.S. (1991) Investigating the stereochemistry of binding to HIV-1 protease with inhibitors containing isomers of 4-amino-3-hydroxy-5-phenylpentanoic acid. *Biochemical and Biophysical Research Communications*, **180**, 187–190.

142. Sham, H.L., Betebenner, D.A., Wideburg, N.E., Saldivar, A.C., Kohlbrenner, W.E., Vasavanonda, S., Kempf, D.J., Norbeck, D.W., Zhao, C., Clement, J.J., Erickson, J.E. and Plattner, J.J. (1991) Potent HIV-1 protease inhibitors with antiviral activities *in vitro*. *Biochemical and Biophysical Research Communications*, **175**, 914–919.

143. Schirlin, D., Baltzer, S., Van Dorsslaer, V., Weber, F., Weill, C., Altenburger, J.M., Neises, B., Flynn, G., Rémy, J.M. and Tarnus, C. (1993) Short and unexpectedly potent difluorostatone type inhibitors of HIV-1 protease. *Bioorganic and Medicinal Chemistry Letters*, **3**, 253–258.

144. Mimoto, T., Imai, J., Tanaka, S., Hattori, N., Takahashi, O., Kisanuki, S., Nagano, Y., Shintani, M., Hayashi, H., Sakikawa, H., Akaji, K. and Kiso, Y. (1991) Rational design and synthesis of a novel class of active-site targeted HIV protease inhibitors containing a hydroxymethylcarbonyl isostere. Use of phenylnorstatine or allophenylnorstatine as a transition-state mimetic. *Chemical and Pharmaceutical Bulletin*, **39**, 2465–2467.

145. Mimoto, T., Imai, J., Tanaka, S., Hattori, N., Kisanuki, S., Akaji, K. and Kiso, Y. (1991) KNI-102, a novel tripeptide HIV protease inhibitor containing allophenylnorstatine as a transition-state mimetic. *Chemical and Pharmaceutical Bulletin*, **39**, 3088–3090.

146. Kageyama, S., Mimoto, T., Murakawa, Y., Nomizu, M., Ford, H., Jr., Shirasaka, T., Gulnik, S., Erickson, J., Takada, K., Hayashi, H., Broder, S., Kiso, Y. and Mitsuya, H. (1993) *In vitro* anti-human immunodeficiency virus (HIV) activities of transition state mimetic HIV protease inhibitors containing allophenylnorstatine. *Antimicrobial Agents and Chemotherapy*, **37**, 810–817.

147. Martin, J.A., Broadhurst, A.V., Duncan, I.B., Galpin, S.A., Handa, B.K., Kinchington, D., Kröhn, A., Lambert, R.W., Machin, P.J., Merrett, J.H., Roberts, N.A., Parkes, K.E.B., Redshaw, S. and Thomas, G.J. (1990) Inhibitors of HIV Proteinase. *Antiviral Research, Supplement I*, Abstract 67, p. 74.

148. Shaw, E. (1967) Site-specific reagents for chymotripsin and trypsin. *Methods in Enzymology*, **11**, 677–686.

149. Segal, D.M., Powers, J.C., Cohen, G.H., Davies, D.R. and Wilcox, P.E. (1971) Substrate binding site in bovine chymotrypsin A$_\gamma$. A crystallographic study using peptide chloromethyl ketones as site-specific inhibitors. *Biochemistry*, **10**, 3728–3738.

150. Luly, J.R., Dellaria, J.F., Plattner, J.J., Soderquist, J.L. and Yi, N. (1987) A synthesis of protected aminoalkyl epoxides from α-amino acids. *Journal of Organic Chemistry*, **52**, 1487–1492.

151. Thompson, W.J., Ghosh, A.K., Holloway, M.K., Lee, H.Y., Munson, P.M., Schwering, J.E., Wai, J., Darke, P.L., Zugay, J., Emini, E.A., Schlief, W.A., Huff, J.R. and Anderson, P.S. (1993) 3'-Tetrahydrofuranylglycine as a novel, unnatural amino acid surrogate for asparagine in the design of inhibitors of HIV protease. *Journal of the American Chemical Society*, **115**, 801–803.

152. Tucker, T.J., Lumma, W.C., Jr., Payne, L.S., Wai, J.M., de Solms, S.J., Giuliani, E.A., Darke, P.L., Heimbach, J.C., Zugay, J.A., Schlief, W.A., Quintero, J.C., Emini, E.A., Huff, J.R. and Anderson, P.S. (1992) A series of potent HIV-1 protease inhibitors containing a hydroxyethyl secondary amine transition state isostere: Synthesis, enzyme inhibition, and antiviral activity. *Journal of Medicinal Chemistry*, **35**, 2525–2533.

153. Ryono, D.E., Free, C.A., Neubeck, R., Samaniego, S.G., Godfrey, J.D. and Petrillo, E.W., Jr. (1985) Potent inhibitors of hog and human renin containing an amino alcohol dipeptide surrogate. *Peptides: Structure and Function. Proceedings of the Ninth American Peptide Symposium*, (Deber, C.M., Hruby, V.J. and Kopple, K.D., Eds.) pp. 739–742. Pierce Chemical Co., Rockford, Il.

154. Arrowsmith, R.J., Dann, J.G., Davies, D.E., Fogden, Y.C., Harris, C.J., Morton, J.A. and Ogden, H. (1988) Inhibitors of human renin; synthesis and comparative SAR of P3-P1 tripeptide-based aminoalcohols. *Peptides, Proceedings of the European Symposium*, 393–395.

155. Rich, D.H., Sun, C.-Q., Prasad, J.V.N.V., Pathiasseril, A., Toth, M.V., Marshall, G.R., Clare, M., Mueller, R.A. and Houseman, K. (1991) Effect of hydroxyl group configuration in hydroxyethylamine dipeptide isosteres on HIV protease inhibition. Evidence for multiple binding modes. *Journal of Medicinal Chemistry*, **34**, 1222–1225.

156. Kröhn, A. unpublished results.

157. Appelt, K. (1993) Crystal structure of HIV-1 protease-inhibitor complexes. *Perspectives in Drug Discovery and Design*, **1**, 23–48.

158. Evans, B.E., Rittle, K.E., Homnick, C.F., Springer, J.P., Hirshfield, J. and Veber, D.F. (1985) A stereocontrolled synthesis of hydroxyethylene dipeptide isosteres using novel, chiral aminoalkyl epoxides and γ-(aminoalkyl) γ-lactones. *Journal of Organic Chemistry*, **50**, 4615–4625.

159. Reetz, M.T. and Binder, J. (1989) Protective group tuning in the stereoselective conversion of α-amino aldehydes into aminoalkyl epoxides. *Tetrahedron Letters*, 5425–5428.

160. Maibaum, J. and Rich, D.H. (1988) A facile synthesis of statine and analogues by reduction of β-keto esters derived from Boc-protected amino acids. HPLC Analyses of their enantiomeric purity. *Journal of Organic Chemistry*, **53**, 869–873.

161. Barton, D.H.R., Crich, D. and Motherwell, W.B. (1983) A practical alternative to the Hunsdiecker reaction. *Tetrahedron Letters*, 4979–4982.

162. Wissner, A. (1979) 2-Hetero substituted silylated ketene acetals: Reagents for the preparation of α-functionalized methyl ketones from carboxylic acid chlorides. *Journal of Organic Chemistry*, **44**, 4617–4622.

163. Caron, M., Carlier, P.R. and Sharpless, K.B. (1988) Regioselective azide opening of 2,3-epoxy alcohols by [Ti(O-*i*-Pr)$_2$(N$_3$)$_2$]: Synthesis of α-amino acids. *Journal of Organic Chemistry*, **53**, 5185–5187.

164. Hanson, R.M. and Sharpless, K.B. (1986) Procedure for the catalytic asymmetric epoxidation of allylic alcohols in the presence of molecular sieves. *Journal of Organic Chemistry*, **51**, 1922–1925.

165. Gao, Y. and Sharpless, K.B. (1988) Vicinal diol cyclic sulfates: like epoxides only more reactive. *Journal of the American Chemical Society*, **110**, 7538–7539.

166. Ghosh, A.K., McKee, S.P., Lee, H.Y. and Thompson, W.J. (1992) A facile and enantiospecific synthesis of 2(*S*)- and 2(*R*)-[1′(*S*)-azido-2-phenylethyl]oxirane. *Journal of the Chemical Society, Chemical Communications*, 273–274.

167. Al-Hakim, A.H., Haines, A.H. and Morley, C. (1985) Synthesis of 1,2-*O*-isopropylidene-*L*-threitol and its conversion to (*R*)-1,2-*O*-isopropylideneglycerol. *Synthesis*, 207–208.

168. Ohno, M., Fujita, K., Nakai, H., Kobayashi, S., Inoue, K. and Nojima, S. (1985) An enantioselective synthesis of platelet-activating factors, their enantiomers, and their analogues from D- and L-tartaric acids. *Chemical and Pharmaceutical Bulletin*, **33**, 572–582.

169. Herdewijn, P. and Van Aerschot, A. (1989) Synthesis of 9-(3-azido-2,3-dideoxy-β-D-erythro-pentofuranosyl)-2,6-diaminopurine (AzddDAP). *Tetrahedron Letters*, 855–858.

170. Houpis, I.N., Molina, A., Reamer, R.A., Lynch, J.E., Volante, R.P. and Reider, P.J. (1993) Towards the synthesis of HIV-protease inhibitors. Synthesis of optically pure 3-carboxyldecahydroisoquinolines. *Tetrahedron Letters*, 2593–2596.

171. Holmes, H.C., Mahmood, N., Karpas, A., Petrik, J., Kinchington, D., O'Connor, T., Jeffries, D.J., Desmyter, J., De Clercq, E., Pauwels, R. and Hay, A. (1991) Screening of compounds for activity against HIV: a collaborative study. *Antiviral Chemistry and Chemotherapy*, **2**, 287–293.

172. Craig, J.C., Duncan, I.B., Hockley, D., Grief, C., Roberts, N.A. and Mills, J.S. (1991) Antiviral properties of Ro 31-8959, an inhibitor of human immunodeficiency virus (HIV) proteinase. *Antiviral Research*, **16**, 295–305.

173. Martin, J.A., Mobberley, M.A., Redshaw, S., Burke, A., Tyms, A.S. and Ryder, T.A. (1991) The inhibitory activity of a peptide derivative against the growth of simian immunodeficiency virus in C8166 cells. *Biochemical and Biophysical Research Communications*, **176**, 180–188.

174. Roberts, N.A., Craig, J.C. and Duncan, I.B. (1992) HIV proteinase inhibitors. *Biochemical Society Transactions*, **20**, 513–516.

175. Vacca, J.P., Guare, J.P., deSolms, S.J., Sanders, W.M., Giuliani, E.A., Young, S.D., Darke, P.L., Zugay, J., Sigal, I.S., Schlief, W.A., Quintero, J.C., Emini, E.A., Anderson, P.S. and Huff, J.R. (1991) L-687,908, a potent hydroxyethylene-containing HIV protease inhibitor. *Journal of Medicinal Chemistry*, **34**, 1225–1228.

176. Lyle, T.A., Wiscount, C.M., Guare, J.P., Thompson, W.J., Anderson, P.S., Darke P.L., Zugay, J.A., Emini, E.A., Schlief, W.A., Quintero, J.C., Dixon, R.A.F., Sigal, I.S. and Huff, J.R. (1991) Benzocycloalkyl amines as novel C-termini for HIV protease inhibitors. *Journal of Medicinal Chemistry*, **34**, 1228–1230.

177. Thompson, W.J., Fitzgerald, P.M.D., Holloway, M.K., Emini, E.A., Darke, P.L., McKeever, B.M., Schlief, W.A., Quintero, J.C., Zugay, J.A., Tucker, T.J., Schwering, J.E., Homnick, C.F., Nunberg, J., Springer, J.P. and Huff, J.R. (1992) Synthesis and antiviral activity of a series of HIV-1 protease inhibitors with functionality tethered to the P_1 or $P_{1'}$ phenyl substituents: X-ray crystal structure assisted design. *Journal of Medicinal Chemistry*, **35**, 1685–1701.

178. Ghosh, A.K., Thompson, W.J., McKee, S.P., Duong, T.T., Lyle, T.A., Chen, J.C., Darke. P.L., Zugay, J.A., Emini, E.A., Schlief, W.A., Huff, J.R. and Anderson, P.S. (1993) 3-Tetrahydrofuran and pyran urethanes as high-affinity P2-ligands for HIV-1 protease inhibitors. *Journal of Medicinal Chemistry*, **36**, 292–294.

179. Dreyer, G.B., Lambert, D.M., Meek, T.D., Carr, T.J., Tomaszek, T.A., Jr., Fernandez, A.V., Bartus, H., Cacciavillani, E., Hassell, A.M., Minnich, M., Petteway, S.R., Jr. and Metcalf, B.W. (1992) Hydroxyethylene isostere inhibitors of human immunodeficiency virus-1 protease: Structure-activity analysis using enzyme kinetics, X-ray crystallography, and infected T-cell assays. *Biochemistry*, **31**, 6646–6659.

180. Getman, D.P., DeCrescenzo, G.A., Heintz, R.M., Reed, K.L., Talley, J.J., Bryant, M.L., Clare, M., Houseman, K.A., Marr, J.J., Mueller, R.A., Vazquez, M.L., Shieh, H.-S., Stallings, W.C. and Stegeman, R.A. (1993) Discovery of a novel class of potent HIV-1 protease inhibitors containing the (R)-(hydroxyethyl)urea isostere. *Journal of Medicinal Chemistry*, **36**, 288–291.

181. Thaisrivongs, S., Tomasselli, A.G., Moon, J.B., Hui, J., McQuade, T.J., Turner, S.R., Strohbach, J.W., Howe, W.J., Tarpley, W.G. and Heinrikson, R.L. (1991) Inhibitors of the protease from human immunodeficiency virus: Design and modeling of a compound containing a dihydroxyethylene isostere insert with high binding affinity and effective antiviral activity. *Journal of Medicinal Chemistry*, **34**, 2344–2356.

182. Thaisrivongs, S., Turner, S.R., Strohbach, J.W., TenBrink, R.E., Tarpley, W.G., McQuade, T.J., Heinrikson, R.L., Tomasselli, A.G., Hui, J.O. and Howe, W.J. (1993) Inhibitors of the protease from human immunodeficiency virus: Synthesis, enzyme inhibition, and antiviral activity of a series of compounds containing the dihydroxyethylene transition-state isostere. *Journal of Medicinal Chemistry*, **36**, 941–952.

183. Kempf, D.J., Norbeck, D.W., Codacovi, L., Wang, X.C., Kohlbrenner, W.E., Wideburg, N.E., Paul, D.A., Knigge, M.F., Vasavanonda, S., Craig-Kennard, A., Saldivar, A., Rosenbrook, W., Jr., Clement, J.J., Plattner, J.J. and Erickson, J. (1990) Structure-based, C_2 symmetric inhibitors of HIV protease. *Journal of Medicinal Chemistry*, **33**, 2687–2689.

184. Kempf, D.J., Codacovi, L., Wang, X.C., Kohlbrenner, W.E., Wideburg, N.E., Saldivar, A., Vasavanonda, S., Marsh, K.C., Bryant, P., Sham, H.L., Green, B.E., Betebenner, D.A., Erickson, J. and Norbeck, D.W. (1993) Symmetry-based inhibitors of HIV protease. Structure-activity studies of acylated 2,4-diamino-1,5-diphenyl-3-hydroxypentane and 2,5-diamino-1,6-diphenylhexane-3,4-diol. *Journal of Medicinal Chemistry*, **36**, 320–330.

185. Kempf, D.J., Marsh, K.C., Paul, D.A., Knigge, M.F., Norbeck, D.W., Kohlbrenner, W.E., Codacovi, L., Vasavanonda, S., Bryant, P., Wang, X.C., Wideburg, N.E., Clement, J.C., Plattner, J.J. and Erickson, J. (1991) Antiviral and pharmacokinetic properties of C_2 symmetric inhibitors of the human immunodeficiency virus type 1 protease. *Antimicrobial Agents and Chemotherapy*, **35**, 2209–2214.

186. Erickson, J.W. (1993) Design and structure of symmetry-based inhibitors of HIV-1 protease. *Perspectives in Drug Discovery and Design*, **1**, 109–128.

187. Humber, D.C., Cammack, N., Coates, J.A.V., Cobley, K.N., Orr, D.C., Storer, R., Weingarten, G.G. and Weir, M.P. (1992) Penicillin derived C_2-symmetric dimers as novel inhibitors of HIV-1 proteinase. *Journal of Medicinal Chemistry*, **35**, 3080–3081.

188. Holmes, D.S., Clemens, I.R., Cobley, K.N., Humber, D.C., Kitchin, J., Orr, D.C., Patel, B., Paternoster, I.L. and Storer, R. (1993) Novel dimeric penicillin derived inhibitors of HIV-1 proteinase: Interaction with the catalytic aspartates. *Bioorganic and Medicinal Chemistry Letters*, **3**, 503–508.

189. Peyman, A., Budt, K.-H., Spanig, J., Stowasser, B. and Ruppert, D. (1992) C2-Symmetrc phosphinic acid inhibitors of HIV protease. *Tetrahedron Letters*, 4549–4552.

190. Sham, H.L., Wideburg, N.E., Spanton, S.G., Kohlbrenner, W.E., Betebenner, D.A., Kempf, D.J., Norbeck, D.W., Plattner, J.J. and Erickson, J.W. (1991) Synthesis of (2S,5S,4R)-2,5-diamino-3,3-difluoro-1,6-diphenylhydroxyhexane: The core unit of a potent HIV proteinase inhibitor. *Journal of the Chemical Society, Chemical Communications*, 110–112.

191. Lam, P.Y.-S., Etermann, C.J., Hodge, C.N., Jadhav, P.K. and DeLucca, G.V. (1993) Cyclic ureas and analogues useful as retroviral protease inhibitors. *World Patent*, WO 93/07128.

192. Chenera, B., DesJarlais, R.L. and Dreyer, G.B. (1992) Carbocyclic and heterocyclic HIV protease inhibitors. *World Patent*, WO 92/21647.

193. DesJarlais, R.L., Seibel, G.L., Kuntz, I.D., Furth, P.S., Alvarez, J.C., Ortiz de Montellano, P.R., DeCamp, D.L., Babé, L.M. and Craik, C.S. (1990) Structure-based design of nonpeptide inhibitors specific for the human immunodeficiency virus 1 protease. *Proceedings of the National Academy of Sciences of the USA*, **87**, 6644–6648.

194. Rutenber, E., Fauman, E.B., Keenan, R.J., Fong, S., Furth, P.S., Ortiz de Montellano, P.R., Meng, E., Kuntz, I.D., DeCamp, D.L., Salto, R., Rosé, J.R., Craik, C.S. and Stroud, R.M. (1993) Structure of a non-peptide inhibitor complexed with HIV-1 protease. *Journal of Biological Chemistry*, **268**, 15343–15346.

195. Friedman, S.H., DeCamp, D.L., Sijbesma, R.P., Srdanov, G., Wudl, F. and Kenyon, G.L. (1993) Inhibition of the HIV-1 protease by fullerene derivatives: Model building studies and experimental verification. *Journal of the American Chemical Society*, **115**, 6506–6509.

5. NONPEPTIDE ANTAGONISTS AT PEPTIDE RECEPTORS

JOHN A. LOWE, III, BRIAN T. O'NEILL and PHILIP A. CARPINO

Central Research Division, Pfizer, Inc. Groton, CT 06340

1. INTRODUCTION

Receptor antagonists have historically played two important roles: as reagents for receptor characterization, especially in the delineation of receptor subtypes, and as novel therapeutic agents. They are thus well suited to make a major contribution to the study of the neuropeptide hormones and their receptors, a rapidly emerging field with a diversity of biological activity and potential for novel therapy. Indeed, selective antagonists have already defined receptor subtypes in the cholecystokinin and angiotensin II areas, and provided strong evidence for the relevant physiological roles of the neurokinin 1 and 2 receptors for the tachykinin peptides. This chapter will discuss the medicinal chemistry which enabled discoveries in each of these three areas. In addition, newly emerging antagonists at several other peptide receptors will be examined. Finally, contributions from the field of molecular biology which have characterized details of the binding of antagonists at peptide receptors and may some day enable design of new antagonists will be considered. Recent reviews on G protein-coupled receptors for peptide hormones[1] and on nonpeptide antagonists at these receptors[2] have appeared.

2. CHOLECYSTOKININ

Although the history of nonpeptide ligands for peptide receptors actually extends back into prehistory with morphine, the recent history of this field begins with the discovery of asperlicin, the first nonpeptide cholecystokinin (CCK) antagonist.[3] CCK is a 33-amino acid polypeptide hormone discovered in 1929 and subsequently shown to control gallbladder function and secretion of digestive enzymes.[4] Compared with the CCK antagonists known at the time of its discovery, the fermentation-derived compound asperlicin was found to be both more potent and more selective for peripheral CCK receptors ($IC_{50} = 1.4$ µM in rat pancreas, $IC_{50} > 100$ µM in rat brain). More importantly, its affinity was subsequently improved, and selectivity tailored for either central or peripheral receptors, by an intensive medicinal chemistry program. The key step in this program was the observation that part of asperlicin's structure could be approximated by the benzodiazepine nucleus of valium, around which there is a rich history of chemical modification. The structural aspects of asperlicin initially excised are illustrated in compound **1**, which has similar affinity for peripheral CCK receptors, $IC_{50} = 0.3$ µM.[5]

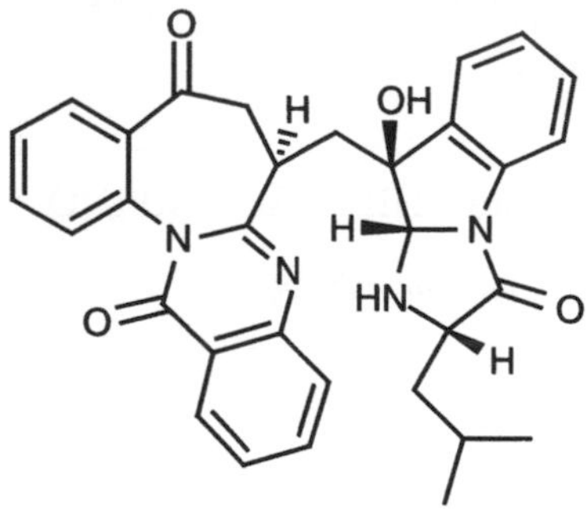

Asperlicin

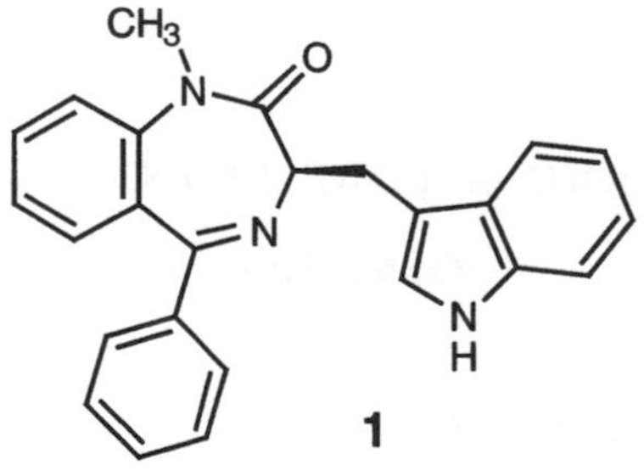

1

Subsequent SAR studies indicated the critical importance of the linking group between the benzodiazepine nucleus and the indole side chain, leading to the insertion of an amide linker to afford L-364,718, with an affinity for the peripheral CCK receptor, $IC_{50} = 80$ pM in rat pancreas, far exceeding that of asperlicin or **1**.[6] Another important discovery was made by extending the side chain SAR of this aminobenzodiazepine series to a urea, L-365,260, selective for central CCK receptors over peripheral CCK receptors ($IC_{50} = 2.0$ nM in guinea pig brain and $IC_{50} = 280$ nM in rat pancreas).

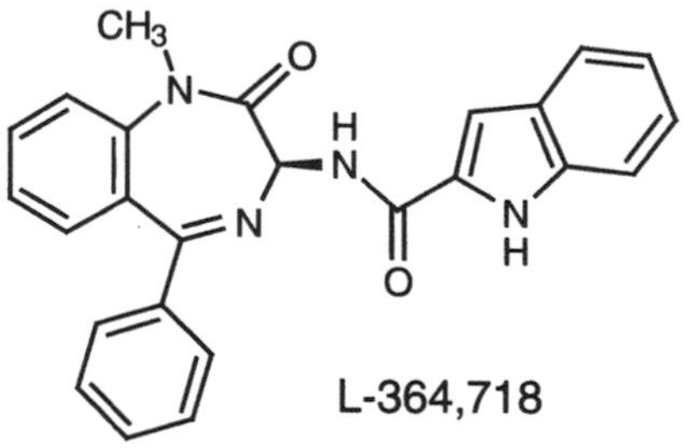

L-364,718

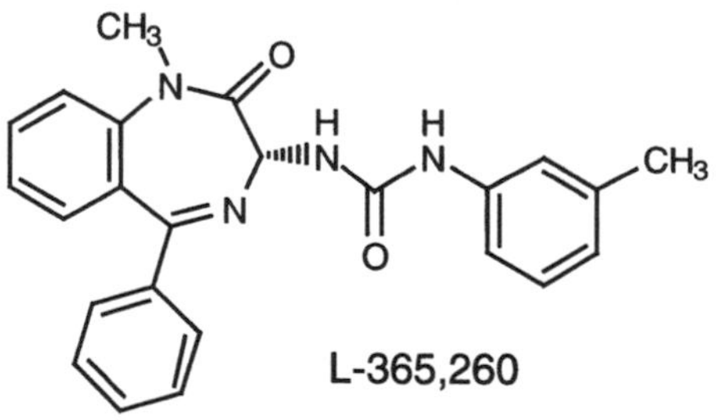

L-365,260

This led to classification of peripheral CCK receptors as CCK-A, and central receptors as CCK-B, identical to the gastrin receptor in the stomach which controls acid secretion.[7] This assignment has been confirmed by cloning and sequencing of these two CCK receptors, the CCK-A from rat[8] and human,[9] and the CCK-B from human brain.[10] In addition, while L-364,718 has confirmed many of the functions of CCK in the digestive tract uncovered by previous work, L-365,260 has revealed a previously unknown role for central CCK activity in panic disorder[11,12] and pain.[13]

Table 1 displays the sensitivity of CCK receptor affinity and selectivity to side chain structure and the stereochemistry at the 3-position of the benzodiazepine nucleus in this series.

Further chemical diversity in this series is illustrated by compound **2**, which has an IC_{50} value of 0.9 nM in rat pancreas and an oral ED_{50} of 0.02 mg/kg at 1 hour for antagonism of CCK-8 inhibition of gastric emptying in mice.[14] Replacement of

Table 1 SAR of benzodiazepine CCK antagonists

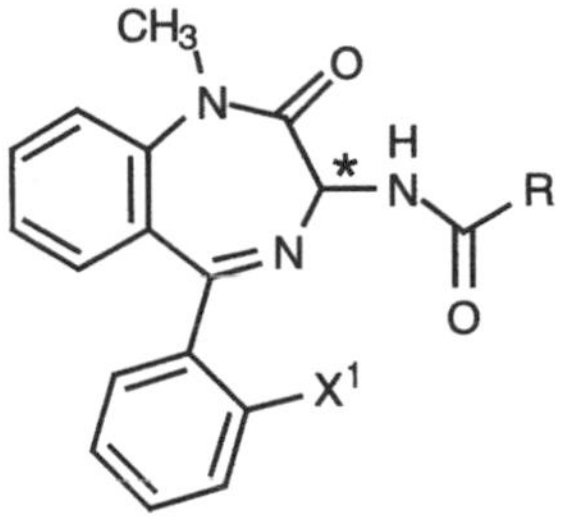

*	X^1	R	CCK-B[#]	CCK-A[#]
S	F	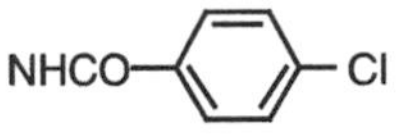 NHCO—⬡—Cl	2900	0.39 ± 0.09
R	H	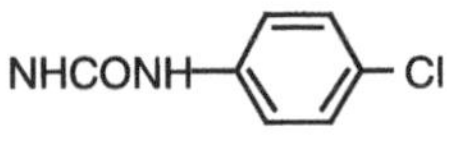 NHCONH—⬡—Cl	5.5	1100
RS	H	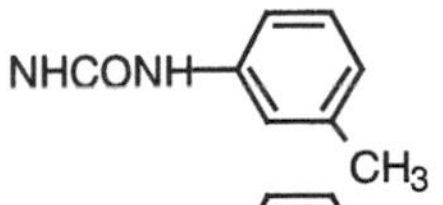 NHCONH—⬡ CH₃	7.1	8.1
R	H	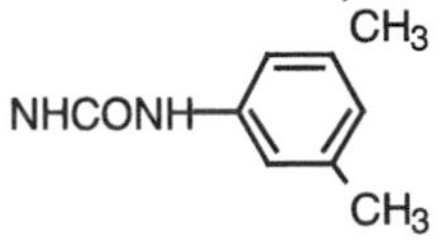 NHCONH—⬡ CH₃	2.0 ± 0.3	280 ± 33

[#] IC_{50} value for inhibition of $[^{125}I]$CCK-8 binding to: A – rat pancreas, B – guinea pig cerebral cortex.

the benzodiazepine nucleus with a 7-membered benzolactam ring afforded compound **3**, with potent affinity for the CCK-A receptor ($IC_{50} = 3$ nM).[15] Extension of this strategy to the urea side chain found to be optimal for CCK-B activity was not as successful, however, since compound **4** afforded weaker affinity and little selectivity (CCK-B $IC_{50} = 110$ nM, CCK-A $IC_{50} = 140$ nM).[16]

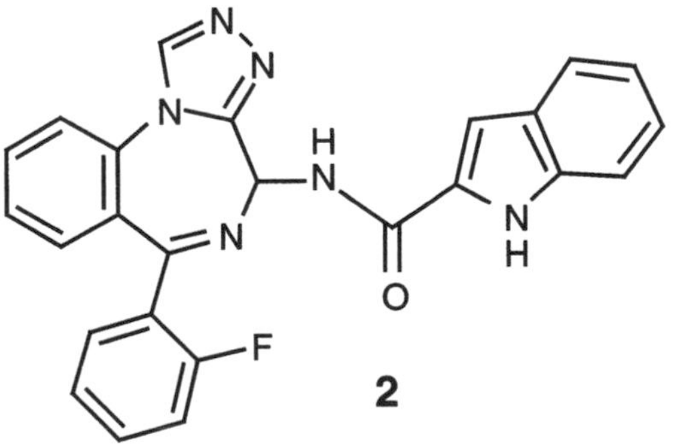

2

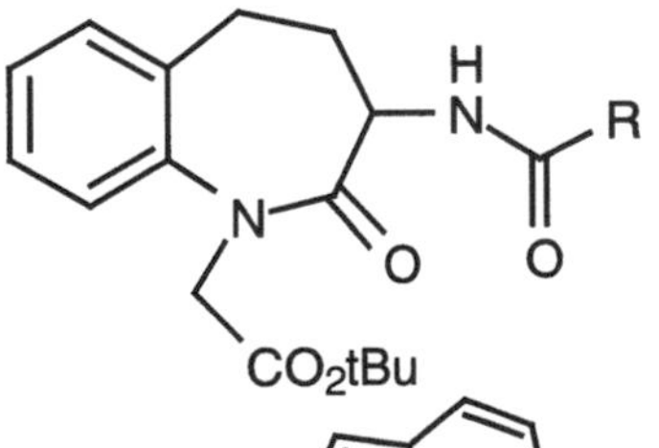

3, R=

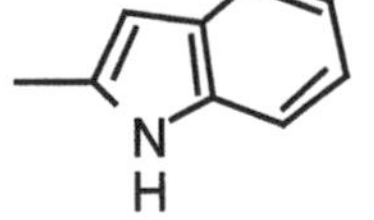

4, R=

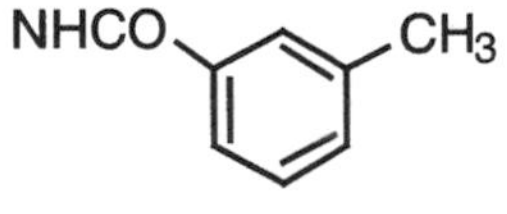

Another surrogate for the benzodiazepine nucleus, the carbazole **5**, was recently reported to be a potent CCK-A antagonist, $IC_{50} = 24$ nM.[17] By excising the quinazolinone portion of asperlicin and exploring substituent SAR, a potent CCK-B antagonist, **6**, $IC_{50} = 9$ nM, was discovered.[18]

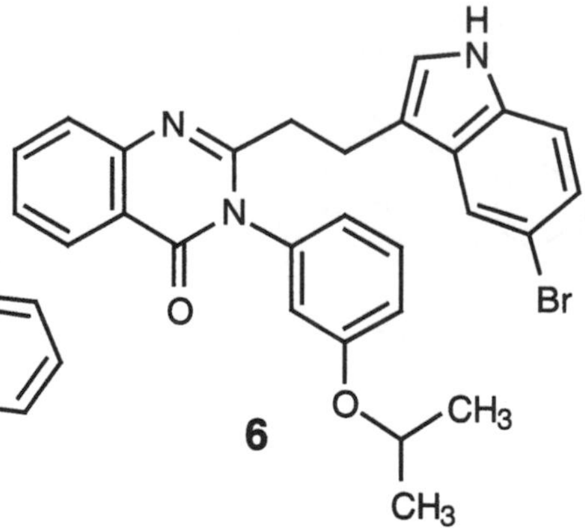

5

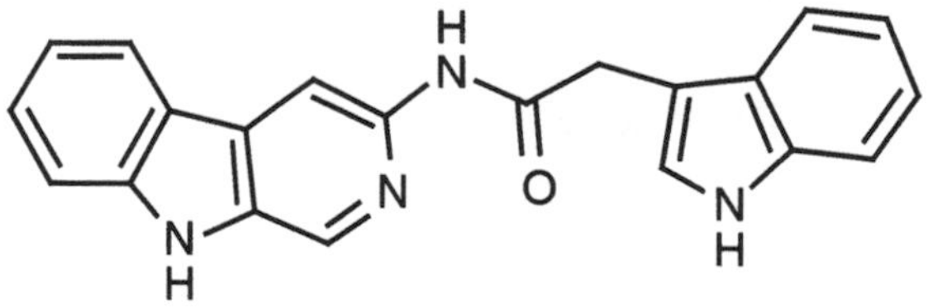

6

The discovery of L-364,718 and L-365,260 inspired the development of several other classes of cholecystokinin antagonists with structural aspects found in the parent compounds. Extending the N-1 substituent of L-365,260 led to YM022, with an IC$_{50}$ value of 0.11 nM at the CCK-B receptor.[19] Molecular modeling studies were recently employed to elucidate a common pharmacophore for L-365,260 and the diphenylpyrazolidinone CCK-B antagonist, LY288513.[20]

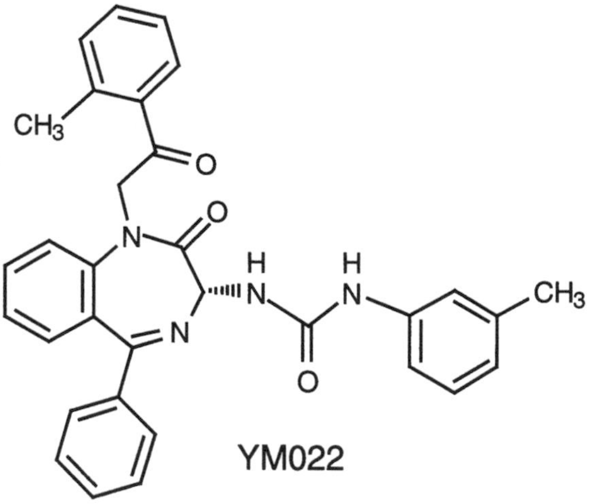

YM022

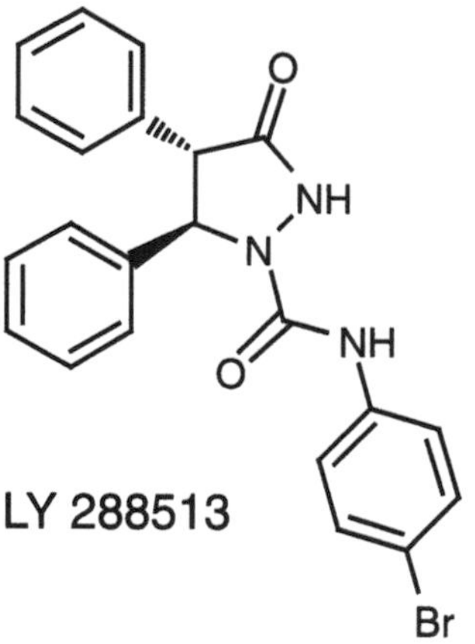

LY 288513

Using the urea side chain of L-365,260 with an "acyclic" nucleus afforded the CCK-B (IC$_{50}$=6 nM) antagonist RP-72540,[21] while the indole side chain of L-364,718 was modified and combined with a phenylthiazole nucleus to produce the potent CCK-A antagonist SR 27897, (IC$_{50}$=0.2 nM in rat pancreas).[22]

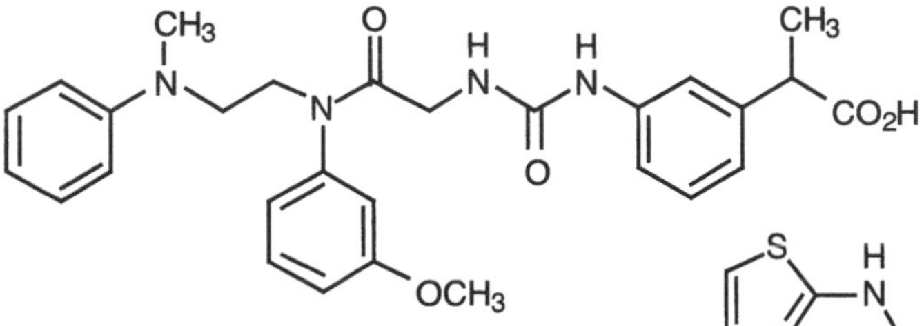

RP-72540

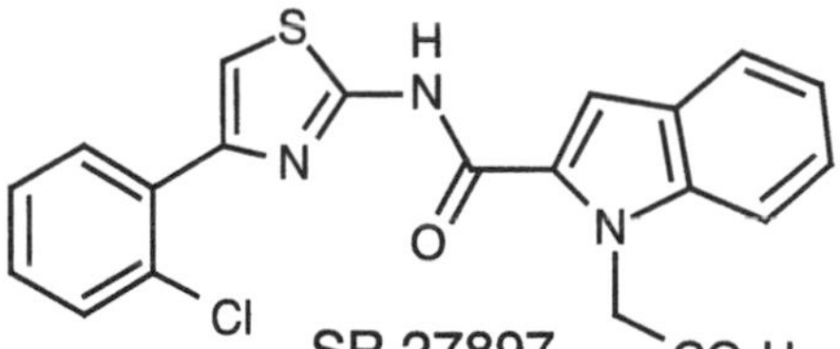

SR 27897

Two novel series of CCK antagonists have been designed by hybridizing the structure of L-364,718 with the peptide-based CCK antagonist lorglumide.[23] The first series modeled the pentyl chains of lorglumide with the fused and pendant benzo rings in L-364,718 and then employed a similar amide side chain to arrive at A-65,186, with a CCK-A IC_{50} value of 5.1 nM for the R isomer.[24] It was subsequently shown that the "glutamate" portion of A-65,186 could be replaced with the indolylmethyl group, affording compound 7, with an IC_{50} value of 23 nM in rat pancreas.[25] Consideration of the structures of lorglumide and L-364,718 suggested the design of an "acyclic" nucleus, which was then combined with a ureido side chain to afford compound 8, with an IC_{50} value of 90 nM at CCK-A receptors in rat pancreas.[26]

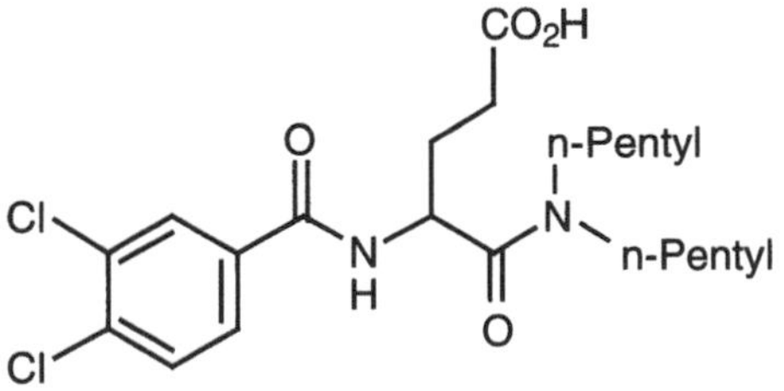

Lorglumide

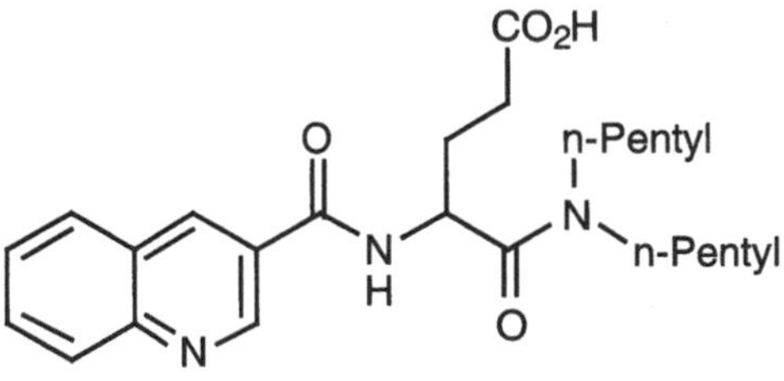

A-65,186

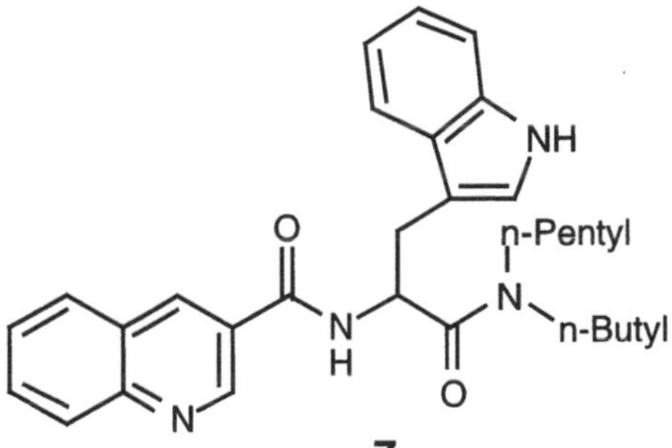

7

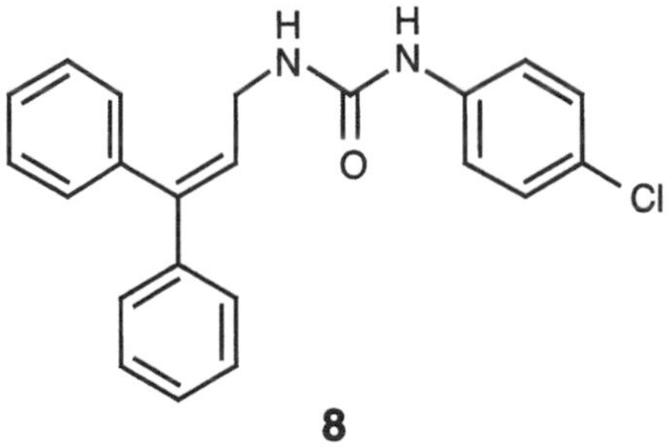

8

JOHN A. LOWE, III, BRIAN T. O'NEILL and PHILIP A. CARPINO

Table 2 SAR of CI-988 and related CCK antagonists

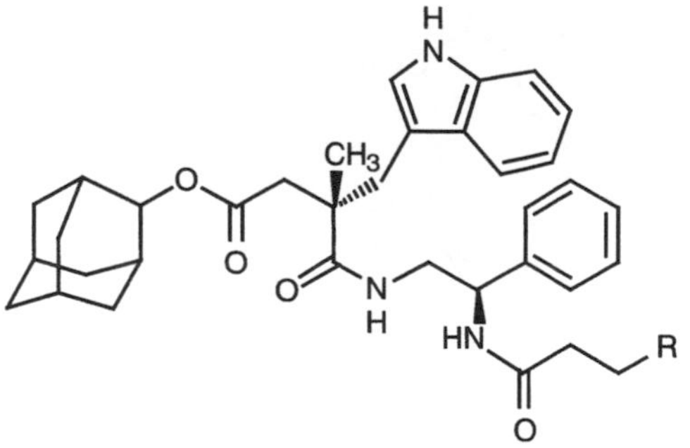

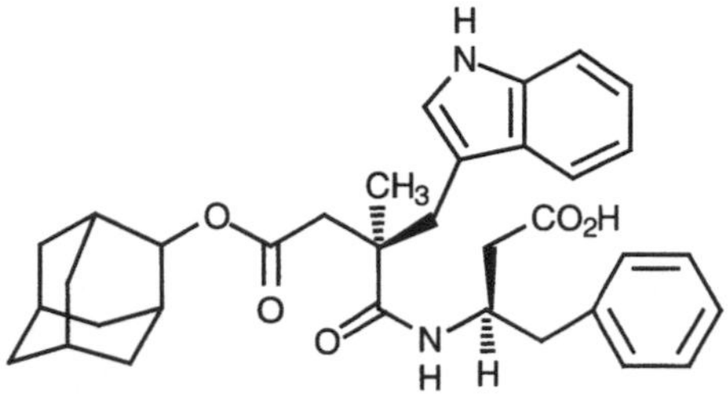

R	CCK-A*	CCK-B*
CO$_2$H#	4300	1.7
Tetrazolyl	1070	5.6
CONH-Tetrazolyl	1200	6.3
#CI-988		
PD 140548	2.8	259

* IC$_{50}$ values of CCK binding in: CCK-A – rat pancreas, CCK-B – mouse cerebral cortex

Another successful approach to CCK antagonists based on modeling of the C-terminal tetrapeptide portion of CCK, Trp-Met-Asp-Phe-NH$_2$, led to the discovery of the CCK-B antagonist CI-988.[27] The salient design principle for these "dipeptoid" CCK antagonists is the rearrangement of the "discontinuous" message Trp-Met-Asp-Phe to the "continuous" message Trp-Phe, with the aromatic rings of Trp and Phe serving as key interaction sites with the receptor, and the remaining peptide backbone serving as a scaffolding element to properly orient the two aromatic rings.[28] This design was based in turn on earlier SAR work on the C-terminal tetrapeptide portion of CCK indicating the proximity of the Trp and Phe residues.[29] Subsequent SAR work has focused on modifications of CI-988 which replace the carboxylic acid with surrogates which retain potent affinity for the CCK-B receptor, such as the tetrazole or tetrazolyl carboxamide,[30] or which alter this side chain and stereochemistry to achieve potent affinity for the CCK-A receptor, as in PD 140548,[31] as outlined in Table 2.

By modeling just the Asp-Phe-NH$_2$ portion of CCK, another carboxylic acid CCK antagonist, designated 2-NAP, was discovered which shows a pK$_b$ value of 6.5 for functional antagonism of CCK activity in a number of peripheral tissue preparations.[32] Modeling the Trp-Asp-Phe side chains of the C-terminal tetrapeptide portion of CCK led to the design of SC-50998, IC$_{50}$ = 16 nM at the CCK-A receptor in rat pancreas, in which the naphthoyl group is a surrogate for the indolyl group as a Trp mimetic.[33]

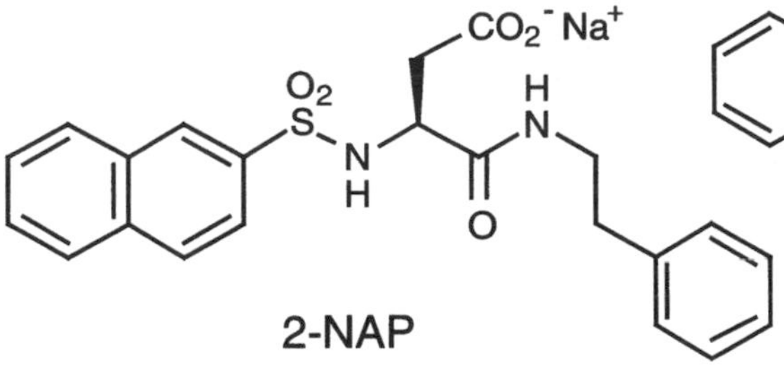

2-NAP

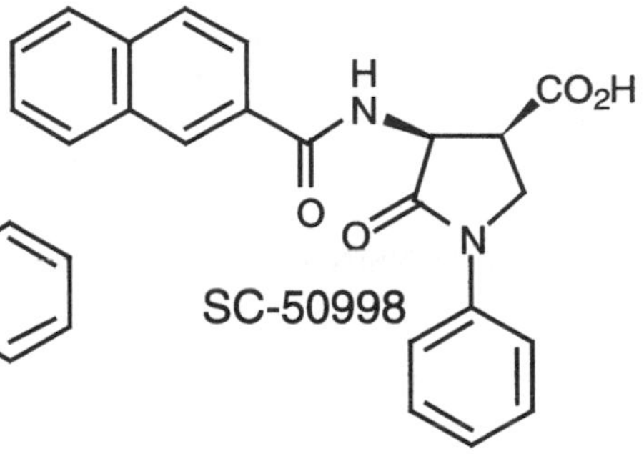

SC-50998

Finally, the search for structurally novel CCK antagonists through fermentation screening continued after the discovery of asperlicin. Two structures, both potent CCK-B antagonists, tetronothiodin, ($IC_{50} = 3.2$ nM),[34] and L-156,586, ($IC_{50} = 7.8$ nM),[35] have been reported.

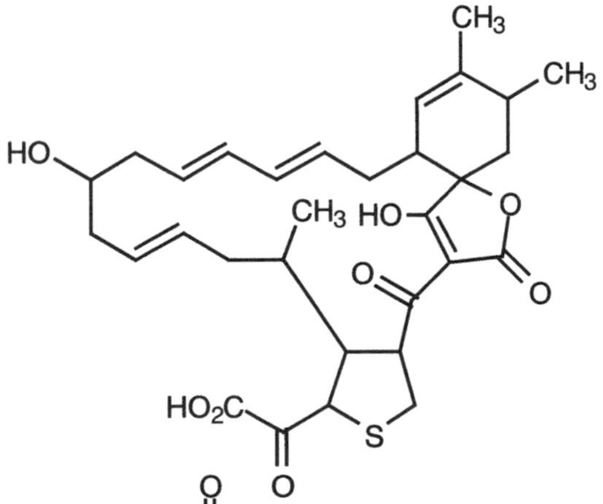

Tetronothiodin

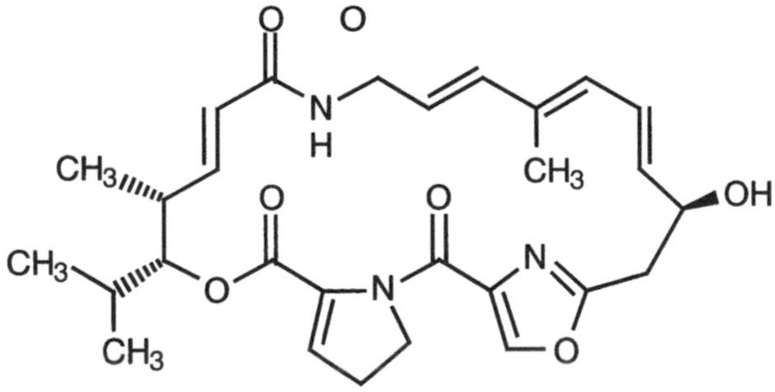

L-156,586

The maturity of the CCK antagonist area is illustrated not only by the variety of structural series displaying potent affinity and selectivity, but also by continued research into potential therapeutic applications of these antagonists. For example, electrophysiological and behavioral studies have been employed to show that both CCK-A[36,37] and CCK-B[38] receptors are involved in regulating the activity of dopaminergic neurons in the central nervous system in an effort to clarify the possible role of CCK in psychosis. With potential applications for CCK antagonists in anxiety, pain, psychosis, and satiety, it is clear that these compounds will continue to provide a rich ground for medicinal chemistry pursuits in the years ahead.

3. ANGIOTENSIN II

Angiotensin II (AII) is an octapeptide (Asp-Arg-Val-Tyr-Ile-His-Pro-Phe) that is biosynthesized from the 60 kD protein angiotensinogen (A_0).[39] The enzyme renin, an aspartyl proteinase, cleaves A_0 in the kidney to give the decapeptide angiotensin I (Asp-Arg-Val-Tyr-Ile-His-Pro-Phe-His-Leu, AI) which is then degraded to AII by the carboxypeptidase — angiotensin converting enzyme (ACE). AII is a potent vasoconstrictor that has been implicated in hypertension and congestive heart failure.[40] AII stimulates the release of aldosterone — another hormone that plays an important role in homeostasis and sodium retention.

The first non-peptidic AII antagonists, reported in 1982, exhibit weak affinity for the receptor with IC_{50} values in the micromolar range.[41,42] S-8307 and S-8308 were shown to be functional antagonists by inhibiting both the AII-induced contraction of rabbit aorta strips and the AII-induced pressor response in conscious rats.

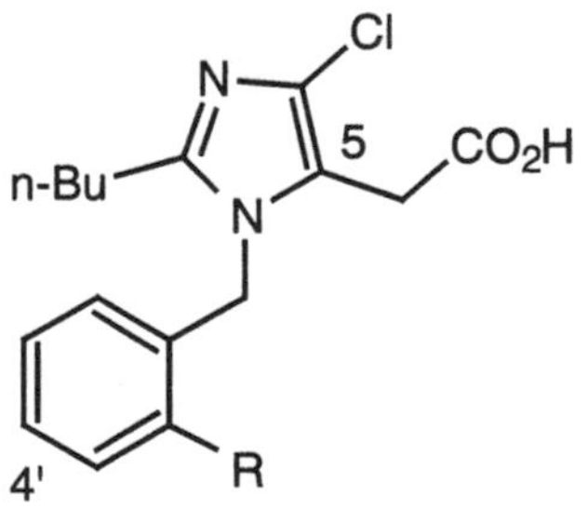

S-8307 R = Cl; IC_{50} = 40 µM (rat adrenal cortex)

S-8308 R = NO_2; IC_{50} = 13 µM (rat adrenal cortex)

The binding affinity of these 2-butyl-4-chloro-1-benzyl-imidazole-5-acetic acid derivatives was improved by a classical medicinal chemistry approach that involved first searching for structural similarities to the angiotensin II octapeptide and then optimizing potentially critical overlapping interactions. Alignment of the imidazole heterocycle in S-8307 with the imidazole ring in the His[6] residue in AII allows the C-5 acetic acid side-chain in S-8307 to point toward the C-terminal Phe[8].[43] A carboxyl moiety was introduced at the C-4' position of the benzyl side chain in S-8307 in order place an ionizable group in the region of the receptor where either the Asp[1] carboxylate side chain or the Tyr[4] hydroxyl group might be situated. The

Table 3 SAR of N-benzyl imidazole angiotensin II antagonists

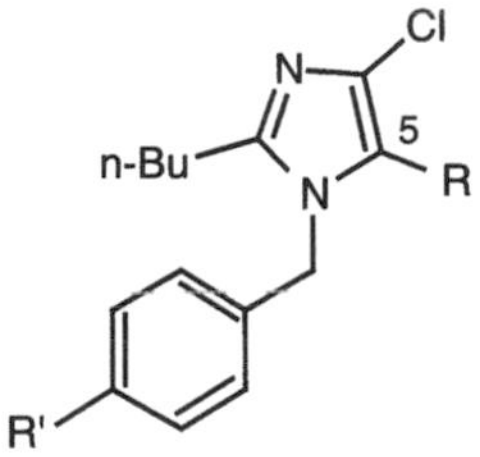

	R	R'	IC_{50} (μM)[*]
EXP6155	CH_2CO_2H	CO_2H	1.2
EXP6803	CH_2CO_2H	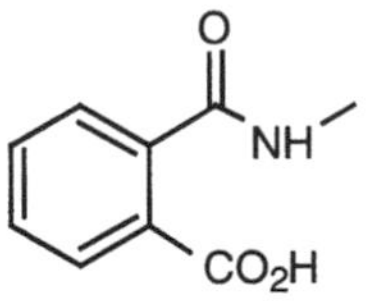	0.12
EXP7711	CH_2OH	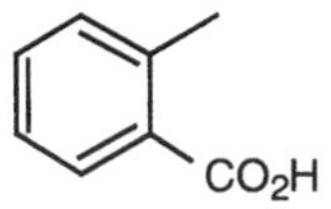	0.23
DuP753	CH_2OH	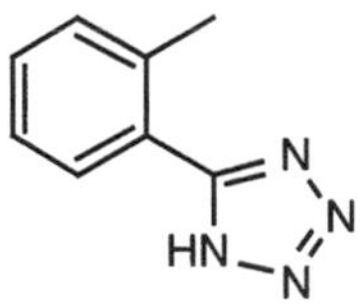	0.019
EXP3174	CO_2H	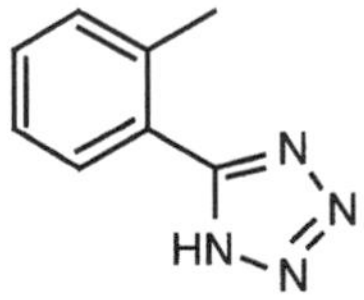	0.0012

[*] Concentration necessary for a 50% reduction in the binding of (^{3}H)-AII (@ 2 nM) from its specific binding sites in rat adrenal cortical microsomes.

binding affinity of EXP6155 (Table 3) was eleven-fold higher than S-8307 which contains no acidic group on its benzyl side-chain. To increase the anionic charge overlap with either the Asp^1 or the Tyr^4 residue, the acidic group was extended further into the receptor using a 2-carboxybenzamido moiety attached to the N-benzyl side chain. This change results in another ten-fold gain in binding affinity (EXP6803, IC_{50} = 0.12 μM). Replacement of the C-5 acetic acid group on the imidazole ring with a hydroxymethyl group and substitution of the amide linkage in the N-benzyl side chain of EXP6803 with a carbon-carbon bond leads to a two-fold loss in binding affinity.[44] Introduction of carboxylic acid isosteres restores the *in vitro* potency of the series. The 2-carboxyphenyl group in EXP7711 was replaced with a 2-tetrazoylphenyl group to give DuP753 which shows a 10× increase in potency ($IC_{50} = 0.019$ μM).[45]

DuP753 (losartan) is the first orally active AII antagonist to advance to the clinic for the treatment of hypertension. It is metabolized in many species (rat, man) to EXP3174, an insurmountable AII antagonist with an IC_{50} value of 1.3 nM.[46] EXP3174 behaves in a non-competitive or insurmountable manner in inhibiting the AII-induced contractions of rabbit aorta strips. Because of its di-acidic character, it is highly protein bound which leads to a long duration of action in many animal species.

A structurally distinct class of AII antagonists was found by randomly screening compounds containing structural elements found on the side chains in the AII octapeptide. PD123,177 and PD123,319 are derivatives of 4,5,6,7-tetrahydroimidazo [4,5-c]pyridine-6-carboxylic acids (spinacine) — the product of the cyclocondensation reaction of histidine with formaldehyde.[47] The structure-activity relationships (SAR) for this series of compounds show that a carboxyl group — preferrably in the (S)-configuration — at C-6 and a bulky acyl group attached to N-5 are required for binding affinity.

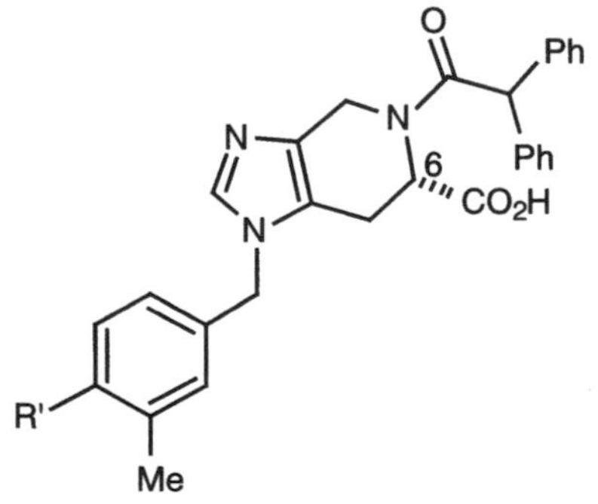

PD123,177 R' = NH_2; IC_{50} = 0.066 μM*

PD123,319 R' = NMe_2; IC_{50} = 0.034 μM*

* Measured in rat adrenal cortical microsomes with dithiothreitol (DTT) and bovine serum albumin (BSA) present

Examination of the *in vitro* and *in vivo* pharmacology of PD123,319 revealed unique receptor binding properties. PD123,319 exhibits no functional antagonism in rabbit aorta strips nor does it lower blood pressure in the renal hypertensive rat (RHR). It was soon discovered that the compound was selective for an angiotensin II receptor site for which DuP753 showed little affinity. This new site is insensitive to the addition of dithiothreitol (DDT) — a disulfide reducing agent used to prevent protein degradation. This distinction led to the identification of two new subtypes of the AII receptor now classified as AT_1 and AT_2.[48] The AT_1 receptor,

Table 4 SAR in acrylic acid series of AII antagonists

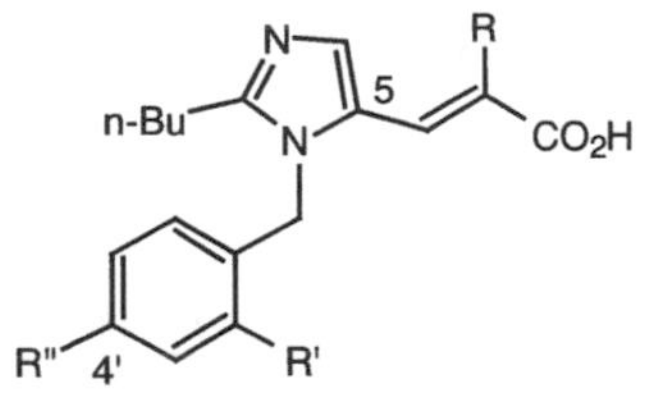

	R''	R'	R	$AT_1 \; IC_{50} \; (\mu M)^*$
9	H	Cl	H	8.9
10	H	Cl	$-CH_2Ph$	2.6
11	H	Cl	$-CH_2$⟨thienyl⟩ 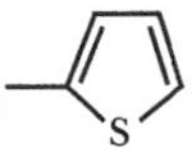	0.44
SK&F 108566	CO_2H	H	$-CH_2$⟨thienyl⟩ 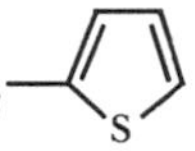	0.001

* Inhibition of [^{125}I]AII specific binding to rat mesenteric arteries

which is specific for ligands such as DuP753, mediates all the known vascular responses of AII. The AT_2 receptor to which PD123,319 specifically binds has as yet no known physiological function.[49] It is found in the adrenal gland, uterus and the brain and it may play a physiological role in ovarian regulation and in fetal development. Both the AT_1 and the AT_2 receptor subtypes have recently been cloned and sequenced.[50]

A second structurally distinct series of AT_1-selective antagonists was developed from the N-benzyl-imidazole-5-acetic acid S-8307 by employing a different overlap of the imidazole heterocycle with the postulated pharmacophoric conformation of the AII octapeptide.[51] In this new model, the imidazole heterocycle is positioned over Pro[7] while the C-5 acetic acid group points toward the C-terminal carboxylate in Phe[8], the C-2 butyl group toward the side chain on Ile[5] and the N-benzyl group toward the lipophilic pocket occupied by the aromatic side-chain of Tyr[4]. It was hypothesized that extension of the C-5 side-chain on the imidazole ring would improve the overlap of the acidic group with the the C-terminal carboxylate in Phe[8] when the N-benzyl group occupies the pocket reserved for the side chain on Tyr[4]. The imidazole acetic acid was changed to an imidazole acrylic acid (Table 4, compound **9**). Addition of a benzyl group on the acrylic acid chain creates a terminal phenylalanine mimic (compound **10**), resulting in a slight improvement in the binding affinity. Replacement of the phenyl group in **10** with a thienyl group increases potency five-fold (**11**, AT_1 IC_{50} = 0.44 nM). Introduction of a carboxyl group at the C-4' position on the N-benzyl side chain led to SK&F 108566, a 1 nM AT_1-selective AII antagonist. This compound antagonizes the AII-induced pressor response in conscious rats with ED_{50} value of 0.08 mg/kg.

DuP753 and SK&F 108566 are the prototypical AT_1-selective angiotensin II antagonists. The discovery of these compounds led to the development of a wide variety of structurally similar AII antagonists — the majority of which are derived from DuP753 by either replacing the imidazole ring (headpiece) or by modifying the (2-tetrazoylphenyl)benzyl moiety (tailpiece).

A wide variety of five-membered heterocycles are satisfactory surrogates for the imidazole ring in DuP753. In SR-47436, a 5-butyl-3-spirocyclopentyl-imidazol-2-one is attached to the (2-tetrazoylphenyl)benzyl tailpiece.[52] This compound has an AT_1 IC value of 1.3 nM (rat liver membrane binding assay) and is orally active in both rats and cynomologous monkeys. Compounds 12[53] and 13[54] are analogs of DuP753 in which either an N^2-alkyl- or an N^2-aryl-1,2,4-triazol-3-one (triazolinones) is attached to the tailpiece. SC-50560 with a 3,5-dibutyl-1H-1,2,4-triazole as the headpiece has an AT_1 IC_{50} value of 5.6 nM (rat uterine membrane binding assay) and a pA_2 value of 8.7 for the inhibition of the AII-induced contraction of rabbit aorta.[55] The pyrazole 14 is an example of a carbon-linked five-membered ring heterocycle connected to the tailpiece instead of the usual nitrogen-linked heterocycle.[56] Compound 14 has a pK_B value of 10.5 (for the inhibition of the contractile effect of AII in rabbit isolated thoracic aorta) and lowers blood pressure at a 1 mg/kg *p.o.* dose in renal-ligated hypertensive rat.

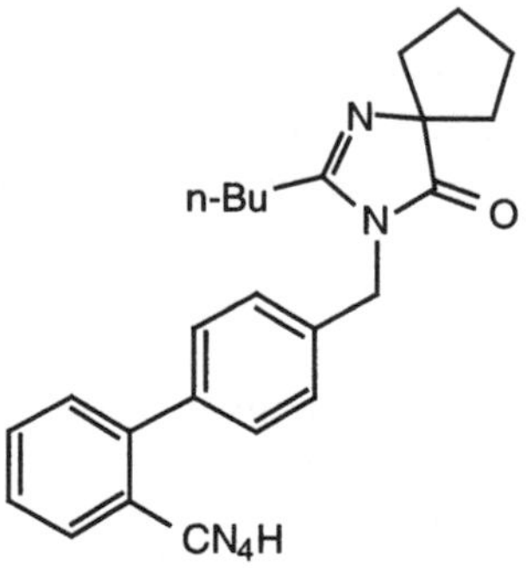

SR-47436

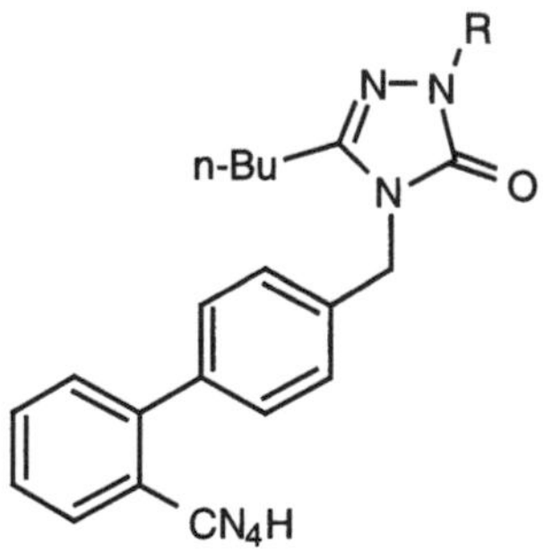

12 R = n-Bu
13 R = 2,6-Cl,Cl-C_6H_3

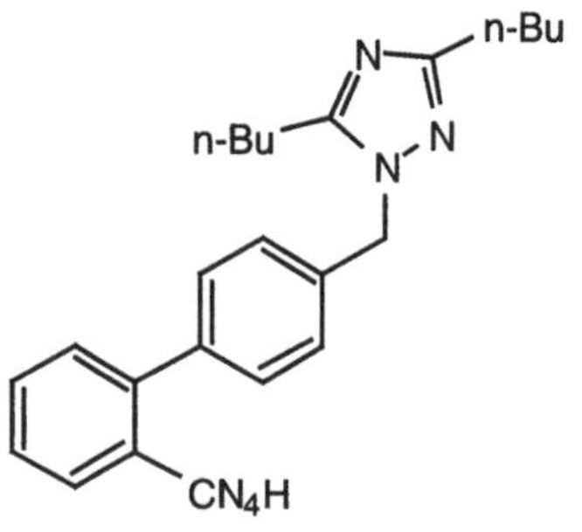

SC-50560 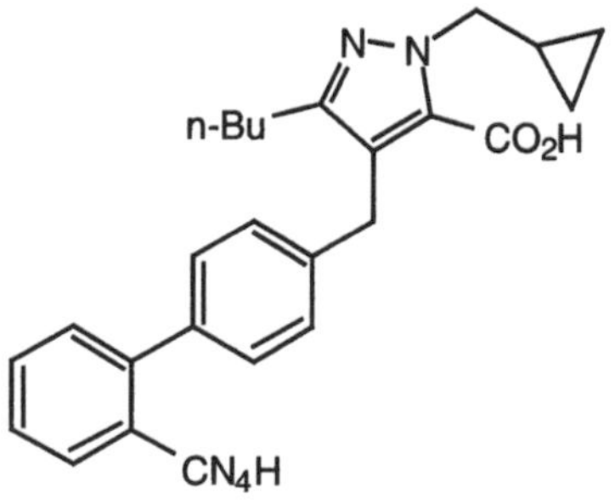   14

Table 5 SAR in benzimidazole series of AII antagonist

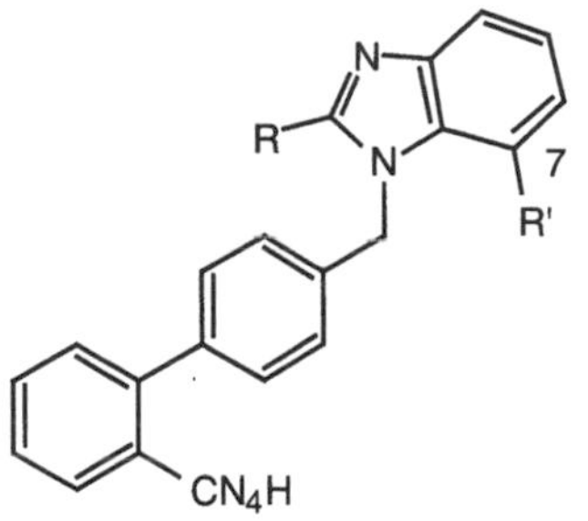

	R	R'	AT_1 IC_{50} (μM)*
15	n-Bu	H	0.9
CV-11194	n-Bu	CO_2H	0.6
CV-11974	EtO	CO_2H	0.1
TCV-116	EtO	$CO_2CH(Me)OCO_2$-c-hexyl	

* Inhibition of specific binding of ^{125}I-AII (0.2 nM) to bovine adrenal cortex.

The replacement of the DuP753 imidazole headpiece with benzimidazoles led to another structural class of AII antagonists. 2-Butyl-1-[2'-(1H-tetrazoyl-5-yl)-biphenyl-4-ylmethyl]-1H-benzimidazole (Table 5, compound **15**) has an AT_1 IC_{50} value of 0.9 μM (bovine adrenal cortex membrane binding assay).[57] This compound exhibits a short duration of action (both *i.v.* and *p.o.*) in inhibiting the pressor response to intraveneously-administered AII in the conscious rat. Introduction of a C-7 carboxy group on the benzimidazole increases the binding affinity two-fold (CV-11194, AT_1 IC_{50}=0.6 μM). More importantly, however, this analog shows significant oral activity. CV-11194 exhibits a dose related (0.3–10 mg/kg) inhibition of the pressor response induced by AII that is superior to that of DuP753. It is postulated that the C-7 carboxyl group induces an unknown secondary conformational change in the receptor that results in an insurmountable antagonist-receptor complex. Replacement of the C-2 butyl group on the benzimidazole subunit with a C-2 ethoxy group (CV-11974) results in a six-fold increase in binding affinity.[58] The oral activity of CV-11974 was enhanced by converting the C-7 carboxyl group into a metabolically labile ester. TCV-116 is the 1-[[(cyclohexyloxy)carbonyl]-oxy]ethyl ester of CV-11974 and is currently in clinical development in Japan for the treatment of hypertension.[59]

Replacement of the benzimidazole heterocycle with an imidazo[4,5-b]pyridine ring gives another structural class of AII antagonists.[60] In an imidazo[4,5-b]pyridine ring, the pyridine nitrogen atom can act as a proton acceptor in a manner similar to the C-5 hydroxymethyl group in the imidazole subunit of DuP753. The SAR for this series of compounds is shown in Table 6. L-158,809 is an orally active AII antagonist with an AT_1 IC_{50} value of 0.3 nM (rabbit aorta membrane binding assay) and *i.v.* ED_{50} of 0.048 mg/kg for the inhibition of the AII-induced pressor response.

Table 6 SAR in an imidazo[4,5-*b*]pyridine series of AII antagonists

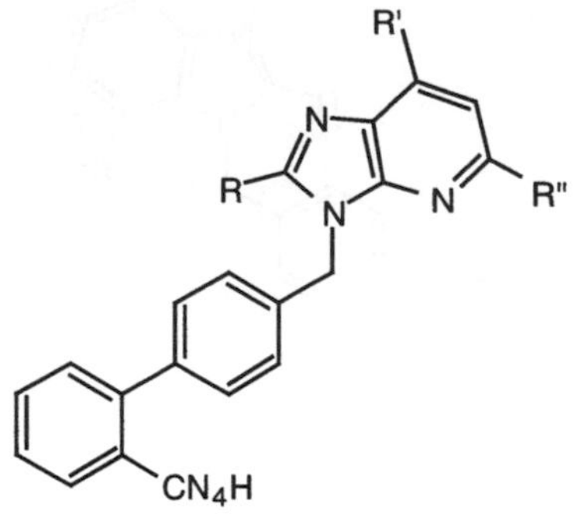

	R	R′	R″	AT_1 IC_{50} (nM)[*]
	n-Pr	H	H	8
	n-Pr	H	Me	2
L-158,338	n-Pr	Me	H	1
	n-Pr	Me	Me	0.3
L-158,809	Et	Me	Me	0.3

[*] Inhibition of ^{125}I-Sar^1, Ile^8-AII specific binding to rabbit aorta membranes.

Table 7 SAR for a series of quinazolinone AII antagonist

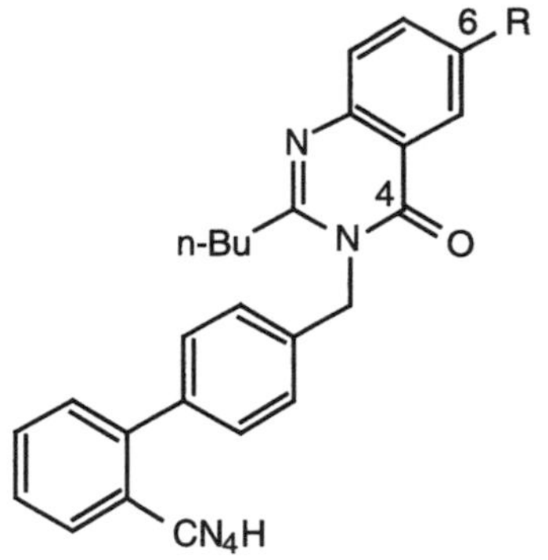

	R	AT_1 IC_{50} (nM)[*]
	H	6.3
	Me	4
	NH$_2$	1.2
	NHCONH*i*-Pr	0.75
L-159,093	NHCON(Me)*i*-Pr	0.1

[*] Inhibition of ^{125}I-Sar^1, Ile^8-AII specific binding to rabbit aorta membranes.

The heterocycle attached to the (2-tetrazoylphenyl)benzyl subunit can also be replaced by various six-membered rings. A series of compounds containing quinazolin-4(1H)-ones was found to be potent AII antagonists.[61] In these structures, the C-4 carbonyl group can act as a hydrogen-bond acceptor similar to the N-4 nitrogen in the imidazo[4,5-b]pyridine series. Shown in Table 7 is the SAR for this series. Substitution at the C-6 position — especially with polar urea derivatives — provides compounds with subnanomolar affinities for the receptor. L-159,093 with an AT_1 IC_{50} value of 0.1 nM (rabbit aorta membrane binding assay) inhibits the AII-induced pressor response in conscious normotensive rhesus monkeys at a 3 mg/kg $p.o.$ dose.

A novel series of AII antagonists contains heterocyclic rings connected via an oxygen or nitrogen atom to the (2-tetrazoylphenyl)benzyl tailpiece. The 4-alkoxy-quinoline D-8371 has an AT_1 IC_{50} value of 30.7 nM (guinea pig adrenal membrane binding assay) and an $i.v.$ ED_{50} value of 1 mg/kg in antagonizing the AII-induced pressure response in the conscious rat.[62] The tetrahydro-4-alkoxyquinoline D-6888 exhibits a six-fold increase in binding affinity (AT_1 IC_{50}=5 nM) which translates into a three-fold increase in the $in\ vivo$ potency ($i.v.$ ED_{50}=0.39 mg/kg).[63] The 3-carboxy-2-amino pyridine A-81988 is an acyclic mimic of the imidazo[4,5-b]pyridine ring with a pA_2 value of 10.6 (for the inhibition of the AII induced contractile effect in rabbit aorta) and an AT_1 K_i value of 0.76 nM (rat liver membrane assay).[64] A-81988 lowers blood pressure in the renal hypertensive rat at a 0.3 mg/kg $p.o.$ dose.

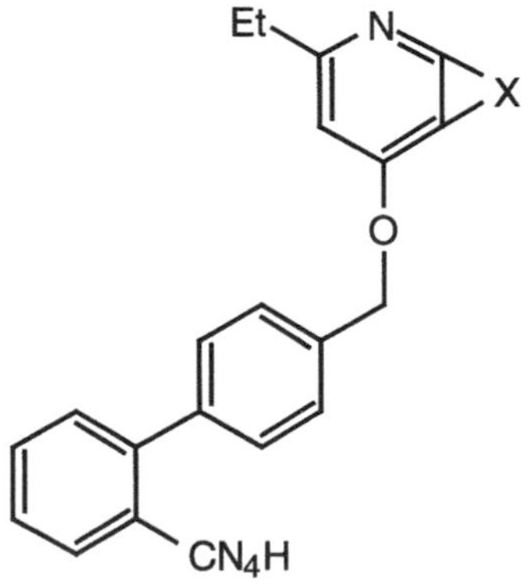

D-8371 X = -CH=CH-CH=CH-

D-6888 X = -(CH$_2$)$_4$-

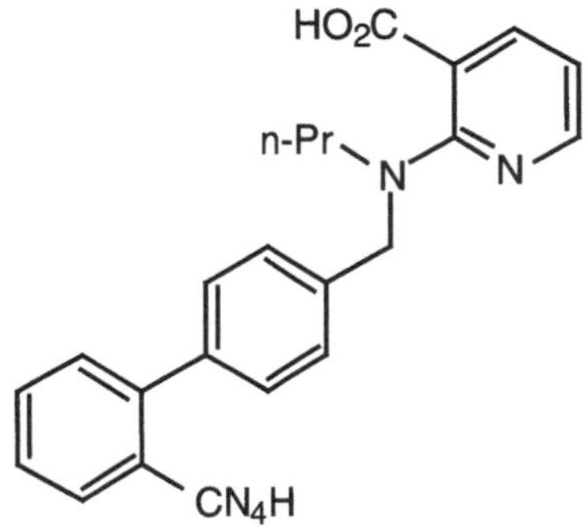

A-81988

Acylated amino acids can also replace the imidazole heterocycle in DuP753. CGP48933 (valsartan) with an N-butyrylvaline residue attached to the 4-(2-tetrazoylphenyl)benzyl tailpiece has an AT_1 IC_{50} value of 2.7 nM (aortic smooth muscle cell membrane binding assay).[65] It lowers blood pressure in the renal hypertensive rat at doses of 1–10 mg/kg $p.o.$ and in the sodium-depleted marmoset at doses of 0.1–3 mg/kg $p.o.$[66]

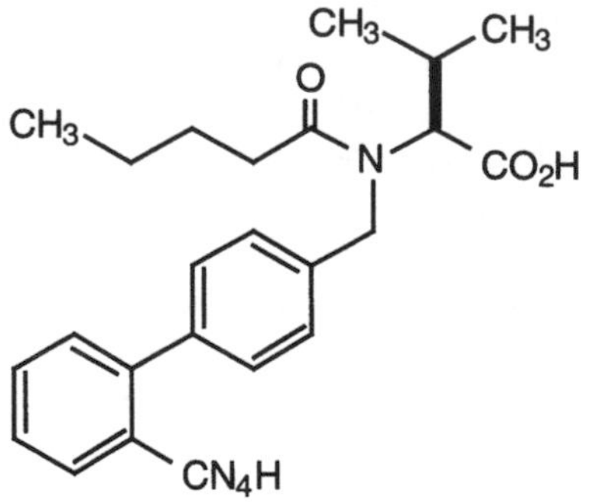

CGP48933 (Valsartan)

The 4-(2-tetrazoylphenyl)benzyl subunit is a common pharmacophore to the many angiotensin II antagonists containing novel ring replacements for the imidazole group in DuP753. It was soon discovered that either one of the aryl rings in the biphenyl system could also be replaced by heterocycles. In a series of compounds with $1H$-1,2,4-triazole headpieces, replacement of the middle phenyl ring with a pyridyl ring affords SC-52458 which has an AT_1 IC_{50} value of 6.9 nM (rat uterine membrane binding assay) and a pA_2 value of 8.18 (antagonism of the AII-induced contractile effect in rabbit aorta).[67] Introduction of a benzofuran ring in the EXP3174 series of compounds with a 2-butyl-4-chloro-imidazole-5-carboxylic acid group as the headpiece yields another potent class of AII antagonists.[68] The SAR for this series of compounds is shown in Table 8. The binding affinity depends on

Table 8   SAR for a series of benzofuran AII antagonists

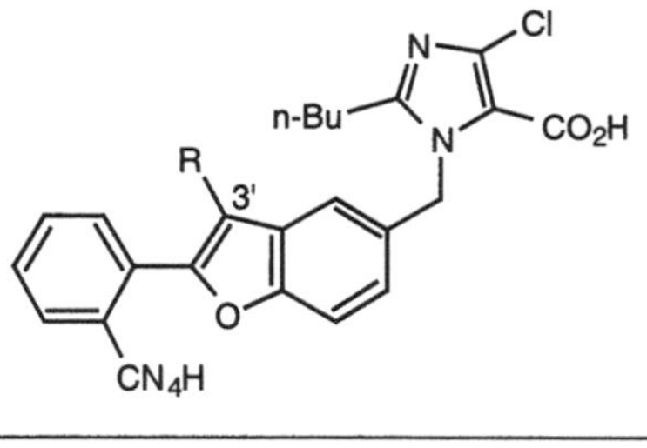

	R	pK_B[*]
	H	8.4
	OMe	9.6
GR117289	Br	9.8
	CF$_3$	10.0
	Cl	10.5

[*] Inhibition of the contractile effect of AII in rabbit isolated thoracic aorta.

the nature of the substituent at the C-3' position of the benzofuran ring — with electron-withdrawing substituents being preferred.[69] Such substituents are postulated to enhance binding affinity by decreasing the electronic charge distribution of the aromatic ring. It is particularly interesting that the replacement of the middle phenyl ring in EXP3174 with a benzofuranyl ring increases the distance between the two critical pharmacophores — the imidazole headpiece and the tetrazoyl group on the distal phenyl ring — yet the potency of this series is similar to that of the EXP3174 biphenyl series. GR117289 has a pK_B value of 9.8 (for the inhibition of the contractile effect of AII in rabbit isolated thoracic aorta); it inhibits the AII pressor response in normotensive rats at a dose of 1 mg/kg administered intra-arterially. It also lowers blood pressure at a dose of 10 mg/kg *p.o.* in the conscious hypertensive rat.[70]

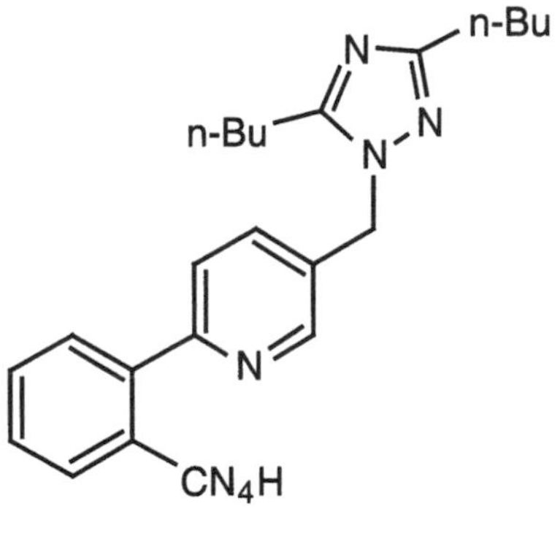

SC-52458

Replacement of the distal phenyl ring also affords potent AII antagonists. In the $1H$-1,2,4-triazole series, the biphenyl ring was replaced with a N-phenylpyrrole to afford compound **16** which has an AT_1 IC_{50} value of 33 nM (rat uterine membrane membrane binding assay) and a pA_2 value of 8.0 for the antagonism of AII-induced contraction in rabbit aorta rings.[71] In a series of compounds with imidazo[4,5-b] pyridine heterocycles as headpieces, the distal phenyl group could be replaced by thienyl or pyridyl rings without significantly affecting the binding affinity. L-159,827 with a 4-(2-tetrazolylthiophen-3-yl)benzyl tailpiece has an AT_1 IC_{50} value of 2.3 nM (rabbit aorta membrane binding assay) and in conscious rats, it antagonizes the AII-induced pressor response at an *i.v.* dose of 1 mg/kg with a duration of action that is greater than 6 h.[72] The binding affinity of compound **17** with a 4-(3-tetrazoyl-pyrid-4-yl)benzyl tailpiece is four-fold less than the corresponding biphenyl analog L-158,338 (see Table 6).[73]

In the biaryl series of AII antagonists, the binding affinity is critically dependent on the presence of an acidic group on the distal ring. The tetrazoyl group (with a pKa value of 5–6) is a common pharmacophore in many potent compounds —

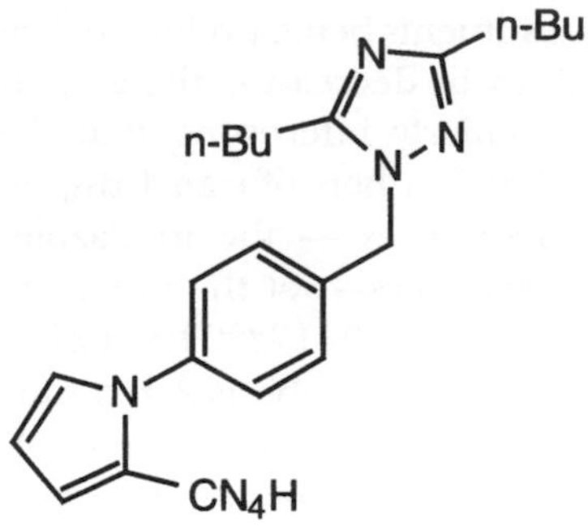

16

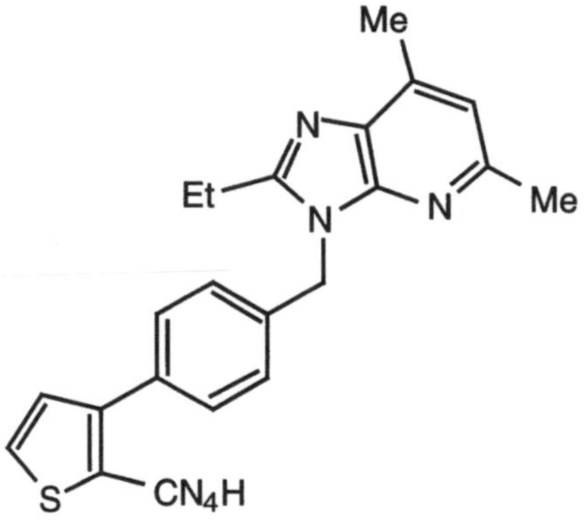

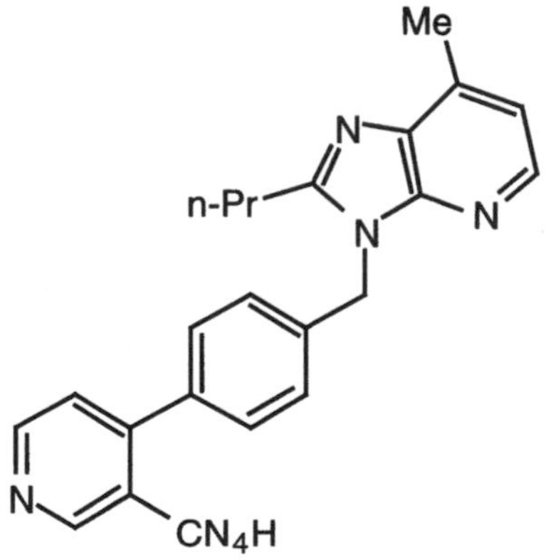

L-159,827

17

usually increasing both the *in vitro* potency and oral efficacy compared to the corresponding carboxylic acid analogs. Several acid isosteres have been discovered to be satisfactory replacements for the tetrazoyl group. The triflamide **18** — with a pK_B value of 8.4 for the inhibition of the contractile effect of AII in rabbit isolated thoracic aorta — lowers blood pressure at a 1 mg/kg *p.o.* dose in the renal hypertensive rat.[74] L-159,282 — the benzoyl sulfonamide analog of the potent tetrazoyl compound L-158,809 (see Table 6) — has an AT_1 IC_{50} value of 0.22 nM (rabbit aorta membrane binding assay) and a pA_2 value of 10.3 in antagonizing AII-induced aldosterone release.[75] This compound is as efficacious as the ACE inhibitor enalapril in lowering blood pressure in the aortic coarcted rat model.

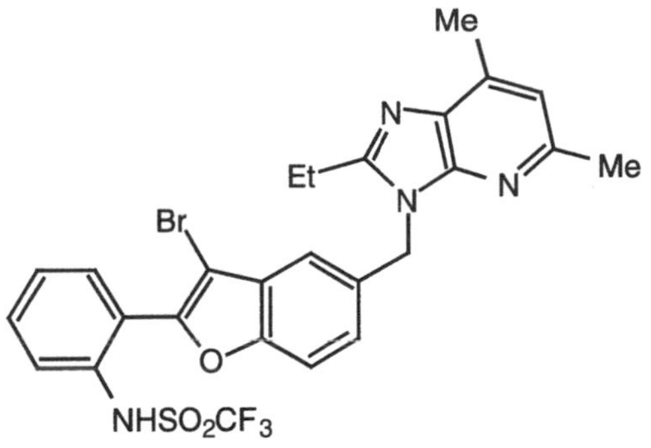

18

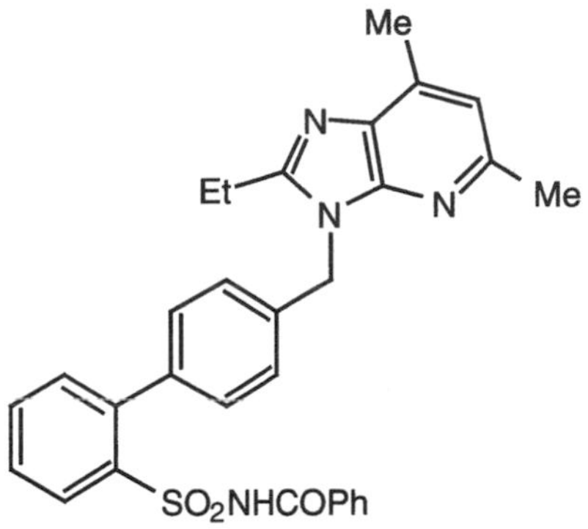

L-159,282

4. TACHYKININS

The principal members of the tachykinin peptide family in mammals are substance P, neurokinin A, and neurokinin B, which bind selectively to the three corresponding members of the neurokinin receptor family, NK_1, NK_2, and NK_3.[76] Extensive pharmacological characterization of substance P (SP) in inflammation, pain, and pulmonary models in animals suggested it plays an important role in human disease. These studies stimulated efforts to find SP antagonists both as reagents for *in vivo* studies and as possible new therapeutants. Although early work with modifications of SP itself did provide peptidic reagents useful in the laboratory,[77] a real breakthrough in this field came with the discovery of nonpeptide antagonists. The first of these, CP-96,345, resulted from SAR studies around a lead structure, compound **19**, discovered by empirical file screening.[78] Thus optimization of three key elements for NK_1 receptor recognition by this class of compounds, the 2-methoxybenzylamine side chain, unsubstituted benzhydryl group, and bridgehead nitrogen[79] resulted in the potent and NK_1 selective CP-96,345, $IC_{50} = 0.77$ nM in human IM-9 cells.

CP-96,345 shows both enantioselectivity in its binding to the NK_1 receptor, with the corresponding 2R,3R enantiomer affording at least 10,000-fold lower NK_1 affinity, and species selectivity, with affinity for the human NK_1 receptor 60-fold greater than in the rat.[80] These properties, and its good oral bioavailability and long plasma half-life, have enabled CP-96,345 to define the role of SP binding to the NK_1 receptor in inflammation, pain, and asthma. A key finding is the importance of NK_1 control of endothelial permeability and hence the access of neutrophils to a proinflammatory site, which provides SP with an important role in neurogenic inflammation.[81] In addition, SP activation of NK_1 receptors plays a role in chronic pain.[82] More recently, CP-96,345 and an analogue CP-99,994 (see below) have been used to delineate the role of SP in angiogenesis[83] and emesis.[84]

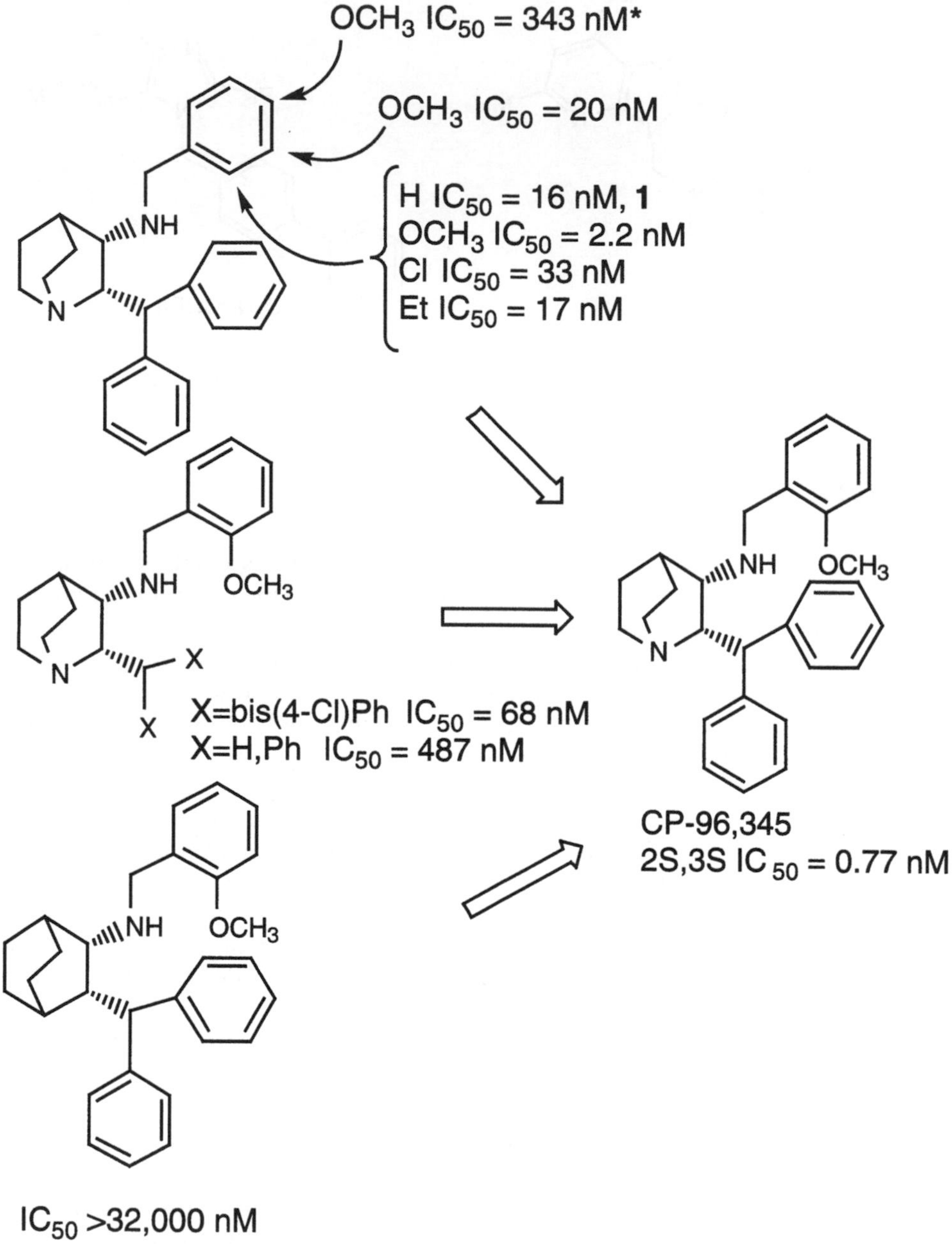

OCH$_3$ IC$_{50}$ = 343 nM*

OCH$_3$ IC$_{50}$ = 20 nM

H IC$_{50}$ = 16 nM, **1**
OCH$_3$ IC$_{50}$ = 2.2 nM
Cl IC$_{50}$ = 33 nM
Et IC$_{50}$ = 17 nM

NH

N

OCH$_3$

NH

N

X

X

X=bis(4-Cl)Ph IC$_{50}$ = 68 nM
X=H,Ph IC$_{50}$ = 487 nM

NH   OCH$_3$

N

CP-96,345
2S,3S IC$_{50}$ = 0.77 nM

OCH$_3$

NH

IC$_{50}$ >32,000 nM

*IC$_{50}$ values in nM units for displacement of [^{3}H]SP binding
in human IM-9 cells.

In the medicinal chemistry work following up on CP-96,345, an important SAR limitation for CP-96,345 was disclosed in the [3.2.2] analogue **20**, with 8-fold lower NK$_1$ affinity.[85] Potent NK$_1$ receptor affinity has been achieved instead with changes in more than one structural parameter. In one study, substitution of the benzylamine nitrogen by an ether oxygen necessitated a new substituent pattern on the side chain phenyl ring, with the 3,5-dimethyl compound **21** affording optimal NK$_1$ affinity.[86] Even more intriguing is the finding that the *trans* 2R,3S enantiomer **22** displays more potent NK$_1$ affinity than its corresponding 2S,3S isomer, in contrast to CP-96,345.[87] Another variation of CP-96,345 is the recently reported CP-99,994, in which the quinuclidine nucleus has been modified to a piperidine ring and the 2-substituent changed to an unsubstituted phenyl ring.[88] CP-99,994 is a potent NK$_1$ antagonist, K$_i$ = 0.25 nM in human IM-9 cells, and shows good oral activity in blocking capsaicin-induced plasma protein extravasation in guinea pig lung, ID$_{50}$ = 4.0 mg/kg.[89] SAR studies again showed the importance of the piperidine nitrogen and 2-methoxy benzylamine side chain for NK$_1$ affinity. These cases illustrate the principle that one modification of a lead structure may sometimes have to be followed by another to achieve potent receptor affinity.

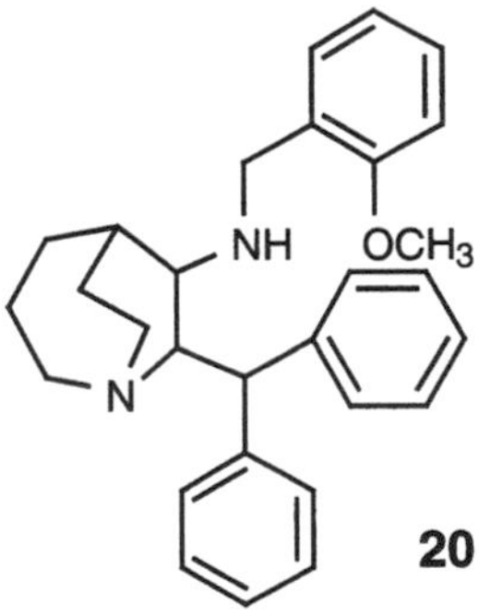

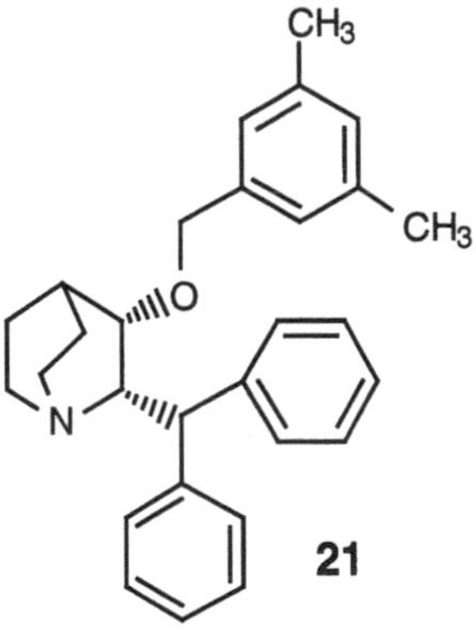

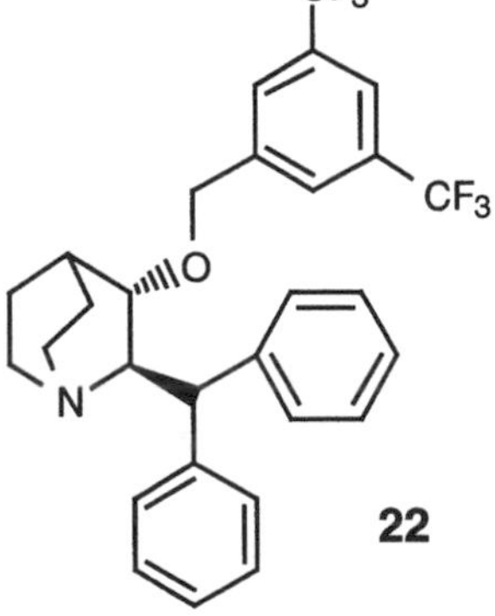

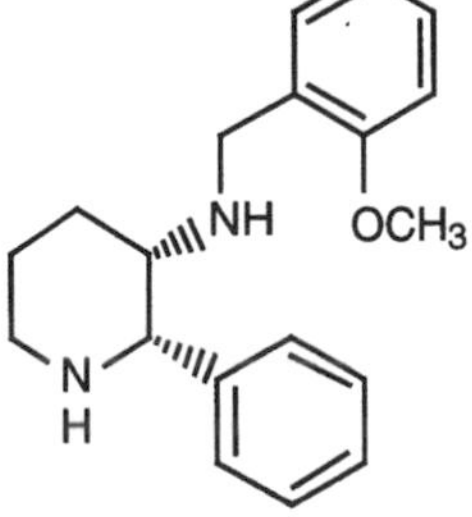

A structurally novel class of SP antagonists, again discovered through SAR optimization of a lead compound derived from empirical screening, is illustrated by RP 67,580,[90] with an IC_{50} value of 4.17 nM for the rat NK_1 receptor.[91] SAR aspects showing the importance of the unsubstituted diphenyl moeity and the amidine and 2-methoxy substituents on the side chain are summarized below.[92] RP 67,580 has also been used extensively for pharmacological characterization of SP, for example, in investigating the role of SP in neurogenic inflammation connected with animal models of headache.[93]

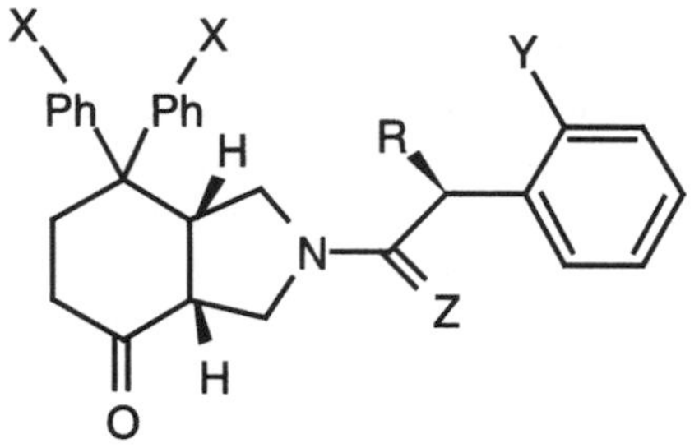

X	Y	Z	R	SP IC_{50}, nM^*
H	H	O	H	181
H	OCH_3	O	H	75
H	$OCH_2CH_2NMe_2$	O	H	40
H	H	O	CH_3	35
H	H	O	OH	113
H	OCH_3	O	CH_3	12
4-CH_3	· H	O	H	393
H	H	NH	H	36
H	OCH_3	NH	H	17#

* IC_{50} value of displacement of $[^3H]$-SP from rat brain membranes. # After resolution, $IC_{50} = 4.1$ nM, RP-67,580.

An extensive SAR effort was required to modify the weak (IC_{50} value of 5,000 nM), screening-derived lead structure **23** into the potent ($IC_{50}=0.02$ nM in rat) and selective NK_1 antagonist SR-140,333.[94] This structural class also serves as the basis for the NK_2 antagonist SR-48,968, described below, illustrating its versatility.

Another screening lead, N-ethyl-L-tryptophan benzyl ester, was modified to the potent NK_1 antagonist **24**, $IC_{50} = 1.7$ nM.[95] The androstane steroid skeleton **25**, uncovered by file screening, has also served as the basis for a series of NK_1 receptor antagonists.[96] SAR studies demonstrated the importance of the groups at the C-17 position of the steroid nucleus, with the most potent compound, **25**, giving an IC_{50} value of 50 nM at the rat brain NK_1 receptor.

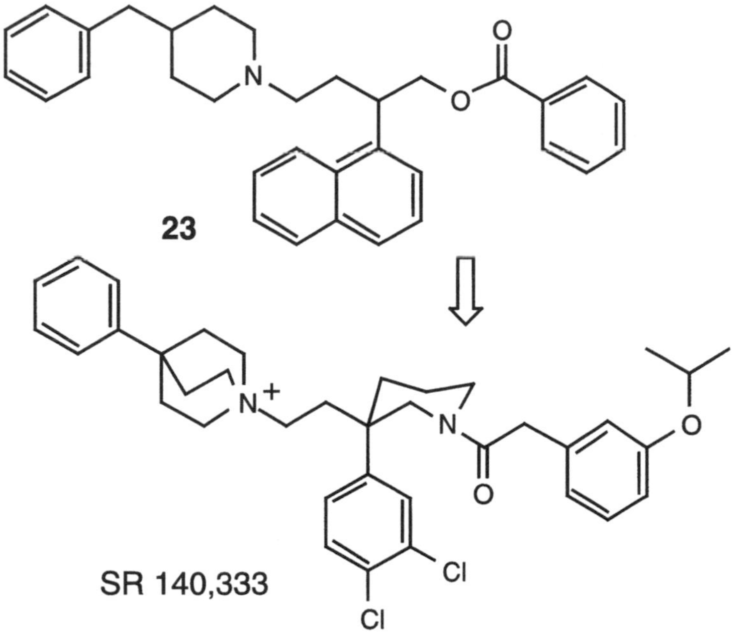

23

SR 140,333

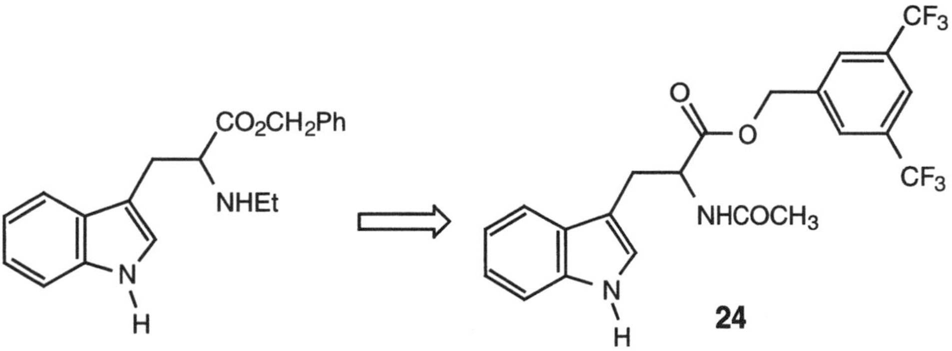

24

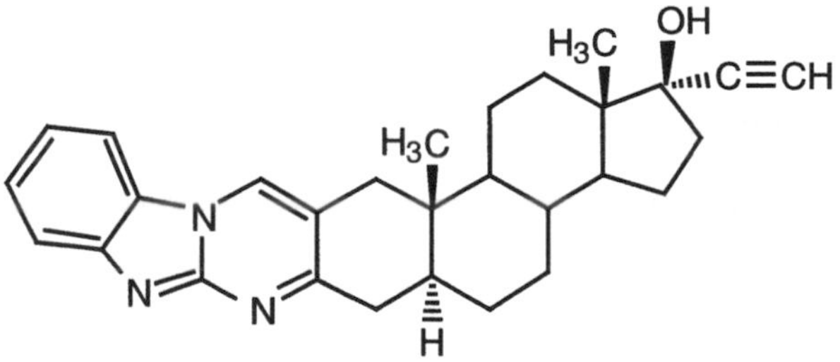

25

The discovery of the NK_1 antagonist CGP-47,899 relied on computer-based modeling of lead structure **26** to understand its relationship to the binding conformation of SP.[97] Thus incorporation of the 2-benzyl substituent on the piperidine ring, based on mimicking Phe-8 in SP, improved NK_1 receptor affinity. Subsequent studies showed that the active enantiomer at the 2-position of the piperidine ring, however, corresponds to the inactive enantiomer at the Phe-8 position of SP. As shown below, recent evidence indicates nonpeptide antagonists occupy a different receptor site from their peptide agonist counterparts, explaining this discrepancy. The application of computer-based design nevertheless helped afford a potent, selective NK_1 antagonist in CGP-47,899, with an IC_{50} value of 10 nM at the bovine receptor.

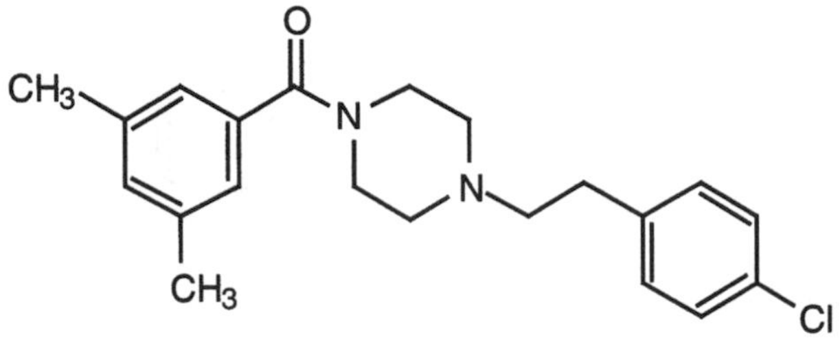

26 , NK_1 IC_{50} = 1,100 nM in bovine retina

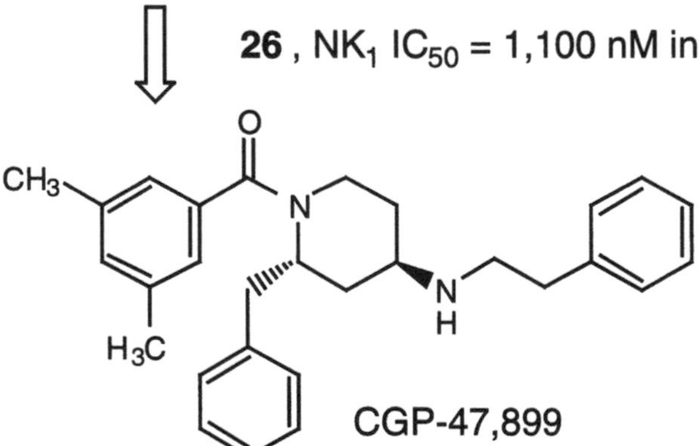

CGP-47,899
NK_1 IC_{50} = 10 nM in bovine retina

Fermentation screening has also afforded structurally novel NK_1 antagonists. The cyclic peptide structure FK 224 affords a mixed spectrum of NK_1 (IC_{50} = 37 nM) and NK_2 (IC_{50} = 72 nM) antagonism.[98] Due to its promising inhibition of non-cholinergic bronchoconstriction in the guinea pig,[99] FK 224 was advanced to clinical trials and demonstrated blockade of bradykinin-induced bronchoconstriction in humans.[100] WIN 64821, a dimeric structure vaguely reminiscent of asperlicin, gives an NK_1 K_i value of 230 nM in human astrocytoma cells.[101]

An alternative to the empirical file screening approach is illustrated by FK-888, discovered through systematic SAR investigation of a peptidic lead structure.[102] In

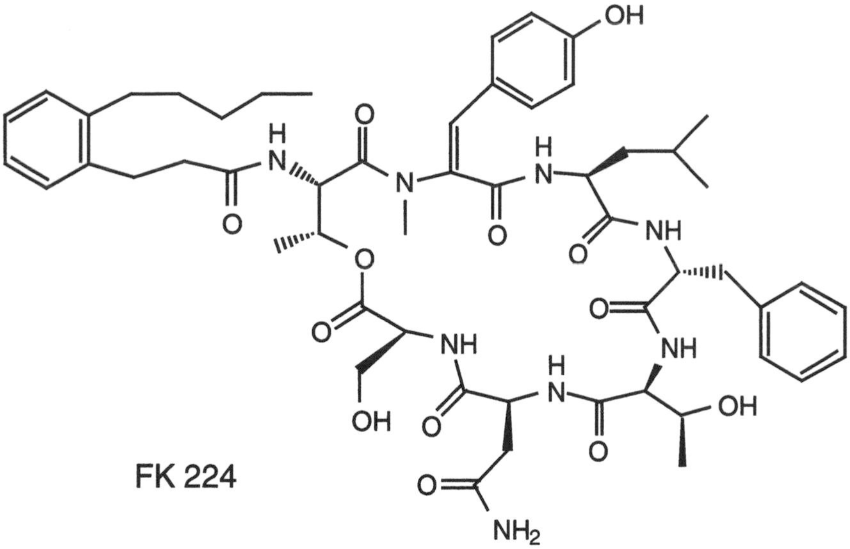

FK 224

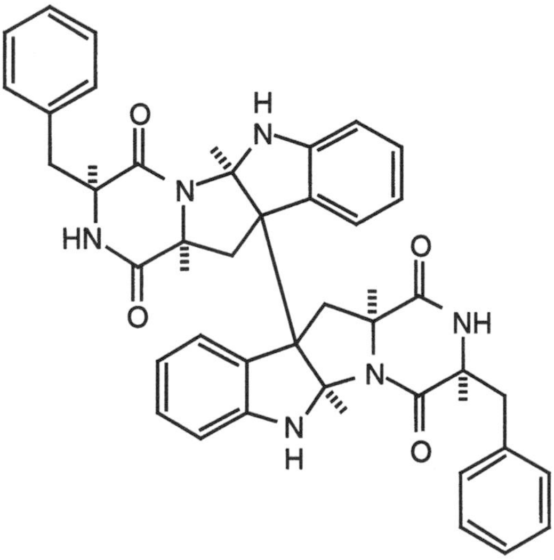

WIN 64821

the first stage of this approach, an octapeptide was reduced to each possible tripeptide fragment to find the essential binding core, leading to FR-106506.[103] Further optimization involved improvement of metabolic stability with the C-terminal N-benzyl, N-methyl amide and N-terminal modification from Boc-Gln to Ac-Thr to afford FR-113680. Finally, extensive modification of the C-terminus, eventually incorporating a hydroxyproline unit for good stability and solubility, and retention of the Trp unit, afforded FK-888.[104] FK-888, with an IC$_{50}$ value at the guinea pig NK$_1$ receptor of 6.9 nM and a pA$_2$ value of 9.29 in guinea pig ileum, is active by the oral route in specifically blocking SP-induced airway edema in the guinea pig.

FR-106506

FR-113680

FK-888

Another compound developed by a design-based program is **27**, in which the β-D-glucose core serves as a scaffold to position the side chains optimally for NK_1 receptor binding, giving an IC_{50} value of 60 nM.[105] Thus a full range of approaches has been successfully applied to the NK_1 antagonist area, from modification of screening-derived lead structures to the use of peptide- or carbohydrate-based rational design, including approaches that combine both strategies.

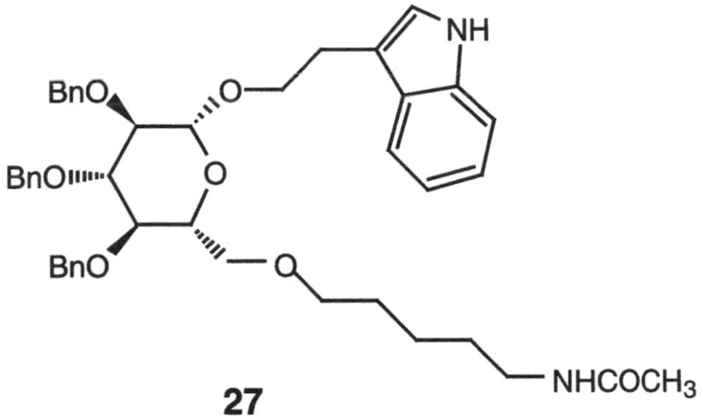

27

Nonpeptide antagonists have also been useful in defining the physiological role of neurokinin A binding to the NK_2 receptor and the potential for therapeutic intervention at this target. The first such compound was SR-48,968, again derived by modification of an empirically-derived lead structure.[106] The N-methyl benzamide is a key structural feature which ensures NK_2 selectivity, in contrast to the N-phenylacetamide group in the potent NK_1 antagonist SR-140,333 discussed above. SR-48,968 shows potent NK_2 affinity with an IC_{50} value of 0.09 nM in the rat.[107]

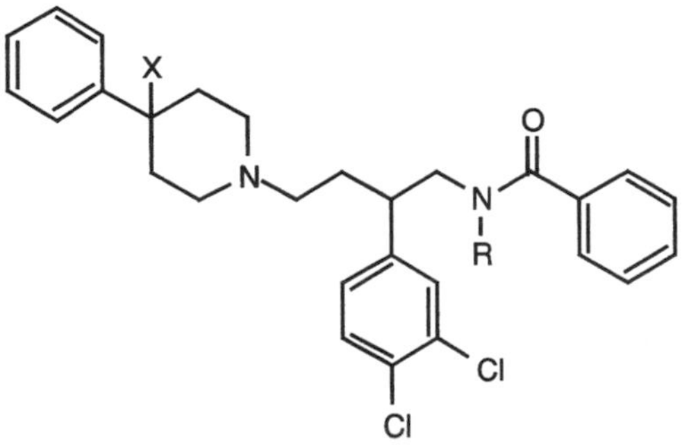

X	R	Rat NK_2 K_i, nM
OH	CH_3	1.0
CH_2OH	CH_3	5.0
NHAc	H	>100
NHAc	CH_3	0.5 (S-isomer, SR 48,968)

SR 48,968 has been used to show that the non-adrenergic non-cholinergic broncho-constriction in guinea pig lung induced by capsaicin challenge or nerve stimulation is mediated mostly by NK_2 receptors, suggesting a role for NK_2 receptors in asthma.[108] In addition, involvement of NK_2 receptors in sensing pain induced by heat has been demonstrated in the rat using SR 48,968.[109] The multitude of potential therapeutic targets indicated by the work with nonpeptide NK_1 and NK_2 antagonists suggests some exciting clinical work ahead for these compounds.

5. MISCELLANEOUS

The recent announcement of the first nonpeptide neurotensin (NT) antagonist, SR 48692, represents an opportunity to evaluate the pharmacological characterization and physiological role of NT.[110] No SAR information has as yet been disclosed around SR 48692, which was discovered by optimization of a lead structure uncovered by empirical screening. Its potent *in vitro* NT receptor affinity, $IC_{50} = 8.7$ nM in human brain, and *in vivo* activity, reversal of NT-induced turning behavior in mice at 80 µg/kg, make it a valuable reagent for further study of NT. A structurally different NT antagonist, CP-73,093, with an IC_{50} value of 5.5 µM at the NT receptor in bovine brain, was recently disclosed.[111]

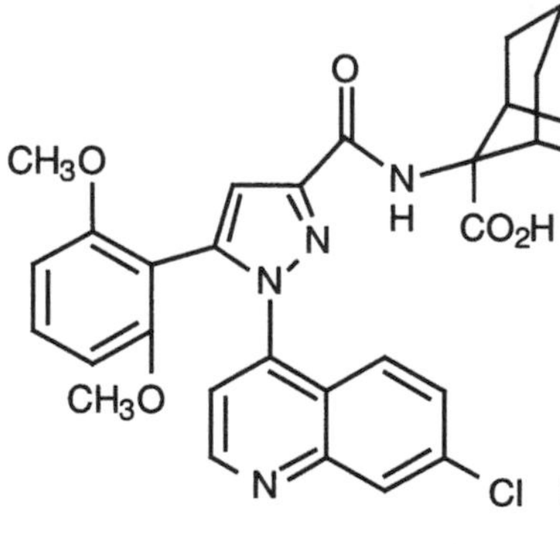

SR 48692

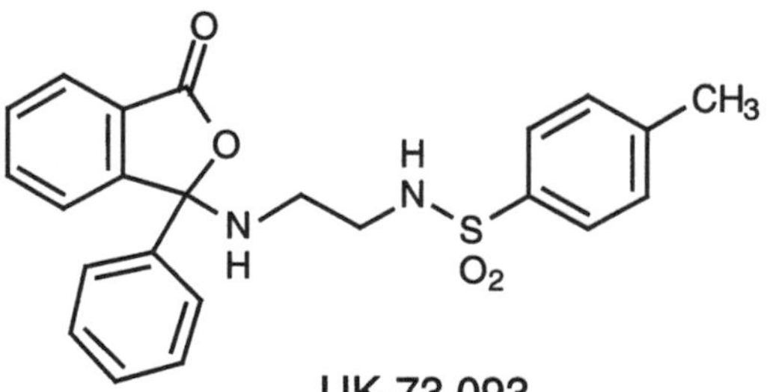

UK-73,093

In the vasopressin area, the discovery of the first nonpeptide V1 receptor antagonist, OPC-21268, ($IC_{50} = 400$ nM for V1 receptors in rat liver)[112] was followed by the first nonpeptide V2 receptor antagonist, OPC-31260, ($IC_{50} = 14$ nM for V2 receptors in rat kidney),[113] both the result of chemical followup of screening leads. The activity of OPC-31260 in blocking the antidiuretic action of a vasopressin agonist and restoring normal sodium levels in rats infused with a vasopressin agonist is helping to clarify the physiological role of the V2 receptor.[114]

While neither of the above vasopressin antagonists shows appreciable affinity for the related oxytocin (OT) receptor, the recently disclosed L-366,948, displays selectivity for the OT receptor, with an IC_{50} value of 370 nM.[115] The key step in the SAR optimization of this compound was the replacement of the tosyl group in lead

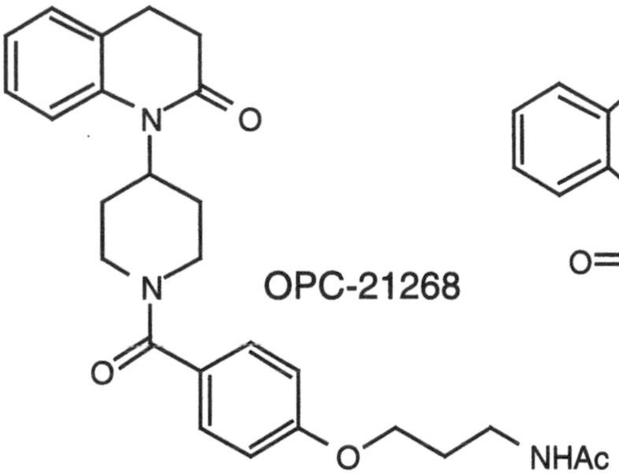

OPC-21268

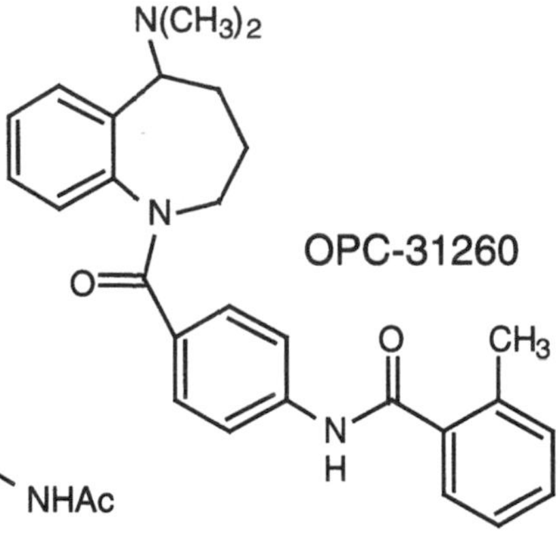

OPC-31260

compound L-674,256, with the camphor group followed by attachment of the acetic acid side chain, increasing both potency and selectivity. Blockade of OT-induced uterine contraction in rat and monkey by L-366,948 suggests a potential therapeutic role in pre-term labor.

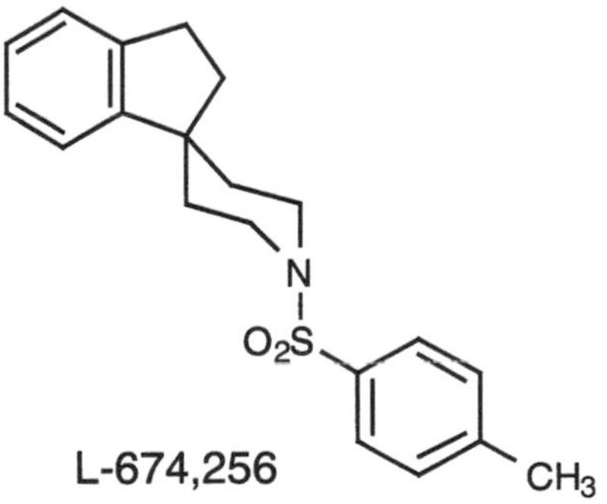

L-674,256

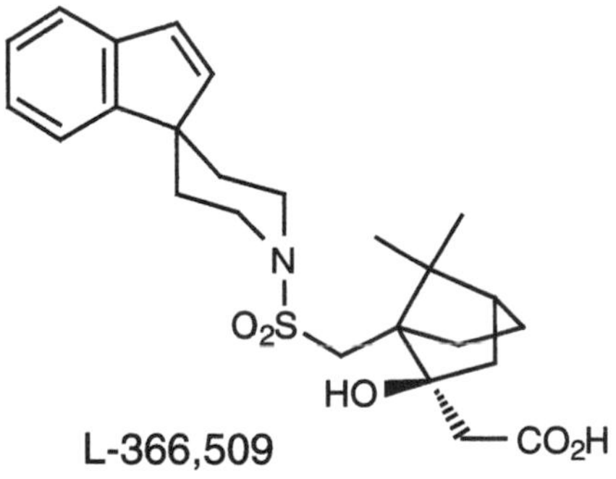

L-366,509

The fermentation screening approach has also borne fruit with a nonpeptide antagonist, the penicillide 28, with an IC_{50} value of 8.4 μM at the OT receptor and at least 10-fold selectivity over the V1 receptor.[116]

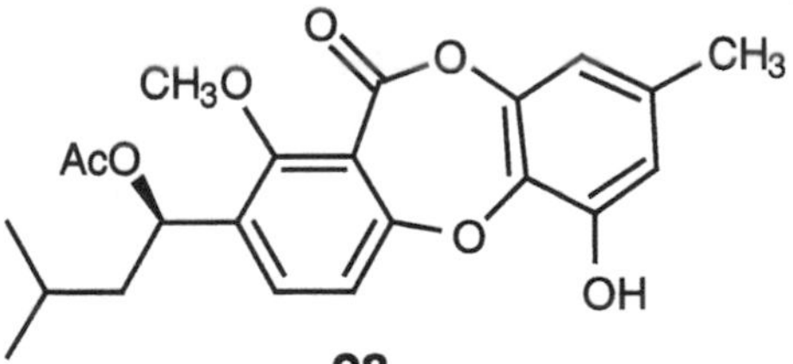

28

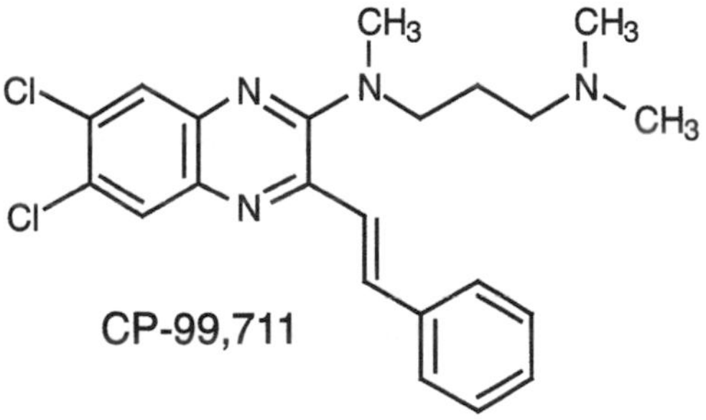

CP-99,711

Other recently announced compounds which promise to open new areas include the glucagon antagonist CP-99,711, ($IC_{50} = 4.1$ μM at the rat liver glucagon receptor),[117] two antagonists of the gastrin-releasing peptide receptor, CP-70,030, and CP-75,998, (IC_{50} values of 1.5–3.0 μM in rat brain, although they show no affinity for the corresponding receptor in human N592 carcinoma cells),[118] and the first reported nonpeptide antagonists of the bradykinin B_2 receptor, compounds 29 and 30 (IC_{50} values of 60 nM and 210 nM at the human IMR 90 fibroblast B_2 receptor).[119]

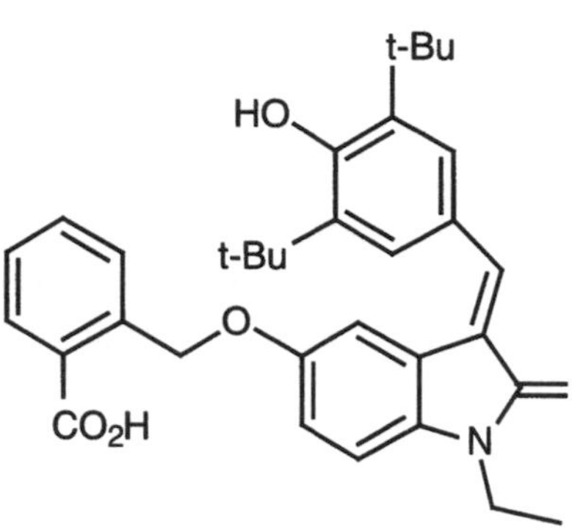

CP-70,030 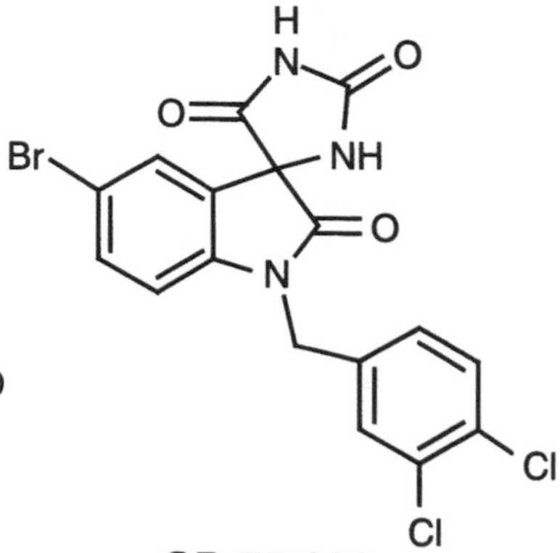   CP-75,998

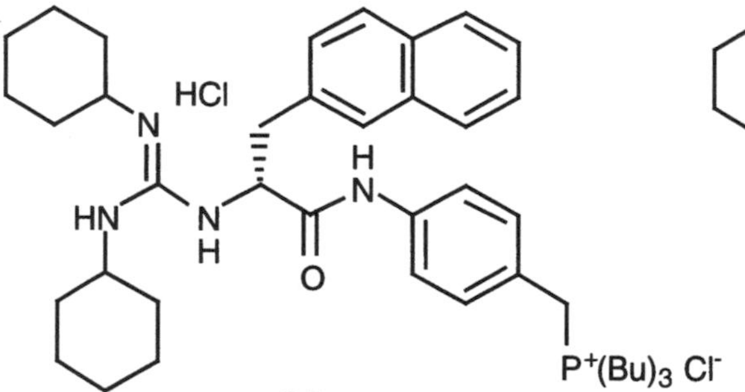

29 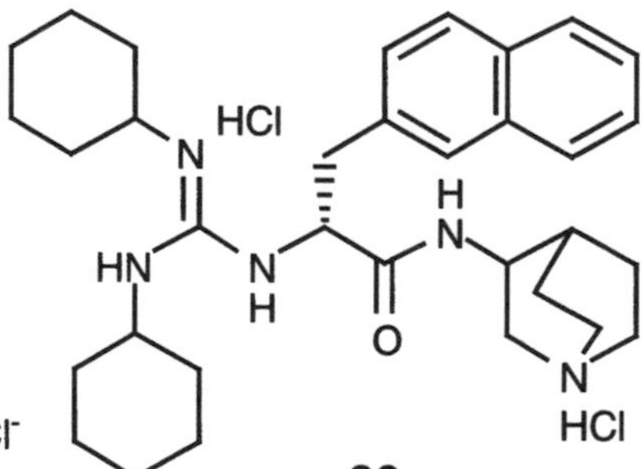   30

·The discovery of L-692,429, a novel growth hormone (GH) secretagogue was recently reported.[120] Selection of structures for screening was based on key SAR aspects of the hexapeptide growth hormone releasing peptide (GHRP-6, His-D-Trp-Ala-Trp-D-Phe-Lys-NH_2), especially the aromatic residues and lysine amine. Lead modification included incorporation of the tetrazole and resolution, affording an EC_{50} value of 60 nM for release of GH. The structurally related L-692,400 shows antagonist activity and thus is a useful reagent for further characterization of this system.

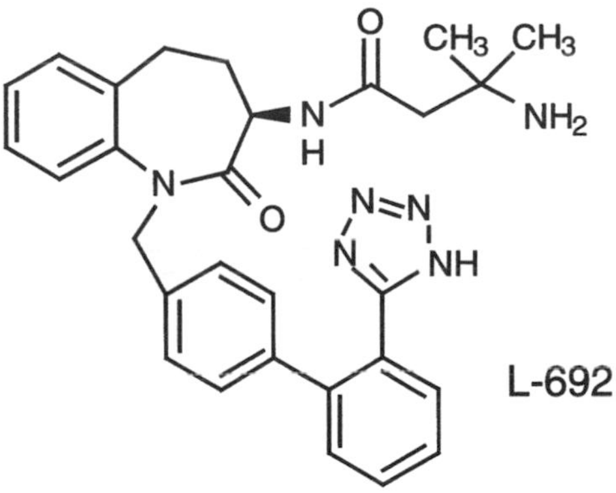

L-692,429 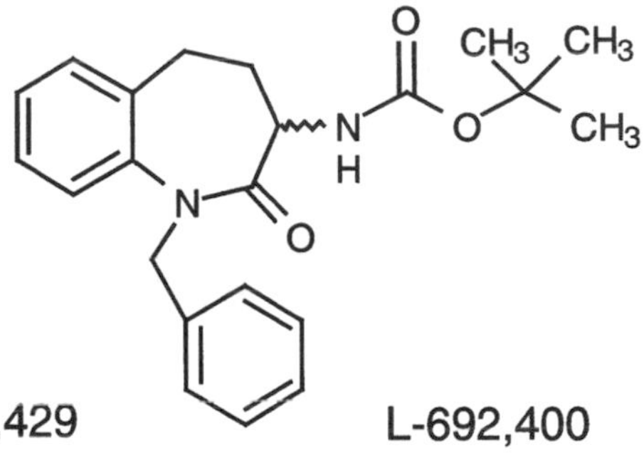   L-692,400

6. CONCLUSION: APPLICATIONS OF MOLECULAR GENETICS TO DESIGN OF RECEPTOR ANTAGONISTS

The nonpeptide antagonists for the cholecystokinin, angiotensin II, and tachykinin receptors reviewed above have enabled delineation of receptor subtypes and evaluation of the relevant physiological roles of these peptides. The potential for similar advances in new areas now exists with recently discovered antagonists at the neurotensin, vasopressin, oxytocin, and bombesin receptors. One of the recurring themes in these medicinal chemistry discovery efforts is the use of the structure of the peptide agonist as a template for molecular interactions to improve the receptor affinity of a lead antagonist structure, which assumes that both agonists and antagonists bind to a common site. Recent advances in our understanding of the mechanism of receptor antagonism provided by molecular genetics studies challenge this assumption by showing that a peptide agonist and its corresponding nonpeptide antagonist occupy different binding sites in the receptor. A brief review of this work in the molecular genetics area is worthwhile in light of its importance to medicinal chemists for new antagonist design.

Molecular genetics studies of the structure and function of receptors use two types of mutant receptors to reveal elements involved in ligand recognition and signal transduction: chimeric receptors, made by substituting sequences from a cognate receptor into the wildtype receptor, and single point mutant receptors, with amino acid changes engineered by the site-directed mutagenesis technique. Chimeric receptors define regions of agonist or antagonist recognition, allowing single amino acid changes to be made in these regions to define molecular contact points. Both techniques have been extensively applied to G protein-coupled receptors for the aminergic neurotransmitters noradrenaline, dopamine, and serotonin.[121] With rapid advances enabling cloning of G protein-coupled receptors for peptides and the discovery of nonpeptide antagonists for these receptors, the stage is set for a powerful combination of molecular genetics with medicinal chemistry to elucidate the mechanism of receptor antagonist action.

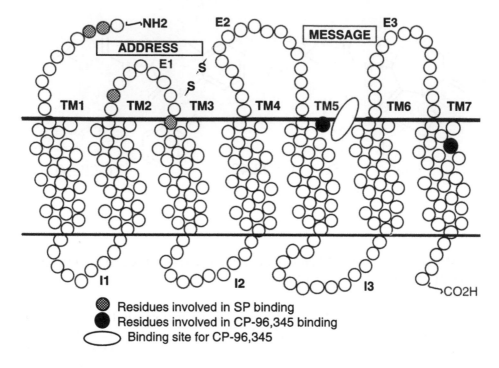

Figure 1 Schematic diagram of the NK$_1$ receptor showing binding sites for SP and CP-96,345.

Cloning, sequencing, and expression of all three neurokinin receptors, NK$_1$, NK$_2$, and NK$_3$ helped initiate application of these new techniques to the tachykinin area.[122] The chimeric receptor technique was first used used to delineate the domains of the NK$_1$ receptor involved in recognizing SP. The chimeric receptors were constructed by substituting sequences from the NK$_3$ receptor, which has poor affinity for SP, into the NK$_1$ receptor and measuring the effect on SP binding. The results were assembled to create a map of the agonist binding site within the NK$_1$ receptor, consisting of two domains: the SP recognition domain delineated by the N-terminal extracellular region and the first two extracellular loops, E1 and E2, and the signal transduction domain, comprised of SP binding in the extracellular epitopes in the C-terminal domain, E2 and E3, located above the the third intracellular loop and C-terminal tail, which interact with the G protein (Figure 1).[123] This model is familiar to medicinal chemists as the "message-address" concept of receptor recognition, with SP using the "address" domain in the N-terminus for recognition, and the "message" domain in the C-terminus for receptor activation.[124,125] Additional studies identified the importance of numerous agonist-binding epitopes clustered around the top of the multi-helix bundle-shaped receptor.[126] Single point mutant receptors then identified residues at positions 23 and 24 in the N-terminal domain and 96 and 108 in the first extracellular loop as important for agonist recognition,[127] while valine-97 is involved in agonist selectivity (see Figure 1).[128]

These chimeric receptors were then used to identify domains involved in recognizing the NK_1 receptor antagonist CP-96,345: two epitopes located at the tops of transmembrane domains V and VI.[129] This antagonist-binding domain was transferred from the NK_1 receptor to the CP-96,345-unresponsive NK_3 receptor, resulting in a chimera with a nearly wildtype NK_1 affinity for CP-96,345. Since changes of receptor sequence in the antagonist binding domain did not effect SP binding, it was concluded that agonist and antagonist binding sites differ. Single point mutant receptors were then used to identify an important interaction between CP-96,345 and histidine-197 in the NK_1 receptor sequence (located at the top of transmembrane domain V), and to provide evidence of an amino-aromatic interaction between the histidine-197 NH and the benzhydryl group of CP-96,345.[130] Isoleucine-290 at the top of transmembrane helix VII was identified using a similar approach to be responsible for a 20-fold portion of the 60-fold loss of NK_1 receptor affinity of CP-96,345 at the rat receptor compared to the human receptor (see Figure 1).[131] In summary, a model is emerging in which antagonists bind to the "message" portion of the receptor, located in the C-terminal domain, to prevent signal transduction.

Extension of this model of antagonist binding to the NK_2 receptor was recently accomplished with chimeric NK_1/NK_2 receptors using both CP-96,345 and the NK_2-selective antagonist SR 48,968.[132] Thus the binding of SR 48,968 to the NK_2 receptor was found to depend on epitopes around the tops of transmembrane domains VI and VII. Studies of the NK_3 receptor await a selective, nonpeptide antagonist. The plethora of structural types of NK_1 antagonists available, however, will no doubt help to provide a more detailed picture of the antagonist binding site in the NK_1 receptor. By encouraging chemists to model antagonist binding independently of overlap with the agonist, these studies are already changing design strategies for new compounds.

Finally, single point mutant receptors were used recently to identify a residue in the C-terminal domain of the CCK receptor, valine 319 near the top of transmembrane helix VI, involved in recognizing the antagonists L-365,260 and L-364,718.[133] If further work on the CCK and other receptors continues to confirm the model of agonist and antagonist binding elucidated for the NK_1 receptor, medicinal chemists will have an important clue for pursuing antagonists at new receptors. Although it is premature to suggest the possibility of *de novo* design of nonpeptide antagonists at peptide receptors, the insights already gained should help in accelerating the rate of progress toward new therapeutic agents which modulate the activity of peptide hormones.

7. REFERENCES

1. Burbach, J.P.H. and Meijer, O.C. (1992) The structure of neuropeptide receptors. *Eur. J. Pharmacol.,* **227**, 1–18.
2. Rees, D.C. (1993) Non-peptide ligands for neuropeptide receptors. *Ann. Rep. Med. Chem.,* **28**, 59–68.
3. Chang, R.S.L., Lotti, V.J., Monaghan, R.L., Birnbaum, J., Stapley, E.O., Goetz, M.A., Albers-Schoenberg, G., Patchett, A.A., Liesch, J.M., Hensens, O.D. and Springer, J.P. (1985) A Potent Nonpeptide Cholecystokinin Antagonist Selective for Peripheral Tissues Isolate from *Aspergillus alliaceus. Science,* **230**, 177–179.

4. Mutt, V. (1980) Cholecystokinin: Isolation, Structure, and Functions. In *Gastrointestinal Hormones*, pp. 169–221. ed. Glass, G.B.J. New York: Raven Press.
5. Evans, B.E., Rittle, K.E., Bock, M.G., DiPardo, R.M., Freidinger, R.M., Whitter, W.L., Gould, N.P., Lundell, G.F., Homnick, C.F., Veber, D.F., Anderson, P.S., Chang, R.S.L., Lotti, V.J., Cerino, D.J., Chen, T.B., King, P.J., Kunkel, K.A., Springer, J.P. and Hirshfeld, J. (1987) Design of Nonpeptidal Ligands for a Peptide Receptor: Cholecystokinin Antagonists. *J. Med. Chem.*, **30**, 1229–1239.
6. Evans, B.E., Bock, M.G., Rittle, K.E., DiPardo, R.M., Whitter, W.L., Homnick, C.F., Veber, D.F., Anderson, P.S. and Freidinger, R.M. (1986) Design of potent, orally effective, nonpeptidal antagonists of the peptide hormone cholecystokinin. *Proc. Natl. Acad. Sci. USA*, **83**, 4918–4922.
7. Wank, S.A., Pisegna, J.R. and de Weerth, A. (1992) Brain and gastrointestinal cholecystokinin receptor family: structure and functional expression. *Proc. Natl. Acad. Sci. USA*, **89**, 8691–8695.
8. Wank, S.A., Harkins, R., Jensen, R.T., Shapira, H., de Weerth, A. and Slattery, T. (1992) Purification, molecular cloning, and functional expression of the cholecystokinin receptor from rat pancreas. *Proc. Natl. Acad. Sci. USA*, **89**, 3125–3129.
9. de Weerth, A., Pisegna, J.R., Huppi, K. and Wank, S.A. (1993) Molecular Cloning, Functional Expression and Chromosomal Localization of the Human Cholecystokinin Type A Receptor. *Biochem. Biophys. Res. Commun.*, **194**, 811–818.
10. Lee, Y.-M., Beinborn, M., McBride, E.W., Lu, M., Kolakowski, L.F. and Kopin, A.S. (1992) The human brain cholecystokinin-B/gastrin receptor. *J. Biol. Chem.*, **268**, 8164–8169.
11. Singh, L., Lewis, A.S., Field, M.J., Hughes, J. and Woodruff, G.N. (1991) Evidence for an involvement of the brain cholecystokinin B receptor in anxiety. *Proc. Natl. Acad. Sci. USA*, **88**, 1130–1133.
12. Harro, J., Vasar, E. and Bradwejn, J. (1993) CCK in animal and human research on anxiety. *Tr. Pharmacol. Sci.*, **14**, 244–249.
13. O'Neill, M.F., Dourish, C.T., Tye, S.J. and Iversen, S.D. (1990) Blockade of CCK-B receptors by L-365,260 induces analgesia in the squirrel monkey. *Brain Res.*, **534**, 287–290.
14. Bock M.G., DiPardo, R.M., Evans, B.E., Rittle, K.E., Veber, D.F., Freidinger, R.M., Chang, R.S.L. and Lotti, V.J. (1988) Cholecystokinin Antagonists. Synthesis and Biological Evaluation of 4-Substituted 4H-[1,2,4]Triazolo[4,3a][1,4]benzodiazepines. *J. Med. Chem.*, **31**, 176–181.
15. Parsons, W.H., Patchett, A.A., Holloway, M.K., Smith, G.M., Davidson, J.L., Lotti, V.J. and Chang, R.S.L. (1989) Cholecystokinin Antagonists. Synthesis and Biological Evaluation of 3-Substituted Benzolactams. *J. Med. Chem.*, **32**, 1681–1685.
16. Bock, M.G., DiPardo, R.M., Veber, D.F., Chang, R.S.L., Lotti, V.J., Freedman S.B. and Freidinger, R.M. (1993) Benzolactams as Nonpeptide Cholecystokinin Receptor Ligands. *Bioorg. Med. Chem. Lett.*, **3**, 871–874.
17. Evans, B.E., Rittle, K.E., Chang, R.S.L., Lotti, V.J., Freedman, S.B. and Freidinger, R.M. (1993) Multipurpose Receptor Ligands: β-Carboline Cholecystokinin Antagonists. *Bioorg. Med. Chem. Lett.*, **3**, 867–870.
18. Yu, M.J., Thrasher, J., McCowan, J.R., Mason, N.R. and Mendelsohn, L.G. (1991) Quinazolinone Cholecystokinin-B Receptor Ligands. *J. Med. Chem.*, **34**, 1505–1508.
19. Satoh, M., Kondoh, Y., Okamoto, Y., Nishida, A., Miyata, K., Ohta, M., Honda, K., Fujikura, T. and Murase, K. (1993) New 1,4-Benzodiazepin-2-one Derivatives as CCK-B/Gastrin Antagonists. 205th American Chemical Society National Meeting, Denver, CO Abstract #158.
20. Howbert, J.J., Lobb, K.L., Britton, T.C., Mason, N.R. and Bruns, R.F. (1993) Diphenylpyrazolidinone and Benzodiazepine Cholecystokinin Antagonists: A Case of Convergent Evolution in Medicinal Chemistry. *Bioorg. Med. Chem. Lett.*, **3**, 875–880.
21. Guyon, C., Dubroeucq, M.C., Barreau, M., Bertrand, P. and Boehme, G.A. (1992) Ureido-acetamides (III): synthesis of optically active derivatives. XXIth International Symposium on Medicinal Chemistry, Basel Switzerland, Poster #P-186.C.
22. Gully, D., Frehel, D., Marcy, C., Spinazze, A., Lespy, L., Neliat, G., Maffrand, J.-P. and LeFur, G. (1993) Peripheral Biological Activity of SR 27897: a new potent non-peptide antagonist of CCK-A receptors. *Eur. J. Pharmacol.*, **232**, 13–19.
23. Makovec, F., Bani, M., Cereda, R., Chiste, R., Revel, L., Rovati, L.C., Setnikar, I. and Rovati, L.A. (1986) Protective Effect of CR 1409 (Cholecystokinin antagonist) on experimental pancreatitis in rats and mice. *Peptides*, **7**, 1159–1164.

24. Kerwin, J.F., Nazdan, A.M., Kopecka, H., Lin, C.W., Miller, T., Witte, D. and Burt, S. (1989) Hybrid cholecystokinin (CCK) antagonists: new implications in the design and modification of CCK antagonists. *J. Med. Chem.*, **32**, 739–742.

25. Kerwin, J.F., Wagenaar, F., Kopecka, H., Lin, C.W., Miller, T., Witte, D., Stashko, M. and Nazdan, A.M. (1991) Cholecystokinin antagonists: (R)-Tryptophan-based hybrid antagonists of high affinity and selectivity for CCK-A receptors. *J. Med. Chem.*, **34**, 3350–3359.

26. van der Bent, A., Blommaert, A.G.S., Melman, C.T.M., IJerman, A.P., van Wijngaarden, I. and Soudijn, W. (1992) Hybrid Cholecystokinin-A antagonist based on molecular modeling of lorglumide and L-364,718. *J. Med. Chem.*, **35**, 1042–1049.

27. Horwell, D.C., Hughes, J., Hunter, J.C., Pritchard, M.C., Richardson, R.S., Roberts, E. and Woodruff, G.N. (1991) Rationally designed "dipeptoid" analogues of CCK. *J. Med. Chem.*, **34**, 404–414.

28. Horwell, D.C., Birchmore, B., Boden, P.R., Higginbottom, M., Yee, P.H., Hughes, J., Hunter, J.C. and Richardson, R.S. (1990) α-Methyl tryptophanylphenylalanines and their arylethylamine "dipeptoid" analogues of the tetrapeptide cholecystokinin (30–33). *Eur. J. Med. Chem.*, **25**, 53–60.

29. Horwell, D.C., Beeby, A., Clark, C.R. and Hughes, J. (1987) Synthesis and binding affinities of analogues of cholecystokinin-(30–33) as probes for central nervous system cholecystokinin receptors. *J. Med. Chem.*, **30**, 729–732.

30. Drysdale, M.J., Pritchard, M.C. and Horwell, D.C. (1992) The synthesis and CCK receptor affinities of selected carboxylic acid mimics of CI-988 — a potent and selective CCK-B antagonist. *Bioorg. Med. Chem. Lett.*, **2**, 45–48.

31. Higginbottom, M., Horwell, D.C. and Roberts, E. (1993) Selective ligands for cholecystokinin receptor subtypes CCK-A and CCK-B within a single structural class. *Bioorg. Med. Chem. Lett.*, **3**, 881–884.

32. Hull, R.A.D., Shankley, N.P., Harper, E.A., Gershkowitch, V.P. and Black, J.W. (1993) 2-Naphthalenesulfonyl L-aspartyl-(2-phenethyl)amide (2-NAP) — a selective cholecystokinin CCK-A-receptor antagonist. *Br. J. Pharmacol.*, **108**, 734–740.

33. Flynn, D.L., Villamil, C.I., Becker, D.P., Gullikson, G.W., Moummi, C. and Yang, D.-C. (1992) 1,3,4-Trisubstituted pyrrolidinones as scaffolds for construction of peptidomimetic cholecystokinin antagonists. *Bioorg. Med. Chem. Lett.*, **2**, 1251–1256.

34. Ohtsuka, T., Kudoh, T., Shimma, N., Kotaki, H., Nakayama, N., Itezono, Y., Fujisaki, N., Watanabe, J., Yokose, K. and Seto, H. (1992) Tetronothiodin, a novel cholecystokinin type-B receptor antagonist produced by Streptomyces sp. *J. Antibiot.*, **45**, 140–143.

35. Lam, Y.K.T., Bogen, D., Chang, R.S., Faust, K.A., Hensens, O.D., Zink, D.L., Schwartz, C.D., Zitano, L., Garrity, G.M., Gagliardi, M.M., Currie, S.A. and Woodruff, H.B. (1991) Novel and potent gastrin and cholecystokinin antagonist from Streptomyces olivaceus. *J. Antibiot.*, **44**, 613–625.

36. Minabe, Y., Ashby, C.R. and Wang, R.Y. (1991) The CCK-A receptor antagonist devazepide but not the CCK-B receptor antagonist L-365,260 reverses the effects of chronic clozapine and haloperidol on midbrain dopamine neurons. *Brain Res.*, **549**, 151–154.

37. O'Neill, M.F., Dourish, C.T. and Iversen, S.D. (1991) Hypolocomotion induced by peripheral or central injection of CCK in the mouse is blocked by the CCK-A receptor antagonist devazepide but not by the CCK-B receptor antagonist L-365,260. *Eur. J. Pharmacol.*, **193**, 203–208.

38. Rasmussen, K., Czachura, J.F., Stockton, M.E. and Howbert, J.J. (1993) Electrophysiological effects of diphenylpyrazolidinone cholecystokinin-B and cholecystokinin-A antagonists on midbrain dopamine neurons. *J. Pharmacol. Exp. Ther.*, **264**, 480–488.

39. (a) Valotton, M.B. (1987) The Renin-Angiotensin System. *Trends Pharmacol. Sci.*, **8**, 69–74. (b) Peach, M.J. (1977) Renin-Angiotensin System: Biochemistry and Mechanisms of Action. *J. Physiol. Rev.*, **57**, 313–370.

40. Ferrario, C.M. (1990) The Renin-Angiotensin System: Importance in Physiology and Pathology. *J. Cardiovasc. Pharmacol.*, **15** (Suppl. 3), S1–S5.

41. (a) Furukawa, Y. and Kishimoto, K. (1982) Hypotensive Imidazole-5-acetic Acid Derivatives. U.S. Patent 4,340,598. (b) Furukawa, Y., Kishimoto, K. and Nishikawa, K. (1982) Hypotensive Imidazole-5-acetic Acid Derivatives. U.S. Patent 4,355,040.

42. (a) Wong, P.C., Chiu, A.T., Price, W.A., Thoolen, M.J., Carini, D.J., Johnson, A.L., Taber, R.I. and Timmermans, P.B.M.W.M. (1988) Nonpeptide Angiotensin II Receptor Antagonists I. Pharma-

cological Characterization of 2-n-Butyl-4-chloro-1-(2-chlorobenzyl)imidazole-5-acetic acid, sodium salt (S-8307). *J. Pharmacol. Exp. Ther.*, **247**, 1–7. (b) Chiu, A.T., Carini, D.J., Johnson, A.L., McCall, D.E., Price, W.A., Thoolen, M.J., Wong, P.C., Taber, R.I. and Timmermans, P.B.M.W.M. (1988) Nonpeptide Angiotensin II Receptor Antagonists II. Pharmacology of S-8308. *Eur. J. Pharmacol.*, **157**, 13–21.

43.   Duncia, J.V., Chiu, A.T., Carini, D.J., Gregory, G.B., Johnson, A.L., Price, W.A., Wells, G.J., Wong, P.C., Calebrese, J.C. and Timmermans, P.B.M.W.M. (1990) The Discovery of Potent Nonpeptide Angiotensin II Receptor Antagonists: A New Class of Potent Antihypertensives. *J. Med. Chem.*, **33**, 1312–1329.

44.   Carini, D.J., Duncia, J.V., Johnson, A.L., Chiu, A.T., Price, W.A., Wong, P.C. and Timmermans, P.B.M.W. (1990) Nonpeptide Angiotensin II Receptor Antagonists: N-[(Benzyloxy)benzyl]imidazoles and Related Compounds as Potent Antihypertensives. *J. Med. Chem.*, **33**, 1330–1336.

45.   Carini, D.J., Duncia, J.V., Aldrich, P.E., Chiu, A.T., Johnson, A.L., Pierce, M.E., Price, W.A., Santella III, J.B., Wells, G.E., Wexler, R.R., Wong, P.C., Yoo, S.-E. and Timmermans, P.B.M.W.M. (1991) Nonpeptide Angiotensin II Receptor Antagonists: The Discovery of a Series of N-(Biphenylylmethyl) imidazoles as Potent, Orally Active Antihypertensives. *J. Med. Chem.*, **34**, 2525–2547.

46.   Wong, P.C., Price, W.A., Chiu, A.T., Duncia, J.V., Carini, D.J., Wexler, R.R., Johnson, A.L. and Timmermans, P.B.M.W.M. (1990) Nonpeptide Angiotensin II Receptor Antagonists. XI. Pharmacology of EXP3174, an Active Metabolite of DUP753 — an Orally Active Antihypertensive Agent. *J. Pharmacol. Exp. Ther.*, **255**, 211–217.

47.   Blankley, C.J., Hodges, J.C., Klutchko, S.R., Himmelsbach, R.J., Chucholowski, A., Connolly, C.J., Neergaard, S.J., van Nieuwenhze, M.S., Sebastian, A., Quin III, J., Essenburg, A.D. and Cohen, D.M. (1991) Synthesis and Structure-Activity Relationships of a Novel Series of Non-peptide Angiotensin II Receptor Binding Inhibitors Specific for the AT$_2$ Subtype. *J. Med. Chem.*, **34**, 3248–3260.

48.   Bumpus, F.M., Catt, K.J., Chiu, A.T., DeGrasparo, M., Goodfriend, T., Husain, A., Peach, M.J., Taylor Jr, D.G. and Timmermans, P.B.M.W.M. (1991) Nomenclature for Angiotensin Receptors. *Hypertension*, **17**, 720–723.

49.   Timmermans, P.B.M.W.M., Benfield, P., Chiu, A.T., Herblin, W.F., Wong, P.C. and Smith, R.D. (1992) Angiotensin II Receptors and Functional Correlates. *Am. J. Hyper.*, **5**, 221S–235S.

50.   For the AT$_1$ Receptor Subtype, see: (a) Murphy, T.J., Alexander, R.W., Griendling, K.K., Runge, M.S. and Bernstein, K.E. (1991) Isolation of a cDNA Encoding the Vascular Type-1 Angiotensin II Receptor. *Nature*, **351**, 233–236. (b) Sasaki, K., Yamano, Y., Bardhan, S., Iwai, N., Murray, J.J., Hasegawa, M., Matsuda, Y. and Inagami, T. (1991) Cloning and Expression of a Complementary DNA Encoding a Bovine Adrenal Angiotensin II Type-1 Receptor. *Nature*, **351**, 230–232. For the AT$_2$ Receptor Subtype, see: Kambayashi, Y., Bardhan, S., Takahashi, K., Tsuzuki, S., Inui, H., Hamakubo, T. and Inagami, T. (1993) Molecular Cloning of a Novel Angiotensin II Receptor Isoform Involved in Phosphotyrosine Phosphatase Inhibition. *J. Biol. Chem.*, **33**, 24543–24546.

51.   (a) Weinstock, J., Keenan, R.M., Samanen. J., Hempel, J., Finkelstein, J.A., Franz, R.G., Gaitanopoulos, D.E., Girard, G.R., Leason, J.G., Hill, D.T., Morgan, T.M., Peishoff, C.E., Aiyar, N., Brooks, D.P., Fredickson, T.A., Ohlstein, E.H., Ruffolo, Jr., R.R., Stack, E.J., Sulpizio, A.C., Weidley, E.F. and Edwards, R.M. (1991) (1-Carboxybenzyl)imidazole-5-acrylic Acids Potent and Selective Angiotensin II Receptor Antagonists. *J. Med. Chem.*, **34**, 1514–1517. (b) Keenan, R.M., Weinstock, J., Finkelstein, J.A., Franz, R.G., Gaitanopoulos, D.E., Girard, G.R., Hill, D.T., Morgan, T.M., Samanen, J.M., Hempel, J., Eggleston, D.S., Aiyar, N., Griffin, E., Ohlstein, E.H., Stack, E.J., Weidley, E.F. and Edwards, R. (1992) Imidazole-5-acrylic Acids: Potent Nonpeptide Angiotensin II Receptor Antagonists Designed Using a Novel Peptide Pharmacophore Model. *J. Med. Chem.*, **35**, 3858–3872. (c) Keenan, R.M., Weinstock, J., Finkelstein, J.A., Franz, R.G., Gaitanopoulos, D.E., Girard, G.R., Hill, D.T., Morgan, T.M., Samanen, J.M., Peishoff, C.E., Tucker, L.M., Aiyar, N., Griffin, E., Ohlstein, E.H., Stack, E.J., Weidley, E.F. and Edwards, R.M. (1993) Potent Nonpeptide Angiotensin II Receptor Antagonists. 2. 1-(Carboxybenzyl)imidazole-5-acrylic Acids. *J. Med. Chem.*, **36**, 1880–1892.

52.   Berhart, C.A., Perreaut, P.M., Ferrari, B.P., Muneaux, Y.A., Assens, J.-L., Clement, J., Haudricourt, F., Muneaux, C.F., Taillades, J.E., Vignal, M.-A., Gougat, J., Guiraudou, P.R., Lacour, C.A., Roccon, A., Cazaubon, C.F., Breliere, J.-C., Fur, G.L. and Nisato, D. (1993) A New Series of Imidazolones: Highly Specific and Potent Nonpeptide AT$_1$ Angiotensin II Receptor Antagonists. *J. Med. Chem.*, **36**, 3371–3380.

53. Colins, G.M., Corpus, V.M., McMahon, E.G., Palomo, M.A., Schuh, J.R., Blehm, D.J., Huang, H.-C., Reitz, D.B., Manning, R.E. and Blaine, E.H. (1992) *In Vitro* Pharmacology of a Nonpeptide Angiotensin II Receptor Antagonists, SC-51316. *J. Pharm. Exp. Ther.*, **261**, 1037–1043.

54. Chang, L.L., Ashton, W.T., Flanagan, K.L., Strelitz, R.A., MacCoss, M., Greenlee, W.J., Chang, R.S., Lotti, V.J., Faust, K.A., Chen, T.-B., Bunting, P., Zingaro, G.J., Kivlighn, S.D. and Siegl, P.K.S. (1993) Triazolinones as Nonpeptide Angiotensin II Antagonists. 1. Synthesis and Evaluation of Potent 2,4,5-Trisubstituted Triazolinones. *J. Med. Chem.*, **36**, 2558–2568.

55. (a) Reitz, D.B. (1992) 1*H*-Substituted-1,2,4-Triazole Compounds for Treatment of Cardiovascular Disorders. U.S. Patent 5,098,920 (b) Reitz, D.B., Penick, M.A., Brown, M.S., Reinhard, E.J., Olins, G.M., Corpus, V.M., MeMahon, E.G., Palomo, M.A., Koepke, J.P., Moore, G.K., Smits, G.J., McGraw, D.E. and Blaine, E.H. *Abs. Pap. American Chemical Society (ACS) National Meeting*, **203**, 189-MEDI.

56. Middlemiss, D., Ross, B.C., Eldred, C., Montana, J.G., Shah, P., Hirst, G.C., Watson, T.A., Panchal, T.A, Paton, J.M.S., Hubbard, T., Drew, G.M., Robertson, M.J., Hilditch, A. and Clark, K.L. (1992) C-linked Pyrazole Biaryl Tetrazoles as Antagonists of Angiotensin II. *Bioorg. Med. Chem. Lett.*, **2**, 1243–1246.

57. Kubo, K., Inada, Y., Kohara, Y., Sugiura, Y., Ojima, M., Itoh, K., Furukawa, Y., Nishikawa, K. and Naka, T. (1993) Nonpeptide Angiotensin II Receptor Antagonists. Synthesis and Biological Activity of Benzimidazoles. *J. Med. Chem.*, **36**, 1772–1784.

58. Kubo, K., Kohara, Y., Imamiya, E., Sugiura, Y., Inada, Y., Furukawa, Y., Nishikawa, K. and Naka, T. (1993) Nonpeptide Angiotensin II Receptor Antagonists. Synthesis and Biological Activity of Benzimidazolecarboxylic Acids. *J. Med. Chem.*, **36**, 2182–2195.

59. Kubo, K., Kohara, Y., Yoshimura, Y., Inada, Y., Shibouta, Y., Furukawa, Y., Kato, T., Nishikawa, K. and Naka, T. (1993) Nonpeptide Angiotensin II Receptor Antagonists. Synthesis and Biological Activity of Potential Prodrugs of Benzimidazolecarboxylic Acids. *J. Med. Chem.*, **36**, 2343–2349.

60. Mantlo, N.B., Chakravarty, P.K., Ondeyka, D.L., Siegl, P.K.S., Chang, R.S., Lotti,, V.J., Faust, K.A., Chen, T.-B., Schorn, T.W., Sweet, C.S., Emmert, S.E., Patchett, A.A. and Greenlee, W.J. (1991) Potent Orally Active Imidazo[4,5-*b*]pyridine-Based Angiotensin II Receptor Antagonists. *J. Med. Chem.*, **34**, 2919–2922.

61. (a) Allen, E.R., de Laszlo, S.E., Huang, S.X., Quagliato, C.S., Greenlee, W.J., Chang, R.S.L., Chen, T.-B., Faust, K.A. and Lotti, V.J. (1993) Quinazolinones 1: Design and Synthesis of Potent Quinazolinone-containing AT$_1$-selective Angiotensin-II Receptor Antagonists. *Bioorg. Med. Chem. Lett.*, **3**, 1293–1298. (b) de Laszlo, S.E., Allen, E.R., Quagliato, C.S., Greenlee, W.J., Patchett, A.A., Nachbar, R.B., Siegl, P.K.S., Chang, R.S., Kivlighn, S.D., Schorn, T.S., Faust, K.A., Chen, T.-B., Zingaro, G.J. and Lotti, V.J. (1993) Quinazolinones as Angiotensin II Antagonists: Part 2. QSAR and *In Vivo* Characterization of AT$_1$-selective AII Antagonists. *Bioorg. Med. Chem. Lett.*, **3**, 1299–1304.

62. Bradbury, R.H., Allott, C.P., Dennis, M., Fisher, E., Major, J.S., Masek, B.B., Oldham, A.A., Pearce, R.J., Rankine, N., Revill, J.M., Roberts, D.A. and Russell, S.T. (1992) New Nonpeptide Angiotensin II Receptor Antagonists. 2. Synthesis, Biological Properties and Structure-Activity Relationships of 2-Alkyl-4-(biphenylyl)methoxyquinoline Derivatives. *J. Med. Chem.*, **35**, 4027–4038.

63. Allot, C.P., Bradbury, R.H., Dennis, M., Fisher, E., Kuke, R.W.A., Major, J.S., Oldham, A.A., Pearce, R.J., Reid, A.C., Roberts, D.A., Rudge, D.A. and Russell, S.T. (1993) Quinoline, 1,5-Naphthyridine and Pyridine Derivatives as Potent, Nonpeptidic Angiotensin II Receptor Antagonists. *Bioorg. Med. Chem. Lett.*, **3**, 899–904.

64. (a) De, B., Winn, M., Zydiwsky, T.M., Kerkman, D.J., DeBernardis, J.F., Lee, J., Buckner, S., Warner, R., Brune, M., Hancock, A., Opgenorth, T. and Marsh, K. (1992) Discovery of a Novel Class of Orally Active Non-peptide Angiotensin II Antagonists. *J. Med. Chem.*, **35**, 3712–3717. (b) Winn, M., De, B., Zydowsky, T., Altenbach, R.J., Basha, F.Z., Boyd, S.A., Brune, M.E., Buckner, S.A., Crowell, D., Drizin, I., Hancock, A.A., Jae, H.-S., Kester, J.A., Lee, J.Y., Mantei, R.A., Marsh, K.G., Novosad, E.I., Oheim, K.W., Rosenberg, S.H.. Shiosaku, K., Sorensen, B.K., Spina, K., Sullivan, G.M., Tasker, A.S., von Geldern, T., Warner, R.B., Opgenorth, T.J., Kerkman, D.J. and DeBernardis, J.F. (1993) 2-(Alkylamino)nicotinic Acid and Analogs. Potent Angiotensin II Antagonists. *J. Med. Chem.*, **36**, 2676–2688.

65. Criscione, L., de Gasparo, M., Buehlmayer, P., Whitebread, S., Ramjoue, H.P. and Wood, J. (1992) Pharmacological Profile of CGP48933, a Novel, Nonpeptide Antagonist of AT$_1$ Angiotensin II Receptor. 14th Scientific Meeting of the International Society of Hypertensin, Madrid/Spain. *J. Hypertension*, **10** (Suppl. 4), 196.

66. Wood, J., Schnell, C., Forgiarini, P., Baum, H.P., Cumin, F. and Buehlmayer, P. (1992) Measurement of Blood Pressure by Telemetry in Conscious, Sodium-depleted Marmosets: Effects of Selective AT$_1$ Subtype Angiotensin II Receptor Antagonism. 14th Scientific Meeting of the International Society of Hypertensin, Madrid/Spain. *J. Hypertension*, **10** (Suppl. 4).

67. (a) Reitz, D.B. (1992) 1-Arylheteroarylalkyl Substituted-1*H*-1,2,4-triazole Compounds for the Treatment of Circulatory Disorders. U.S. Patent 5,155,117. (b) Reitz, D.B., Garland, D.J., Norton, M.B., Collins, J.T., Reinhard, E.J., Manning, R.E., Olins, G.M., Chen, S.T., Palomo, M.A. and McMahon, E.G. (1993) N$_1$-Sterically Hindered 2H-imidazol-2-one Angiotensin II Receptor Antagonists: The Conversion of Surmountable Antagonists to Insurmountable Antagonists. *Bioorg. Med. Chem. Lett.*, **3**, 1055–1060.

68. Middlemiss, D., Drew, G.M., Ross, B.C., Robertson, M.J., Scopes, D.I.C., Dowle, M.D., Akers, J., Cardwell, K., Clark, K.L., Coote, S., Eldred, C.D., Hamblett, J., Hilditch, A., Hirst, G.C., Jack. T., Montana, J., Panchal, T.A., Paton, J.M.S., Shah, P., Stuart, G. and Travers, A. (1991) Bromo-benzofurans: A New Class of Potent, Nonpeptide Antagonists of Angiotensin II. *Bioorg. Med. Chem. Lett.*, **1**, 711–716.

69. Middlemiss, D., Watson, S.P., Ross, B.C., Dowle, M.D., Scopes. D.I.C., Montana, J.G., Shah, P., Hirst, G.C., Panchal, T.A., Paton, J.M.S., Pass, M., Hubbard, T., Hamblett, J., Cardwell, K.S., Jack, T.I., Stuart, G., Coote, S., Bradshaw, J., Drew, G.M., Hilditch, A., Clark, K.L., Robertson, M.J., Bayliss, M.K., Donnelly, M., Palmer, E. and Manchee, G.R.M. (1993) Benzofuran Based Angiotensin II Antagonists Related to GR117289: Enhancement of Potency *In Vitro* and Oral Activity. *Bioorg. Med. Chem. Lett.*, **3**, 589–594.

70. Hilditch, A., Akers, J.S., Travers, A., Hunt, A.A.E., Robertson, M.J., Drew, G.M., Middlemiss, D. and Ross, B.C. (1991) Cardiovascular Effects of the Angiotensin II Receptor Antagonist, GR117289, in Conscious Renal Hypertensive and Normotensive Rats. *Brit. J. Pharmacol.*, **104** (proc. Suppl), 423P.

71. Bovy, P.R., Reitz, D.B., Collins, J.T., Chamberlain, T.S., Olins, G.M., Corpus, V.M., McMahon, E.G., Palomo, M.A., Koepke, J.P., Smits, G.J., McGraw, D.E. and Gaw, J.F. (1993) Nonpeptide Angiotensin II Antagonists: N-Phenyl-1H-pyrrole Derivatives are Angiotensin II Receptor Antagonists. *J. Med. Chem.*, **36**, 101–110.

72. Rivero, R.A., Kevin, N.J. and Allen, E.E. (1993) New Potent Angiotensin II Receptor Antagonists Containing Phenylthiophenes and Phenylfurans in Place of the Biphenyl Moiety. *Bioorg. Med. Chem. Lett.*, **3**, 1119–1124.

73. Mantlo, N.B., Chang, R.S.L. and Siegl, P.K.S. (1993) Angiotensin II Receptor Antagonists Containing a Phenylpyridine Element. *Bioorg. Med. Chem. Lett.*, **3**, 1693–1696.

74. (a) Judd, D.B., Cardwell, K.S., Panchal, T.A., Jack, T.I., Pass, M., Hubbard, T., Dean, A.W., Butt, A.U., Hobson, J.E., Heron, N.M., Watson, S.P., Currie, G.S., Middlemiss, D., Aston, N.M., Drew, G.M., Hilditch, A., Gallacher, D., Bayliss, M. and Donnelly, M. (1993) Non-peptide Antagonists of Angiotensin II: Imidazopyridinyl Benzofuranyl Triflamides. *Abs. Pap. American Chemical Society (ACS) National Meeting*, **206**, 83-MEDI. (b) Judd, D.B., Cardwell, K.S., Panchal, T.A., Jack, T.I., Pass, M., Hubbard, T., Dean, A.W., Currie, G.S. and Middlemiss, D. (1993) Non-peptide Antagonists of Angiotensin II: Imidazopyridinylbromobenzofurans. *Abs. Pap. American Chemical Society (ACS) National Meeting*, **206**, 84-MEDI.

75. Naylor, E.M., Chakravarty, P.K., Chen, A., Strelitz, R.A., Chang, R.S., Chen, T.-B., Faust, K.A. and Lotti, V.J. (1993) Potent Imidazopyridine Angiotensin II Antagonists: Acyl Sulfonamides as Tetrazole Replacements. *Abs. Pap. American Chemical Society (ACS) National Meeting*, **206**, 76-MEDI.

76. Guard, S. and Watson, S.P. (1991) Tachykinin Receptor Types: Classification and Membrane Signalling Mechanisms. *Neurochem. Int.*, **18**, 149–165.

77. Folkers, K., Rosell, S., Hakanson, R., Chu, J.-Y., Lu, L.-A., Leander, S., Tang, P.F.L. and Ljungqvist, A. (1985) Chemical Design of Antagonists of Substance P. In *Tachykinin Antagonists* pp. 259–266. Hakanson, R. and Sundler, F. Amsterdam: Elsevier.

78. Snider, R.M., Constantine, J.W., Lowe, III, J.A., Longo, K.P., Lebel, W.S., Woody, H.A., Drozda, S.E., Desai, M.C., Vinick, F.J., Spencer, R.W. and Hess, H.-J. (1991) A Potent Nonpeptide Antagonist of the Substance P (NK₁) Receptor. *Science*, **251**, 435–437.

79. Lowe, J.A., III, Drozda, S.E., Snider, R.M., Longo, K.P., Zorn S.H., Morrone, J., Jackson, E.R., McLean, S., Bryce, D.K., Bordner, J., Nagahisa, A., Kanai, Y., Suga, O. and Tsuchiya, M. (1992) The Discovery of (2S,3S)-cis-2-(Diphenylmethyl)-N-[(2-methoxyphenyl)methyl]-1-azabicyclo[2.2.2]octan-3-amine as a Novel, Nonpeptide Substance P Antagonist. *J. Med. Chem.*, **35**, 2591–2600.

80. Beresford, I.J.M., Birch, P.J., Hagan, R.M. and Ireland, S.J. (1991) Investigation into species variants in tachykinin NK₁ receptors by use of the non-peptide antagonist, CP-96,345. *Br. J. Pharmacol.*, **104**, 292–293.

81. Lembeck, F., Donnerer, Tsuchiya, M. and Nagahisa, A. (1992) The non-peptide tachykinin antagonist, CP-96,345, is a potent inhibitor of neurogenic inflammation. *Br. J. Pharmacol.*, **105**, 527–530.

82. Radhakrishnan, V. and Henry, J.L. (1991) Novel substance P antagonist, CP-96,345, blocks responses of cat spinal dorsal horn neurons to noxious cutaneous stimulation and to substance P. *Neurosci. Lett.*, **132**, 39–43.

83. Fan, T.-P.D., Hu, D.-E., Guard. S., Gresham, A. and Watling, K.J. (1993) Stimulation of angiogenesis by substance P and interleukin-1 in the rat and its inhibition by NK₁ or interleukin-1 receptor antagonists. *Br. J. Pharmacol.*, **110**, 43–49.

84. Bountra, C., Bunce, K., Dale, T., Gardner, C., Jordan, C., Twissell, D. and Ward, P. (1993) Anti-emetic profile of a non-peptide neurokinin NK₁ receptor antagonist, CP-99,994, in ferrets. *Eur. J. Pharmacol.*, **249**, R3–R4.

85. Lowe, J.A., III, Drozda, S.E., Snider, R.M., Longo, K.P. and Rizzi, J.P. (1993) Nuclear variations of quinuclidine substance P antagonists: 2-diphenylmethyl-1-azabicyclo[3.2.2]nonan-3-amines. *Bioorg. Med. Chem. Lett.*, **3**, 921–924.

86. Seward, E.M., Swain, C.J., Merchant, K.J., Owen, S.N., Sabin, V., Cascieri, M.A., Sadowski, S., Strader, C. and Baker, R. (1993) Quinuclidine-based NK-1 antagonists I: 3-benzyloxy-1-azabicyclo[2.2.2] octanes. *Bioorg. Med. Chem. Lett.*, **3**, 1361–1366.

87. Swain, C.J., Seward, E.M., Sabin, V., Owen, S., Baker, R., Cascieri, M.A., Sadowski, S., Strader, C. and Ball, R.G. (1993) Quinuclidine-based NK-1 antagonists 2: determination of the absolute stereo-chemical requirements. *Bioorg. Med. Chem. Lett.*, **3**, 1703–1706.

88. Desai, M.C., Lefkowitz, S.L., Thadeio, P.F., Longo, K.M. and Snider, R.M. (1992) Discovery of a potent substance P antagonist: recognition of the key molecular determinant. *J. Med. Chem.*, **35**, 4911–4913.

89. McLean, S., Ganong, A., Seymour, P.A., Snider, R.M., Desai, M.C., Rosen, T., Bryce, D.K., Longo, K.P., Reynolds, L.S., Robinson, G., Schmidt, A.W., Siok, C. and Heym, J. (1993) Pharmacology of CP-99,994: a nonpeptide antagonist of the tachykinin neurokinin-1 receptor. *J. Pharmacol. Exp. Ther.*, **267**, 472–479.

90. Peyronal, J.-F., Truchon, A., Moutonnier, C. and Garret, C. (1992) Synthesis of RP-67,580, a new potent nonpeptide substance P antagonist. *Bioorg. Med. Chem. Lett.*, **2**, 37–40.

91. Garret, C., Carruette, A., Fardin, V., Moussaoui, S., Peyronel, J.-F., Blanchard, J.-C. and Laduron, P.M. (1991) Pharmacological properties of a potent and selective nonpeptide substance P antagonist. *Proc. Natl. Acad. Sci. USA*, **88**, 10208–10212.

92. Tabart, M., Peyronel, J.F., Truchon, A., Moutonnier, C., Dubroeucq, M.C., Fardin, V., Carruette, A. and Garret, C. (1992) Perhydroisoindolone Derivatives: A New Class of Potent and Selective Nonpeptide NK₁ Antagonists. XIIth International Symposium on Medicinal Chemistry, Basel Switzerland, P-190.A.

93. Shepheard, S., Williamson, D.J., Hill, R.G. and Hargreaves, R.J. (1993) The non-peptide neurokinin₁ receptor antagonist, RP 67580, blocks neurogenic plasma extravasation in the dura mater of rats. *Br. J. Pharmacol.*, **108**, 11–12.

94. Emonds-Alt, X., Doutremepuich, J.D., Jung, M., Proietto, E., Santucci, V., Van Broeck, D., Vilain, P., Soubrie, Ph., LeFur, G. and Breliere, J.C. (1993) SR 140,333, a non-peptide antagonist of substance-P (NK₁) receptor. *Neuropeptides*, **24**, 231, C18.

95. MacLeod, A.M., Merchant, K.J., Cascieri, M.A., Sadowski, S., Ber, E., Swain, C.J. and Baker, R. (1993) N-Acyl-L-tryptophan benzyl esters: potent substance P receptor antagonists. *J. Med. Chem.*, **36**, 2044–2045.

96. Venepalli, B.R., Aimone, L.D., Appell, K.C., Bell, M.R., Dority, J.A., Goswami, R., Hall, P.L., Kumar, V., Lawrence, K.B., Logan, M.E., Scensny, P.M., Seelye, J.A., Tomczuk, B.E. and Yanni, J.M. (1992) Synthesis and substance P receptor binding activity of androstano[3,2b]pyrimido[1,2a]-benzimidazoles. *J. Med. Chem.*, **35**, 374–378.

97. Schilling, W., Bittiger, H., Brugger, F., Criscione, L., Hauser, K., Ofner, S., Olpe, H.R., Vassout, A. and Veenstra, S. (1992) Approaches Towards the Design and Synthesis of Nonpeptidic Substance P Antagonists. XIIth International Symposium on Medicinal Chemistry, Basel Switzerland, ML-11.3.

98. Hashimoto, M., Hayashi, K., Murai, M., Fujii, T., Nishikawa, M., Kitoyoh, S., Okuhara, M., Kohsaka, M. and Imanaka, H. (1992) WS9326A, a novel tachykinin antagonist isolated from Streptomyces violaceusniger no. 9326. *J. Antibiot.*, **45**, 1064–1070.

99. Hirayama, Y., Lei, Y.-H., Barnes, P.J. and Rogers, D.F. (1993) Effects of two novel tachykinin antagonists, FK224 and FK888, on neurogenic airway plasma exudation, bronchoconstriction and systemic hypotension in guinea-pigs *in vivo. Br. J. Pharmacol.*, **108**, 844–851.

100. Ichinose, M., Nakajima, N., Takahashi, T., Yamauchi, H., Inoue, H. and Takishima, T. (1992) Protection against bradykinin-induced bronchoconstriction in asthmatic patients by neurokinin receptor antagonist. *Lancet*, **340**, 1248–1251.

101. Barrow, C.J., Cai, P., Snyder, J.K., Sedlock, D.M., Sun, H.H. and Cooper, R. (1993) WIN 64821, a new competitive antagonist to substance P, isolated from an *Aspergillus* species: structure determination and solution conformation. *J. Org. Chem.*, **58**, 6016–6021.

102. Hagiwara, D., Miyake, H., Morimoto, H., Murai, M., Fujii, T. and Matsuo, M. (1991) The discovery of a tripeptide substance P antagonist and its structure-activity relationships. *J. Pharmacobiodyn.*, **14**, 5104.

103. Hagiwara, D., Miyake, H., Morimoto, H., Murai, M., Fujii, T. and Matsuo, M. (1992) Studies on Neurokinin Antagonists. 1. The Design of Novel Tripeptides Possessing the Glutaminyl-D-tryptophanylphenylalanine Sequence as Substance as Antagonists. *J. Med. Chem.*, **35**, 2015–2025.

104. Hagiwara, D., Miyake, H., Igari, N., Morimoto, H., Murai, M., Fujii, T. and Matsuo, M. (1992) Design of a Novel Dipeptide Substance P Antagonist FK888. *Regul. Peptides*, Supplement 1, S66.

105. Hirschmann, R., Nicolaou, K.C., Pietranco, S., Salvino, J., Leahy, E.M., Sprengler, P.A., Furst, G., Smith, A.B., III, Strader, C.D., Cascieri, M.A., Candelore, M.R., Donaldson, C., Vale, W. and Maechler, L. Nonpeptidal Peptidomimetics with a β-D-glucose Scaffolding. A Partial Somatostatin Agonist Bearing a Close Structural Resemblance to a Potent, Selective Substance P Antagonist. *J. Am Chem. Soc.*, **114**, 9217–9218.

106. Emonds-Alt, X., Proietto, V., Van Broeck, D., Vilain, P., Advenier, C., Neliat, G., LeFur, G. and Breliere, J.C. (1993) Pharmacological profile and chemical synthesis of SR 48968, a non-peptide antagonist of the neurokinin A (NK₂) receptor. *Bioorg. Med. Chem. Lett.*, **3**, 925–930.

107. Emonds-Alt, X., Vilain, P., Goulaouic, P., Proietto, V., Van Broeck, D., Advenier, C., Naline, E., Neliat, G., LeFur, G. and Breliere, J.C. (1992) A potent and selective non-peptide antagonist of the neurokinin A (NK₂) receptor. *Life Sci.*, **50**, PL101–PL106.

108. Lou, Y.-P., Lee, L.-Y., Satoh, H. and Lundberg, J.M. (1993) Postjunctional inhibitory effect of the NK₂ receptor antagonist, SR 48968, on sensory NANC bronchoconstriction in the guinea-pig. *Br. J. Pharmacol.*, **109**, 765–773.

109. Santucci, V., Gueudet, C., Emonds-Alt, X., Breliere, J.C., Soubrie, Ph. and LeFur, G. (1993) The NK₂ receptor antagonist SR 48968 inhibits thalamic responses evoked by thermal but not mechanical nociception. *Eur. J. Pharmacol.*, **237**, 143–146.

110. Gully, D., Canton, M., Boigegrain, R., Jeanjean, F., Molimard, J.-C., Poncelet, M., Gueudet, C., Heaulme, M., Leyris, R., Brouard, A., Delaprat, D., Labbe-Jullie, C., Mazella, J., Soubrie, P., Mafrand, J.-P., Rostene, W., Kitabgi, P. and LeFur, G. (1993) Biochemical and pharmacological profile of a potent and selective nonpeptide antagonist of the neurotensin receptor. *Proc. Natl. Acad. Sci. USA*, **90**, 65–69.

111. Snider, R.M., Pereira, Longo, K.P., Davidson, R.E., Vinick, F.J., Laitinen, K., Genc-Sehitoglu, E. and Crawley, J.N. (1992) UK-73,093: A non-peptide neurotensin receptor antagonist. *Bioorg. Med. Chem. Lett.*, **2**, 1535–1540.

112. Yamamura, Y., Ogawa, H., Chihara, K., Kondo, K., Onogawa, S., Nakamura, T., Tominaga, M. and Yabuuchi, Y. (1991) OPC-21268, an orally effective, nonpeptide vasopressin V1 receptor antagonist. *Science*, **252**, 572–574.

113. Yamamura, Y., Ogwaw, H., Yamashita, H., Chihara, T., Miyamoto, H., Nakamura, S., Onogawa, T., Yamashita, T., Hosokawa, T., Mori, T., Tominaga, M. and Yabuuchi, Y. (1992) Characterization of a novel aquaretic agent, OPC-31260, as an orally effective non-peptide vasopressin V2 receptor antagonist. *Br. J. Pharmacol.*, **105**, 787–791.

114. Saito, T., Fujisawa, G., Tsuboi, Y., Okada, K. and Ishikawa, S. (1993) Nonpeptide vasopressin antagonist and its application in the correction of experimental hyponatremia in rats. *Regul. Peptides*, **45**, 295–298.

115. Pettibone, D.J., Clineschmidt, B.V., Bock, M.G., Evans, B.E., Freidinger, R.M., Veber, D.F. and Williams, P.D. (1993) Development and pharmacological assessment of novel peptide and nonpeptide oxytocin antagonists. *Regul. Peptides*, **45**, 289–293.

116. Salituro, G.M., Pettibone, D.J., Clineschidt, B.V., Williamson, J.M. and Zink, D.L. (1993) Potent, non-peptidic oxytocin receptor antagonists from a natural source. *Bioorg. Med. Chem. Lett.*, **3**, 337–340.

117. Collins, J.L., Dambek, P.J., Goldstein, S.W. and Faraci, W.S. (1992) CP-99,711: A non-peptide glucagon receptor antagonist. *Bioorg. Med. Chem. Lett.*, **2**, 915–918.

118. Valentine, J.J., Nakanishi, S., Hageman, D.L., Snider, R.M., Spencer, R.W. and Vinick, F.J. (1992) CP-70,030 and CP-75,998: The first non-peptide antagonists of bombesin and gastrin releasing peptide. *Bioorg. Med. Chem. Lett.*, **2**, 333–338.

119. Salvino, J.M., Seoane, P.R., Douty, B.D., Awad, M.M.A., Dolle, R.E., Houck, W.T., Faunce, D.M. and Sawutz, D.G. (1993) Design of potent non-peptide competitive antagonists of the human bradykinin B2 receptor. *J. Med. Chem.*, **36**, 2583–2584.

120. Smith, R.G., Cheng, K., Schoen, W.R., Pong, S.-S., Hickey, G., Jacks, T., Butler, B., Chan, W.W.-S., Chaung, L.-Y.P., Judith, F., Taylor, J., Wyvratt, M.J. and Fisher, M.H. (1993) A nonpeptidyl growth hormone secretagogue. *Science*, **260**, 1640–1643.

121. See for example Garon, M.G. and Lefkowitz, R.J. (1993) Catecholamine receptors: structure, function, and regulation. *Recent Prog. Horm. Res.*, **48**, 277–291.

122. Nakanishi, S. (1991) Mammalian tachykinin receptors. *Ann. Rev. Neurosci.*, **14**, 123–136.

123. Yokota, Y., Akazawa, C., Ohkubo, H. and Nakanishi, S. (1992) Delineation of structural domains involved in the subtype specificity of tachykinin receptors through chimeric formation of substance P/substance K receptors. *EMBO J.*, **11**, 3585–3591.

124. Schwyzer, R. (1977) ACTH: a short introductory review. *Ann. N.Y. Acad. Sci.*, **297**, 3–26.

125. Portoghese, P.S. (1989) Bivalent ligand and the message-address concept in the design of selective opioid receptor antagonists. *Tr. Pharmacol. Sci.*, **10**, 230–235.

126. Gether, U., Johansen, T.E. and Schwartz, T.W. (1993) Chimeric NK_1 (Substance P)/NK_3 (Neurokinin B) receptors. *J. Biol. Chem.*, **268**, 7893–7898.

127. Fong, T.M., Huang, R.-R.C. and Stader, C.D. (1992) Localization of agonist and antagonist binding domains of the human neurokinin-1 receptor. *J. Biol. Chem.*, **267**, 25664–25667.

128. Fong, T.M., Yu, H., Huang, R.-R.C. and Strader, C.D. (1992) The extracellular domain of the neurokinin-1 receptor is required for high-affinity binding of peptides. *Biochem.*, **31**, 11806–11811.

129. Gether, U., Johansen, T.E., Snider, R.M., Lowe III, J.A., Nakanishi, S. and Schwartz, T.W. (1993) Different binding epitopes on the NK1 receptor for substance P and a non-peptide antagonist. *Nature*, **362**, 345–348.

130. Fong, T.M., Cascieri, M.A., Yu, H., Bansal, A., Swain, C. and Strader, C.D. (1993) Amino-aromatic interaction between histidine 197 of the neurokinin-1 receptor and CP 96345. *Nature*, **362**, 350–353.

131. Sachais, B., Snider, R.M., Lowe III, J.A. and Krause, J.E. (1993) Molecular basis for the species selectivity of the substance P antagonist CP-96,345. *J. Biol. Chem.*, **268**, 2319–2323.

132. Gether, U., Yokota, Y., Emonds-Alt, X., Breliere, J.-C., Lowe III, J.A., Snider, R.M., Nakanishi, S. and Schwartz, T.W. (1993) Two nonpeptide tachykinin antagonists act through epitopes on corresponding segments of the NK1 and NK2 receptors. *Proc. Natl. Acad. Sci. USA*, **90**, 6194–6198.

133. Beinborn, M., Lee, Y.-M., McBride, E.W., Quinn, S.M. and Kopin, A.S. (1993) A single amino acid of the cholecystokinin-B/gastrin receptor determines specificity for non-peptide antagonists. *Nature*, **362**, 348–350.

6. CONFORMATIONS AND BINDING INTERACTIONS OF THYROID HORMONES AND ANALOGUES

DAVID J. CRAIK,* BRENDAN M. DUGGAN** AND SHARON L.A. MUNRO**

*Centre for Drug Design and Development, University of Queensland, Brisbane 4072, Queensland, Australia
**School of Pharmaceutical Chemistry, Victorian College of Pharmacy, Monash University, 381 Royal Parade, Parkville, Victoria 3052, Australia

Thyroid hormones are important in a wide range of metabolic processes. This article describes the conformations of the hormones and their binding interactions with various receptor and transport proteins, and examines the structures and conformations of a range of other molecules which interact with thyroid hormone binding sites.

1. INTRODUCTION

The structures of the two major thyroid hormones, thyroxine (T_4) and $3',3,5$-triiodothyronine (T_3) are shown below.

Each consists of a diphenyl ether nucleus which is iodinated and further substituted with an alanyl sidechain at one end of the molecule and a hydroxyl group at the other. These hormones regulate a wide range of metabolic events, including heat output and oxygen consumption, as well as cell growth and differentiation.[1] The hormones are synthesised in the thyroid gland, which concentrates iodide from the bloodstream and incorporates it into the protein thyroglobulin, which acts as a storage reservoir for the hormones. Upon a signal from the pituitary gland via the messenger TSH (thyroid-stimulating hormone, also known as thyrotropin), the thyroid gland releases T_4, and smaller quantities of T_3, for transport to peripheral tissue. In the bloodstream the hormones are mostly (>99%) protein-bound to the

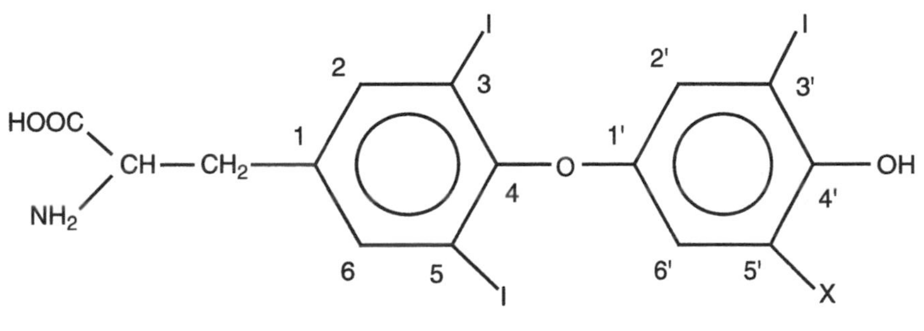

Figure 1 Structures of T_4 (X = I) and T_3 (X = H).

255

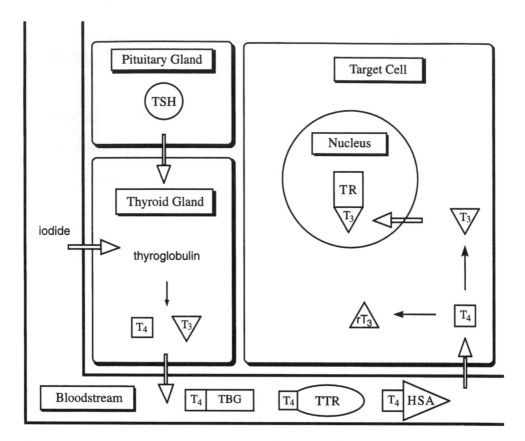

Figure 2   Schematic representation of thyroid hormone release, transport and binding interactions. The symbols are defined in the text, except for TR, which is the thyroid receptor.

three transport proteins, thyroxine-binding globulin (TBG), transthyretin (TTR) and albumin (HSA). In peripheral tissue, unbound hormone passes into the cell and eventually exerts its action at a nuclear receptor. Aspects of hormone action are summarised in Figure 2.

T_4 is the major circulating thyroid hormone; however, T_3 is more active at the nuclear receptor. As indicated in Figure 2, T_4 is converted to T_3 in peripheral tissue by deiodinase enzymes. These enzymes are also responsible for conversion of T_4 to the inactive derivative, reverse-T_3 (rT_3, 3,3'-5'-triiodothyronine), and thus provide a local control mechanism. There are various other feedback control loops, including regulation of TSH by thyrotropin-releasing hormone (TRH) from the hypothalamus (positive feedback). TSH is also regulated by levels of the circulating hormones in a negative feedback loop. Together these various mechanisms finely regulate the amount of secreted hormones.

Because of the importance of the thyroid hormones, there have been many studies on their structures and mechanism of action.[1] The structures of a wide range of hormones and analogues in the solid state have been determined by X-ray crystallography.[2,3] Much is also known about the binding affinities of hormones and analogues to the transport proteins and to the nuclear receptor, as well as on the thyromimetic activities of these analogues.[4-9] These studies have been extremely useful in defining molecular features necessary for binding and activity. X-ray crystallographic studies of protein-hormone complexes have provided detailed three-dimensional representations of the interactions between hormones and the transport protein TTR.[8,9] Such studies have revealed that even quite subtle changes in structure may significantly affect the mode of binding to target macromolecules. It has therefore become apparent that much remains to be determined about the precise nature of functional group interactions present in the thyroid system. In this article, we have attempted to summarise the current state of knowledge of structures, binding interactions, and activities of the hormones and other compounds which interact with thyroid binding sites. As will be seen from the following discussion, there are potentially a number of areas in which the medicinal chemist can contribute to improved therapy for diseases associated with thyroid malfunction and in the clinical evaluation of thyroid function.

There are many disease states associated with malfunction of the thyroid system.[10] These range from myxedema to thyrotoxicosis, goitre and thyroid tumours. Within each of these categories there is a range of causes and contributing factors. Two examples of malfunction are the undersupply and oversupply of thyroid hormone to the body, defined respectively as hypothyroidism and hyperthyroidism. Treatment of the former is readily accomplished by hormone replacement therapy with synthetic T_4. Treatment of the latter is more complicated, depending on the extent and cause of the hormone excess. Current antithyroid drugs are relatively slow-acting, as they exert their effects on biosynthesis, rather than directly at the nuclear receptor. The development of a potent thyroid hormone antagonist acting directly at the nuclear receptor would be extremely useful.

Novel thyroid hormone analogues may have potential uses in other disease states. For example, it has been suggested that thyroid hormone analogues with organ-selective activity may have potential as anti-atherosclerotic drugs. It has been known for some time that T_4 lowers circulating cholesterol levels[11] but T_4 cannot be used as a treatment for hypercholesterolaemia because of its unfavourable cardiac effects when delivered in doses sufficient to lower serum cholesterol. Since the cholesterol lowering effect is manifested mainly by thyromimetic action on the liver, an organ-selective compound (ie, one with high activity in the liver but not in the heart)[12-14] would have potential clinical application.

In addition to the drugs used to treat thyroid disease, a number of other drugs affect the thyroid system indirectly. These include such diverse compounds as benzodiazepines, non-steroidal anti-inflammatories and the anti arrhythmic drug, amiodarone.[15-45] There are also many other chemical compounds which bind to proteins in the thyroid system.[46-74] The range of different structures that interact with thyroid hormone binding sites provides an opportunity to further define the molecular shapes of the binding sites. Such studies are described in Section 4 of this review.

The fact that many non-thyroidal compounds can bind to thyroid binding sites has important implications in thyroid function tests.[18,19] This also provides the opportunity for the medicinal chemist to design novel and selective thyroid-binding compounds for use in such tests.

2. STRUCTURE OF THYROID HORMONES

In addition to the basic hormones T_4 and T_3, there is a range of other naturally occurring hormones of similar structure which differ in the degree of iodination.[75] These are summarised in Figure 3.

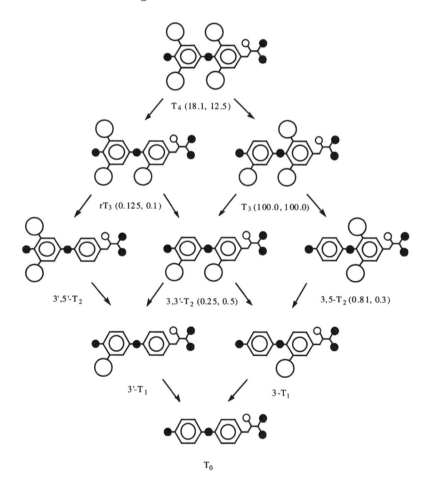

Figure 3 Deiodination cascade of the nine thyroid hormones. The large open circles represent iodine atoms, the small open circles nitrogen and small closed circles, oxygen. Protons have been omitted to simplify the figure. The numbers in parentheses are *in vivo* activity and *in vitro* nuclear receptor affinity, relative to T_3.[76]

All nine of the thyroid hormones are present *in vivo* in at least trace amounts[75] and have significantly reduced potency relative to T_3. The activity of the thyroid hormones is closely related to their binding affinity for the nuclear receptor, which in turn depends on both the structure of the compounds and their conformations.

2.1 Structure-Activity Relationships

Extensive investigation of the structure-activity relationships of the thyroid hormones and chemically synthesised analogues[6,76–81] has shown that hormonal activity requires:

(i)   a central hydrophobic nucleus composed of two mutually perpendicular aromatic rings linked by a bridging group;
(ii) bulky lipophilic substituents ortho to the bridging group which maintain the active conformation of the aromatic moiety;
(iii) a hydroxyl group para to the bridging group;
(iv) a charged sidechain at the opposite end of the molecule to the hydroxyl group.

In the case of the thyroid hormones themselves, the lipophilic substituents are iodines, an ether link is the bridging group and the sidechain is an alanyl moiety. The hormones are remarkably well optimised, and while many analogues with different substitutions show hormonal activity, very few are more potent than the native hormones. In studies of analogues, factors which may increase potency include:

(i)   increasing the compound's affinity for the nuclear receptor,
(ii) slowing the compound's metabolism, or
(iii) increasing the efficiency with which the compound is transported to cell nuclei.

Figure 4 summarises the structural features of the thyroid hormones which affect their activity. The importance of these various features is briefly described in the

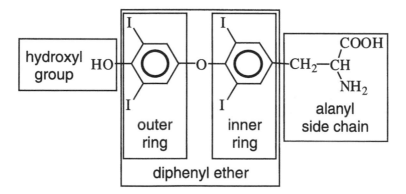

Figure 4   Structural features of the thyroid hormones.

following sections. In this discussion we adopt the commonly used nomenclature in which the aromatic ring adjacent to the alanyl sidechain is referred to as the inner ring and the hydroxyl-bearing ring as the outer ring.

2.1.1 Aromatic nucleus

Two phenyl rings linked by an ether bridge form the core of the thyroid hormone structure. Bridging groups other than oxygen do lead to active compounds[6] but the diphenyl ether moiety is the preferred framework.

2.1.2 The iodine atoms

A major function of the iodine atoms is to maintain the thyronine nucleus in a conformation suitable for binding. The iodine atoms achieve this simply by restricting the motion of the two rings about the bridging group. Substituents smaller than iodine cannot restrict the motion as efficiently and thus molecules with these substituents generally do not bind as well and are less active.

On the inner ring, halogen atoms have been found to be more effective substituents than other similarly sized groups.[6] The iPr group is approximately the same size as an iodine atom, but molecules carrying this alkyl group have very little activity. Halogens smaller than iodine are better substituents than larger alkyl groups, while substituents larger or more polar than iodine render the molecule inactive. Thus, the order of activity follows the series $I > Br > Cl = Me \gg iPr > sBu$. The formation of hydrogen bonds and charge transfer complexes between protein binding sites and the iodine atoms have been proposed[77] to explain why iodine atoms increase activity so much more than alkyl groups.

As with the inner ring substituents, activity increases with the size of the outer ring substituents. The iPr group gives the maximum activity ($iPr > I > Et > Br > Me > Cl > F > H$). Substituents larger than the iPr group cause a decrease in activity ($iPr > sBu > nPr > tBu > iBu > Ph$). Unlike the substituents on the inner ring, the outer ring substituent need not be a halogen for maximum activity. Indeed, the highly active thyroid-hormone analogue 3,5-diiodo-3'-isopropyl-L-thyronine, carries an alkyl group on the outer ring. It is thought that the greater activity of the alkyl substituted compound is due to the lack of an iodine atom on the outer ring, thereby preventing its metabolism by 5'-deiodinases.[6]

Hormonal activity decreases in proportion to the size of a second substituent on the outer ring. However, disubstitution of the outer ring increases the molecule's affinity for the transport proteins which appear to act as a reservoir of T_4 which is metabolised to T_3 as required. The nuclear receptor and the transport proteins also differ in that the receptor's affinity is unaffected by the nature of the outer ring substituent, whereas the transport proteins bind more tightly to halogen-bearing molecules.

2.1.3 The hydroxyl group

The hydroxyl group is thought to contribute to the affinity of the thyroid hormones for the nuclear receptor by forming a hydrogen bond with the receptor. However, some compounds in which the hydroxyl group is blocked still show activity. These compounds are presumably metabolised *in vivo* to give a free hydroxyl function capable of forming a hydrogen bond. Compounds which cannot be converted to give a free hydroxyl group are inactive, as are those in which the hydroxyl group is in a position other than para to the bridging group.[6]

Substituents which increase the acidity of the hydroxyl function (ie, halogens) also increase the affinity of the hormone for the transport proteins, but not for the nuclear receptor. This is consistent with the proposal that the transport proteins interact with the phenoxide ion, whereas the nuclear receptor forms a hydrogen bond with the ligand.[6,76,82]

2.1.4   The amino-acid sidechain

The thyroid hormones require an amino-acid sidechain para to the bridging group linking the two aromatic rings. Analogues with the amino-acid sidechain at different positions show virtually no activity. The length of the sidechain is also important. Sidechains longer than two carbon atoms show decreased activity, as does the shorter formic acid derivative. The carboxyl function is particularly important for activity; removal of this group decreases activity by an order of magnitude.

Nuclear receptor affinity is greater for analogues without the amino group than for those with it. However, the thyromimetic activity of the acid sidechain analogues is less than that of the amino-acid sidechain analogues. Similarly, the activity of compounds with the D-configuration is much less than one would expect from the affinity of these molecules for the nuclear receptor. Both these discrepancies have been rationalised by proposing rapid metabolism and less efficient transport of the less active compounds.[6]

2.2   Conformation

Most of the available information on conformations of the thyroid hormones has come from X-ray studies of the crystalline state and has been extensively reviewed by Cody.[3] NMR spectroscopy[83–93] and theoretical calculations[94–101] have also provided valuable information.

The thyroid hormones possess five single bonds about which rotation may take place. The torsion angles about these bonds are ψ, χ_1, χ_2, ϕ and ϕ', and are shown in Figure 5. The first three control the conformations of the amino-acid sidechain, while the remaining two define the diphenyl ether conformation.

2.2.1   The amino-acid sidechain

To minimise steric clashes with the inner ring, the amino-acid sidechain adopts a conformation which places the amino and carboxyl groups out of the plane of the ring. The summary in Figure 6 of observed values of χ_2 in a number of crystal structures of thyroid-hormone analogues[3,82] confirms the preference for $\chi_2 \approx 90°$. Figure 6 also shows that there is generally a preference for $\psi = 0°$. By contrast, with

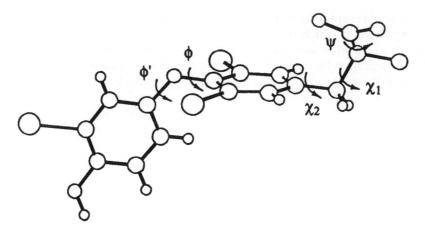

Figure 5   Structure of T_3 showing the torsion angles ψ, χ_1, χ_2, ϕ and ϕ'. The structure is shown in the twist-skewed, transoid, distal conformation.

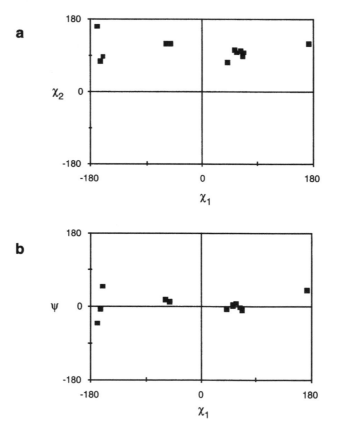

Figure 6   Plot of torsion angles found in the crystal structures of thyroid hormones and analogues:[3,82,102] (a) χ_2 vs χ_1 and (b) ψ vs χ_1.

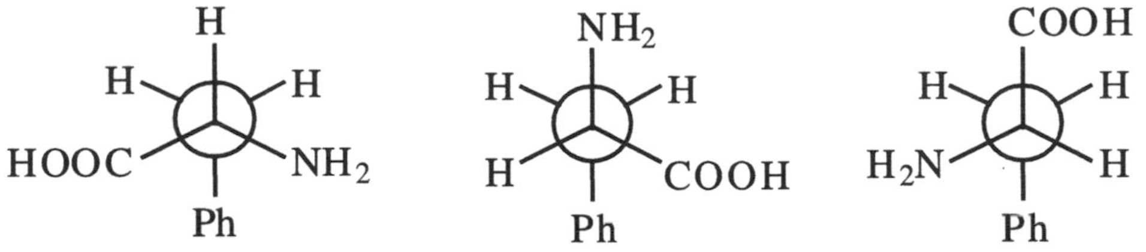

Figure 7   Three staggered conformers corresponding to χ_1 angles of 60° (left), 180° (centre) and –60° (right).

the relative lack of variation in χ_2 and ψ, each of the three staggered conformers about χ_1 (Figure 7) is seen in the crystal state. This means that any one of the three groups attached to the α-carbon may project away from the aromatic rings, while the other two point back towards them. An analysis of the ^{1}H NMR coupling constants indicates that all three conformers are present in solution.[91]

2.2.2   The diphenyl ether moiety

It has been recognised for some time[103,104] that the bulky iodine atoms restrict the movement of the thyronine nucleus, thereby creating preferred conformations. The symmetrical conformation in which the ether linkage lies in the plane of the outer ring and bisects the inner ring ($\phi = 90°$, $\phi' = 0°$) is known as the skewed structure. Extensive crystallographic investigations[3] have shown that nearly all the thyroid hormones and their analogues crystallise in a slight variation of the skewed conformation known as twist-skewed (see Figure 5), in which the outer ring is slightly off perpendicular to the inner ring and no longer edge-on to it. Thyroxine itself adopts the conformation in which $\phi = 108°$ and $\phi' = -30°$.[82] This arrangement of the diphenyl ether nucleus results in a characteristic three-dimensional shape for the thyroid hormones which is important for their interactions with target proteins.

Another conformational factor that affects the overall three-dimensional shape of the thyroid hormones is the relationship of the alanyl sidechain to the outer (hydroxyl bearing) aromatic ring. When these two moieties are on opposite faces of the inner aromatic ring, the structures are referred to as transoid, whereas if they are on the same face, the structure is cisoid. The two forms are of similar energy and, indeed, in the salt of thyroxine crystallised from diethanolamine[82] both forms are seen in the same crystal. Figure 8 shows values of ϕ and ϕ' for a large number of thyroid-hormone analogues[3,82,102] and illustrates that in most structures the ϕ/ϕ' angles correspond to a twist-skewed orientation with the angles clustered into two groups corresponding to cisoid and transoid.

A third conformational aspect of particular importance to those analogues with monosubstitution on the outer ring, such as T_3, is the orientation of the 3'- or 5'-position with respect to the inner aromatic ring. Conformers in which the 3'-iodine is over the inner ring are referred to as proximal, while those in which it is rotated

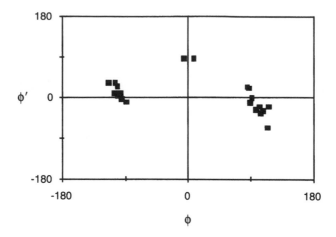

Figure 8 Observed torsion angles ϕ and ϕ' in the crystal structures of thyroid hormones and analogues.[3,82,102] The cluster centred on $\phi = 108°$, $\phi' = -28°$ is due to the transoid conformer, while the cluster around $\phi = -108°$, $\phi' = 28°$ is due to the cisoid form. The points near $\phi = 0°$, $\phi' = 90°$ are the two antiskewed conformers observed in the unit cell of rT_3.[102]

away from the rest of the molecule are called distal conformers. Both forms have been observed in crystals; however, for T_3 analogues the distal conformation has been shown to be the biologically active one.[103]

Only one thyroid hormone, rT_3, has been reported to crystallise in a conformation other than twist-skewed. In the antiskewed conformation adopted by rT_3,[102] the ether linkage bisects the outer ring and is in the same plane as the inner ring. As would be expected, the outer ring straddles the protonated, rather than the iodonated, side of the inner ring. The biological activity of rT_3 is very low and actually increases if one of the iodines on the outer ring is removed (Figure 3).

2.3 Mobility

The substituents on the thyroid hormones impose preferred conformations on the compounds, but they are not static, rigid molecules. The outer ring protons would give rise to different ^1H NMR signals if the molecules were held in a rigid twist-skewed conformation like that shown in Figure 5 because of the substantially different ring-current effects on the H2' and H6' positions. The fact that only one signal is observed[83] indicates that rapid averaging of the H2' and H6' environment takes place. Lowering the temperature below $-80°C$ slows the interconversion sufficiently that a separate signal is observed for each proton.[85] Such low-temperature NMR experiments have been used to determine the energy barriers to interconversion between proximal and distal forms of thyroid hormone analogues[85] and for T_3 and T_4.[89] The experimental values for the analogues and the hormones were similar (~35 kJ/mol) and matched well with the values obtained from molecular orbital calculations.[95]

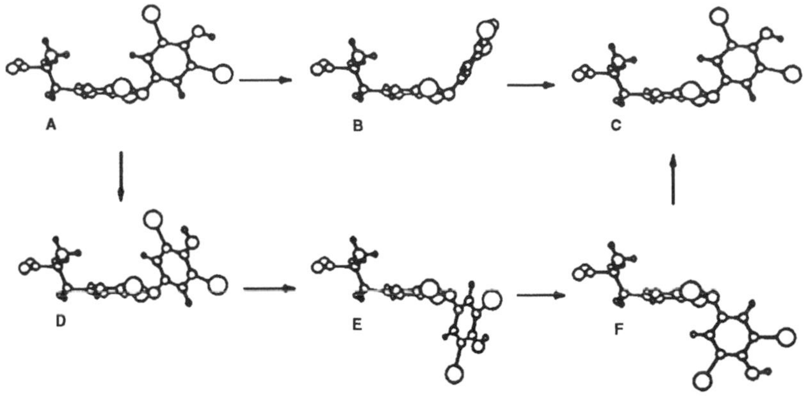

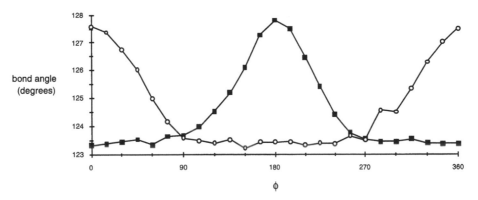

Figure 9 Two possible pathways for the interconversion of environments of H2′ and H6′. In pathway 1 [(A) → (C)], ϕ' alone is varied from 0° to 180°. In pathway 2 [(A) → (D) → (E) → (F)], co-operative rotation of ϕ and ϕ' occurs. Transoid conformer (F) may be converted into the cisoid structure (C) through rotation of the sidechain about χ_2. One of the protons of the outer (hydroxyl-bearing) aromatic ring is shaded to highlight the change in environment that occurs during the conformational changes.

The mechanism by which the distal conformer converts to the proximal conformer has been investigated in a recent modelling study[101] using molecular orbital calculations. Two pathways are possible in principle (Figure 9). The first involves simple rotation about ϕ', while the second involves a concerted rotation of both ϕ and ϕ'. The theoretically determined energy barriers for both pathways are similar[101] and in reasonable agreement with experimental values.[89] The theoretical calculations for the second pathway showed that there is a degree of co-operativity of motion of several bond angles near the diphenyl ether linkage. For example, as the outer ring passes the iodine atoms attached to the inner ring the I3-C3-C4 and I5-C5-C4 bond angles open out, allowing the ring easier passage (see Figure 10).

Figure 10 Theoretically predicted[101] variation in the iodine bond angles with rotation about the torsion angle ϕ following pathway 2. Filled squares correspond to the I3-C3-C4 angle and open circles to the I5-C5-C4 bond angle.

3. BINDING INTERACTIONS OF THYROID HORMONES

The thyroid hormones bind to many different proteins throughout the body.[4,5,10] These proteins carry out a variety of functions, some of which are yet to be fully understood. In the blood, thyroid hormones are bound to proteins which transport them from the thyroid gland, where they are synthesised, to the cells where they have their effects. The three major transport proteins, thyroxine-binding globulin, transthyretin and human serum albumin, carry virtually all of the thyroid hormones. A small fraction is transported by a variety of other proteins. At the cell, other thyroid-hormone binding proteins are found in the cell membrane, in the cytosol, on the surface of mitochondria and inside the cell nucleus. Each of these thyroid-hormone binding proteins will be briefly described and the variations in their binding characteristics discussed.

3.1 Thyroxine-Binding Globulin

Thyroxine-binding globulin (TBG) has the highest affinity for the thyroid hormones of all the transport proteins and the lowest serum concentration. It carries about 75% of the total serum T_4 and about 50% of the T_3. TBG is a glycosylated single polypeptide chain of approximately 54 kDa[105] with a single thyroid-hormone binding site. Its three-dimensional structure has not yet been experimentally determined. This has limited the amount of structural information that can be derived for ligand binding to this protein, although the situation has been improved by some recent modelling studies. The amino-acid sequence of TBG shows 42% homology with α_1-antitrypsin,[106] identifying TBG as a member of the serpin family of proteins. The crystal structure of α_1-antitrypsin has been solved[107] and used as the basis of two models of TBG.[108,109] Terry and Blake's model[109] uses a slightly different sequence alignment and a different orientation of the ligand from that of Jarvis et al.[108] and proposes that the high affinity of TBG is due to an abundance of phenylalanine residues surrounding the iodine atoms of the thyroid hormones. Otherwise, both models are similar and place the thyroid-hormone binding site in a β-barrel structure near charged residues which could interact with the amino-acid sidechain.

Other information on the binding sites has been obtained from spin-label studies[110] and, indirectly, from studies of various TBG mutants.[108,111] The lack of a crystal structure for TBG, however, limits the structural interpretation which may be placed on these data.

An indication of some of the factors which influence thyroid-hormone binding to TBG is presented in Table 1. This shows relative binding affinities for a number of hormones and analogues to TBG, TTR and HSA. The data show that reductions in binding occur for deiodinated derivatives. The chirality of the sidechain does not have a great influence on TBG binding (reducing it by only a factor of two in going from the L to D form of T_4), but the charge on the sidechain appears to be important. The compound T_4-acetic acid, commonly referred to as tetrac, is a useful probe of binding, as it shows the effect of shortening, and removing the NH_2 group from the amino-acid sidechain. Triac is the corresponding analogue of T_3. In both cases, there is a substantial loss in affinity at TBG (>30-fold) on removal of the NH_2 group.

Table 1 Relative binding affinities of selected thyroid-hormone analogues[5,6]

	TBG	TTR	HSA
T_4	100	100	100
T_3	9	4.8	55
rT_3	38	8	100
$3',5'-T_2$	0.1	2.3	85
$D-T_4$	54	2.7	100
T_4-acetic acid (tetrac)	1.7	275	—
T_3-acetic acid (triac)	0.3	20	—

3.2 Transthyretin

Transthyretin (TTR) has a lower affinity for T_4 and T_3 than TBG but its serum concentration is much higher. It carries roughly 15% of the total serum T_4 and T_3, and is also responsible for the transport of retinol through an interaction with retinol-binding protein.[112]

The relationship between structure and binding affinity of many hormones and their analogues has been extensively reviewed.[1-10] From the summary of binding data for a few key compounds in Table 1 it can be seen that TTR binding differs from that of TBG in terms of its stereoselectivity for T_4 (the selectivity for the L form over the D form is greater for TTR) and in its affinity for the deaminated T_4 and T_3 derivatives triac and tetrac (these bind more avidly to TTR than T_4 and T_3 respectively, but less avidly to TBG).

TTR is a tetrameric protein of approximately 55 kDa, composed of identical 127-residue subunits. The crystal structure of TTR was first published in 1974.[113] Since then several other crystallographic studies of the protein complexed with its ligands have appeared.[114-117] Consequently, TTR is the most well studied thyroid-hormone binding protein. The TTR tetramer consists of four curved β-sheets stacked on top of each other. A cylindrical channel, roughly 8 Å in diameter and 50 Å long, passes straight through the centre of the molecule between the sheets.[115] Crystal structures of the T_4-TTR complex show that the thyroid hormones bind in this central channel. Near the centre of the protein, paired serine and threonine sidechains project into the channel, dividing it into two identical binding sites 21 Å deep.[115] Spin-labelled T_4 has been used to confirm the depth of the binding site.[110]

Despite the crystal structures showing two identical binding sites, binding studies at high concentrations of T_4 have found that TTR binds the first thyroid hormone with high affinity and the second with an affinity two orders of magnitude less.[118] This negative co-operativity has yet to be satisfactorily explained, although evidence for conformational change in the protein upon binding the first T_4 molecule has been detected by both fluorescence[119,120] and ${}^1H$ NMR[92] techniques.

The original model[114] of T_4-TTR binding placed the hormone in the binding site with the hydroxyl group at the centre of the protein, where it could interact with the hydrophilic Ser117 and Thr119 residues. In this orientation, the amino-acid sidechain of T_4 may interact with Lys15 and Glu54 near the binding channel entrance. The iodine atoms are then surrounded by hydrophobic valine and leucine residues. However, molecular modelling and recent crystallographic studies now suggest that the thyroid hormones exhibit multiple modes of binding. The thyroid hormones carry a maximum of four iodines but there are six pockets into which they can fit. This redundancy of iodine binding sites, coupled with the flexibility of the charged sidechains near the channel entrance and the ability of water molecules to fill vacant pockets, are factors which apparently allow the thyroid hormones to bind to TTR in a variety of ways.

Molecular modelling studies of the T_4-TTR interaction[100,121] support the existence of a variety of binding modes, including:

(i)   the original model, known as the forward mode;
(ii)   the reverse mode, in which the hydroxyl is at the entrance to the binding site and the amino-acid sidechain is at the centre of the protein;
(iii) variations of the forward mode in which the ligand is either closer to the centre of the protein or closer to the binding site entrance.

A recent crystallographic investigation of the binding of thyroid hormones to TTR[8] proposes multiple modes of binding to explain the electron density observed in the binding site. As well as the forward and reverse modes of binding, some ligands are proposed to bind deeper inside the molecule. This is supported by the crystal structure of the $3,3'$-T_2-TTR complex[117] which shows the hormone to be 3.5 Å closer to the centre of the protein than T_4.[113] The diiodothyronine binds in such a way that its 3-iodine occupies a pocket filled by the $3'$-iodine of T_4 and the $3'$-iodine fills a pocket occupied by a water molecule in the T_4-TTR complex.

Each of the TTR binding sites contains three pairs of pockets formed by the pleats in the β-sheets.[8] The pockets are approximately 6 Å, 9 Å and 13 Å from the centre of the molecule, and accommodate the iodine atoms when thyroid hormones bind to TTR. An indication of the location of these pockets may be seen from Figure 11, which schematically represents electron density at different depths in the channel, summed for a range of different thyroid-hormone analogues.

While most of the available information on the structure and ligand-binding interactions of TTR has come from X-ray crystallographic studies, NMR spectroscopy has provided some information. The molecular weight of the TTR tetramer is too large to allow a ^1H NMR structure determination. However, some resonances have been assigned by using the crystal structure and a ring-current shift prediction program.[93] ^{13}C NMR techniques have also been used to investigate the protein,[122] as the greater dispersion of ^{13}C chemical shifts reduces peak overlap compared to ^1H NMR. Figure 12 shows that the use of spectral editing techniques to obtain separate subspectra of CH, CH_2 and CH_3 groups provides considerable simplification of the ^{13}C spectrum. In principle relaxation time measurements on such spectra may allow information on protein dynamics in solution to be obtained.

Recently, variants of TTR involving point mutations have been identified.[123–164] Some examples where altered binding affinities for TTR have been well characterised

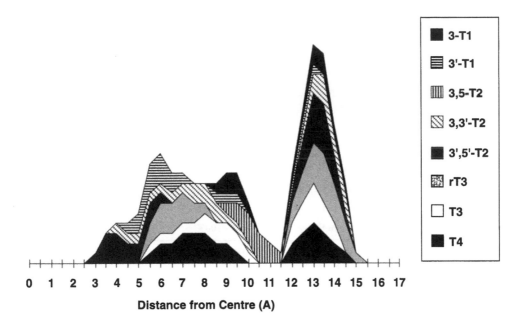

Figure 11 Cumulative electron density within the TTR binding site of eight different thyroid hormone-TTR complexes.[8]

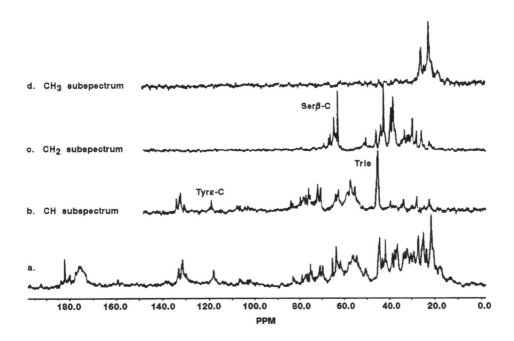

Figure 12 ^{13}C subspectra of human TTR.[122]

Table 2 Effect of transthyretin mutations on binding affinity

Mutation	Location	Relative binding affinity[a]	Reference
Gly6 $\rightarrow$ Ser	channel	348%	Lalloz et al.[123]
Val30 $\rightarrow$ Met	buried	17%	Refetoff et al.[124]
Leu58 $\rightarrow$ His	buried	33%	Murrell et al.[125]
Thr60 $\rightarrow$ Ala	surface	78%	Refetoff et al.[124]
Ser77 $\rightarrow$ Tyr	surface	<100%	Murrell et al.[126]
Ile84 $\rightarrow$ Ser	channel	19%	Refetoff et al.[124]
Ala109 $\rightarrow$ Thr	channel	317%	Moses et al.[127]
Thr119 $\rightarrow$ Met	channel	213%	Barlow et al.[128]
Val122 $\rightarrow$ Ile	channel	<100%	Murrell et al.[126]

[a] Relative binding affinity was calculated using the binding affinity of the wild-type protein determined at the same time as the affinity of the mutant.

are listed in Table 2.[123–128] Crystal structures of two of these mutants have been determined. The Met30 variant has decreased affinity for the thyroid hormones and the crystal structure[162,163] shows that this is caused by the bulky methionine residue forcing a β-strand into the central channel, thereby narrowing the binding site. The other variant whose crystal structure has been solved, Ala109,[164] has an increased affinity for T_4. The crystal structure of this variant shows the opposite effect to the Met30 variant. The substitution has allowed a β-strand to move away from the binding site, making it wider than in the native molecule. Further investigation of these mutants and identification of new variants will give a more detailed understanding of the TTR binding site. In principle, similar information could be derived from TTRs from different species, but in general these are highly homologous and have similar binding affinities.[165–167] The binding site residues are particularly highly conserved.

It is interesting to note that TTR may have important physiological roles other than thyroid-hormone and retinol transport, as evidenced by its high concentration in the choroid plexus,[168–173] although the significance of this finding is not yet fully understood.

3.3 Human Serum Albumin

The affinity of T_4 for human serum albumin (HSA) is two orders of magnitude less than that for TTR but, because of the high serum concentration of HSA, it transports almost as much of the thyroid hormones as TTR. HSA has six binding sites for T_4, the first of higher affinity than the others.[5] It binds a variety of hormones, drugs, fatty acids, dyes, inorganic ions and long-chain alcohols, and is considered to be more of a universal carrier than a specific binder of thyroid hormones.

HSA consists of a single polypeptide chain of 65 kDa. The crystal structure[174] shows it to consist of 28 α-helices held in place by 17 disulphide bridges. The protein is composed of three structurally similar domains based on two smaller subdomains.

Tryptophan and T_4 are known to bind to the same site on HSA.[175] The crystal structure shows that the high-affinity tryptophan binding site is in subdomain IIIA, in a cavity between four α-helices. One side of the binding site is lined with hydrophobic residues, while on the other side are a tyrosine and two arginine residues. A second binding site has been located in sub-domain IIA. Studies using spin-labelled thyroxine found binding sites of high and low affinity. The high-affinity site was determined to be more than 23 Å deep. The lower affinity site is at least 21 Å deep, but is wider than the high-affinity site. HSA is the only thyroxine-binding protein in which the binding site is constructed from α-helices, rather than β-sheets. As with the other thyroxine-transporting proteins, a number of variant proteins are known,[176] although only limited affinity data are available.

3.4   Other Thyroid Hormone Binding Serum Proteins

A small proportion of the thyroid hormones in the blood is bound by proteins other than TBG, TTR or HSA. One of these proteins, known as 27K protein, exists as a dimer of 66 kDa and binds two molecules of T_4 with an affinity slightly less than that of TTR.[177] Many of the lipoproteins have been found to bind T_4 with affinities greater than that of TTR.[178–182] They show similar affinity for L-T_4, D-T_4, rT_3 and triac, and lower affinity for L-T_3. The T_4 binding sites of the lipoproteins have been localised to regions of the proteins predicted to have β-sheet structure.[182]

3.5   Cell Uptake Site

The passage of thyroid hormones into the cell was originally thought to occur by a process of passive diffusion through the cell membrane.[183] More recently, substantial evidence has accumulated that thyroid-hormone uptake is the result of an active transport process, as reviewed by Davis.[184] Recent evidence also suggests that thyroid-hormone uptake may be similar to that of certain classes of amino-acid uptake.[185,186] The transport of T_3 across cell membranes is energy-dependent and saturable, and L-T_3 is transported in preference to L-T_4, other L-iodothyronines and D-T_3.[27,184,187] These findings imply that thyroid hormones bind to a structurally specific site for transport into cells.

3.6   Deiodinase Enzymes

These enzymes are responsible for the deiodination cascade shown in Figure 3. They are associated with microsomal or cellular membranes and function in most tissues of the body, although their activities in individual organs vary widely. There are two general types of deiodination reactions, 5- and 5'-deiodination, and at least three enzymes types responsible for these reactions.[188,189] These are 5-deiodinase, type I 5'-deiodinase and type II 5'-deiodinase. The two isotypes of 5'-deiodinase are

Table 3 Summary of some properties of deiodinase enzymes

Property	5-Deiodinase	Type I 5'-Deiodinase	Type II 5'-Deiodinase
Substrate preference	T_3 (sulphate) > T_4	rT_3 > T_4 > T_3	T_4 > rT_3
Molecular weight	unknown	55,000	200,000
Tissue distribution	almost every tissue	thyroid, kidney, liver, CNS	CNS, pituitary, brown adipose tissue

discriminated by differences in their substrate specificity, degree of susceptibility to inhibition by 6-n-propyl-2-thiouracil (PTU), and response to physiological perturbations.[189] A summary of some properties of the three deiodination enzymes is given in Table 3. They have been extensively reviewed by Koehrle *et al.*[189] recently and only aspects relevant to medicinal chemistry are summarised below.

3.6.1 5'-Deodination

Type I 5'-deiodinase is found in most tissues. In humans, the highest levels of activity are in the thyroid, kidney and liver. It is thought that type I 5'-deiodinase plays an important role in the production of T_3, the elimination of sulphate-conjugated T_3 and $3,3'$-T_2, and the disposal of rT_3 under normal conditions.[189] Purification has been hampered because most detergents used to isolate the enzyme also inhibit catalytic activity. Type I substrates include iodothyronines with a free or sulphate-conjugated phenolic hydroxyl, but the glucuronide conjugate of the 4'-hydroxyl cannot be 5'-deiodinated. The order of affinity of the naturally occurring iodothyronines for rat liver microsomal fractions is rT_3 > $3',5'$-T_2 > $3,3'$-T_2 > $3'$-T_1 > T_4 > T_3,[189] with rT_3 being the preferred substrate. Sulphation of the 4'-hydroxyl of T_3 and $3,3'$-T_2, but not of rT_3, markedly increases the catalytic efficiency.[189,190] Type I 5'-deiodinase is completely blocked by micromolar concentrations of the antithyroid drug PTU.

The essential features of iodothyronine substrates include:

(i)   a negatively charged sidechain,
(ii) a negatively charged substituent in the 4'-position,
(iii) substituents with the ionic diameter of iodine in the 3- or 3- and 5-positions of the inner (tyrosyl) ring,
(iv) 3'- and 5'- or 3'-iodine substituents.[189]

Type I 5'-deiodinase does not catalyse dehalogenation of brominated or chlorinated iodothyronines. It shows an absolute preference for the L-enantiomers of iodothyronines, with both D-T_4 and D-T_3 undergoing predominantly 5-deiodination.[191] Metabolites such as tetrac and triac are type I substrates with higher affinities than those of the parent iodothyronine.[191] On the other hand, decarboxylated analogues have no affinity for type I deiodinase. Molecular modelling studies suggest that the iodothyronine site of the type I deiodinase closely resembles the iodothyronine site of TTR, differing considerably from other thyroid-hormone binding sites.[191,192]

Type II 5'-deiodinase is restricted to fewer tissues than type I 5'-deiodinase and appears to generate T_3 for local use. In euthyroid rats, the highest activities occur in the CNS, pituitary gland, placenta and brown adipose tissue.[189] Like type I 5'-deiodinase, type II 5'-deiodinase is an integral component of cellular membranes, and soluble active enzyme preparations can only be obtained with ionic and some non-ionic detergents. Type II 5'-deiodinase has a preference for T_4 over rT_3 but both iodothyronines act as competitive ligands *in vivo*. No data are available on the stereoselectivity or the affinity of type II 5'-deiodinase for other iodothyronines or their analogues.

3.6.2 5-Deiodination

5-deiodination reactions are major routes for degradation of T_4, T_3, the diiodothyronines and the monoiodothyronines. The metabolites of T_4 generated by 5-deiodination reactions are devoid of calorigenic or thyromimetic activity and this reaction has therefore been called a *bioinactivating pathway*.[189] It is critical in the recycling of iodide and also yields rT_3, which is a competitive substrate for 5'-deiodination of T_4, but less is known about 5-deiodination than 5'-deiodination.

5-Deiodination occurs in almost every tissue except the anterior pituitary, and high levels of activity are present in liver, brain and placenta. 5-deiodinase has not been fully characterised, but it appears that several isoenzymes exist. Their subcellular location has not been established due to enzyme instability, but they appear to be membrane-bound. No clear-cut structure-activity relationships have been reported for 5-deiodinase. *In vitro*, T_4, T_3, 3,3'-T_2 and 3,5-T_2 all have similar K_m values. Because of the rapid 5'-deiodination of rT_3, no reliable data have been reported for the apparent K_m and V_{max} values of the 5-deiodination of T_4.[189]

3.7 Other Cellular Interaction Sites

Several other physiological sites which interact with T_4 and T_3 have been identified. These include cytoplasmic, endoplasmic reticulum and mitochondrial binding sites.[184,193]

3.7.1 Cytoplasmic binding sites

Thyroid hormones interact with soluble cytoplasmic proteins in a variety of tissues with up to 50% of extranuclear thyroid hormone being localised in the cytoplasm.[184] The low binding affinity and high capacity to hormone suggests that these proteins are an intracellular reservoir of hormones. Unlike steroid hormones, thyroid hormone action at the nuclear level is not facilitated by cytoplasmic binding proteins, which can actually restrict the thyroid hormone's access to the nuclear receptor sites.[184]

3.7.2   Mitochondrial thyroid-hormone binding proteins

High-affinity binding of T_3 to the inner membranes of mitochondria was first reported by Sterling and Milch.[194] Sterling *et al.*[195] subsequently showed that T_3 stimulated mitochondrial oxidative phosphorylation within two minutes of its addition *in vitro*, presumably via a direct action on the mitochondria. The thyroid-hormone binding site was then purified and shown to be similar to mitochondrial adenine translocase.[196] It has since been surmised that thyroid hormones have dual regulatory effects on mitochondria, one being a direct action through the mitochondrial receptor, and the other secondary to the effect mediated by the thyroid hormone via the nuclear receptors.[194]

3.7.3   Endoplasmic reticulum

The membrane fraction of cells includes the plasma membrane, endoplasmic reticulum and sarcoplasmic reticulum. The endoplasmic reticulum in various tissues includes a thyroid-hormone binding site, the p55 protein. This protein was originally thought to be part of the plasma membrane site but has since been localised to the endoplasmic reticulum.[197] The bovine liver protein has since been cloned and is highly homologous in structure with rat and human protein disulphide isomerase.[197] This enzyme is important in post-translational modifications of secretory and membrane-associated proteins, and the interaction of hormone and enzymes at, or in, the cell membrane may affect protein traffic in the cell. The biological roles of these interactions remain unclear.

3.8   Nuclear Receptor Site

High-affinity, limited capacity nuclear binding sites for T_3 were first identified in 1972 in rat liver and kidney nuclei,[198] and subsequently in many other tissues.[199] These sites are 50 kDa, non-histone nucleoproteins and their characterisation as true receptors, responsible for the initiation of thyroid-hormone action, is based on the limitation of biological response resulting from their full occupation, and the correlation between the *in vivo* biological response and nuclear affinity of a wide range of T_3 analogues,[200] as reviewed by Oppenheimer.[80] Structure-activity relationships of many thyroid hormone analogues have been reported, and the order of their nuclear receptor affinity and *in vivo* biological response is identical when account is taken of metabolism and distribution.[80] The receptor has an affinity for T_3 of $6 \times 10^9 \, M^{-1}$ and a ten-fold lower affinity for T_4.[201] The relative analogue-binding hierarchy for the nuclear receptor is triac > 3'-isopropyl 3,5-diiodothyronine > T_3 > T_4 > reverse T_3 > diiodothyronine > mono-iodothyronine.[80] The binding order of these analogues to receptors derived from different tissues is the same.

   Genes coding for two thyroid-hormone receptors have been identified in the human genome.[202,203] Known as $T_3R\alpha$ and $T_3R\beta$, these receptors were identified by their structural similarity to the steroid-hormone nuclear receptors. Receptors for retinoic acid, vitamin D_3 and several others of unknown function also belong to the steroid/thyroid-hormone superfamily.[204,205] It is interesting that, like the thyroid hormones, the steroid hormones and retinoic acid are involved in the regulation of body morphogenesis and homeostasis.

Upon hormone binding, the receptor protein undergoes a conformational change, evidenced by a difference in partition coefficient and circular dichroism spectroscopy between the protein itself and the protein-hormone complex.[206,207] Only after hormone binding occurs can the receptor protein bind to DNA. This occurs at DNA sites[208] flanking genes which code for specific proteins, for example growth hormone, α_{2u}-globulin and malic enzyme, and the production of these proteins increases as a result.[80] There are other protein and enzyme systems, however, for which thyroid hormone attenuates or inhibits transcription.[205] Virtually all thyroid-hormone-responsive genes (eg, rat growth hormone, thyrotropin, S14, malic enzyme, and the α- and β-myosin heavy-chain genes) are regulated in a cell-specific or tissue-specific fashion.[209]

Because of their importance in endocrinology and developmental biology, these nuclear receptors have been vigorously studied in recent years. As the cDNA for the glucocorticoid receptor was the first isolated, cloned and expressed,[210,211] this receptor is probably the best characterised member of the nuclear receptor super-family. Three-dimensional structures of the DNA-binding domains of both the glucocorticoid and oestrogen receptor have been determined by NMR,[212,213] and a crystallographic study of the glucocorticoid receptor DNA-binding domain complexed with a glucocorticoid response element (GRE) has also been reported.[214] The X-ray study shows two protein molecules bound to a palindromic sequence of DNA within the GRE. The proteins bind to DNA as a dimer, with each subunit interacting with successive major grooves of the DNA. Each of the DNA binding domains has two zinc coordinating sites where the Zn ions are each tetrahedrally coordinated by four cysteines to stabilise two peptide loops.[215] The residues coordinating the Zn and supporting the fold of the domain are conserved throughout the family. Like the glucocorticoid receptor, the thyroid-hormone receptor binds to DNA as a dimer. The substantial homology between the DNA binding domains of the two proteins means that this structure should also be a good model for the interaction of the thyroid-hormone receptor with DNA.

Site-directed mutagenesis and deletion mapping studies have shown that all of the nuclear hormone receptors can be dissected into various functional domains.[204] This common structure within the receptor superfamily is shown in Figure 13. Transcription signals reside on the A/B domain (or maximum activity region), which varies considerably in size among the receptor family. The highly conserved C-domain is essential, being sequence-specific for DNA binding. The D-region is a small linker region which also varies considerably in size among the family. The large E-region is responsible for hormone binding. Although this ligand-binding domain of the receptor proteins has a lower sequence homology than the DNA binding domain, its function is retained throughout the superfamily. The dissimilarity in this region probably reflects the structural diversity of the ligands for the different receptor proteins. It has proved possible, using molecular biology, to produce biologically active hybrid receptors which consist of the ligand-binding domain from

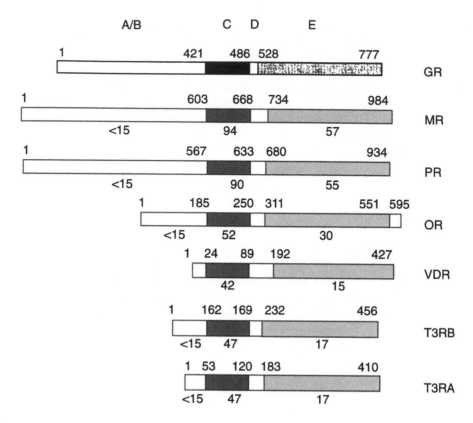

Figure 13 Comparison of the amino-acid sequences of the steroid-hormone receptor superfamily. Numbers above each bar relate to the amino-acid sequence; numbers below are the percentage homology to the gluco-corticoid receptor. The domains identified are the transcription region (A/B), the highly conserved DNA region (C), the linkage sequence (D) and the hormone-binding region (E). The code for receptor naming is: GR – glucocorticoid receptor, MR – mineralocorticoid receptor, PR – progesterone receptor, OR – oestrogen receptor, VDR – vitamin D_3 receptor, $T_3R\alpha$ and $T_3R\beta$ – the thyroid-hormone receptors. Figure adapted from Evans.[204]

one member of the family combined with the DNA-binding region of another.[216] These proteins respond to the presence of one hormone but produce the response of another.[204]

4.   INTERACTIONS OF NON-THYROIDAL COMPOUNDS WITH THYROID-BINDING SITES

4.1   Introduction

Many non-thyroidal compounds interact with the various thyroid-hormone binding sites. These include drugs used to treat non-thyroid diseases as well as a range of other exogenous compounds. Although the majority of these interactions are relevant only at high physiological concentrations of the various competitors, these compounds play a role in understanding the medicinal chemistry of the various hormone binding sites because many of the interactions are highly specific. There are, of course, other compounds which bind non-specifically to thyroid hormone binding sites or alter the physiological balance of the thyroid hormone system indirectly.

Ways in which pharmacological agents may affect thyroid hormone metabolism include:[15]

(i) altering the serum concentrations of thyroid-hormone binding proteins, their binding affinity for the thyroid hormones or their metabolism;
(ii) inducing changes which alter the binding of thyroid hormones to the transport proteins;
(iii) altering the cellular uptake of the thyroid hormones or their intracellular metabolism.

Because the focus of this review is on potential medicinal chemistry applications, only specific interactions of ligands with the respective thyroid-hormone binding sites, in particular those competitive interactions which have been characterised by *in vitro* binding assays, are described.

4.2 Polychlorinated Biphenyls, Dibenzodioxins and Dibenzofuran Analogues

4.2.1 TTR

The binding of polychlorinated biphenyls (PCBs), dibenzodioxins and dibenzofurans to isolated TTR has been studied by McKinney and coworkers.[46–49] This work was initiated on the hypothesis that TTR may be useful as a model for the cytosolic dioxan or Ah receptor, which has been implicated in the toxic action of these compounds. Molecular modelling of 2,3,7,8-tetrachlorodibenzo-p-dioxin (TCDD) and 2,3,7,8-tetrachlorodibenzofuran (TCDF) binding to the X-ray crystal coordinates of TTR indicated that the hypothesis was plausible. Subsequent measurement of the relative binding affinities of the adipamide derivatives of TCDD and TCDF showed that these molecules have a moderate affinity for TTR.[46] Testing of selected PCBs revealed that several also have very high affinities for TTR[47] (Figure 14).

The most potent compound, 3,5-dichloro-4-hydroxybiphenyl, has an affinity for TTR 8.5 times that of T_4. Interestingly, biphenyl itself shows no affinity for TTR, but addition of chlorine and hydroxy groups at specific positions increases binding markedly. Maximum affinity results from chlorine substituents at the 3- and 5-positions and hydroxyl substitution at the 4-position. Substitution at positions 2 and 6 appears to reduce binding, as shown by 2,2',6,6'-tetrachlorobiphenyl, which has no detectable binding to TTR.[47] Theoretical differential free energies for the formation of a complex between TTR and several PCBs, dibenzodioxins and dibenzofurans were calculated using the AMBER forcefield and found to correctly predict strong, intermediate and non-binders.[48]

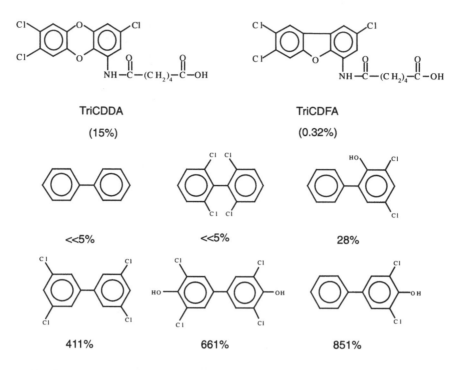

Figure 14 Structures and binding affinities for TTR (relative to T_4, 100%) of the adipamide derivatives of TCDD and TCDF, and several PCBs.[46,47]

Subsequent *in vitro* studies showed that only hydroxylated metabolites of 3,4,3′,4′-tetrachlorobiphenyl (TCB), but not the parent compound, could compete with T_4 for binding to TTR.[50] Similar results were obtained for halogenated benzenes and their phenolic metabolites,[51,52] where it was shown that a hydroxyl group on the parent compound was essential for TTR binding. A more extensive range of polychlorinated hydroxybiphenyls, dibenzo-p-dioxins and dibenzofurans was subsequently tested for competitive binding of T_4 with isolated TTR,[53] and some representative PCBs reported in that study are shown in Figure 15.

The most potent of the dioxan and furan derivatives (relative to T_4, 100%) were 3-hydroxy-2,6,7,8-tetrachlorodibenzofuran (450%) and 2-hydroxy-1,3,7,8-tetrachlorodibenzo-p-dioxin (437%). While there were several differences in the binding results of Rickenbacher *et al.*[47] and Lans *et al.*[53] for the PCB, dioxan and furan analogues assayed, overall, the structural requirements for competitive binding of the three classes of aromatic analogues appears to resemble those observed for the natural TTR ligand, T_4,.

4.2.2 *Nuclear receptor*

The binding of PCBs to a nuclear extract was also investigated by McKinney *et al.*[49] It is difficult, however, to relate these studies to other investigations of binding to

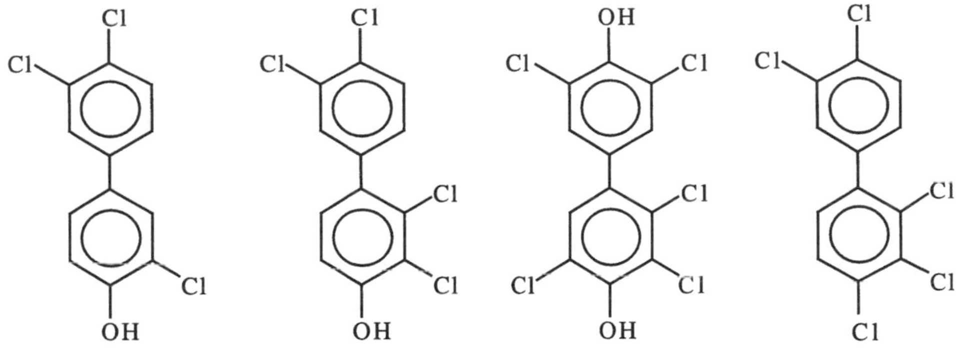

840%   1020%   1360%   <1%

Figure 15   Selected PCBs and their binding affinities for TTR, relative to T_4 (100%).[53]

the thyroid-hormone nuclear receptor because the assay used T_4 as a reference compound. Most other studies of nuclear receptor binding use T_3 as the reference ligand. Overall however, this study identified specific binding interactions of the PCBs and related compounds to thyroxine-specific binding sites in rat liver nuclear extracts. The structure-binding relationship for this series of compounds was found to be similar to that reported at the T_4-TTR site.[47]

4.3   Flavonoids

The flavonoids are a group of natural and semi-synthetic compounds which possess a wide range of pharmacodynamic properties. They are present in the extracts from plants such as *Lycopus virginicus*, *Melissa officinale* and *Lithospermum officinale*, which have been used in folk medicine for the treatment of thyroid hormone related diseases.[54] A broad survey of the active constituents of these extracts found that several phenolic metabolites, especially the flavonoids, were able to inhibit rat liver microsomal iodothyronine deiodinase (type I). This class of compounds includes the chalcones, aurones and flavones, shown in Figure 16.

4.3.1   TTR

SAR data of thyroid hormone analogues for the rat liver microsomal type I iodothyronine deiodinase shows strong correlations with the binding requirements of hormone analogues to serum TTR.[54] A model was therefore developed using the crystal structure of TTR which suggested that functionally specific ligands for inhibiting peripheral thyroid-hormone metabolism were possible.[55] The relative binding affinities of an aurone derivative (4',4,6-trihydroxyaurone, Figure 16) and phloretin were assayed and the compounds were found to have comparable binding (relative to T_4) at TTR and the deiodinase sites.[56]

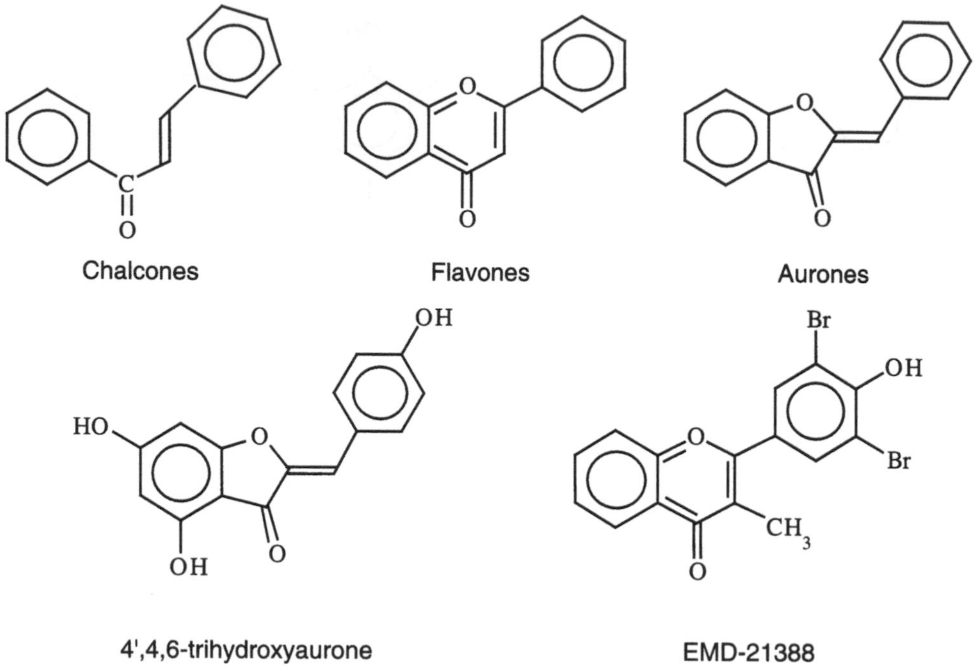

Figure 16   Structures of the classes of flavonoids which inhibit T_4 binding and two of the more potent members of these classes.

However, a study of the *in vitro* competitive binding of several clinical drugs for the T_4 site of isolated human TTR, identified several compounds (eg, amiodarone, ipodate and iopanoic acid) which are potent inhibitors of type I 5'-deiodinase,[15,57] yet have very low potency as competitors for T_4 binding to TTR.[17] This finding suggests that parallel inhibition of T_4 binding to TTR and type I 5'-deiodinase may be a particular feature of the flavonoids, rather than a general reflection of homology between these binding sites.

By systematically varying substitution around the flavonoid nucleus and comparing the derived structures with the thyroid hormones, a synthetic flavonoid (EMD-21388, Figure 16) was designed which has a binding affinity 2.5 times that of T_4 binding to TTR.[58]

4.3.2 Deiodinase enzymes

Several natural, polycyclic compounds (for example, dicoumarol, the secondary plant metabolites rosmarinic acid, flavonoids and auronoids) competitively inhibit type I 5'-deiodinase and 5-deiodinase activity. Those flavonoids and auronoids with somewhat rigid conformations resembling that of T_4 are very active deiodinase inhibitors, whereas chalcones, with a similar molecular structure but with a flexible conformation, do not interfere with deiodinase activity.[54,59,60] The most potent naturally occurring compounds are 4,4′,6-trihydroxyaurone and 2,4,4′,6-tetrahydroxyaurone. Chemical modification of the 4′-hydroxyl group reduces their activity, as does hydrogenation of the exocyclic double bond.[59] The inhibitory potency was increased by introduction of an iodine atom in the ortho-position of 4,4′,6-trihydroxyaurone. The structure-activity relationships and of naturally occurring or synthetic flavonoids and aurones were similar in rat-liver microsomal membranes and intact hepatocytes.[60]

4.4 Therapeutic Agents

There are many drugs used to treat a range of non-thyroidal illnesses which affect thyroid hormone binding sites.[15–45] Many of these interactions were first discovered as a result of clinical thyroid hormone assays which gave unclear or misleading results for patients taking these drugs. Examples of discoveries in this way include fenclofenac[19,20] and furosemide.[21,22] As a result of these findings, a number of studies of the binding of these compounds to isolated thyroid hormone binding proteins were initiated. In this section we examine binding data for selected examples of commonly used drugs which displace thyroid hormones from their binding sites.

The antiarrhythmic agent amiodarone is examined separately in Section 4.5, as unlike many of the drugs which displace hormone from binding sites, amiodarone also displays thyroid hormone antagonist properties.

4.4.1 TTR and TBG

An *in vitro* study of drug binding to TTR and TBG identified about 25 compounds which displace T_4 from these binding sites to various extents.[17] The more potent compounds are shown in Figure 17 and their relative binding affinities are in Table 4.

The most potent compound at the TTR binding site is flufenamic acid, which has an affinity about twice that of T_4 itself, while the closely related compounds mefenamic acid and meclofenamic acid have lower affinities (*ca.* 20–26% of T_4 binding). The hierarchy of drug binding to TBG differs markedly from that found at TTR, as is particularly evident for the compounds furosemide and milrinone. Furosemide is the most potent binder at TBG but shows no displacement of T_4 from TTR at a concentration as high as 10^{-4} M. Conversely, milrinone has moderate affinity for TTR but extremely low affinity for TBG. These trends are consistent with earlier *in vitro* studies for these two drugs.[21,23]

The compounds which bind to TTR and TBG belong to several structural classes and have different drug actions. Flufenamic acid, mefenamic acid, diflunisal, meclofenamic acid, diclofenac, fenclofenac, indomethacin and sulindac are non-steroidal anti-inflammatory drugs (NSAIDs), milrinone is a positive cardiac inotropic agent, and ethacrynic acid and furosemide are diuretics. There appear to be broad structural similarities for these compounds: many of the drugs are acidic; most contain two aromatic moieties linked by a rotatable linkage; and the aromatic rings are connected by a single bond (eg, diflunisal) or by two bonds and a bridging atom

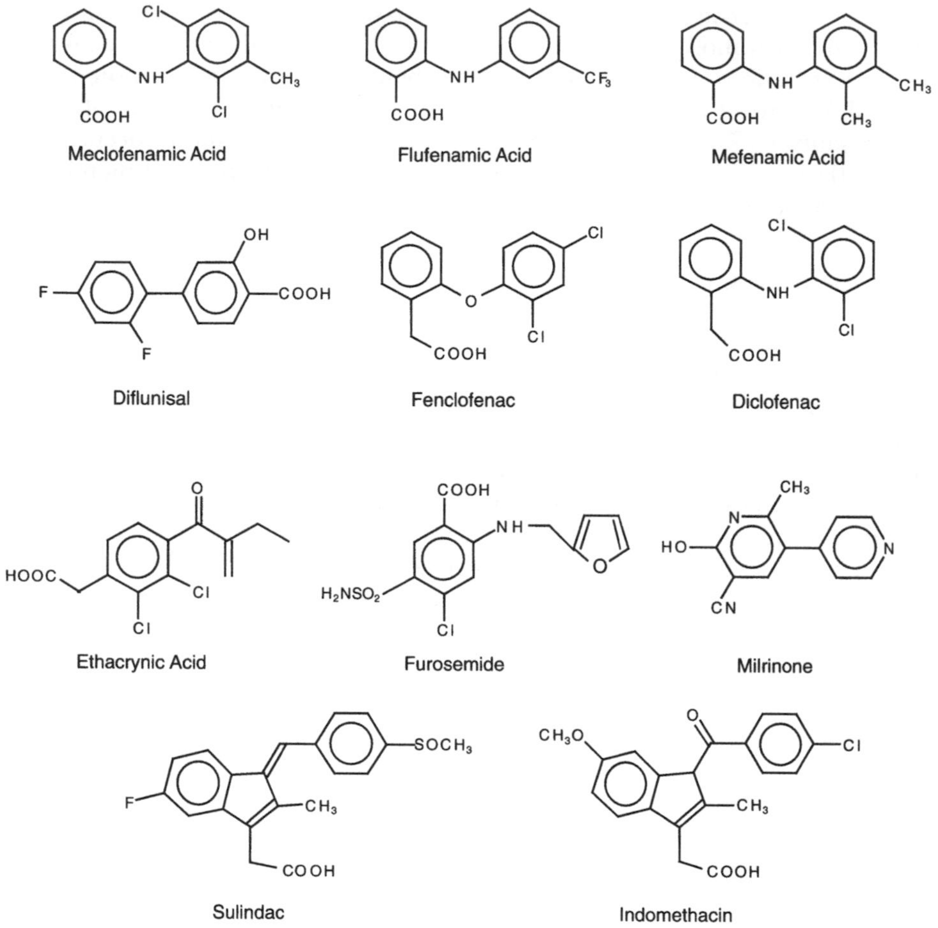

Figure 17   Drugs which displace T_4 from the thyroid-hormone transport proteins TTR and TBG.[17]

(eg, meclofenamic acid, sulindac). The crystal structure of milrinone complexed with TTR has recently been determined[24] and shows that milrinone binds deep within the central cavity of TTR, with the hydroxyl group innermost.

4.4.2   Cell-uptake site

Cell membrane binding sites for T_3, which may represent part of the T_3 uptake mechanism, have been identified, but these sites have not been completely characterised and at present their roles are uncertain.[184] Several groups have identified compounds which inhibit T_3 uptake at cell membrane sites at concentrations ranging from 1 to 100 μM. These agents have a diverse range of chemical structures and biological properties and include inhibitors of cytoskeleton function, radio-

Table 4 Relative affinities of selected drugs for TTR and TBG[17]

Compound	Affinity for TTR (%)	Affinity for TBG (%)
T_4	100	100
Flufenamic acid	174.6	6.3×10^{-3}
Mefenamic acid	26.5	2.7×10^{-2}
Diflunisal	22.4	7.1×10^{-4}
Meclofenamic acid	20.4	8.4×10^{-2}
Milrinone	4.8	$<3.2 \times 10^{-5}$
Ethacrynic acid	2.1	7.6×10^{-4}
Diclofenac	2.0	3.9×10^{-2}
Indomethacin	2.0	8.4×10^{-3}
Sulindac	1.8	9.1×10^{-4}
Fenclofenac	0.8	1.5×10^{-2}
Tolmetin	0.004	7.6×10^{-5}
Furosemide	<0.024	0.11

graphic agents, compounds normally taken up by the non-bile acid cholephil system (bromosulphopthalein, lopanoic acid, indocyanine green), benzodiazepines[25,26] and some non-steroidal anti-inflammatory drugs.[27] The thyroid-hormone uptake system is also inhibited at high concentrations by phenylalanine and tryptophan, but not by other, non-aromatic amino acids.[28] Several calcium channel blockers (dihydropyridines), calmodulin antagonists and β-blockers have also been shown to inhibit T_3 uptake.[29,30]

Although the benzodiazepines are better known as anxiolytics, studies by Kragie and Doyle[25,26] show that many are also inhibitors of thyroid-hormone uptake. The most potent of these compounds, triazolam and lormetazepam (Figure 18) inhibit T_3 accumulation in human liver cell lines with K_i values of 4.1×10^{-8} M and 4.4×10^{-8} M, respectively. Analysis of structure-activity relationships for this set of compounds found that several substituents contributed favourably to binding. Specifically, binding was enhanced by methylation at N1, the presence of a carbonyl group at position 2 and by fluorine or chlorine substituents at the 2'-position. Although it could not be determined whether the benzodiazepines interacted directly with the transport site, it was found that little of the added benzodiazepines entered the cells.[25] On this basis, it was suggested that the benzodiazepines do not act as false substrates for the putative transport protein.

Binding of some non-steroidal anti-inflammatory drugs including mefanamic acid, meclofenamic acid and flufenamic acids to cell uptake sites have also been reported.[31] A discussion of their binding data is given in Section 4.6, where they form part of a series of phenylanthranilic acid derivatives synthesized to investigate the mechanism of thyroid-hormone uptake and obtain further evidence for the existence of a thyroid-hormone membrane transport protein.

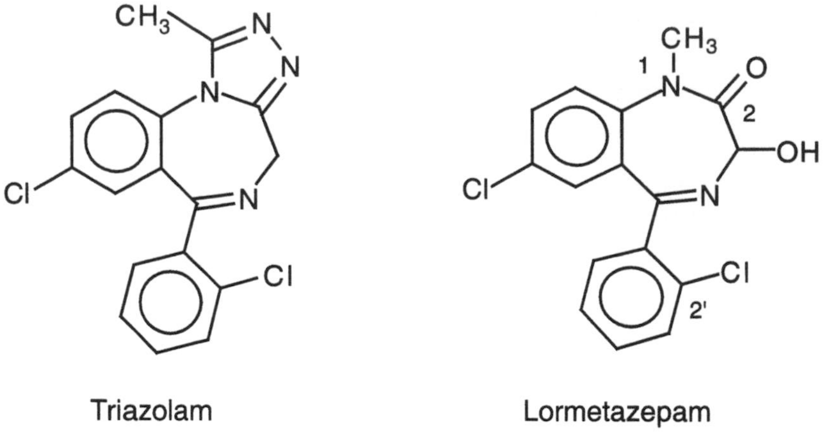

Triazolam   Lormetazepam

Figure 18   Structures of triazolam and lormetazepam.

4.4.3 Deiodinase enzymes

Deiodination of the thyroid hormones is inhibited by a variety of drugs which may be divided into two classes: those which are iodinated and those which are not. These agents generally result in a decrease in serum T_3 levels. The general difference between these two classes is that the non-iodinated drugs do not usually cause changes in serum T_4 levels, while the iodinated ones tend to increase serum T_4.[15]

The most important drugs in the non-iodinated class are propylthiouracil, propranolol and dexamethasone. In several species, including humans, propyl-thiouracil is the only known specific inhibitor of type I 5'-deiodinase.[189] The effects of both dexamethasone and propranolol are not well-characterised and the precise mechanism by which they reduce serum T_3 levels is not yet clear.[15]

The iodinated drugs which reduce monodeidodination of the thyroid hormones include the iodinated radiographic contrast agents and amiodarone. The overall physiological effects of these compounds are complex and they cannot all be attributed solely to inhibiting 5'-monodeiodination.[15] The effects of amiodarone on the thyroid-hormone system are more fully described in the following section.

4.5   Amiodarone

Amiodarone (Figure 19) is an important antiarrhythmic agent whose mechanism of action is not yet fully understood, although it has long been known to have side effects that involve the thyroid hormone system.[32-36]

Interest in the interactions of amiodarone at the various thyroid hormone binding sites increased recently, with the proposal that it may be the first thyroid hormone antagonist to be identified.[37] This suggestion arose as a consequence of amiodarone's ability to inhibit T_3-induced increases in growth hormone mRNA levels in cultured rat pituitary GC cells.[37] There are however, several problems with using amiodarone

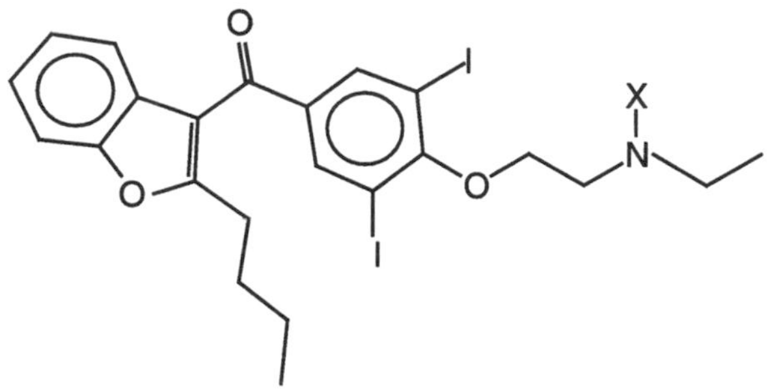

Figure 19 Amiodarone (X = Et) and desethylamiodarone (X = H).

clinically as an antagonist for thyroid hormone action.[36,37] Its role is thus that of a lead compound for the design of more specific antagonists rather than as an antithyroid drug itself.

4.5.1 TTR and TBG

An *in vitro* binding assay for competitive binding of amiodarone to isolated TTR yielded a relative binding affinity of <0.05% relative to T_4.[17] The same study also reported a relative binding of <0.00003% to TBG relative to T_4 (100%). These data show that amiodarone interacts weakly, if at all, with these two T_4-binding sites in plasma.

4.5.2 Nuclear receptor

Disruption of thyroid hormone levels by amiodarone is common and, in severe cases, can result in induction of either hypothyroidism or thyrotoxicosis.[35] This can be partly explained by amiodarone's ability to bind to the thyroid-hormone nuclear receptor and competitively displace T_3. The binding affinity for this site is low and has been variously reported as 'modest',[38] ranging from 6.3×10^6 M^{-1} to 7.4×10^5 M^{-1} in different tissues.[37] The major metabolite of amiodarone, desethylamiodarone, also binds to the receptor,[38] but with ten-fold greater affinity.

Recently, Chalmers *et al.*[39] undertook a conformational analysis of amiodarone using molecular mechanics and semi-empirical molecular orbital methods. The molecular mechanics calculations revealed that the low-energy conformers of amiodarone can be divided into four distinct groups. These depend on the orientation of the butyl sidechain and carbonyl group in relation to the benzofuran ring. The minimum-energy structures were then fitted to an extension of Jorgensen and Andrea's thyroid-hormone receptor model.[78] By superimposition of functional groups with similar steric and electronic properties, four possible binding modes were devised. This study was different from two previous reports which speculated on the mode of amiodarone binding to the nuclear receptor,[37,40] as the low-energy structures were derived from systematic conformational searches and key functional groups

Table 5   TTR and cell-uptake binding data for phenylanthranilic acids[31,41]

Compound	% T_4 at TTR site	% T_3 at cell-uptake site[a]
VCP-2	300	51.4
VCP-3	480	51.0
VCP-5	459	39.5
VCP-6	523	60.6
VCP-16	290	—
VCP-20	439	40.0
VCP-27	205	—

[a] Inhibition of ^{125}I[T_3] by H4 rat hepatoma cells by 0.1 mM solutions of the compounds listed.

were used to guide the superimpositions developed. Although none of the four models[39] could be discounted, the preferred one has the butylbenzofuran moiety of amiodarone superimposed on the inner ring of T_3 and the diiodophenyl group fitted on the outer ring. This mode of superimposition provides the best overlap of molecular volumes, and gives the closest match of functional groups for the two molecules.

4.6   Phenylanthranilic Acid Analogues

Of the therapeutic agents reported by Munro et al.[17] as interacting with TTR, three of the four compounds with the greatest affinity for TTR are derivatives of phenylanthranilic acid. These compounds were subsequently used as starting structures for the development of novel thyroid-hormone analogues. The structural relationship between phenylanthranilic acids and the thyroid hormones was investigated and three pharmacophore models for the binding of phenylan-thranilic acid to TTR developed. To test the various binding models, 33 analogues were synthesized.[41] Relative binding affinities at TTR and the cell-uptake site for representative compounds are given in Table 5, and the structures are in Figure 20.

4.6.1   TTR

All the analogues shown above bind to TTR more strongly than T_4 and some have a substantially higher affinity than tetrac. Removal of the COOH group from the phenylanthranilic acid framework results in a total loss of competitive binding for the T_4-TTR site. The most potent compounds have a 3-chloro or 3-iodo substituent or 3,5-dichloro, or 3,5-bis(trifluoromethyl) substituents. Although a direct comparison has not been made, VCP-6 appears to be approximately twice as potent as the flavonoid EMD-23188 referred to in Section 4.3.1.

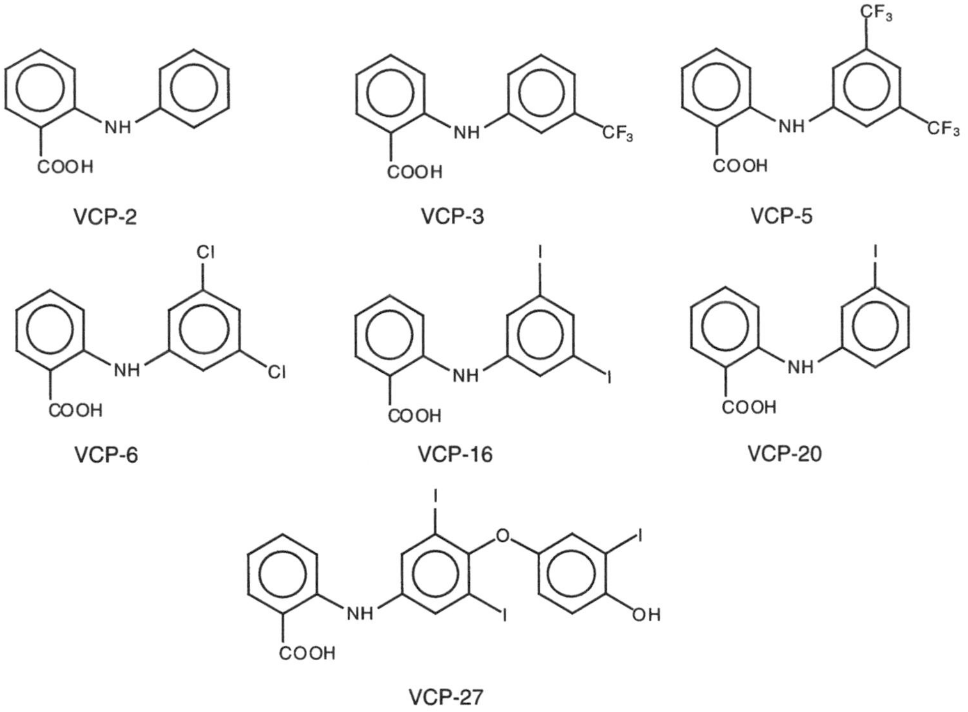

Figure 20 Structures of the analogues shown in Table 5.

4.6.2 Cell-uptake site

The compounds listed in Table 5 are representative of the series tested at the cell uptake site.[31] From the full series of compounds, it was deduced that the carboxylic acid group is not an absolute prerequisite for inhibition, unlike for binding at the T_4-TTR site. The binding data for all analogues were analysed for quantitative structure-activity relationships, resulting in the derivation of an equation to describe affinity at the cell-uptake site:

$$\text{logit}(\%\ I) = 3.92\ c\log P - 0.346\ (c\log P)^2 - 11.0$$

In this equation, $c\log P$ is the logarithm of the calculated octanol-water partition coefficient of the un-ionized molecule and $\text{logit}(I\%)$ is a logarithmically transformed measure of the uptake inhibition. Lipophilicity is the parameter of prime importance in the inhibition of T_3 uptake by the phenylanthranilic acids. It is interesting to note that the parabolic relationship between $c\log P$ and inhibition is independent of the position of substitution and the size of the substituents on the second ring (ie, the ring not bearing the carboxylic acid group); this suggests that steric interactions do not play a large part in the interactions. The finding that hydrophobicity plays an important role in the inhibition of thyroid-hormone uptake reflects the hydrophobic nature of the thyroid hormones themselves, and may represent a basic requirement of inhibitors/binders at the T_3 cell-uptake site.

4.7 Nonesterified Fatty Acids

The nonesterified fatty acids (NEFAs) have been associated with changing the levels of circulating thyroid-hormone,[64] particularly following the administration of heparin.[15] These effects have recently been characterised and are described below.

4.7.1 Plasma binding proteins: TTR, TBG and albumin

It has been suggested that NEFAs may inhibit protein binding of circulating thyroid-hormones, particularly in some critically ill patients.[65] This report is at variance with data suggesting that physiological NEFA concentrations have only minor effects on serum binding of T_4.[66,67] In order to clarify the effect of NEFAs on circulating levels of thyroid hormone, the effect of various long-chain NEFAs on T_4 binding in whole serum and diluted serum was examined.[68] Oleic acid was reported to be the most potent of the NEFA's studied, with both T_3 and T_4 levels increasing in serum. Results for the other NEFAs showed that palmitic and stearic acids had no effect, but addition of 15 mM linoleic, linolenic and arachidonic acids increased the free T_4 fraction by $11 \pm 10\%$, $23 \pm 16\%$ and $26 \pm 5\%$, respectively.[68]

Recently the competitive binding of several NEFAs was studied *in vitro* relative to T_4 binding to TBG[69] and TTR[70], with different hierarchies for the competitors emerging at the two sites. Only arachidonic acid was able to displace $[^{125}I]$-T_4 from TTR, albeit weakly (relative binding affinity of 0.49%). Inhibition of binding to TBG was more potent, as shown in Table 6.[69] The saturated NEFAs, lauric (12:0), myristic (14:0), palmitic (16:0) and stearic (18:0) acids had no effect on $[^{125}I]$-T_4 binding to TBG.[69]

Table 6 Relative affinities of NEFA for T_4-TBG binding[69]

NEFA	Ligand concentration at 50% bound/free of labelled T_4 (M)	Relative affinity
(T_4)	5.0×10^{-11}	100
Palmitoleic acid (16:1)	3.2×10^{-6}	0.0016
Oleic acid (18:1)	1.0×10^{-5}	0.0005
Linoleic acid (18:2)	1.6×10^{-6}	0.0032
Linolenic acid (18:3)	2.0×10^{-6}	0.0025
Arachidonic acid (20:4)	1.0×10^{-6}	0.005
Eicosapentaenoic acid (20:5)	1.6×10^{-6}	0.0032
Docosahexaenoic acid (22:6)	1.0×10^{-6}	0.005

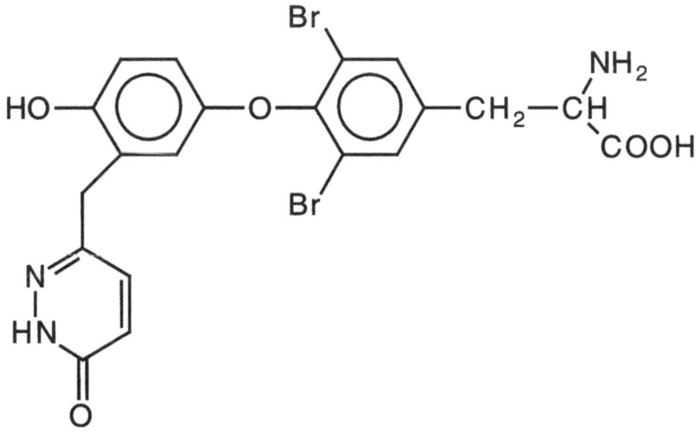

Figure 21   Structure of SK&F-L94901.

4.8   Selective Thyromimetics

It is well established that the thyroid hormones lower plasma cholesterol, probably via increased liver metabolism. However, due to the adverse consequences of some of their other actions, particularly on the heart, the hormones cannot be used to treat hypercholesterolaemia. In the mid 1980s, a group at Smith Kline & French reported 3'-substituted 3,5-diiodo-L-thyronines which appeared to interact selectively with thyroid-hormone receptors in the liver but not in the heart.[12] The binding of the most interesting of this class of molecules, SK&F-L94901 (structure shown in Figure 21), was subsequently assayed at a number of thyroid binding sites. This class of compounds is included here as another example of compounds which affect the thyroid system but are not specifically targeted at the treatment of thyroid disease.

4.8.1   Plasma sites: TTR, TBG and HSA

Barlow *et al.*[74] studied *in vitro* the interactions of SK&F-L94901 with the primary plasma binding proteins of thyroid hormones. Relative to T_4, the affinity of the analogue for TBG was 0.0035%, for TTR 1.66% (similar to T_3) and for albumin 1.26%. Although the interaction of SK&F-L94901 with other serum proteins cannot be ruled out, these data suggest that at equimolar concentrations of T_3 and L-94901, tissues will be exposed to a higher concentration of free SK&F-L94901 than free T_3.[74]

4.8.2   Nuclear sites

The analogue SK&F-L94901 was developed from analysis of many thyronine analogues. Initially, a group of 29 3'-substituted 3,5-diiodo-L-thyronines was synthesised to explore the size and shape of the 3'-substituent pocket of the thyroid hormone nuclear receptor. By analysing the SAR of these analogues, quantified by their ability to induce GPDH (mitochondrial 3-phosphoglycerate oxidoreductase) in heart and liver of hypothyroid rats, it was shown that the 3'-substituent binding pocket on the thyroid-hormone receptor is lipophilic and substantially larger than the natural iodo-substituent. Furthermore, the free 4'-hydroxyl is necessary for receptor binding. Analysis of the correlations observed for full agonists revealed some tissue differences. Activity in the heart was readily explained by nuclear binding alone, whereas an additional 3'-substituent coefficient was required to obtain a similarly significant correlation in liver. Since these compounds do not differentiate between heart and liver receptors *in vitro*, it was suggested that the tissue differences *in vivo* result from pharmacokinetic effects controlling analogue delivery to the nuclear receptors.[13]

Following the initial studies, a second line of analogues based on 3'-(arylmethyl)-3,5-dihalo-L-thyronines was developed.[14] Members of this group were shown to be hypocholesterolaemic and have high thyromimetic activity in the liver and little or no direct action on the heart. Although the novel compounds showed no difference in relative *in vitro* affinities at the two tissue sites, the observed selective thyromimetic activities were consistent with *in vivo* binding data.[14] SAR of the 3,5-diiodo members of the 3'-arylmethyl series showed that hormonal activity is positively correlated with 3'-substituted hydrophobicity, while liver selectivity is negatively correlated with this property.[14] However, the 3,5-dibromo and 3,5-dichloro analogues of the same compounds were shown to be 4–8 times more potent as liver thyromimetics, while also displaying selective *in vivo* liver nuclear receptor binding and were orally active (in rats). Therefore, while it was established that the 3'-arylmethyl groups interact with the same size-limited, hydrophobic pocket for 3'-halogen and alkyl groups in non-selective thyromimetics (with increased affinity when electronegative substituents are in the para position of the 3'-arylmethyl ring), 3,5-substituents and ether link substitutions were necessary to increase potency and gain oral bioavailability. Replacement of the 3,5-iodo groups by bromine proved crucial in meeting both objectives, with SK&F-L94901 being equipotent with T_3 as a liver thyromimetic and hypocholesterolaemic agent while possessing only one-thousandth the activity of T_3 in the heart.[12] Studies with alanine sidechain replacements showed the amino group was essential for activity; however, introduction of acetic acid and butyric acid substituents resulted in retention of selectivity but reduced potency.

The binding of SK&F-L94901 to the thyroid-hormone nuclear receptor in whole human HeLa cells and with nuclear extracts from this line was investigated by Barlow *et al.*[74] In nuclei of whole cells, the analogue was 0.8% as potent as T_3, whereas in experiments with nuclear extracts, the analogue was 7.7% as potent as T_3. These results suggest that an extranuclear component may be involved in restricting access of SK&F-L94901 to the nucleus. Whether or not such mechanisms account for the observed differences in its effects on different tissues, as shown by Underwood *et al.*,[12] remains unclear.[52]

4.9 Miscellaneous

There are a number of other non-thyroidal molecules which interact with the thyroid hormone system but do not fall under any of the previous classes.

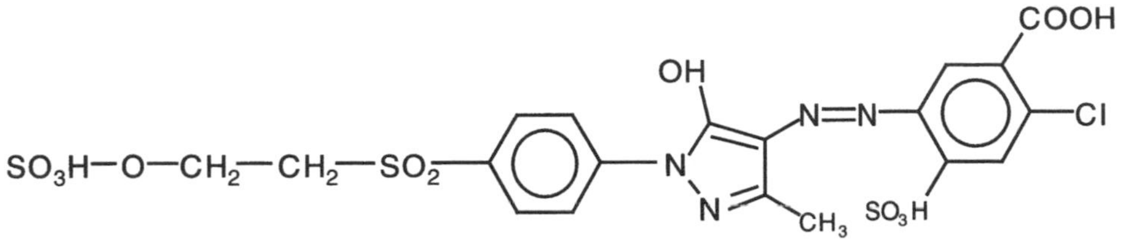

Figure 22 Structure of Remazol Yellow GGL.

The interaction of Remazol Yellow GGL (Figure 22) and several other dyes with isolated TTR has been studied by affinity phase partitioning, difference spectroscopy, and equilibrium dialysis.[63] Of those dyes studied, only Remazol Yellow GGL has a specific interaction with TTR. The complex has a ratio of Remazol Yellow: TTR of 4:1, and using Scatchard analysis, two classes of binding site were described: two with a $K_d = 3.3 \times 10^{-6}$ M, and two lower affinity states with a K_d of 258×10^{-6} M. This finding has more relevance to the isolation of TTR than direct chemical implications.

A series of technetium-99m-labelled phosphine and isocyanide cationic complexes has recently been investigated due to their ability to allow imaging of the myocardium in animals but not in humans. In particular, it was found that [^{99m}TC (DEPE)$_2$ (CNR$_2$)], where R = t-butyl isocyanide, images the myocardium in three animal species very well but remains primarily in the blood pool of humans.[73] The reason for this is that it binds almost exclusively to TTR in humans but non-specifically to other proteins in rabbits. The interaction with TTR can be blocked by sodium salicylate and appears to involve the T$_4$ binding sites.

5. CONCLUSIONS

There have been many studies of structure-activity relationships for thyroid hormones and their analogues. In particular, there are many binding data available for the plasma transport proteins and the nuclear receptors. By far the best characterised binding protein at a structural level is TTR, for which X-ray structures of a range of complexed hormones are available. With the increasing characterisation of other hormone binding proteins, the number of sites at which a molecular understanding is achieved will significantly increase over the next few years. Developments include the proposal of structural models for TBG and the characterisation of the various deiodinase enzymes and other cellular binding proteins. The cloning and expression of the thyroid hormone receptors will also lead to greater structural characterisation of this binding site.

These developments open new opportunities for the design of novel thyroid-active drugs. A potent and specific thyroid hormone antagonist is one important design target. The molecular information obtained from structural studies on the thyroid nuclear receptor will be extremely valuable for this purpose. Given the link between the thyroid system and other important physiological functions, compounds for the treatment of other, non-thyroidal, diseases may also develop from these studies.

The binding of many clinically used drugs and other exogenous compounds to hormone binding proteins has important implications in thyroid hormone physiology. While the fact that such interactions occur has been known for more than two decades, the amount of quantitative information on relative affinities and conformations of non-thyroidal ligands has increased significantly over the last five years. Thus the binding sites of thyroid-binding proteins are being increasingly understood not only by characterisation of the proteins themselves but also by reference to various probe ligands.

6. ACKNOWLEDGMENTS

Some of the studies from our laboratory reported herein were supported by a grant from the Australian Research Council. We thank David Chalmers and Chen-Fee Lim for providing unpublished data, Jackie Jarvis and Spiro Pavlopoulos for helpful comments, and Jacqui King for typing the manuscript.

7. REFERENCES

1. Oppenheimer, J.H. (1983) *The Molecular Basis of Thyroid Hormone Action.* New York: Academic Press.
2. Cody, V. (1978) Thyroid hormones: crystal structure, molecular conformation, binding, and structure-function relationships. *Rec. Prog. Horm. Res.,* **34,** 437–475.
3. Cody, V. (1980) Thyroid hormone interactions: molecular conformation, protein binding, and hormone action. *Endocr. Rev.,* **1,** 140–166.
4. Bartalena, L. (1990) Recent achievements in studies on thyroid-hormone-binding proteins. *Endocr. Rev.,* **11,** 47–64.
5. Robbins, J. (1991) Thyroid hormone transport proteins and the physiology of hormone binding. In *Werner and Ingbar's The Thyroid: A Fundamental and Clinical Text,* (eds. Braverman, L.E. and Utiger, R.D.), 6th edn, pp. 111–125. Philadelphia: J.B. Lippincott Co.
6. Jorgensen, E.C. (1978) Thyroid hormones and analogs. II. Structure-activity relationships. In *Hormonal Proteins and Peptides,* (ed. Li, C.H.), pp. 108–204. New York: Academic Press.
7. Jorgensen, E.C. (1981) Thyromimetic and antithyroid drugs. In *Burger's Medicinal Chemistry,* (ed. Wolf, M.F.), 4th edn, Part III, pp. 103–146. New York: John Wiley & Sons.
8. de la Paz, P., Burridge, J.M., Oatley, S.J. and Blake, C.C.F. (1992) Multiple modes of binding of thyroid hormones and other iodothyronines to human plasma transthyretin. In *The Design of Drugs to Macromolecular Targets,* (ed. Beddell, C.R.), pp. 119–172. Chichester: John Wiley & Sons.
9. Oatley, S.J., Blake, C.C.F., Burridge, J.M. and de La Paz, P. (1984) The structure of human serum prealbumin and the nature of its interactions with the thyroid hormones. In *X-Ray Crystallography and Drug Action,* (eds. Horn, A.S. and De Ranter, C.J.), pp. 207–221. Oxford: Clarendon Press.
10. Braverman, L.E. and Utiger, D. (Eds.) (1991) *Werner and Ingbar's The Thyroid: A Fundamental and Clinical Text,* 6th edn. Philadelphia: J.B. Lippincott Co.
11. Boyd, G.S. and Oliver, M.F. (1960) The effect of certain thyroxine analogues on the serum lipids in human subjects. *J. Endocrinol.,* **21,** 33–43.
12. Underwood, A.H., Emmett, J.C., Ellis, D., Flynn, S.B., Leeson, P.D., Benson, G.M., Novelli, R., Pearce, N.J. and Shah, V.P. (1986) A thyromimetic that decreases plasma cholesterol levels without increasing cardiac activity. *Nature,* **324,** 425–429.

13. Leeson, P.D., Ellis, D., Emmett, J.C., Shah, V.P., Showell, G.A. and Underwood, A.H. (1988) Thyroid hormone analogues: synthesis of 3-substituted 3,5-diiodo-L-thyronines and quantitative structure-activity studies of *in vitro* and *in vivo* thyromimetic activities in rat liver and heart. *J. Med. Chem.*, **31**, 37–54.

14. Leeson, P.D., Emmett, J.C., Shah, V.P., Showell, G.A., Novelli, R., Prain, H.D., Benson, M.G., Ellis, D., Pearce, N.J. and Underwood, A.H. (1989) Selective thyromimetics. Cardiac-sparing thyroid hormone analogues containing 3-arylmethyl substituents. *J. Med. Chem.*, **32**, 320–336.

15. Burger, A.G. (1991) Effects of pharmacologic agents on thyroid hormone metabolism. In *Werner and Ingbar's The Thyroid: A Fundamental and Clinical Text*, (eds. Braverman, L.E. and Utiger, R.D.), 6th edn, pp. 335–347. Philadelphia: J.B. Lippincott Co.

16. Cavalieri, R.R. and Pitt-Rivers, R. (1981) The effects of drugs on the distribution and metabolism of thyroid hormones. *Pharmacol. Rev.*, **33**, 55–80.

17. Munro, S.L.A., Lim, C.-F., Hall, J.G., Barlow, J.W., Craik, D.J., Topliss, D.J. and Stockigt, J.R. (1989) Drug competition for thyroxine binding to transthyretin (prealbumin): comparison with effects on thyroxine-binding globulin. *J. Clin. Endocrinol. Metab.*, **68**, 1141–1147.

18. Wenzel, K.W. (1981) Pharmacological interference with *in vitro* tests of thyroid function. *Metabolism*, **30**, 717–732.

19. Ratcliffe, W.A., Pearson, D.W.M. and Thomson, J.A. (1980) Fenclofenac and thyroid function tests. *Br. Med. J.*, **281**, 1282.

20. Capper, S.J., Humphrey, M.J. and Kurtz, A.B. (1981) Inhibition of thyroxine binding to serum proteins by fenclofenac and related compounds. *Clin. Chem. Acta*, **112**, 77–83.

21. Stockigt, J.R., Lim, C.-F., Barlow, J.W., Stevens, V., Topliss, D.J. and Wynne, K.N. (1984) High concentrations of furosemide inhibit serum binding of thyroxine. *J. Clin. Endocrinol. Metab.*, **59**, 62–66.

22. Stockigt, J.R., Lim, C.-F., Barlow, J.W., Wynne, K.N., Mohr, V.S., Topliss, D.J., Hamblin, P.S. and Sabto, J. (1985) Interaction of furosemide with serum thyroxine-binding sites: *in vivo* and *in vitro* studies and comparison with other inhibitors. *J. Clin. Endocrinol. Metab.*, **60**, 1025–1031.

23. Davis, P.J., Cody, V., David, F.B., Warnick, P.R., Schoenl, M. and Edwards, L. (1987) Competititon of milrinone, a non-iodinated cardiac inotropic agent, with thyroid hormone for binding sites on human serum prealbumin (TBPA). *Biochem. Pharmacol.*, **36**, 3635–3640.

24. Wojtczak, A., Luft, J. and Cody, V. (1993) Structural aspects of inotropic bipyridine binding — crystal structure determination to 1.9 Å of the human serum transthyretin-milrinone complex. *J. Biol. Chem.*, **268**, 6202–6206.

25. Kragie, L. and Doyle, D. (1992) Benzodiazepines inhibit temperature-dependent L-[^{125}I] triiodothyronine accumulation into human liver, human neuroblast and rat pituitary cell lines. *Endocrinology*, **130**, 1211–1216.

26. Kragie, L. (1992) Requisite structural characteristics for benzodiazepine inhibition of triiodothyronine uptake into a human liver cell line. *Life Sci.*, **51**, 83–94.

27. Topliss, D.J., Kolliniatis, E., Barlow, J., Lim, C.-F. and Stockigt, J.R. (1989) Uptake of 3,5,3'-triiodothyronine by cultured rat hepatoma cells is inhibitable by non-bile acid cholephils, diphenylhydantoin and non-steroidal anti-inflammatory drugs. *Endocrinology*, **124**, 980–986.

28. Topliss, D.J., Kolliniatis, E., Barlow, J.W. and Stockigt, J.R. (1991) Mutual inhibition of cellular uptake of T_3 and system T amino acids in the rat hepatoma cells. In *Proc. 65th Meeting of the American Thyroid Association, Thyroid 1991*, **1**, 79.

29. Krenning, R., Docter, R., Bernard, B., Visser, T. and Hennemann, G. (1981) Characteristics of active transport of thyroid hormone into rat hepatocytes. *Biochim. Biophys. Acta*, **676**, 314–320.

30. Topliss, D.J., Scholz, G.H., Kolliniatis, E., Barlow, J.W. and Stockigt, J.R. (1993) Influence of calmodulin antagonists and calcium channel blockers on triiodothyronine uptake by rat hepatoma and myoblast cell lines. *Metabolism*, **42**, 376–380.

31. Chalmers, D.K., Scholz, G.H., Topliss, D.J., Kolliniatis, E., Munro, S.L.A., Craik, D.J., Iskander, M.N. and Stockigt, J.R. (1993) Thyroid hormone uptake by hepatocytes: structure-activity relationships of phenylanthranilic acids with inhibitory activity. *J. Med. Chem.*, **36**, 1272–1277.

32. Burger, A., Dinichert, D., Nicod, P., Jenny, M., Lemarchand-Beraud, T. and Vallotton, M.B. (1976) Effect of amiodarone on serum triiodothyronine, reverse triiodothyronine, thyroxine and thyrotropin: a drug influencing peripheral metabolism of thyroid hormones. *J. Clin. Invest.*, **58**, 255–259.

33. Melmed, S., Nademee, K., Reed, A.W., Hendrickson, J., Singh, B.N. and Hershman, J.M. (1981) Hyperthyroxinaemia with bradycardia and normal thyrotropin secretion following chronic amiodarone administration. *J. Clin. Endocrinol. Metab.*, **83**, 597–100.

34. Singh, B.N. and Nademee, K. (1983) Amiodarone and thyroid function: clinical implications during antiarrhythmic therapy. *Am. Heart J.*, **106**, 857–869.

35. Wiersinga, W.M. and Trip, M.D. (1986) Amiodarone and thyroid hormone metabolism. *Postgrad. Med. J.*, **62**, 909–914.

36. Mason, J.W. (1987) Amiodarone. *New Engl. J. Med.*, **316**, 455–466.

37. Norman, M.F. and Lavin, T.N. (1989) Antagonism of thyroid hormone action by amiodarone in rat pituitary tumor cells. *J. Clin. Invest.*, **83**, 306–313.

38. Latham, K.R., Selliti, D.F. and Goldstein, R.E. (1987) Interaction of amiodarone and desethylamiodarone with solubilized nuclear thyroid hormone receptors. *J. Am. Coll. Cardiol.*, **9**, 872–876.

39. Chalmers, D.K., Munro, S.L.A., Iskander, M.N. and Craik, D.J. (1992) Models for the binding of amiodarone at the thyroid hormone receptor. *J. Comput.-Aided Mol. Design*, **6**, 19–31.

40. Cody, V. and Luft, J. (1989) Structure-activity relationships of antiarrhythmic agents: crystal structure of amiodarone hydrochloride and two derivatives, and their conformational comparison with thyroxine. *Acta Cryst.*, **B45**, 172–178.

41. Chalmers, D.K. (1993) *The Design and Synthesis of Novel Thyroid Hormone Analogues*, PhD Thesis. Melbourne: University of Melbourne.

42. Larsen, P.R., Atkinson, A.J., Wellman, H.N. and Goldsmith, R.E. (1970) The effect of diphenyl-hydantoin on thyroxine metabolism in man. *J. Clin. Invest.*, **49**, 1266–1279.

43. Smith, P.J. and Surks, M.I. (1984) Multiple effects of 5,5'-diphenylhydantoin on the thyroid hormone system. *Endocr. Rev.*, **5**, 514–524.

44. Larsen, P.R. (1972) Salicylate-induced increases in free triiodothyronine in human serum. *J. Clin. Invest.*, **51**, 1125–1134.

45. Surks, M.I. and Oppenheimer, J.H. (1963) Effect of penicillin on thyroxine binding by plasma proteins. *Endocrinology*, **72**, 567–573.

46. McKinney, J.D., Chae, K., Oatley, S.J. and Blake, C.C.F. (1985) Molecular interactions of toxic chlorinated dibenzo-p-dioxins and dibenzofurans with thyroxine binding prealbumin. *J. Med. Chem.*, **28**, 375–381.

47. Rickenbacher, U., McKinney, J.D., Oatley, S.J. and Blake, C.C.F. (1986) Structurally specific binding of halogenated biphenyls to thyroxine transport protein. *J. Med. Chem.*, **29**, 641–648.

48. Pederson, L.G., Darden, T.A., Oatley, S.J. and McKinney, J.D. (1986) A theoretical study of the binding of polychlorinated biphenyls (PCBs), dibenzodioxins and dibenzofurans to human plasma prealbumin. *J. Med. Chem.*, **29**, 2451–2457.

49. McKinney, J.D., Fannin, R., Jordan, S., Chae, K., Rickenbacher, U. and Pederson, L. (1987) Polychlorinated biphenyls and related compounds. Interactions with specific binding sites for thyroxine in rat liver nuclear extracts. *J. Med. Chem.*, **30**, 79–86.

50. Brouwer, A., Klasson-Wehler, E., Bokdam, M., Morse, D.C. and Traag, W.A. (1990) Competitive inhibition of thyroxine binding to transthyretin by monohydroxy metabolites of 3,4,3',4'-tetrachlorobiphenyl. *Chemosphere*, **20**, 1257–1262.

51. Den Besten, C., Vet, J.J.R.M., Besselink, H.T., Kiel, G.S., Ban Berkel, B.J.M., Beems, R. and Van Bladeren, P.J. (1991) The liver, kidneys and thyroid toxicity of chlorinated benzenes. *Toxicol. Appl. Pharmacol.*, **111**, 69–81.

52. Van den Berg, K.J. (1990) Interaction of chlorinated phenols with thyroxine binding sites of human transthyretin, albumin and thyroid binding globulin. *Chem.-Biol. Interact.*, **76**, 63–75.

53. Lans, M.C., Klasson-Wehler, E., Willemsen, M., Meuseen, E., Safe, S. and Brouwer, A. (1993) Structure-dependent, competitive interaction of hydroxy-polychlorbiphenyl, dibenzo-p-dioxins and dibenzofurans with human transthyretin. *Chem. Biol. Interact.*, **88**, 7–21.

54. Auf'mkolk, M., Koehrle, J., Gumbinger, H., Winterhoft, H. and Hesch, R.-D. (1984) Antihormonal effects of plant extracts: iodothyronine deiodinase of rat liver is inhibited by extracts and secondary metabolites of plants. *Horm. Metab. Res.*, **16**, 136–141.

55. Cody, V., Koehrle, J., Auf'mkolk, M. and Hesch, R.-D. (1986) Structure-activity relationships of flavonoid deiodinase inhibitors and enzyme active-site models. In *Plant Flavonoids in Biology and*

Medicine: Biochemical, Pharmacological and Structure-Activity Relationships, (eds. Cody, V., Middleton, E. and Harborne, J.B.), pp. 373–382. New York: Alan R. Liss, Inc.

56. Auf'mkolk, M., Koehrle, J., Hesch, R.-D. and Cody, V. (1986) Inhibition of rat liver iodothyronine deiodinase. *J. Biol. Chem.*, **261**, 11623–11630.

57. Kaplan, M.M. and Utiger, R.D. (1978) Iodothyronine metabolism in rat liver homogenates. *J. Clin. Invest.*, **61**, 459–471.

58. Koehrle, J., Spanka, M., Irmscher, K. and Hesch, R.-D. (1988) Flavonoid effects on transport, metabolism and action of thyroid hormones. In *Plant Flavonoids in Biology and Medicine. II. Biochemical, Cellular and Medicinal Properties*, (eds. Cody, V., Middleton, E., Harborne, J.B. and Beretz, A.), pp. 323–340. New York: Alan R. Liss, Inc.

59. Koehrle, J., Auf'mkolk, M., Spanka, M., Irmscher, K., Cody, V. and Hesch, R.-D. (1986) Iodothyronines deiodinase is inhibited by plant flavonoids. In *Plant Flavonoids in Biology and Medicine. Biochemical, Pharmacological, and Structure-Activity Relationships*, (eds. Cody, V., Middleton, E. and Harborne, J.B.), pp. 359–371. New York: Alan R. Liss, Inc.

60. Spanka, M., Hesch, R.-D., Irmscher, K. and Koehrle, J. (1990) 5'-Deiodination in rat hepatocytes: effects of specific flavonoid inhibitors. *Endocrinology*, **126**, 1660–1667.

61. Ciszak, E., Cody, V. and Luft, R. (1992) Crystal structure determination at 2.3 Å resolution of human transthyretin-3',5'-dibromo-2',4,4',6-tetrahydroxy-aurone complex. *Proc. Natl. Acad. Sci. USA*, **89**, 6644–6648.

62. Cheng, S.-Y., Pages, R.A., Saroff, H.A., Edelhoch, H. and Robbins, J. (1977) Analysis of thyroid hormone binding to human serum prealbumin by 8-anilinonaphthalene-1-sulfonate fluorescence. *Biochemistry*, **16**, 3707–3713.

63. Birkenmeier, G. and Kopperschlager, G. (1987) Interaction of the dye Remazol Yellow GGL to prealbumin and albumin studied by affinity phase partition difference spectroscopy and equilibrium dialysis. *Mol. Cell Biol.*, **73**, 99–110.

64. Tabachnick, M., Hao, Y.-L. and Korcek, L. (1972) Effect of oleate, diphenylhydantoin and heparin on the binding of ^{125}I thyroxine to purified thyroxine-binding globulin. *J. Clin. Endocrinol. Metab.*, **36**, 392–394.

65. Chopra, K.J., Teco, G.N.C., Mead, J.F., Huang, T-S., Beredo, A. and Solomon, D.H. (1985) Relationship between serum free fatty acids and thyroid hormone binding inhibitor in nonthyroid illnesses. *J. Clin. Endocrinol. Metab.*, **60**, 980–984.

66. Mendel, C.M., Frost, P.H. and Cavalieri, R.R. (1986) Effect of free fatty acids on the concentration of free thyroxine in human serum: the role of albumin. *J. Clin. Endocrinol. Metab.*, **63**, 1394–1399.

67. Braverman, L.E., Arky, R.A., Foster, A.E. and Ingbar, S.H. (1969) Effect of physiological variations in free fatty acid concentration on the binding of thyroxine in the serum of euthyroid and thyrotoxic subjects. *J. Clin. Invest.*, **48**, 878–884.

68. Lim, C.-F., Bai, Y., Topliss, D.J., Barlow, J.W. and Stockigt, J.R. (1988) Drug and fatty acid effects on serum thyroid hormone binding. *J. Clin. Endocrinol. Metab.*, **67**, 682–688.

69. Lim, C.-F. (1989) *Inhibitors of Thyroid Hormone Binding in Human Serum, PhD Thesis*. Melbourne: Monash University.

70. Munro, S.L.A. (1990) *Studies of Thyroid Hormone Binding to Transthyretin, PhD Thesis*. Melbourne: Monash University.

71. Tabachnick, M. (1964) Thyroxine-protein interactions. III. Effect of fatty acids, 2,4-dinitrophenol and other anionic compounds on the binding of thyroxine by human serum albumin. *Arch. Biochim. Biophys.*, **106**, 415–421.

72. Tabachnick, M. and Korcek, L. (1986) Effect of long-chain fatty acids on the binding of thyroxine and triiodothyronine to human thyroxine-binding globulin. *Biochim. Biophys. Acta*, **881**, 292–296.

73. Zanelli, G.D., Cook, N., Lahiri, A., Ellison, D., Webbon, P. and Woolley, G. (1988) Protein binding studies of Technetium-99m-labelled phosphine and isocyanide cationic complexes. *J. Nucl. Med.*, **29**, 62–67.

74. Barlow, J.W., Raggatt, L.E., Lim, C.-F., Munro, S.L., Topliss, D.J. and Stockigt, J.R. (1989) The thyroid hormone analogue SKF L-94902: nuclear occupancy and serum binding studies. *Clin. Sci.*, **76**, 495–501.

75. Chopra, I.J. (1981) *Triiodothyronines in Health and Disease*. Berlin: Springer-Verlag.

76. Dietrich, S.W., Bolger, H.B., Kollman, P.A. and Jorgensen, E.C. (1977) Thyroxine analogues. 23. Quantitative structure-activity correlation studies of *in vivo* and *in vitro* thyromimetic activities. *J. Med. Chem.*, **20**, 863–880.

77. Andrea, T.A., Cavalieri, R.R., Goldfine, I.D. and Jorgensen, E.C. (1980) Binding of thyroid hormones and analogues to the human plasma protein prealbumin. *Biochemistry*, **19**, 55–63.

78. Andrea, T.A., Dietrich, S.W., Murray, W.J., Kollman, P.A., Jorgensen, E.C. and Rothenberg, S. (1979) A model for thyroid hormone interactions. *J. Med. Chem.*, **22**, 221–232.

79. Somack, R., Andrea, T.A. and Jorgensen, E.C. (1982) Thyroid hormone binding to human prealbumin and rat liver nuclear receptor: kinetics and contribution of the hormone phenolic hydroxyl group and accommodation of hormone sidechain bulk. *Biochemistry*, **21**, 163–170.

80. Oppenheimer, J.H. (1991). Thyroid hormone action at the molecular level. In *Werner and Ingbar's The Thyroid: a Fundamental and Clinical Text*, (eds. Braverman, L.E. and Utiger, R.D.), 6th edn., pp. 204–224. Philadelphia: J.B. Lippincott Co.

81. Cody, V. (1991) Thyroid hormone structure-function relationships. *Werner and Ingbar's The Thyroid: A Fundamental and Clinical Text*, (eds. Braverman, L.E. and Utiger, R.D.), 6th edn., pp. 225–229. Philadelphia: J.B. Lippincott Co.

82. Cody, V. (1981) Structure of thyroxine: role of thyroxine hydroxyl in protein binding. *Acta Crystallogr.*, **B37**, 1685–1689.

83. Lehman, P.A. and Jorgensen, E.C. (1965) Thyroxine analogs. XIII. NMR evidence for hindered rotation in diphenyl ethers. *Tetrahedron*, **21**, 363–380.

84. Camerman, N., Fawcett, J.K., Reynolds, W.F. and Camerman, A. (1975) Thyroid hormone conformation from NMR studies of triiodothyropropionic acid. *Nature*, **253**, 50–51.

85. Emmett, J.C. and Pepper, E.S. (1975) Conformation of thyroid hormone analogues. *Nature*, **257**, 334–336.

86. Fawcett, J.K., Camerman, N. and Camerman, A. (1976) Thyroid hormone stereochemistry. IV. Molecular conformation of 3'-isopropyl-3,5-diiodo-L-thyronine in the crystal and in solution. *J. Am. Chem. Soc.*, **98**, 587–591.

87. Mazzocchi, P.H., Ammon, H.L., Colicelli, E., Hohokaba, Y. and Cody, V. (1976) NMR studies of triiodothyropropionic acid in ethanol-HCl. *Experientia*, **32**, 419–420.

88. Mazzochi, P.H., Ammon, H.L. and Colicelli, E. (1978) Carbon-13 NMR studies of thyroid hormones and model compounds. The conformational analysis of diphenyl ethers. *Org. Magn. Reson.*, **11**, 143–149.

89. Gale, D.J., Craik, D.J. and Brownlee, R.T.C. (1988) Variable temperature NMR studies of thyroid hormone conformations. *Magn. Reson. Chem.*, **26**, 275–280.

90. Mazzocchi, P.H., Ammon, H.L., Liu, L., Colicelli, E., Ravi, P. and Burrows, E. (1981) Synthesis of diphenyl ether models of thyroid hormones. Diphenyl ethers linked to the 3-oxo-2-azabicyclo[2.2.1]-heptane ring system as substrates for conformational analysis. *J. Org. Chem.*, **46**, 4530–4536.

91. Andrews, P.R., Craik, D.J. and Munro, S.L.A. (1988) The relationship between NMR, X-ray and theoretical methods of conformational analysis: application to thyroid hormones. In *NMR Spectroscopy in Drug Research*, (ed. Kofod, H.), pp. 20–36. Copenhagen: Munksgaard.

92. Reid, D.G., MacLachlan, L.K., Voyle, M. and Leeson, P.D. (1989) A proton and fluorine-19 nuclear magnetic resonance and fluorescence study and the binding of some natural and synthetic thyromimetics to prealbumin (transthyretin). *J. Biol. Chem.*, **264**, 2013–2023.

93. Reid, D.G. and Saunders, M.R. (1989) A proton nuclear magnetic resonance and nuclear Overhauser effect (NOE) study of human plasma prealbumin, including the development and application to spectral assignment of a combined ring current shift and NOE prediction program. *J. Biol. Chem.*, **264**, 2003–2012.

94. Kier, L.B. and Hoyland, J.R. (1970) A molecular orbital study of 3,3',5-trihalothyronine analogs. *J. Med. Chem.*, **13**, 1182–1184.

95. Kollman, P.A., Murray, W.J., Nuss, M.E., Jorgensen, E.C. and Rothenberg, S. (1973) Molecular orbital studies of thyroid hormone analogues. *J. Am. Chem. Soc.*, **95**, 8518–8125.

96. Benjamins, H., Dar, F.H. and Chandler, W.D. (1974) Conformations of bridged diphenyls. VI. Substituent effects and internal rotation in triply ortho-substituted diphenyl ethers. *Can. J. Chem.*, **52**, 3297–3302.

97. Lehmann, P.A. (1972) Intramolecular aryl-iodine π complex formation and its relation to thyromimetic activity. *J. Med. Chem.*, **15**, 404–409.

98. Blaney, J.M., Weiner, P.K., Dearing, A., Kollman, P.A., Jorgensen, E.C., Oatley, S.J., Burridge, J.M. and Blake, C.C.F. (1982) Molecular mechanics simulation of protein-ligand interactions: binding of thyroid hormone analogues to prealbumin. *J. Am. Chem. Soc.*, **104**, 6424–6434.

99. Blaney, J.M., Jorgensen, E.C., Connolly, M.L., Ferrin, T.E., Langridge, R., Oatley, S.J., Burridge, J.M. and Blake, C.C.F. (1982) Computer graphics in drug design: molecular modelling of thyroid hormone-prealbumin interactions. *J. Med. Chem.*, **25**, 785–796.

100. Kuntz, I.D., Blaney, J.M., Oatley, S.J., Langridge, R. and Ferrin, T.E. (1982) A geometric approach to macromolecule — ligand interactions. *J. Mol. Biol.*, **161**, 269–288.

101. Craik, D.J., Andrews, P.R., Border, C. and Munro, S.L.A. (1990) Conformational studies of thyroid hormones. I. The diphenyl ether moiety. *Aust. J. Chem.*, **43**, 923–936.

102. Okabe, N., Fujiwara, T., Yamagata, Y. and Tomita, K.I. (1982) The crystal stucture of a major metabolite of thyroid hormone: 3,3',5'-triiodo-L-thyronine. *Biochim. Biophys. Acta*, **717**, 179–181.

103. Zenker, N. and Jorgensen, E.C. (1959) Thyroxine analogs. I. Synthesis of 3,5-diodo-4(2'-alkylphenoxy)-DL-phenylalanines. *J. Am. Chem. Soc.*, **81**, 4643–4647.

104. Jorgensen, E.C., Lehman, P.A., Zenker, N. and Greenberg, C. (1962) Thyroxine analogues. VII. Antigoitrogenic, calorigenic, and hypo-cholesteremic activities of some aliphatic, alicyclic, and aromatic ethers of 3,5-diiodotyrosine in the rat. *J. Biol. Chem.*, **237**, 3832–3838.

105. Flink, I.L., Bailey, T.J., Gustafson, T.A., Markham, B.E. and Morkin, E. (1986) Complete amino-acid sequence of human thyroxine-binding globulin deduced from cloned DNA: close homology to the serine proteases. *Proc. Natl. Acad. Sci. USA*, **83**, 7708–7712.

106. Carrell, R.W., Pemberton, P.A. and Boswell, D.R. (1987) The serpins: evolution and adaption in a family of proteinase inhibitors. *Cold Spring Harbor Symp. Quant. Biol.*, **52**, 527–535.

107. Loebermann, H., Tokuoka, R., Diesenhofer, J. and Huber, R. (1984) Human alpha-1-proteinase inhibitor. Crystal structure analysis of two crystal modifications: molecular model and preliminary analysis of the implications for function. *J. Mol. Biol.*, **177**, 531–555.

108. Jarvis, J.A., Munro, S.L.A. and Craik, D.J. (1992) Homology model of thyroxine-binding globulin and elucidation of the thyroid hormone binding site. *Protein Eng.*, **5**, 61–67.

109. Terry, C.J. and Blake, C.C.F. (1992) Comparison of the modelled thyroxine binding site in TBG with the experimentally determined site in transthyretin. *Protein Eng.*, **5**, 505–510.

110. Cheng, S.-Y., Rakhit, G., Erard, F., Robbins, J. and Chignell, C.F. (1981) A spin label study of the thyroid hormone binding sites in human plasma thyroxine transport proteins. *J. Biol. Chem.*, **256**, 831–836.

111. Refetoff, S. (1989) Inherited thyroxine-binding globulin abnormalities in man. *Endocr. Rev.*, **10**, 275–293.

112. Fex, G., Albertsson, P.A. and Hansson, B. (1979) Interaction between prealbumin and retinol binding protein studied by affinity chromatography, gel filtration and two-phase partition. *Eur. J. Biochem.*, **99**, 353–360.

113. Blake, C.C.F., Geisow, M.J., Swan, I.D.A., Rerat, C. and Rerat, B. (1974) Structure of human plasma prealbumin at 2.5 Å resolution. *J. Mol. Biol.*, **88**, 1–12.

114. Blake, C.C.F. and Oatley, S.J. (1977) Protein-DNA and protein-hormone interactions in prealbumin: a model of the thyroid hormone nuclear receptor?. *Nature*, **268**, 115–120.

115. Blake, C.C.F., Geisow, M.J., Oatley, S.J., Rerat, B. and Rerat, C. (1978) Structure of prealbumin: secondary, tertiary and quaternary interactions determined by Fourier refinement at 1.8 Å. *J. Mol. Biol.*, **121**, 339–356.

116. Blake, C.C.F., Burridge, J.M. and Oatley, S.J. (1978) X-ray analysis of thyroid hormone binding to prealbumin. *Biochem. Soc. Trans.*, **6**, 1114–1118.

117. Wojtczak, A., Luft, J. and Cody, V. (1992) Mechanism of molecular recognition. Structural aspects of 3,3'-diiodo-L-thyronine binding to human serum transthyretin. *J. Biol. Chem.*, **267**, 353–357.

118. Ferguson, R.N., Edelhoch, H., Saroff, H.A. and Robbins, J. (1975) Negative cooperativity in the binding of thyroxine to human serum prealbumin. *Biochemistry*, **14**, 282–289.

119. Gonzalez, G. (1989) Fluorescent derivative of cysteine-10 reveals thyroxine-dependent conformational modifications in human serum prealbumin. *Arch. Biochem. Biophys.*, **271**, 200–205.

120. Gonzalez, G. and Tapia, G.A. (1992) Fluorescence study of the thyroxine-dependent conformational changes in human serum transthyretin. *FEBS Lett.*, **297**, 253–256.

121. DesJarlais, R.L., Sheridan, R.P., Dixon, J.S., Kuntz, I.D. and Venkataraghawan, R. (1986) Docking flexible ligands to macromolecular receptors by molecular shape. *J. Med. Chem.*, **29**, 2149–2153.

122. Craik, D.J., Hall, J.G. and Higgins, K.A. (1987) C-13 NMR studies of the thyroid hormone transport protein, transthyretin, and the pancreatic insulin storage moiety, the zinc-insulin hexamer. *Biochem. Biophys. Res. Commun.*, **143**, 116–125.

123. Lalloz, M.R.A., Byfield, P.G.H., Goel, K.M., Loudon, M.M., Thomson, J.A. and Himsworth, R.L. (1987) Hyperthyroxinemia due to the coexistence of two raised affinity thyroxine-binding proteins (albumin and prealbumin) in one family. *J. Clin. Endocrinol. Metab.*, **64**, 346–352.

124. Refetoff, S., Dwulet, F.E. and Benson, M.D. (1986) Reduced affinity for thyroxine in two of three structural thyroxine-binding prealbumin variants associated with familial amyloidotic polyneuropathy. *J. Clin. Endocrinol. Metab.*, **63**, 1432–1437.

125. Murrell, J.R., Schoner, R.G., Liepnieks, J.J., Rosen, H.N., Moses, A.C. and Benson, M.D. (1992) Production and functional analysis of normal and variant recombinant human transthyretin proteins. *J. Biol. Chem.*, **267**, 16595–16600.

126. Murrell, J., Schoner, R., Moses, A., Rosen, H. and Benson, M.D. (1990) Structure and function of recombinant human transthyretin. In *Proc. VIth International Symposium on Amyloidosis*, pp. 647–650. Dordrecht, Netherlands: Kluwer Academic Publishers.

127. Moses, A.C., Rosen, H.N., Moller, D.E., Tsuzaki, S., Haddow, J.E., Lawlor, J., Liepnieks, J.J., Nichols, W.C. and Benson, M.D. (1990) A point mutation in transthyretin increases affinity for thyroxine and produces euthyroid hyper-thyroxinemia. *J. Clin. Invest.*, **86**, 2025–2033.

128. Barlow, J.W., Curtis, A.J., Scrimshaw, B.J., Topliss, D.J., George, P.M. and Stockigt, J.R. (1993) Rapid isolation of transthyretin (TTR) from human serum: mutations at position 119, but not 54, increase binding affinity for T_4 and T_3. In *Proc. 36th Annual Scientific Meeting*, Vol. 36, p. 93. Melbourne: Endocrine Society of Australia.

129. Almeida, M.R., Altland, K., Rauh, S., Gawinowicz, M., Moreira, P., Costa, P.P. and Saraiva, M.J. (1991) Characterization of a basic transthyretin variant — TTR Arg 102 — in the German population. *Biochim. Biophys. Acta*, **1097**, 224–226.

130. Almeida, M.R., Gawinowicz, M., Costa, P.P., Salvi, F., Ferlini, A., Plasmati, R., Tassinari, C. and Saraiva, M.J. (1990) A new transthyretin variant associated with familial amyloidotic poly-neuropathy in an Italian kindred. In *Proc. VIth International Symposium on Amyloidosis*, pp. 599–602. Dordrecht, Netherlands: Kluwer Academic Publishers.

131. Almeida, M.R., Hesse, A., Steinmetz, A., Maisch, B., Altland, K., Linke, R.P., Gawinowicz, M.A. and Saraiva, M.J. (1991) Transthyretin Leu 68 in a form of cardiac amyloidosis. *Basic Res. Cardiol.*, **86**, 567–571.

132. Dwulet, F.E. and Benson, M.D. (1986) Characterization of a transthyretin (prealbumin) variant associated with familial amyloidotic polyneuropathy type II (Indiana/Swiss). *J. Clin. Invest.*, **78**, 880–886.

133. Fitch, N.J.S., Akbari, M.T. and Ramsden, D.B. (1991) An inherited non-amyloidogenic transthyretin variant, [Ser[6]]-TTR, with increased thyroxine-binding affinity, characterized by DNA sequencing. *J. Endocrinol.*, **129**, 309–313.

134. Gorevic, P.D., Prelli, F.C., Wright, J., Pras, M. and Frangione, B. (1989) Systemic senile amyloidosis. Identification of a new prealbumin (transthyretin) variant in cardiac tissue: immunologic and biochemical similarity to one form of familial amyloidotic polyneuropathy. *J. Clin. Invest.*, **83**, 836–843.

135. Harding, J., Skare, J. and Skinner, M. (1991) A second transthyretin mutation at position 33 (Leu/Phe) associated with familial amyloidotic polyneuropathy. *Biochim. Biophys. Acta*, **1097**, 183–186.

136. Harrison, H.H., Gordon, E.D., Nichols, W.C. and Benson, M.D. (1991) Biochemical and clinical characterization of prealbumin[CHICAGO]: an apparently benign variant of serum prealbumin (transthyretin) discovered with high-resolution two-dimensional electrophoresis. *Am. J. Med. Genet.*, **39**, 442–452.

137. Ii, S., Minnerath, S., Ii, K., Dyck, P.J. and Sommer, S.S. (1991) Two-tiered DNA-based diagnosis of transthyretin amyloidosis reveals two novel point mutations. *Neurology*, **41**, 893–898.

138. Izumoto, S., Younger, D., Hays, A.P., Martone, R.L., Smith, R.T. and Herbert, J. (1992) Familial amyloidotic polyneuropathy presenting with carpal tunnel syndrome and a new transthyretin mutation, asparagine 70. *Neurology*, **42**, 2094–2102.

139. Jacobson, D.R., McFarlin, D.E., Kane, I. and Buxbaum, J.N. (1992) Transthyretin Pro55, a variant associated with early-onset aggressive, diffuse amyloidosis with cardiac and neurologic involvement. *Hum. Genet.*, **89**, 353–356.

140. Jones, L.A., Skare, J.C., Harding, J.A., Cohen, A.S., Milunsky, A. and Skinner, M. (1991) Proline at position 36: a new transthyretin mutation associated with familial amyloidotic polyneuropathy. *Am. J. Hum. Genet.*, **48**, 979–982.

141. Jones, L.A., Skare, J.C., Cohen, A.S., Harding, J.A., Milunsky, A. and Skinner, M. (1992) Familial amyloidotic polyneuropathy: a new transthyretin position-30 mutation (alanine for valine) in a family of German descent. *Clin. Genet.*, **41**, 70–73.

142. Lalloz, M.R.A., Byfield, P.G.H. and Himsworth, R.L. (1984) A prealbumin variant with an increased affinity for T_4 and reverse T_3. *Clin. Endocrinol.*, **21**, 331–338.

143. Murakami, T., Maeda, S., Yi, S., Ikegawa, S., Kawashima, E., Onodera, D., Shimada, K. and Araki, S. (1992) A novel transthyretin mutation associated with familial amyloidotic polyneuropathy. *Biochem. Biophys. Res. Commun.*, **182**, 520–526.

144. Nakazato, M., Kangawa, K., Minamino, N., Tawara, S., Matuso, H. and Araki, S. (1984) Revised analysis of amino acid replacement in a prealbumin variant (SKO-III) associated with familial amyloidotic polyneuropathy of Jewish origin. *Biochem. Biophys. Res. Commun.*, **123**, 921–928.

145. Nakazato, M., Tanaka, M., Yamamura, Y., Kurihara, T., Matsukura, S., Kangawa, K. and Matsuo, H. (1988) Abnormal transthyretin in asymptomatic relatives in familial amyloidotic polyneuropathy. *Arch. Neurol.*, **44**, 1275–1278.

146. Nakazato, M., Ikeda, S., Shiomi, K., Matsukura, S., Yoshida, K., Shimizu, H., Atsumi, T., Kangawa, K. and Matsuo, H. (1992) Identification of a novel transthyretin variant (Val30 → Leu) associated with familial amyloidotic polyneuropathy. *FEBS Lett.*, **306**, 206–208.

147. Nichols, W.C., Liepnieks, J.J., McKusick, V.A. and Benson, M.D. (1989) Direct sequencing of the gene for Maryland/German familial amyloidotic polyneuropathy type II and genotyping by allele-specific enzymatic amplification. *Genomics*, **5**, 535–540.

148. Nordlie, M., Sletten, K., Husby, G. and Ranløv, P.J. (1988) A new prealbumin variant in familial amyloid cardiomyopathy of Danish origin. *Scand. J. Immunol.*, **27**, 119–122.

149. Saeki, Y., Ueno, S., Yorifuji, S., Sugiyama, Y., Ide, Y. and Matsuzawa, Y. (1991) New mutant gene (transthyretin Arg 58) in cases with hereditary polyneuropathy detected by non-isotope method of single-strand conformation polymorphism analysis. *Biochem. Biophys. Res. Commun.*, **180**, 380–385.

150. Saeki, Y., Ueno, S., Takahashi, N., Soga, F. and Yanagihara, T. (1992) A novel mutant (transthyretin Ile-50) related to amyloid polyneuropathy. Single-strand conformation polymorphism as a new genetic marker. *FEBS Lett.*, **308**, 35–37.

151. Saraiva, M.J., Almeida, M.R., Alves, I.L., Moreira, P., Gawinowicz, M., Costa, P.P., Rauh, S., Bahnzoff, A. and Altland, K. (1991) Molecular analyses of an acidic transthyretin Asn 90 variant. *Am. J. Hum. Genet.*, **48**, 1004–1008.

152. Saraiva, M.J., Almeida, M.R., Sherman, W., Gawinowicz, M., Costa, P.P. and Goodman, D.S. (1992) A new transthyretin mutation associated with amyloid cardiomyopathy. *Am. J. Hum. Genet.*, **50**, 1027–1030.

153. Saraiva, M.J.M., Birken, S., Costa, P.P. and Goodman, D.S. (1984) Family studies of the genetic abnormality in transthyretin (prealbumin) in Portuguese patients with familial amyloidotic polyneuropathy. *Ann. N.Y. Acad. Sci.*, **435**, 86–100.

154. Skinner, M., Harding, J., Skare, I., Jones, L.A., Cohen, A.S., Milunsky, A. and Skare, J. (1992) A new transthyretin mutation associated with amyloidotic vitreous opacities. Asparagine for isoleucine at position 84. *Ophthalmology*, **99**, 503–508.

155. Strahler, J.R., Rosenblum, B.B. and Hanash, S.M. (1987) Identification and characterisation of a human transthyretin variant. *Biochem. Biophys. Res. Commun.*, **148**, 471–477.

156. Uemichi, T., Murrell, J.R., Zeldenrust, S. and Benson, M.D. (1992) A new mutant transthyretin (Arg 10) associated with familial amyloid polyneuropathy. *J. Med. Genet.*, **29**, 888–891.

157. Ueno, S., Uemichi, T., Yorifuji, S. and Tarui, S. (1990) A novel variant of transthyretin (Tyr[114] to Cys) deduced from the nucleotide sequences of gene fragments from familial amyloidotic polyneuro-pathy in Japanese sibling cases. *Biochem. Biophys. Res. Commun.*, **169**, 143–147.

158. Ueno, S., Uemichi, T., Takahashi, N., Soga, F., Yorifuji, S. and Tarui, S. (1990) Two novel variants of transthyretin identified in Japanese cases with familial amyloidotic polyneuropathy: transthyretin (Glu42 to Gly) and transthyretin (Ser50 to Arg). *Biochem. Biophys. Res. Commun.*, **169**, 1117–1121.

159. Wallace, M.R., Dwulet, F.E., Conneally, P.M. and Benson, M.D. (1986) Biochemical and molecular genetic characterisation of a new variant prealbumin associated with hereditary amyloidosis. *J. Clin. Invest.*, **78**, 6–12.

160. Wallace, M.R., Dwulet, F.E., Williams, E.C., Conneally, P.M. and Benson, M.D. (1988) Identification of a new hereditary amyloidosis prealbumin variant, Tyr-77, and detection of the gene by DNA analysis. *J. Clin. Invest.*, **81**, 189–193.

161. Benson, M.D., Julien, J., Liepnieks, J., Zeldenrust, S. and Benson, M.D. (1993) A transthyretin variant (alanine 49) associated with familial amyloidotic polyneuropathy in a French family. *J. Med. Genet.*, **30**, 117–119.

162. Hamilton, J.A., Steinrauf, L.K., Braden, B.C., Liepnicks, J., Benson, M.D., Holmgren, G., Sandgren, O. and Steen, L. (1993) The X-ray crystal structure refinements of normal human transthyretin and the amyloidogenic Val-30 → Met variant to 1.7 Å resolution. *J. Biol. Chem.*, **268**, 2416–2424.

163. Terry, C.J., Damas, A.M., Oliveira, P., Saraiva, M.J.M., Alves, I.L., Costa, P.P., Matias, P.M., Sakaki, Y. and Blake, C.C.F. (1993) Structure of Met30 variant of transthyretin and its amyloidogenic implications. *EMBO J.*, **12**, 735–741.

164. Steinrauf, L.K., Hamilton, J.A., Braden, B.C., Murrell, J.R. and Benson, M.D. (1993) X-ray crystal structure of the Ala-109 → Thr variant of human transthyretin which produces euthyroid hyper-thyroxinemia. *J. Biol. Chem.*, **268**, 2425–2430.

165. Dickson, P.W., Howlett, G.J. and Schreiber, G. (1985) Rat transthyretin (prealbumin). *J. Biol. Chem.*, **260**, 8214–8219.

166. Larsson, M., Pettersson, T. and Carlstrom, A. (1985) Thyroid hormone binding in serum of 15 vertebrate species: isolation of thyroxine-binding globulin and prealbumin analogs. *Gen. Comp. Endocrinol.*, **58**, 360–375.

167. Sundelin, J., Melhaus, H. and Das, S. (1985) The primary structure of rabbit and rat prealbumin and a comparison with the tertiary structure of human prealbumin. *J. Biol. Chem.*, **260**, 6481–6487.

168. Aleshire, S.L., Bradley, C.A., Richardson, L.D. and Parl, F.F. (1983) Localization of human prealbumin in choroid plexus epithelium. *J. Histochem. Cytochem.*, **81**, 608–612.

169. Dickson, P.W., Aldred, A.R., Marley, P.D., Bannister, D. and Schreiber, G. (1986) Rat choroid plexus specializes in the synthesis and the secretion of transthyretin. *J. Biol. Chem.*, **261**, 3475–3478.

170. Dickson, P.W., Aldred, A.R., Menting, J.G.T., Marley, P.D., Sawyer, W.H. and Schreiber, G. (1987) Thyroxine transport in the choroid plexus. *J. Biol. Chem.*, **262**, 13907–13915.

171. Herbet, J., Wilcox, J.N., Phan, K.-T., Fremeau, R.T., Zeviani, M., Dwork, A., Soprano, D.R., Makover, A., Goodman, D., Zimmerman, E.A., Roberts, J.L. and Schon, E.A. (1986) Transthyretin: a choroid plexus-specific transport protein in human brains. *Neurology*, **36**, 900–911.

172. Soprano, D.R., Herbet, J., Soprano, K.J., Schon, E.A. and Goodman, D.S. (1985) Demonstration of transthyretin mRNA in the brain and other extra-hepatic tissues in the rat. *J. Biol. Chem.*, **260**, 11793–11798.

173. Weisner, B. and Kauerz, U. (1983) The influence of the choroid plexus on the concentration of prealbumin in cerebrospinal fluid. *J. Neurol. Sci.*, **61**, 27–35.

174. Min He, X. and Carter, D.C. (1992) Atomic structure and chemistry of human serum albumin. *Nature*, **358**, 209–215.

175. Tritsch, G.L. and Tritsch, N.E. (1963) Thyroxine binding. II. The nature of the binding site of human serum albumin. *J. Biol. Chem.*, **238**, 138–142.

176. Docter, R., Bod, G., Krenning, E.P., Fekkes, D., Visser, T.J. and Hennemann, G. (1981) Inherited thyroxine excess: a serum abnormality due to an increased affinity for modified albumin. *Clin. Endocrinol.*, **15**, 363–371.

177. Grimaldi, S., Bartalena, L., Carlini, F. and Robbins, J. (1986) Purification and partial characterization of a novel thyroxine-binding protein (27K protein) from human plasma. *Endocrinology*, **118**, 2362–2369.

178. Benvenga, S., Gregg, R.E. and Robbins, J. (1988) Binding of thyroid hormones to human plasma lipoproteins. *J. Clin. Endocrinol. Metab.*, **67**, 6–16.

179. Benvenga, S., Cahnmann, H.J., Gregg, R.E. and Robbins, J. (1989) Characterization of the binding of thyroxine to high density lipoproteins and apolipoproteins A-I. *J. Clin. Endocrinol. Metab.*, **68**, 1067–1072.

180. Benvenga, S., Cahnmann, H.J., Rader, D., Kindt, M. and Robbins, J. (1992) Thyroxine binding to the apolipoproteins of high density lipoproteins HDL_2 and HDL_3. *Endocrinology*, **131**, 2805–2811.

181. Benvenga, S., Cahnmann, H.J. and Robbins, J. (1993) Characterization of thyroid hormone binding to apolipoprotein-E: localization of the binding site in the exon 3-coded domain. *Endocrinology*, **133**, 1300–1305.

182. Benvenga, S. and Robbins, J. (1993) Lipoprotein-thyroid hormone interactions. *Trends Endocrinol. Metab.*, **6**, 194–198.

183. Lein, A. and Dowben, R.M. (1961) Uptake and binding of thyroxine and triiodothyronine by the rat diaphragm *in vitro*. *Am. J. Physiol.*, **200**, 1029–1032.

184. Davis, P.J. (1991) Cellular actions of thyroid hormones. In *Werner and Ingbar's The Thyroid: A Fundamental and Clinical Text*, (eds. Braverman, L.E. and Utiger, R.D.), 6th edn., pp. 190–203. Philadelphia: J.B. Lippincott Co.

185. Osty, J., Jego, L., Francon, J. and Blondeau, J.-P. (1988) Characterization of triiodothyronine transport and accumulation in rat erythrocytes. *Endocrinology*, **123**, 2303–2311.

186. Topliss, D.J., Kolliniatis, E., Barlow, J.W. and Stockigt, J.R. (1990) Interaction between the cellular uptake systems for T_3 and amino acids. In *Proc. 33rd Annual Scientific Meeting*, Vol. 33, p. 51. Melbourne: Endocrine Society of Australia.

187. Blondeau, J.-P., Osty, J. and Francon, J. (1988) Characterization of the thyroid hormone transport system of isolated hepatocytes. *J. Biol. Chem.*, **236**, 2685–2692.

188. Leonard, J.L. and Visser, T.J. (1986) Biochemistry of deiodination. In *Thyroid Hormone Metabolism*, (ed. Hennemann, G.) p. 189. New York: Marcel Dekker.

189. Koehrle, J., Hesch, R.-D. and Leonard, J.L. (1991) Intracellular pathways of iodothyronine metabolism. In *Werner and Ingbar's The Thyroid: A Fundamental and Clinical Text*, (eds. Braverman, L.E. and Utiger, R.D.,), 6th edn., pp. 144–189. Philadelphia: J.B. Lippincott Co.

190. Visser, T.J., van Buuren, J.C.J., Rutgers, M., Rooda, S.J.E. and de Herder, W.W. (1990) The role of sulfation in thyroid hormone metabolism. *Trends Endocrinol. Metab.*, **1**, 211–218.

191. Koehrle, J. and Hesch, R.-D. (1984) Biochemical characteristics of iodothyronine monodeiodination by rat liver microsomes: the interaction between iodothyronine substrate analogues and the ligand binding site of the iodo-thyronine deiodinase resembles that of TBPA-iodothyronine ligand binding. *Horm. Metab. Res. Suppl.*, **14**, 42–55.

192. Koehrle, J., Auf'mkolk, M., Rokos, H., Hesch, R.-D. and Cody, V. (1986) Rat liver iodothyronine monodeiodinase — evaluation of the iodothyronine ligand-binding site. *J. Biol. Chem.*, **261**, 11613–11622.

193. Ichikawa, K. and Hashizume, K. (1991) Cellular binding proteins of thyroid hormones. *Life Sci.*, **49**, 1513–1522.

194. Sterling, K. and Milch, P.O. (1975) Thyroid hormone binding by a component of mitochondrial membrane. *Proc. Natl. Acad. Sci. USA*, **72**, 3225–3229.

195. Sterling, K., Milch, P.O., Brenner, M.A. and Lazarus, J.H. (1977) Thyroid hormone action: the mitochondrial pathway. *Science*, **197**, 996–999.

196. Sterling, K. (1986) Direct thyroid hormone activation of mitochondria: the role of adenine nucleotide translocase. *Endocrinology*, **119**, 292–295.

197. Obata, T., Kitagawa, S., Gong, Q.-H., Pastan, I. and Cheng, S.-Y. (1988) Thyroid hormone down-regulates p55, a thyroid hormone-binding protein that is homologous to protein disulfide isomerase and the β-subunit of prolyl-4-hydroxylase. *J. Biol. Chem.*, **263**, 782–785.

198. Oppenheimer, J.H., Koerner, D. and Schwartz, H.L. (1972) Specific nuclear triiodothyronine binding sites in rat liver and kidney. *J. Clin. Endocrinol. Metab.*, **35**, 330–333.

199. Oppenheimer, J.H., Schwartz, H.L. and Surks, M.L. (1974) Tissue differences in the concentration of triiodothyronine nuclear binding sites in the rat liver, kidney, pituitary, heart, brain, spleen and testis. *Endocrinology*, **95**, 897–903.

200. Samuels, H.H., Stanley, F. and Casanova, I. (1979) Relationship of receptor affinity to the modulation of thyroid hormone nuclear receptor levels and growth hormone synthesis by L-triiodothyronine and iodothyronine analogs in cultured GH_1 cells. *J. Clin. Invest.*, **63**, 1229.

201. Samuels, H.H. and Tsai, J.S. (1974) Thyroid hormone action. Demonstration of similar receptors in nuclei of rat liver and cultured GH_1 cells. *J. Clin. Invest.*, **53**, 656–659.

202. Sap, J., Munoz, A., Damm, K., Goldberg, Y., Ghysdael, J., Leutz, A., Beurg, H. and Vennstrom, B. (1986) The c-*erb* A protein is a high affinity receptor for thyroid hormone. *Nature*, **324**, 635–640.

203. Weinberger, C., Thompson, C.C., Ong, E.S., Lebo, R., Gruit, D.J. and Evans, R.M. (1986) The c-*erb* A gene encodes a thyroid hormone receptor. *Nature*, **324**, 641–646.

204. Evans, R.M. (1988) The steroid and thyroid hormone receptor superfamily. *Science*, **240**, 889–895.

205. Glass, C.K. and Holloway, J.M. (1990) Regulation of gene expression by the thyroid hormone receptor. *Biochim. Biophys. Acta*, **1032**, 157–176.

206. Ichikawa, K., Hashizume, K., Miyamatot, T., Nishii, Y., Yamauchi, K., Ohtsuka, H. and Yamada, T. (1988) Conformational transition of thyroid hormone receptor upon hormone binding: demonstration by aqueous two-phase partitioning. *J. Endocrinol.*, **119**, 431–437.

207. Toney, J.H., Wu, L., Summerfield, A.E, Sanyal, G., Forman, B.M., Zhu, J. and Samuels, H.H. (1993) Conformational changes in chicken thyroid hormone receptor $\alpha1$ induced by binding to ligand or to DNA. *Biochemistry*, **32**, 2–6.

208. Lavin, T.N., Norman, M.F., Eberhardt, N. and Baxter, J.D. (1989) Thyroid hormone receptor interactions with DNA. In *The Steroid/Thyroid Hormone Receptor Family and Gene Regulation*, (eds. Carlstedt-Duke, J., Eriksson, H. and Gustafsson, J.-A.) Vol. 4, pp. 69–81. Basel: Birkhauser Verlag.

209. Samuels, H.H., Forman, B.M., Horowitz, Z. and Ye, Z.-S. (1989) Regulation of gene expression by thyroid hormone. *Ann. Rev. Physiol.*, **51**, 623–639.

210. Miesfeld, R., Okret, S., Wikstrom, A.C., Wrange, O., Gustafsson, J.A. and Yamamoto, K.R. (1984) Characterization of a steroid hormone receptor gene and mRNA in wild-type and mutant cells. *Nature*, **312**, 779–781.

211. Hollenberg, S.M., Weinberger, C., Ong, E.S., Cerelli, G., Oro, A., Lebo, R., Thomson, E.B., Rosenfeld, M.G. and Evans, R.M. (1985) Primary structure and expression of a functional human glucocorticoid receptor cDNA. *Nature*, **318**, 635–641.

212. Härd, T., Kellenbach, E., Boelens, R., Maler, B.A., Dahlman, K., Freedman, L.P., Carlstedt-Duke, J., Yamamoto, K.R., Gustafsson, J.A. and Kaptein, R. (1990) Solution structure of the glucocorticoid receptor DNA binding domain. *Science*, **249**, 157–160.

213. Swabe, J.W.R., Neuhaus, D. and Rhodes, D. (1990) Solution structure of the DNA-binding domain of the oestrogen receptor. *Nature*, **348**, 458–461.

214. Luisi, B.F., Xu, W.X., Otwinowskia, Z., Freedman, L.P., Yamamoto, K.R. and Sigler, P.B. (1991) Crystallographic analysis of the interaction of the glucocorticoid receptor with DNA. *Nature*, **352**, 497–505.

215. Freedman, L.P. and Luisi, B.F. (1993) On the mechanism of DNA binding by nuclear hormone receptors: a structural and function perspective. *J. Cell. Biochem.*, **51**, 140–150.

216. Evans, R.M. (1989) The contribution of the steroid receptor superfamily to development, physiology and medicine. In *The Steroid/Thyroid Hormone Receptor Family and Gene Regulation*, (eds. Carlstedt-Duke, J., Eriksson, H. and Gustafsson, J.-A.) Vol. 4, pp. 69–81. Basel: Birkhauser Verlag.

7. CHEMICAL AND COMPUTER ASSISTED DEVELOPMENT OF AN INCLUSIVE PHARMACOPHORE FOR THE BENZODIAZEPINE RECEPTOR

WEIJIANG ZHANG,[†] KONRAD F. KOEHLER,[‡] HERNANDO DIAZ-ARAUZO,[†] MICHAEL S. ALLEN,[†] and JAMES M. COOK[†,*]

[†] *Department of Chemistry, University of Wisconsin-Milwaukee, Milwaukee, WI 53201, U.S.A.*
[‡] *Istituto di Ricerche di Biologia Molecolare P. Angeletti (IRBM), Via Pontina Km. 30,600; 00040 Pomezia (Roma), Italy*

A unified pharmacophore model of the benzodiazepine receptor (BzR) has been developed using the techniques of chemical synthesis, radioligand binding, and receptor mapping. This model is based on 136 different ligands spanning ten structurally diverse classes of compounds and qualitatively accounts for the relative affinities, efficacies, and functional effects (agonism vs. antagonism vs. inverse agonism) displayed by various ligands at the BzR. In addition, the model is expanded to account for the pharmacology of a recently discovered BzR receptor subtype termed the 'Diazepam-Insensitive' (DI) BzR. Moreover, the unified model described in this chapter is compared and contrasted with other published pharmacophore models (Table 13). The synthesis of both partial agonists and partial inverse agonists has been achieved *via* this model. The former compounds should provide useful leads for treatment of anxiety disorders, while the partial inverse agonists may furnish important clues for the treatment of age-associated memory impairment and of the symptoms of hepatic encephalopathy.

1. INTRODUCTION

1.1 GABA$_A$/BzR/Chloride Ion Channel Complex

The coupling of the gamma-aminobutyric acid (GABA), benzodiazepine receptor (BzR) and TBPS binding sites to a chloride ion channel in the mammalian central nervous system (CNS)[1-4] endows the GABA/Bz/Cl supramolecular complex with important regulatory functions in many neurological processes. The GABA$_A$ receptor has recently been shown to be a hetero-oligomeric family of ligand-gated ion channels which constitutes the major inhibitory neurotransmitter system in the CNS.[5] This membrane-bound protein complex plays a central role in the molecular mechanisms which underlie anxiety,[6,7] sleep,[8] memory-learning,[9] and consequently it represents an important target for the design of selective agents to treat specific disease states in the CNS. The inhibitory GABA neurotransmitter acts by binding to its receptor which is followed by the opening of an intrinsic chloride channel. The increase in chloride flux results in hyperpolarization of the neuronal cell membrane with a concomitant decrease in neuronal transmission. A number of agents have been found to modulate GABA$_A$ receptors including benzodiazepines, barbiturates, neurosteriods, and ethanol.[10]

1.2 BzR Ligands and their *in vivo* Functional Effects

Activation of the $GABA_A$ receptor system *via* the modulatory benzodiazepine receptor site by various structural classes of ligands has resulted in markedly different pharmacological effects. Agonist ligands, including the classical 1,4-benzodiazepines, have wide therapeutic applications as anxiolytics, anticonvulsants, sedative-hypnotics and myorelaxants.[11,12] In contrast, compounds which elicit pharmacological properties opposite those of the 1,4-benzodiazepines are known as inverse agonists and can exhibit anxiogenic, somnolytic, convulsant and proconvulsant actions.[7,13] In addition, there are many series of structurally related ligands whose pharmacological profiles lie in a continuum ranging from full agonist to full inverse agonist. Clinically, there is a need for partial agonists which exhibit anxiolytic/anticonvulsant activity in the absence of myorelaxant and sedative-hypnotic activity. Moreover, there is a need for partial (or selective) inverse agonists which enhance neuronal firing in the CNS but are devoid of proconvulsant/convulsant activity. This latter series would be useful for treatment of barbiturate-alcohol induced CNS depression, hepatic encephalopathy, and for cognition enhancement among others.

1.3 BzR Molecular Biology

In 1987 Seeburg, Schofield, and coworkers reported the cloning and functional expression of a GABA/Bz receptor complex from bovine brain.[14] It was initially proposed that the complex was composed of α- and β-subunits, the latter of which contains the $GABA_A$ receptor site. Subsequent to this report, several other subunits of the $GABA_A$ receptor have been characterized including γ- and δ-subunits.[15–17] The identification of multiple α-, β-, and γ-subunits is consistent with the pharmacological evidence of multiple $GABA_A$ receptor isoforms in the CNS.[18] Moreover, recent molecular biological studies have established that expression of at least two (α, γ) of three (α, β, γ) subunits is necessary to constitute a functional receptor which mimics many of the pharmacological, biochemical and electrophysiological properties of native receptors.[19–21] Previous studies of site directed mutagenesis have shown that changes in the α-subunit altered the activity of the benzodiazepine receptor-mediated response.[9,22–24] It is now known that changes in the γ-subunit[25–28] also alter this response. However, neither the stoichiometry nor the composition of native $GABA_A$ receptors in the CNS has been established. It appears now, based on preliminary evidence,[20,21,26–29] that the BzR binding site lies between the α- and γ-subunits.

1.4 BzR Pharmacology

Initially it was proposed that the type-I BzR was responsible for the anxiolytic and anticonvulsant effects[30] of the benzodiazepines while the type-II BzR mediated the muscle-relaxant, sedative-hypnotic properties of these ligands. SAR data from a variety of recently synthesized BzR ligands,[10,21,31,32] however, demonstrate that the actual pharmacological situation is more complicated. Bz type-I selective ligands such as the triazolopyridazine CL 218872 and the imidazopyridine zolpidem have also been shown to possess sedative effects[10,33] in contrast to previous reports.[30] Receptor isoforms whose pharmacology resembles that of previously reported Bz-I and Bz-II receptors have been expressed: the type-I BzR was constructed from an $\alpha1\beta2\gamma2$ combination of subunits while the type-II zolpidem-insensitive[32] BzR was comprised of $\alpha2\beta2\gamma2$, $\alpha3\beta2\gamma2$, and $\alpha5\beta2\gamma2$ receptor isoforms. Moreover, photo-labeling experiments[34] with [^{3}H]Ro 15-4513 have identified a site termed the 'diazepam insensitive' (DI) receptor which is made up of the subunits designated $\alpha4\beta2\gamma2$ and $\alpha6\beta2\gamma2$,[34,35] the latter of which has been studied extensively. The physiological activity of agonists and inverse agonists at benzodiazepine receptors in the CNS is therefore a consequence of action at one or more of these receptor subtypes.

1.5 BzR Pharmacophore Models

1.5.1 Models from other laboratories

An enormous amount of structure-activity relationship (SAR) data available for a large number of diverse structural classes of ligands has resulted in the formulation of several different models of the pharmacophore for benzodiazepine receptor binding.[36] Models which attempt to explain ligand efficacy as a function of ligand-receptor interactions at the molecular level have been put forth by Loew,[37] Crippen,[38,39] Codding,[40] Fryer,[41] Gilli and Borea,[42] Wermuth,[43] and Gardner.[44]

1.5.2 Skolnick/Cook model

Our previously published model rationalized various aspects of BzR ligand affinity and efficacy.[45–52] Inverse agonist/antagonist, agonist, and DI sites were originally treated as separate entities in these studies and therefore it is not surprising that the models which resulted were somewhat different in nature. In brief, our previously reported inverse agonist/antagonist pharmacophore/receptor model contains three major structural components:

(1) a hydrogen bond acceptor site A_2,
(2) a hydrogen bond donor site H_1,
(3) a lipophilic pocket (L_1) that accommodates substituents at position-3 of β-carbolines which have chain lengths less than five non-hydrogen atoms (Figure 1).[47,48]

The agonist model contains one additional hydrogen bonding site of interaction (H_2) as well as two additional areas of lipophilic interaction (L_2 and L_3). An area of negative steric interaction (S_1) between the ligand and the receptor-binding protein is also defined (Figure 2).[50,51] These models have resulted in the design and synthesis of several pharmacologically interesting ligands. From the agonist model, 6PBC, [6-(n-propoxy)-4-(methoxymethyl)-β-carboline-3-carboxylic acid ethyl ester][51] was conceived and subsequently shown to be a partial agonist which elicited anxiolytic

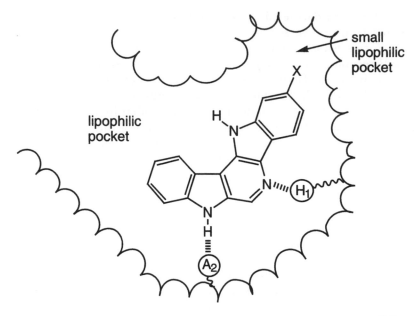

Figure 1   Schematic representation of the inverse agonist pharmacophore.[47,48] Receptor hydrogen bond donating and accepting sites of interaction are designated as H_1 and A_2, respectively.

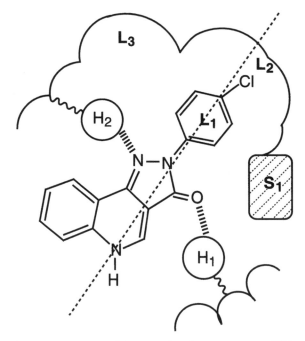

Figure 2   Schematic representation of the agonist pharmacophore.[50,51] Receptor hydrogen bond donating sites of interaction are designated as H_1 and H_2, respectively. Lipophilic sites of interaction are labeled as L_1, L_2, and L_3. S_1 is an area of negative interaction.

and anticonvulsant properties devoid of the myorelaxant effects of the classical 1, 4-benzodiazepines.[51] Likewise the design of a long-lived partial inverse agonist, 3-ethoxy-β-carboline[45,47,48] was based on the inverse agonist model.

1.5.3   Arguments for a unified model

Since various benzodiazepine receptor ligands can display markedly different pharmacological properties (i.e., agonism vs. inverse agonism), it could be argued that each of these effects is mediated by a different receptor on the BzR/GABA$_A$/ chloride ion channel complex. However, since these ligands:

(1) display a continuum of pharmacological effects ranging from agonist to anta-gonist to inverse agonist,[53]
(2) small modifications in the structures of these ligands can cause a shift from one pharmacological class to another,
(3) each functional class of BzR ligands competitively inhibits the binding of the other two classes to the BzR,
(4) inverse agonists and agonists are functional antagonists of each other.[54]

It is reasonable to assume that all three classes of BzR ligands (agonist, antagonist, and inverse agonist) bind to the same location in the receptor complex.

However, the binding site corresponding to each of these functional states of the receptor may overlap only partially. Furthermore, different functional states of the complex (i.e., agonistic vs. antagonistic vs. inverse agonistic) correspond to different conformations of the protein complex (open vs. closed chloride channel), hence one cannot rule out the possibility that the conformation of the receptor may also differ between the functional states. While agonists, antagonists, and inverse agonists may share common pharmacophoric descriptors consistent with a single binding domain, by definition, some of the descriptors must differ to allow for different functional effects.[55]

Recent studies of site directed mutagenesis by Pritchett[56] and Seeburg[57] have implicated two amino acid residues (Glu-199 and His-101 respectively; α1-numbering) as responsible, in part, for the pharmacological heterogeneity of the α-subunit. The implication of these studies is that these amino acid residues directly interact with bound benzodiazepine receptor ligands. However, one cannot rule out the possibility that mutations at one or both of these sites indirectly influence ligand binding through an overall reorganization of the tertiary structure of the protein.

If one assumes that all three pharmacological classes of BzR ligands are recognized in the same binding domain on the receptor, it follows that the pharmacophores for each of these classes (agonist, antagonist and inverse agonist) possess similar dimensions. Although Haefely earlier suggested that differences in efficacy between ligands may simply be a consequence of spare receptors,[58] it now appears more likely that efficacy is a consequence of interaction (full or partial) at different (BzR) GABA$_A$ isoforms.[21]

1.5.4   Objectives of a unified model

Described below is the development of an inclusive agonist/antagonist/inverse agonist pharmacophore/receptor model which unifies our previously published inverse agonist/antagonist[47,48] and agonist[50] pharmacophores and compares them to the recently developed model for the 'diazepam-insensitive' site.[52] A unified pharmacophore model which accounts for agonist, antagonist, and inverse agonist ligand binding to a common receptor yet which is consistent with the observed differences in pharmacology is a prerequisite for a better understanding of the structure-function relationships of these BzR ligands. More importantly, this will assist in the design of more selective ligands to treat disease states.

2.   METHODS

2.1   Computer Modeling

The pharmacophore modeling was carried out on a Silicon Graphics Personal Iris 4D/35 workstation with SYBYL version 5.5 (Tripos Associates. Inc. St. Louis, MO, USA). The core structures of the ligands were taken from available X-ray crystallographic coordinates[40,52,59-67] or generated using CONCORD[68] and modified where necessary using the fragment library of SYBYL. All bond lengths and valence angles of these structures were in turn fully optimized using Gaussian-90[69] or 92[70] (Gaussian Inc., Carnegie-Mellon University, Pittsburgh, PA, USA) at the *ab initio* 3-21G level. Rotational isomers of flexible side chains were generated using the MULTIC option of MacroModel[71] version 3.5 or 4.0 (Columbia University, New York, NY, USA). The geometries of the side chain conformers were optimized (holding the heterocyclic core structure fixed) using MacroModel BatchMin. Calculations of ring centroids, least squares fitting, and excluded/included[72,73] volume analyses were also carried out using SYBYL. In certain cases, it was necessary not only to fit points corresponding to ligand heteroatoms and ring centroids, but also presumed hydrogen bonding sites located in the receptor. This was accomplished through the addition of hydrogen bond extension vectors attached to ligand hetero or hydrogen atoms which are capable of forming a hydrogen bond with the receptor. The geometries of these vectors were chosen to mimic ideal hydrogen bond orientations.[74-79] The receptor modeling strategy employed here has been applied earlier for the inverse agonist/ antagonist and agonist pharmacophores.[47,48,50] The structures used in the receptor modeling represent ten different structural families (Tables 1-12)[80-91] that bind to the BzR with high affinities; ligands with poor affinities were employed as negative controls.

2.2   Synthetic Chemistry Required for SAR

2.2.1   β-carbolines

Although many different syntheses of β-carbolines have been reported, one of the most versatile is depicted in Scheme I.[92,93] This route employs a Pictet-Spengler reaction and is general since most ring-A substituted tryptophans can be reacted with various aldehydes to provide a number of substituted β-carbolines for SAR studies.[92–97] The synthesis of inverse agonists BCCE (16) and BCCM (17) can be accomplished in this fashion, as well as of the partial inverse agonist/antagonist BCCT (8).[98] The β-carboline DMCM (15) illustrated in Scheme I, prepared by Neef and coworkers[97] is a potent convulsant which also acts *via* the GABA$_A$/BzR/Cl system. These β-carbolines have been employed to study many pharmacological processes including anxiety, convulsions, sleep, and cognition enhancement.

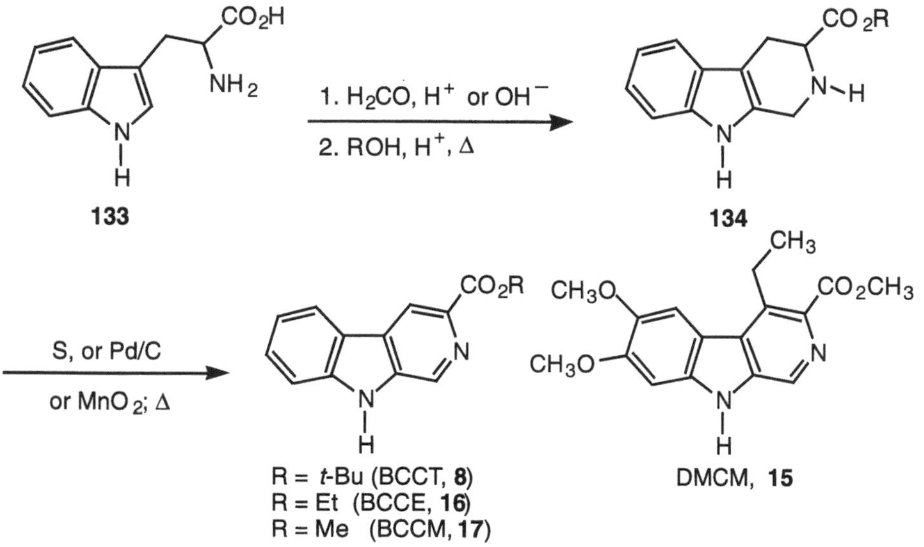

R = t-Bu (BCCT, 8)
R = Et (BCCE, 16)
R = Me (BCCM, 17)

DMCM, 15

Scheme I

Since β-carboline-3-carboxylic acid alkyl esters are generally short-lived *in vivo*, and nearly insoluble in water [DMCM (15) and BCCE (16) both have solubilities less than 1 mg/mL H$_2$O],[83] the synthesis of β-carboline-3-alkyl ethers was pursued. This was accomplished by conversion of the 3-amino-β-carboline (3ABC) 135 into the corresponding diazonium salt followed by reaction with an alcoholic solvent (Scheme II). For example, 3ABC (135) was stirred with isoamyl nitrite in anhydrous

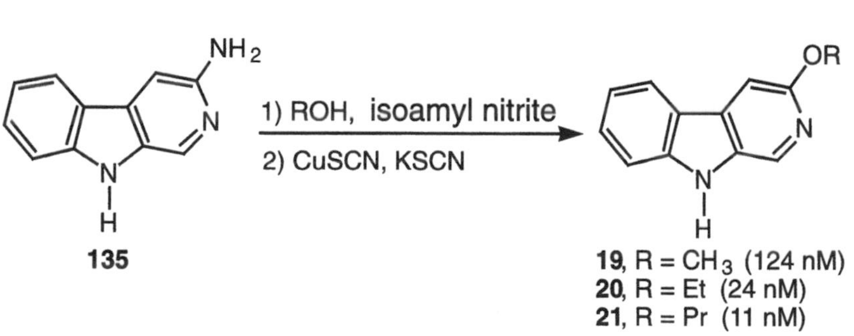

135

19, R = CH$_3$ (124 nM)
20, R = Et (24 nM)
21, R = Pr (11 nM)

Scheme II

ethanol in the presence of a copper I salt to provide 3-ethoxy-*β*-carboline **20**.[45] The other analogs (**19–29**) represented in Scheme II and Table 1 were prepared under conditions analogous to that of **20** with the exception of the neopentyl ether **26**. The preparation of the latter ether was achieved by heating the diazonium salt in neopentyl alcohol in the presence of sulfuric acid.[82]

The synthesis of 6-(benzyloxy)-4-(methoxymethyl)-*β*-carboline-3-carboxylic acid ethyl ester **1** (ZK-93423) is shown in Scheme III. The 5-benzyloxyindole **136** was allowed to react with the nitroester **137** to provide the corresponding indolylnitroester **138**, according to the method of Neef *et al.*[97] Reduction of the nitro group over Raney nickel furnished 3-(5-benzyloxy)-indol-3-yl-2-propanoic acid ethyl ester **139** in reasonable overall yield from **136** (Scheme III). Cyclization of amine **139** with glyoxylic acid was achieved *via* a modification of the Pictet-Spengler reaction[99] to furnish the 1,2,3,4-tetrahydro-*β*-carboline intermediate. Oxidation of the tetrahydro-*β*-carboline over Pd/C in mixed xylenes[97] at reflux or over activated manganese dioxide provided the fully aromatic *β*-carboline **1**. The other *β*-carbolines (**4**, **10–12**, **31**, and **33**) depicted in Table 1 were synthesized *via* a modification of the route illustrated in Scheme III.[49]

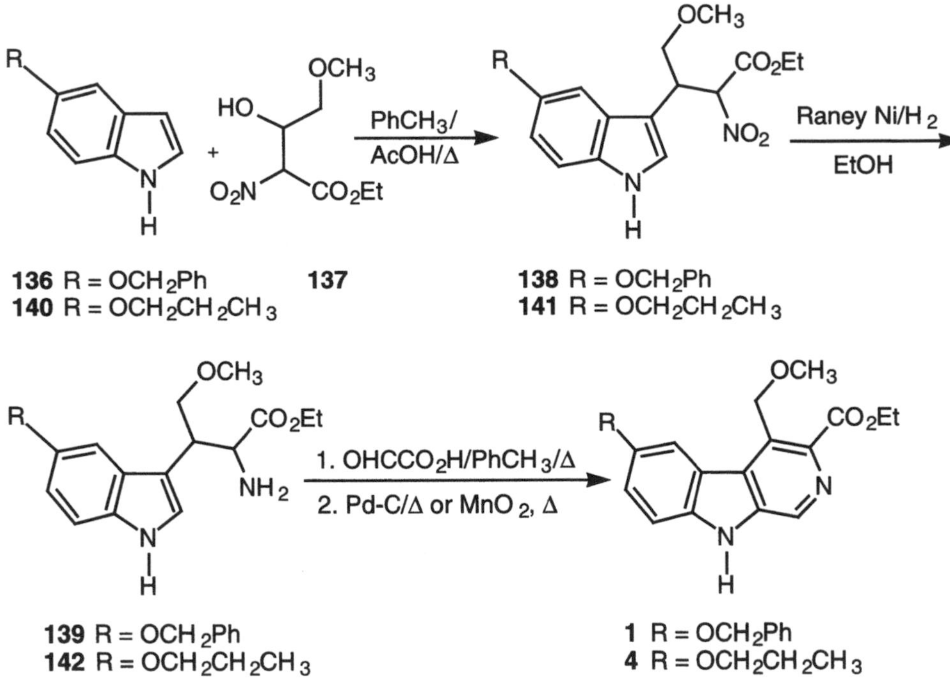

R
HO
OCH3
+
N
H
O2N
CO2Et
PhCH3/
AcOH/Δ
OCH3
CO2Et
NO2
Raney Ni/H2
EtOH
136 R = OCH2Ph
140 R = OCH2CH2CH3
137
138 R = OCH2Ph
141 R = OCH2CH2CH3
NH2
1. OHCCO2H/PhCH3/Δ
2. Pd-C/Δ or MnO2, Δ
139 R = OCH2Ph
142 R = OCH2CH2CH3
1 R = OCH2Ph
4 R = OCH2CH2CH3

Scheme III

The 5-propyloxyindole **140** was reacted with the nitroester **137** under analogous conditions to those employed for ZK-93423 (**1**)[49,97] to provide the tryptophan derivative **141**. This material was reduced to the amine **142** and subsequently converted into 6PBC (**4**), as illustrated in Scheme III.

Table 1 β-Carbolines[46,47,49,50,58,80]

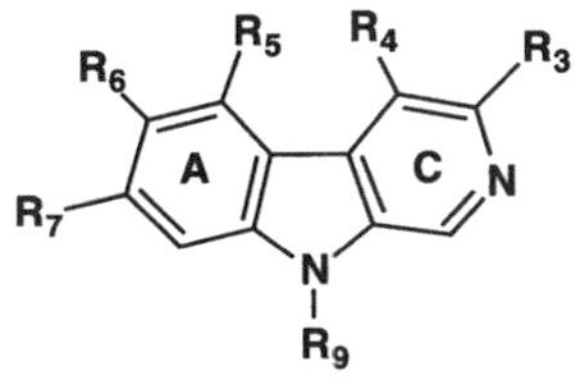

		R_3	R_4	R_5	R_6	R_7	R_9	IC_{50} (nM)	Profile	Ref.
ZK-93423	1	CO_2Et	CH_2OMe	H	OCH_2Ph	H	H	1.0	full agonist	49
ZK-91296	2	CO_2Et	CH_2OMe	OCH_2Ph	H	H	H	0.4	partial agonist	50
abecarnil	3	CO_2-i-Pr	CH_2OMe	H	OCH_2Ph	H	H	0.82	partial agonist	50
6-PBC	4	CO_2Et	CH_2OMe	H	O-n-Pr	H	H	5.0	partial agonist	50
BCCBu	5	CO_2-n-Bu	H	H	H	H	H	29	antagonist	80
BC3OP	6	O-n-Pr	H	H	H	H	H	11	antagonist	47
BCCP	7	CO_2-n-Pr	H	H	H	H	H	3.0	antagonist	46
BCCtBu	8	CO_2-t-C_4H_9	H	H	H	H	H	10	antagonist	47
ZK-93426	9	CO_2Et	Me	O-i-Pr	H	H	H	0.4	antagonist	81
BC6OM	10	CO_2Et	CH_2OMe	H	OMe	H	H	1.0	antagonist	50
BC6H	11	CO_2Et	CH_2OMe	H	H	H	H	2.3	antagonist	49
BC6OBz	12	CO_2Et	H	H	OCH_2Ph	H	H	8.9	inactive	49
BC4Et	13	CO_2Et	CH_2CH_3	H	OCH_2Ph	H	H	22	inactive	49
BC6NA	14	CO_2Et	CH_2OMe	H	O-α-naphthyl	H	H	73	inactive	*
DMCM	15	CO_2Me	Et	H	OMe	OMe	H	1.0	full inverse agonist	47
BCCE	16	CO_2Et	H	H	H	H	H	5.0	full inverse agonist	47
BCCM	17	CO_2Me	H	H	H	H	H	5.0	full inverse agonist	47
BC3COP	18	CO-n-Pr	H	H	H	H	H	2.8	inverse agonist	82

Table 1 β-Carbolines *(Continued)*

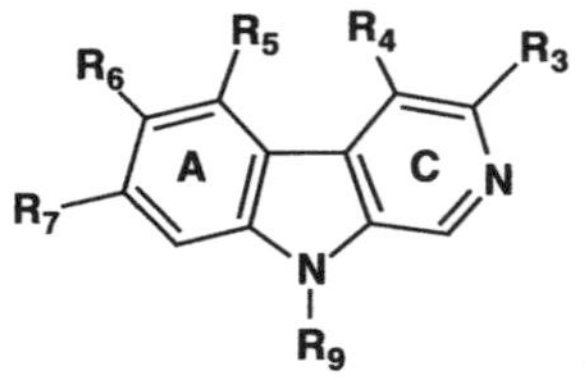

		R_3	R_4	R_5	R_6	R_7	R_9	IC_{50} (nM)	Profile	Ref.
BC3OMe	19	OCH_3	H	H	H	H	H	124	partial inverse agonist	45
BC3OE	20	OCH_2CH_3	H	H	H	H	H	24	partial inverse agonist	45
BC3OnPr	21	$OCH_2CH_2CH_3$	H	H	H	H	H	11	antagonist	82
BC3O2Bu	22	$OCH(CH_3)CH_2CH_3$	H	H	H	H	H	471	NT	82
BC3OiBu	23	$OCH_2CH(CH_3)_2$	H	H	H	H	H	93	NT	82
BC3OiP	24	$OCH_2CH_2CH(CH_3)_2$	H	H	H	H	H	535	NT	82
BC3OBz	25	OCH_2Ph	H	H	H	H	H	>1000	NT	82
BC3ONP	26	$OCH_2C(CH_3)_3$	H	H	H	H	H	104	NT	82
BC3OnBu	27	$OCH_2CH_2CH_2CH_3$	H	H	H	H	H	98	NT	82
BC3OiPr	28	$OCH(CH_3)_2$	H	H	H	H	H	500	NT	82
BC3EtOH	29	OCH_2CH_2OH	H	H	H	H	H	220	NT	*
BC6MOM	30	CO_2Et	CH_2OCH_3	H	$O\text{-}n\text{-}Pr$	H	H	309	NT	*
ZK4H9M	31	CO_2Et	H	H	OCH_2Ph	H	Me	>5000	negative control	49
BC3M9M	32	CO_2Me	H	H	H	H	Me	>5000	negative control	47
BC9M6P	33	CO_2Et	CH_2OCH_3	H	OCH_2Ph	H	Me	945	negative control	49
BC4E6OP	34	CO_2Et	CH_2CH_3	H	OCH_2Ph	H	Me	>5000	negative control	49

* E. Cox, A.J. Laloggia, B. Harris, P. Skolnick, and J.M. Cook unpublished results. NT = not tested.

Table 2 Dihydropyrido[3,4-*b*,5,4-*b′*]diindoles[45–48,83]

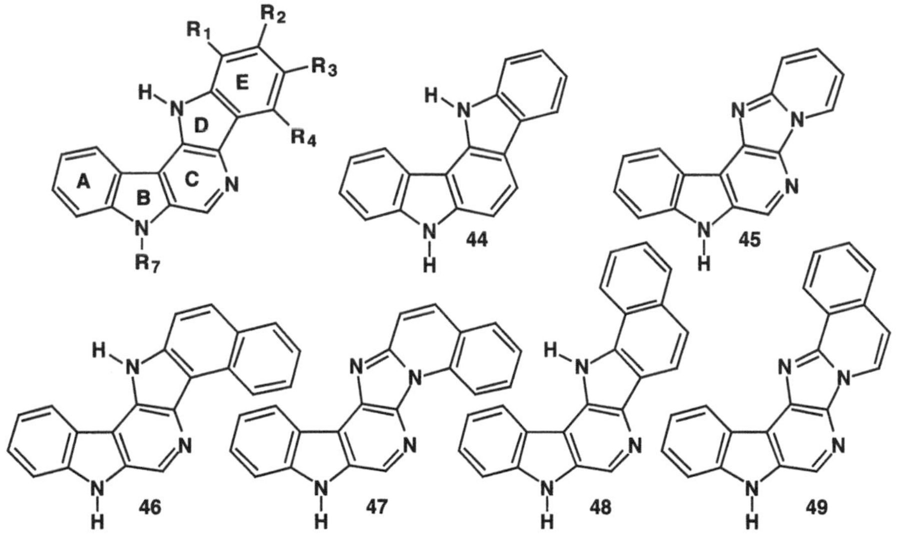

		R_1	R_2	R_3	R_4	R_7	IC_{50} (nM)	Profile	Ref.
PRI2Cl	35	H	Cl	H	H	H	10	antagonist	[48]
PRI3F	36	H	H	F	H	H	6	antagonist	[48]
PRI1Me	37	Me	H	H	H	H	83	ND	[47]
PRI2Me	38	H	Me	H	H	H	10	antagonist	[47]
PRI	39	H	H	H	H	H	4	inverse agonist	[45,46]
PRI2MeO	40	H	OMe	H	H	H	8	inverse agonist	[48]
PRI7Me	41	H	H	H	H	Me	1163	negative control	[47]
PRI3Cl	42	H	H	Cl	H	H	2150	negative control	[48]
PRI4Cl	43	H	H	H	Cl	H	715	negative control	[47]
PRICZ	44						1970	negative control	[45,46]
PRIMD	45						39	not active	[83]
PRI3BZ	46						>>125	negative control	[83]
PRIMD3BZ	47						3220	negative control	[83]
PRIPY1BZ	48						>1000	negative control	[83]
PRIMD1BZ	49						>2000	negative control	[83]

2.2.2 Pyridodiindoles

In 1987 Trudell reported both the synthesis and inverse agonist activity of the novel heterocycle, 7,12-dihydropyrido[3,2-*b*:5,4-*b′*]diindole **39** (Table 2).[100] The preparation of this molecule was achieved by heating 4-oxo-2-benzoyl-1,2,3,4-tetrahydro-*β*-carboline **143** with phenylhydrazine for a period of several hours. It was soon discovered that the parent diindole **39** bound to BzR *in vitro* with an IC_{50} of 4 nM and exhibited proconvulsant activity in the pentylenetetrazole paradigm.[100] The preparation of pyridodiindole analogs for SAR was achieved, as illustrated in Scheme IV.[101]

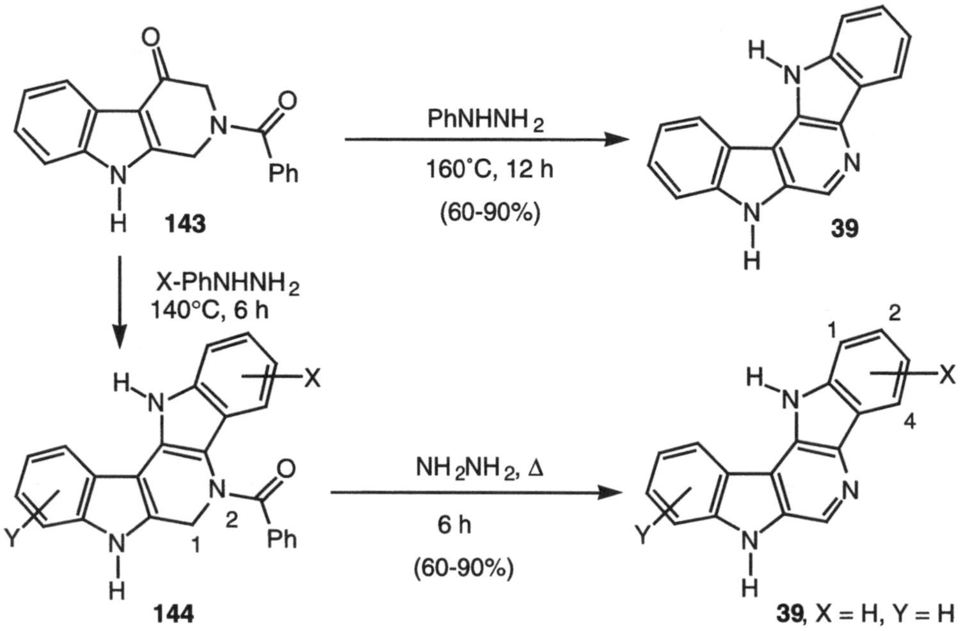

Scheme IV

The ketobenzamide **143** was heated with the substituted phenylhydrazine for six hours after which hydrazine was added and heating continued for an additional six hours. In this manner only one or two equivalents of the substituted phenylhydrazine were required for the Fischer-indole cyclization to provide benzamide **144**. The hydrazine-mediated cleavage of the benzamide function was followed by oxidation across the C1-N2 bond to furnish the desired diindole represented by **39**. This is the only class of BzR ligands discovered, to date, which is completely rigid and possesses a planar topography.

2.2.3 Benzofused pyridodiindole molecular yardsticks

This same approach has been employed to prepare a number of planar "molecular yardsticks" **45–49** designed to probe the dimensions of the binding cleft at BzR. These agents depicted in Scheme V were available from the 4-oxo-tetrahydro-β-carboline **143** by way of a Fischer-indole cyclization with the appropriate arylhydrazines. It is important to note that both the parent diindole **39** and the parent imidazoindole **45** bound tightly to the BzR while the benzofused ligands **46, 47, 48,** and **49** exhibited no appreciable affinity at the BzR. The latter heterocycles are

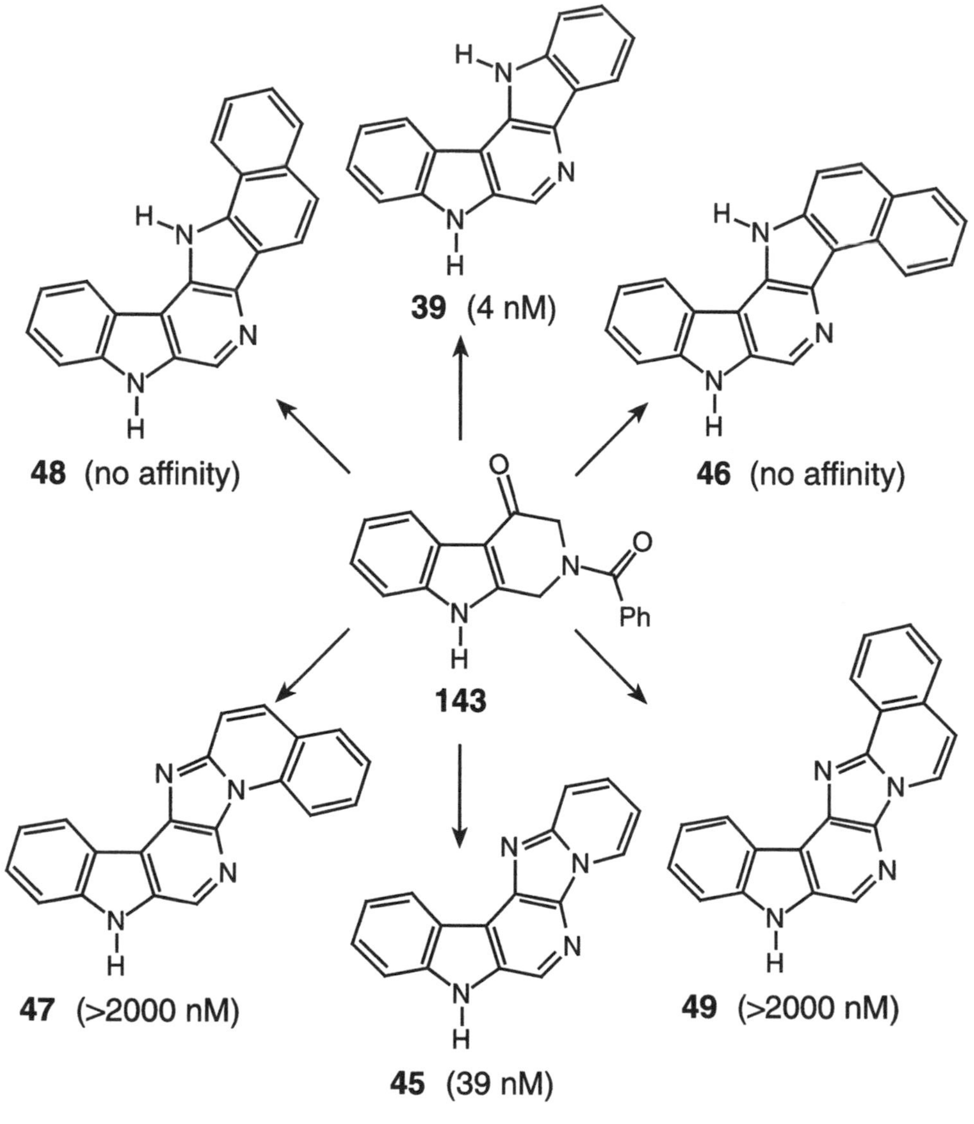

39 (4 nM)

48 (no affinity)

46 (no affinity)

143

47 (>2000 nM)

45 (39 nM)

49 (>2000 nM)

Scheme V Molecular Yardsticks.

important as negative controls in receptor modeling at the BzR. The molecular yardsticks depicted in Scheme V[83] when combined with molecular graphics (SYBYL) were important in defining the regions of steric repulsion (S_1 and S_2) between ligands and the receptor protein at the BzR.[83] As a direct consequence of this approach, more accurate dimensions of lipophilic pockets (L_2 and L_3) at the BzR have been elucidated.[91]

2.2.4   Benzofused benzodiazepines

In order to further delineate the dimensions of the BzR, the benzofused 1,4-benzodiazepine derivatives **123, 127**, and **131** (Table 12) were synthesized by methods outlined in the review by Fryer.[102,103] In this sequence, the preparation of the necessary α-aminobenzophenones **145–147** is illustrated in Scheme VI.[104]

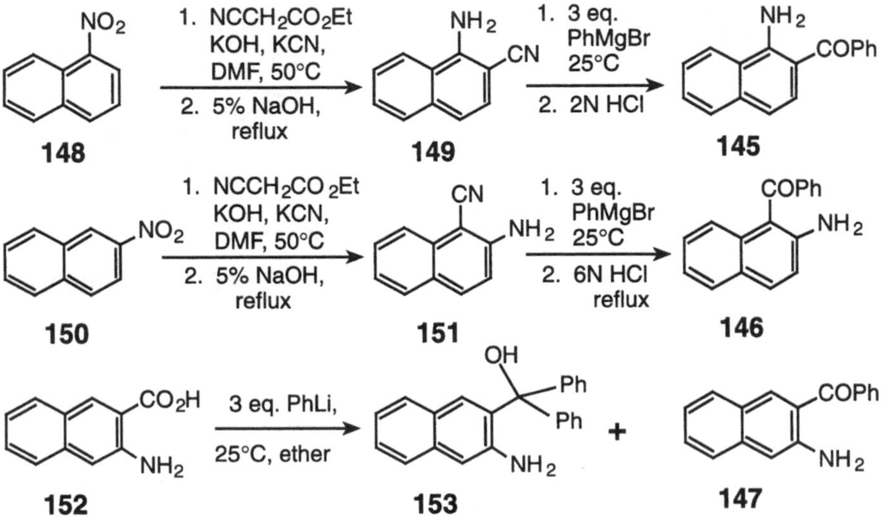

Scheme VI

Conversion of 1-nitronaphthalene **148** into 1-amino-2-naphthonitrile **149** was accomplished, according to the method of Tomioka *et al.*[105,106] as shown. This novel transformation constitutes a one pot sequence to convert an arylnitro compound into an *ortho* substituted aminoarylnitrile. The 2-nitro analog **150** was converted into 2-amino-1-naphthylnitrile **151** under conditions analogous to those described above in 56% yield.

The 1-amino-2-naphthonitrile **149** was stirred with three equivalents of phenyl-magnesium bromide and this was followed by hydrolysis to provide 1-amino-2-benzoylnaphthalene **145** in 85% yield. The reaction of 2-amino-1-naphthonitrile **151** with three equivalents of phenylmagnesium bromide, as illustrated in Scheme VI, gave 2-amino-1-benzoylnaphthalene **146** in 86% yield. The 2-amino-3-benzoyl-naphthalene derivative **147** was synthesized from commercially available 3-amino-2-naphthoic acid **152** and phenyllithium at 25°C. This process also generated 10% of the by-product, diphenyl tertiary alcohol **153**, which was readily separable from **147** by flash chromatography.[91] The three α-aminobenzophenones (**145–147**) were converted into the target benzofused benzodiazepines **123, 127**, and **131**, according to the methods of Sternbach[107] and Fryer.[102,103]

The synthesis of the fluorosubstituted analogs **157, 161**, and **164** proved to be more difficult. Although a number of synthetic routes to the required orthoaminofluoro ketones **157**, **161**, and **164** were attempted,[91,104] only the most successful one will be described here. The amino acid **152** was stirred with benzoyl chloride at 170°C to provide the 2-phenyl-4*H*-naphtho[2,3-*d*]-1,3-oxazin-4-one **154** in 91% yield, according to the procedure of Clemence *et al.*[108] Treatment of the oxazin-4-one **154** with one equivalent of *o*-fluorophenyllithium at −78°C furnished the desired 2′-fluorophenylketone **155** in 51% yield,[109] accompanied by 14% of the bis(2′-fluorophenyl)alcohol **156**. Hydrolysis of benzamide **155** under acidic conditions furnished the desired α-aminoketone **157** in excellent yield (see Scheme VII).

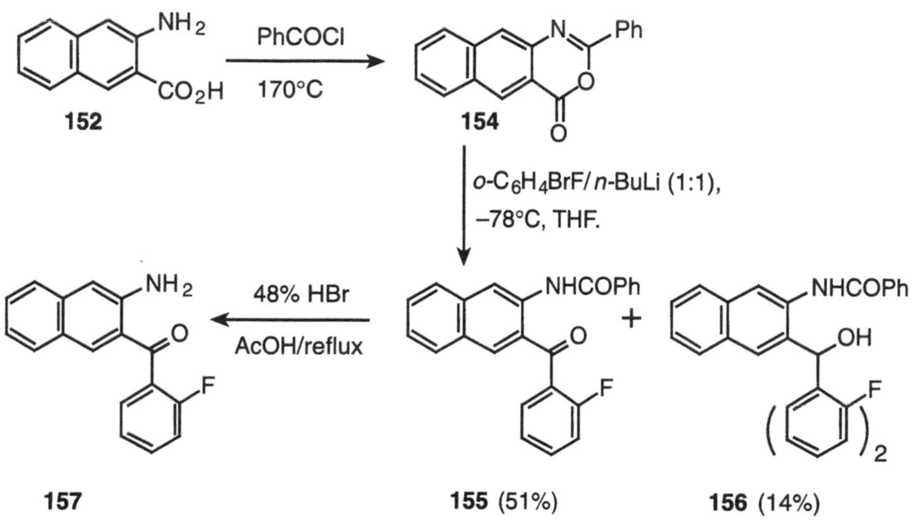

Scheme VII

In order to prepare the α-amino-2′-fluorobenzophenone required for the synthesis of benzofused 1,4-benzodiazepine **128**, 1-amino-2-naphthonitrile **149** was hydrolyzed to provide the corresponding 1-amino-2-naphthoic acid **158** in 85% yield. This amino acid **158** was heated with benzoyl chloride at 170°C to furnish the corresponding naphthoxazinone **159** in 91% yield. The naphthoxazinone **159** was stirred with one equivalent of *o*-fluorophenyllithium at −78°C to provide the ketoamide intermediate **160** as the sole product. This process was followed by hydrolysis to furnish the desired amine **161** in excellent yield (Scheme VIII).

For the synthesis of target **132**, the starting amino acid related to **151** was not easily obtained, a related intermediate **162** from the process developed by Tomioka *et al.*[105] was synthesized (Scheme IX). Hydrolysis of the nitrile **162** in sulfuric acid provided the 2-biscyanonaphthoxazinone **163**. This naphthoxazinone was then stirred with two equivalents of *o*-fluorophenyllithium at −78°C and this step was followed by hydrolysis to furnish the desired fluorine substituted 1-(2′-fluorobenzoyl)-2-aminonaphthalene **164** in 71% yield. The conversion of the orthoamino-2′-fluorophenylketones into the desired 2′-fluorosubstituted benzofused 1,4-benzodiazepines **124**, **128**, and **132** was again accomplished by the methods of Sternbach[107] and Fryer.[102,103]

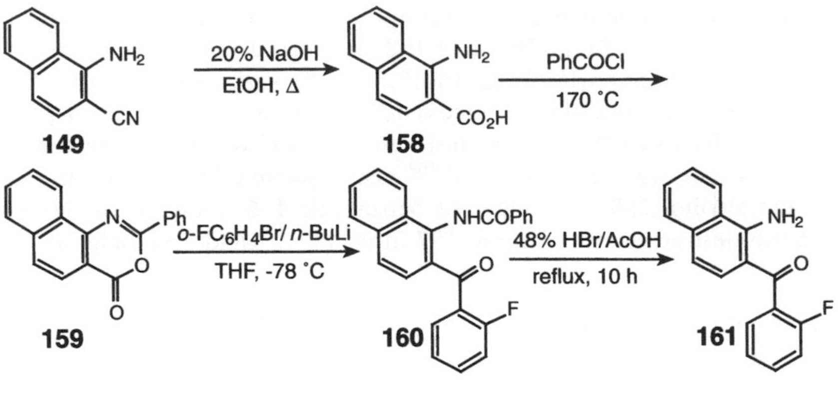

Scheme VIII

3. INVERSE AGONIST/ANTAGONIST PHARMACOPHORE

3.1 Superposition of Rigid BzR Ligands and Interactions with Receptor Hydrogen Bonding Sites H₁ and A₂

In order to define the inverse agonist/antagonist pharmacophore, rigid high affinity ligands were employed. The diindole series of ligands were synthesized and employed (as described in Section 2.2.2) in our laboratories for this purpose. Ligands chosen for the initial fit were BCCE (**16**), the pyridodiindole **39** (Table 2), phenylpyrazolo-quinolone **52** (Table 3), and thienylpyrazoloquinolone **63** (Table 4), all of which exhibited inverse agonist activity when their D- or E-rings, respectively, were unsubstituted (see Figure 3a).

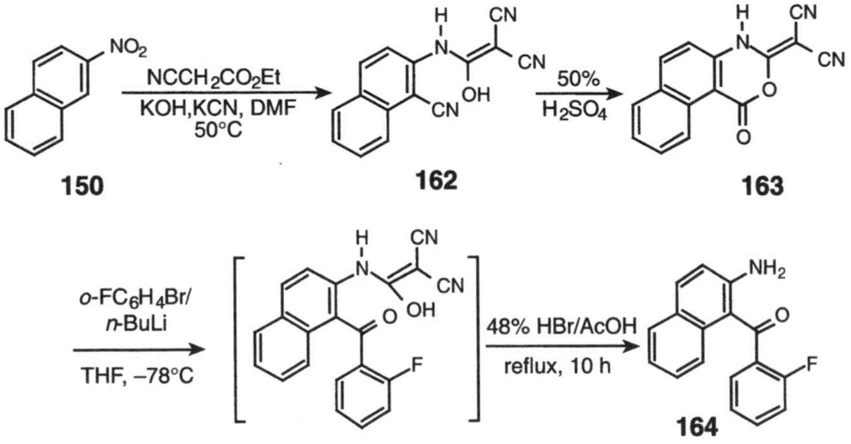

Scheme IX

Table 3   2-Aryl-pyrazolo[3,4-c]quinolin-3-ones (APQ)[41,84–86]

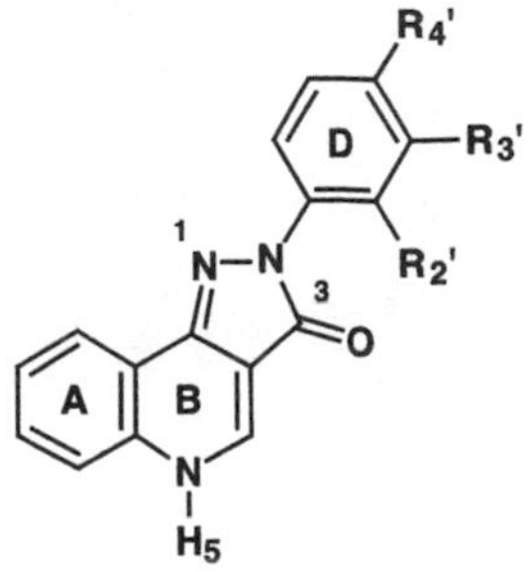

		R'_4	R'_3	R'_2	IC_{50} (nM)	Profile	Ref.
CGS-9896	**50**	Cl	H	H	0.6	partial agonist	84
CGS-9895	**51**	OMe	H	H	0.1	partial agonist	84
CGS-8216	**52**	H	H	H	0.4	inverse agonist	84
APQ3'OMe	**53**	H	OMe	H	0.5	agonist/antagonist	86
APQ3'Cl	**54**	H	Cl	H	3.9	agonist	86
APQ2'Cl	**55**	H	H	Cl	70	antagonist	86
APQ2'OMe	**56**	H	H	OMe	990	negative control	86

Table 4   Thien-2-yl-pyrazolo[3,4-c]quinolin-3-ones (2TPQ)[85,87]

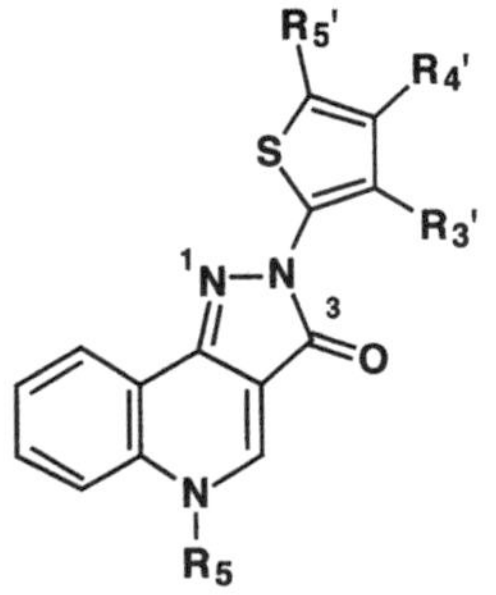

		R_5	R'_5	R'_4	R'_3	K_i (nM)	Profile	Ref.
2TPQ5'Me	**57**	H	Me	H	H	0.3	partial agonist	85
2TPQ5'Et	**58**	H	Et	H	H	1.0	partial agonist	85
2TPQ5'nBu	**59**	H	n-Bu	H	H	1.5	partial agonist	85
2TPQ4'nBu	**60**	H	Cl	H	H	1.9	inverse agonist	85
2TPQ4'5'diMe	**61**	H	Me	Me	H	0.7	partial agonist	87
2TPQ3'Et5'Me	**62**	H	Me	H	CO_2Et	9.8	antagonist	87
2TPQ	**63**	H	H	H	H	0.4	inverse agonist	85
2TPQ4'Me	**64**	H	H	Me	H	0.5	inverse agonist	85
2TPQ5Me	**65**	CH_3	H	H	H	>1023	negative control	87

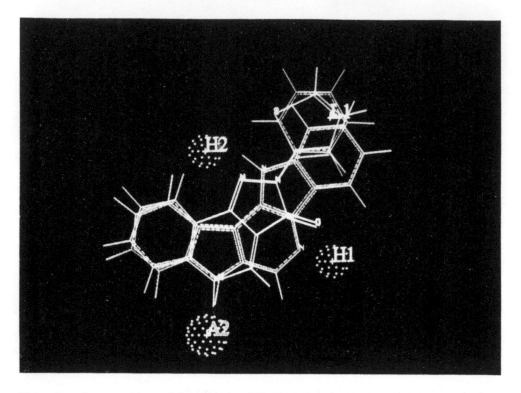

Figure 3a   Superposition of 3-carboethoxy-β-carboline (**16**), dihydropyridodiindole (**39**), pyrazoloquinolinone (**52**), and thienylpyrazoloquinolinone (**63**).

The potent affinity of **39** was immediately employed to formulate a model for the inverse agonist/antagonist pharmacophore at BzR.[45,47,100,101] The rigid nature of the structure of the diindole **39** permitted the effective superposition (using a template approach) of the active analogs, *via* the method of Marshall,[72,73] to arrive at the topography of the inverse agonist/antagonist site in the binding cleft. These results supported the earlier claim[92,96] that inverse agonists and agonists which can assume a planar or pseudoplanar topography exhibit the most potent activity *in vitro* at the BzR. The 2-methoxypyridodiindole (**40**) was the most active inverse agonist in the series and the corresponding 2-pivaloyl ester analog (a 2-hydroxydiindole prodrug)[101] also exhibited proconvulsant activity. Substitution of the parent diindole **39** with halogen atoms (especially 2-chloro) resulted in a ligand **35** with antagonist activity. Based on a limited number of examples, it appears that substitution of ring-E with electron releasing groups provides an agent with inverse agonist properties while substitution with halogen substituents results in an antagonist profile of activity (see Table 2).[101]

In the pyridodiindole series, both the N5 pyridine nitrogen atom and the hydrogen atom attached to the N7 position appear necessary for high affinity. If either of these positions are blocked, a marked decrease in affinity is observed. For example, the parent pyridodiindole **39** possesses high affinity for the BzR ($IC_{50} = 4$ nM); however, if the N5 nitrogen atom is replaced with a methine moiety (*e.g.*, **44**; $IC_{50} = 1970$ nM) or if the hydrogen atom attached to the pyrrole N7 nitrogen atom is replaced with a methyl group (*e.g.*, **41**; $IC_{50} = 1163$ nM), a drastic reduction in affinity for the BzR is observed. These observations demonstrate that both the N5 and N7 sites of the pyridodiindoles are critical for high affinity and it is, therefore, proposed that these ligand sites interact with receptor hydrogen bond donor and acceptor sites labeled as H_1 and A_2, respectively (Figure 3b). As mentioned above, pyridodiindoles which possess strong electron releasing substituents at C2 exhibit enhanced inverse agonist activity [*e.g.*, **40** ($R_2 = OCH_3$)][101] while diindoles with other substituents at position-1, -2, or -3 [*e.g.*, **35** ($R_2 = Cl$), **36** ($R_3 = F$), **37** ($R_1 = CH_3$), and **38** ($R_2 = CH_3$)], exhibit antagonist activity. As expected, ligands which elicit an antagonistic profile are compounds whose structures lack one or more features required for a potent interaction in the agonist and inverse agonist pharmacophore/receptor models (Figures 1 and 2).

The hydrogen bond donor group H_1 on the receptor protein interacts with the carbonyl moiety at position-3 of the pyrazoloquinolinones and the pyridine (N5) nitrogen atom of the pyridodiindoles. The donor group H_1 in the new inclusive model corresponds to the hydrogen bond donor group H_1 of the inverse agonist/antagonist pharmacophore developed earlier.[45–48] In agreement with the previously proposed inverse agonist/antagonist model, ligand interaction with the hydrogen bond acceptor group A_2 and donor site H_1, as well as with the specific lipophilic area L_1 is required to elicit potent inverse agonist activity (see Figure 3a).

3.2 Evidence from 3D-QSAR for BzR Sites H_1, A_2, and H_2

The existence of the receptor hydrogen bonding sites H_1 and A_2 is further supported by a 3D-QSAR analysis of BzR ligands[47,82] which revealed significant correlations between enhanced binding affinity and increases in the positive electrostatic potential about H_1 and decreases in the negative potential about A_2. These 3D-QSAR analyses also demonstrated that high affinity strongly correlated with increases in the negative electrostatic potential in the vicinity of the N1 nitrogen atom of the pyrazoloquinolinone series (Tables 3–5).[47,82] This observation has led in part to the proposal of a third hydrogen bonding site in the receptor which we have designated as H_2 (see Figure 3b).[47]

3.3 β-Carbolines

3.3.1 Discovery

β-Carboline-3-carboxylic acid ethyl ester [BCCE (**16**); Table 1] was isolated from human urine by Braestrup *et al.*[110] and subsequently shown to bind to benzodiazepine receptors (BzR) *in vitro* and *in vivo*.[111] Although the origin of this β-carboline was shown to be an artifact of the isolation process,[112] nevertheless the potent anxiogenic and proconvulsant properties (inverse agonism) displayed by this ligand have

Table 5   Thien-3-yl-pyrazolo[3,4-*c*]quinolin-3-ones (3TPQ)[85,87]

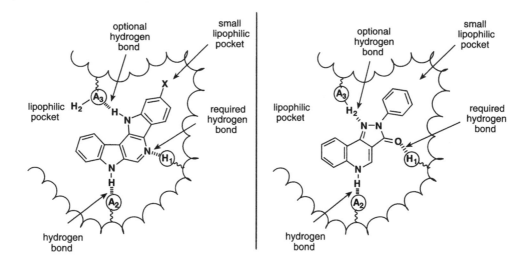

		R'_5	R'_4	R'_2	K_i (nM)	Profile	Ref.
3TPQ5'nBu	**66**	*n*-Bu	H	H	1.86	partial agonist	85
3TPQ5'Me2'Br	**67**	Me	H	Br	27.2	antagonist	87
3TPQ5'Et	**68**	Et	H	H	0.41	antagonist	85
3TPQ	**69**	H	H	H	0.73	inverse agonist	85
3TPQ5'Me	**70**	Me	H	H	0.32	inverse agonist	85
3TPQ4'Me	**71**	H	Me	H	6.86	inverse agonist	87

Figure 3b   Proposed model of the benzodiazepine receptor inverse agonist/antagonist active site shown with: (left) 7,12-dihydropyridol[3,2-*b*:5,4-*b'*]diindole (IC$_{50}$ = 4 nM) and (right) 2,5-dihydro-2-phenyl-3H-pyrazolo[4,3-*c*]quinolin-3-one (CGS-8216; IC$_{50}$ = 0.4 nM)[82,83] interacting with the hydrogen bond donor sites (H$_1$ and H$_2$) and hydrogen bond acceptor sites (A$_2$ and A$_3$) on the binding protein. Binding to H$_2$/A$_3$ is not necessary for inverse agonist activity; however, interaction with H$_2$/A$_3$ and especially H$_2$ does appear to enhance affinity. The development of the current pharmacophore was based on the superposition of ligands to the planar template **39**.

stimulated a number of investigations into the synthesis and biological properties of β-carbolines. It is now known that such inverse agonists act through the $GABA_A/$ BzR/Cl ion channel by retarding the flow of chloride ions through this ionophore. This results in a decrease in the inhibitory effects of the neurotransmitter gamma aminobutyric acid (GABA) providing then an excitatory response in the CNS.[5]

3.3.2   Therapeutic potential

Even though DMCM (15), BCCE (16), and BCCM (17) have served as important pharmacological tools with which to study benzodiazepine receptors, the potent proconvulsant/convulsant activity of these ligands coupled with poor water solubility (<1 mg/mL in H_2O)[83] have impeded the realization of a clinical use for such agents. Likewise BCCT (8) is an excellent antagonist of the CNS depressant effects of diazepam,[98] moreover, this β-carboline has been shown to be devoid of proconvulsant/ convulsant activity. However, the low water solubility and short shelf-life of this ligand have precluded its use in the clinic, to date. With respect to drug design, there is a need for inverse agonists which are devoid of the proconvulsant/convulsant activity of the typical β-carboline-3-alkyl esters. These new agents should also be long-lived in vivo and exhibit enhanced water solubility, as compared to the simple β-carbolines of Scheme I. Such drugs might be important in the treatment of hepatic encephalopathy,[113,114] the reversal of the effects of barbiturate-alcohol[115] induced CNS depression (overdose), or in the enhancement of cognition.[9,22] It is important to note that DMCM (15) has been shown to reverse the effects of pentobarbital-induced CNS depression (lethality) in mice, while Ro 15-4513 has reversed some of the coma-like effects[115] in a rat model of hepatic encephalopathy.[113,114] Agents targeted against age-associated memory impairment as well as enhancement of general memory-learning have also attracted a great deal of attention. Sarter has recently shown that partial (or selective) inverse agonists do indeed enhance cognition in some paradigms.[23,24]

3.3.3   Definition of active conformation

The rigid pyridodiindole and pseudoplanar pyrazoloquinoline ligands in turn served as templates to define the active conformation of flexible BzR ligands. The 2-methylpyridodiindole 38 was especially useful for defining the active conformation of flexible analogs, since it is the largest rigid ligand known which retains significant affinity ($IC_{50} = 10$ nM) for the BzR and therefore maps out a significant portion of the sterically allowed regions of the receptor. In contrast, substitution at the 1-, 3-, or 4-positions (e.g., pyridodiindoles 37, 42, and 43) is not compatible with high affinity.[101] Hence the region of the BzR in the vicinity of the 2-methyl group of bound 38 defines a sterically tolerated region in the receptor. The terminal methyl group of the longer chain 3-substituted ether and ester β-carbolines (6 and 7) was "threaded" through this sterically allowed region (see Figure 4) which resulted in the conclusion that the N2—C3—C=O torsional angle is syn (τ=0°) and the remainder of the torsional angles of the side chains are in the extended trans conformation

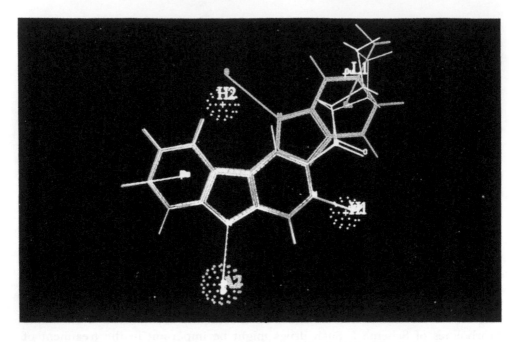

Figure 4 Assignment of the active conformation of the flexible side chains of
3-*n*-propoxy- (**6**) and 3-*n*-carbopropyl-*β*-carboline (**18**) by superposition with the rigid
2-methyldihydropyrido[3,2-*b*:5,4-*b′*]diindole (**38**).

(Figure 4). The *syn* conformation of the 3-ester group permits a three centered
hydrogen bonding interaction between the carbonyl oxygen and N2 nitrogen atom
of the *β*-carbolines with the receptor site at H_1. The three centered hydrogen
bonding interaction is apparently a very stable arrangement for it is found in the
X-ray crystallographic structure of BCCM (**17**).[63] In principle the *anti* conformation
(N2—C3—C=O torsional angle = 180°) could also form a three-centered hydrogen
bond with receptor site at H_1; however, this would involve interaction between H_1
and the less electronegative ether oxygen of the ester group and therefore would
be less favorable energetically.

3.3.4   *In vitro pharmacology*

The *in vitro* potency of 3-substituted *β*-carboline alkyl ethers appears to increase with
increasing chain length [*e.g.*, methoxy **19** (IC_{50} = 124 nM), ethoxy **20** (IC_{50} = 24 nM),
and *n*-propoxy **21** (IC_{50} = 11 nM)]. Consistent with this hypothesis, replacement of the
propyl group of **21** (IC_{50} = 11 nM) with a hydroxyethyl group **29** (IC_{50} = 220 nM)[116]
resulted in a two fold drop in binding potency *in vitro* at the BzR (see Table 1 for
details).

3.3.5 L_1/L_2 *receptor mapping*

Analogs (**19–29**) illustrated in Table 1 whose synthesis is described in Section 2.2.1

were employed both to probe the dimensions of the hydrophobic pocket (region L_1/L_2) in the BzR and to test the predictive ability of the previously reported 3D-QSAR regression model.[47] Examination of the IC_{50} values depicted in Table 1 indicate that γ-branched analogs (**23** and **26**) possess significantly higher affinity for the BzR than either β- (**22** and **28**) or δ- (**24**) branched derivatives. Furthermore, β-carboline 3-alkyl ethers with substituents at position-3 whose chain length is in excess of five heavy (i.e., non-hydrogen) atoms display poor affinity in this series. These results further indicate that β-carbolines which possess substituents of shorter length which are constrained to be in the plane of the aromatic ring tend to display inverse agonist activity while β-carbolines with longer substituents (*e.g.*, **8** and **21**) which can access regions of space above and below the plane of the aromatic rings are likely to elicit antagonist activity.[47] This antagonistic region which appears to be above and/or below receptor site L_1 is reminiscent of the general "accessory receptor antagonist binding sites" first proposed by Ariens.[117,118] In principle, β-carbolines with side chains longer than five atoms could occupy receptor region L_2 and, therefore, bind with high affinity. However, occupation of L_1 with a rigid aromatic ring appears as a necessary condition for interaction of substituents in L_2.

The β-carboline analogs in Table 1 were also used to test the predictive ability of a previously derived 3D-QSAR regression model.[47] These molecules were synthesized and tested after the regression model was developed. Gratifyingly, an excellent correlation between predicted and experimentally measured binding affinities was obtained.[82]

3.3.6 In vivo activity

The first two members of the congeneric series **19** and **20** are inverse agonists, while **21** has been shown to exhibit an antagonist profile of activity at the BzR.[82] This result suggests that smaller lipophilic substituents at position-3 which lie close to the plane of the β-carboline ring system tend to be inverse agonists [*e.g.*, BCCE (**16**), BCCM (**17**), **19**, and **20**], while compounds which carry bulkier substituents at the same position are antagonists [*e.g.*, BCCT (**8**)[98] and **21**]. It is possible that favorable steric/lipophilic interactions between substituents at position-3 and the lipophilic pocket (L_1) are responsible for an antagonist profile of activity for BCCT (**8**) and propylether **21**. A strong lipophilic interaction in this region may compromise the hydrogen bonds between the ligand and receptor at H_1 and A_2 (see Figure 4).

In vivo, both 3-methoxy-β-carboline **19** ($IC_{50} = 124$ nM) and 3-ethoxy β-carboline **20** ($IC_{50} = 24$ nM) exhibited proconvulsant activity characteristic of inverse agonists. Since **20** was the more potent of the two ethers *in vitro*, it was chosen for further study. The 3EBC **20** was shown to be proconvulsant with an ED_{50} of 7 mg/kg and inhibited stress-induced ulcer formation in the rat;[119] moreover, even at doses up to 40 mg/kg **20** did not produce convulsions. This result is significant for it demonstrates that **20** is a partial inverse agonist and may be useful for the reversal of barbiturate overdose,[115] as well as treatment of hepatic encephalopathy.[113,114] This ethyl ether **20** exhibits a longer duration of action *in vivo* than the corresponding β-carboline-3-alkylesters [BCCE (**16**), BCCM (**17**), etc.], moreover, the 3-alkylethers **19–23** depicted in Table 1 are soluble in water as the hydrochloride salts [compare

BCCE•HCl (1 mg/mL in H_2O) vs. **20**•HCl (13 mg/mL in H_2O)].[83] The remaining ether derivatives listed in Table 1 may exhibit a partial inverse agonist spectrum of activity and are therefore worthy of additional investigation.

4. AGONIST PHARMACOPHORE

4.1 Agonist Pyrazoloquinolines and Definition of Receptor Site L_2

In contrast to the inverse agonists mentioned above, phenylpyrazoloquinolinone **50** and the thienylpyrazoloquinolinone **57** bear chloro- or methyl-substituents, respectively, on ring-D and display agonist activity (Tables 3–5). As with inverse agonists, the agonist ligands also interact with the receptor protein through the common hydrogen bond donor site H_1 *via* the carbonyl oxygen at position-3 in **50** and **57** but require an additional electrostatic interaction with a hydrogen bond donor group at H_2 to exhibit agonist activity (Figure 2).[50]

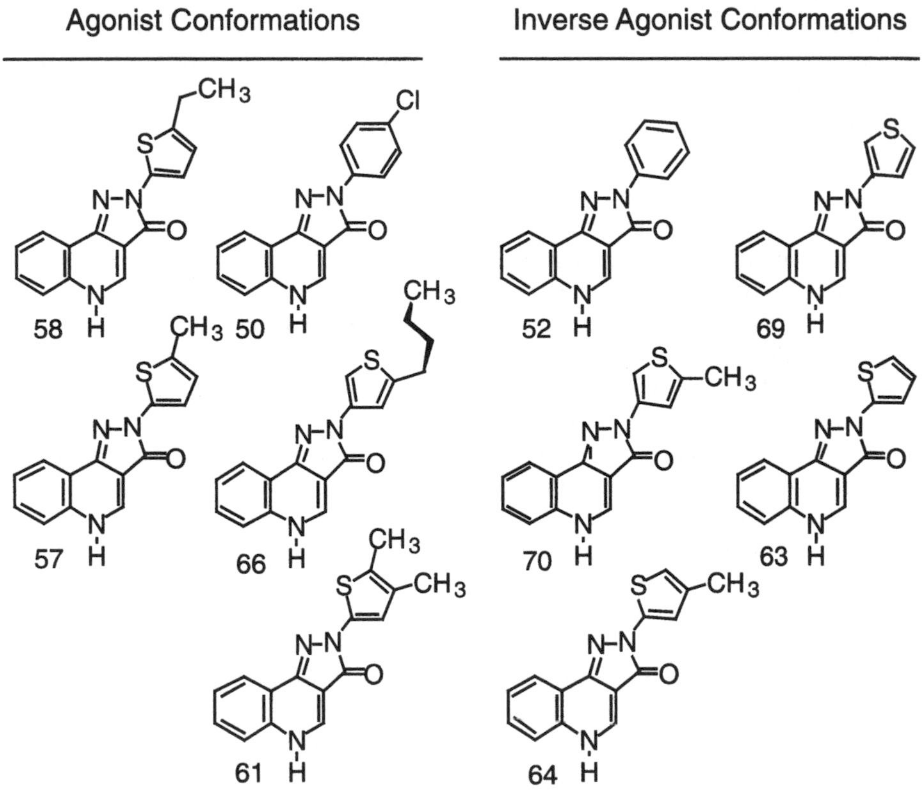

Figure 5a Proposed active conformations of agonist and inverse agonist pyrazolo[3,4-*c*] quinolin-3-ones.

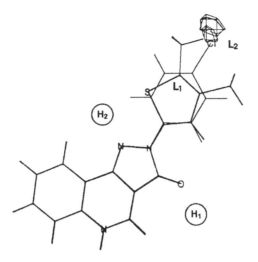

Figure 5b Agonist essential volume analysis [agonist volume (intersection of **50, 57, 58, 61**, and **66**) minus inverse agonist volume (intersection of **52, 63, 64, 69, 70**)] of pyrazolo[3,4-c]quinolin-3-ones CGS-9896 (**50**) and thien-2-yl-pyrazolo[3,4-c]quinolin-3-one (**61**) used to define the position of lipophilic receptor region L_2 required for agonist activity.

The functional SAR for the pyrazolo[3,4-c]quinolin-3-ones (Tables 3–5) appears very complex. However on closer inspection, each of these agonists have substituents which are located in a region of space which corresponds to the para position of the 2-aryl (CGS) series (see Figure 5a). A receptor essential volume analysis[72,73] has been employed to identify a volume element which is common to all agonists but which is not occupied by any of the inverse agonists (Figure 5b). We designate this region which correlates with agonist activity as lipophilic area L_2. The L_2 steric model presented here is not fully consistent with Fryer's latest report[86] which demonstrates that the *meta*-substituted phenyl-pyrazoloquinolines (i.e., **53** and **54**) are agonists. A preliminary 3D-QSAR analysis (3-21G//6-31G*//ESPFIT charges) revealed no significant correlation between either *in vivo* functional effects (n = 10; agonist = 1.0, partial agonist = 0.5, antagonist = 0.0, partial inverse agonist = –0.5, inverse agonist = –1.0) or GABA shift ratios (n = 5) and the molecular steric and electrostatic potentials. Additional experiments and analyses are underway to resolve this puzzling SAR and will be reported in due course.

4.2 Alignment of 1,4-Benzodiazepines in the Inclusive Pharmacophore

4.2.1 *Alignment with receptor descriptors H_1 and H_2*

The next question which must be addressed is how to align the agonist 1,4-benzodiazepines with respect to the pyrazoloquinolines. The benzodiazepines possess two potential hydrogen bond accepting sites of interaction: (1) the C2=O carbonyl oxygen atom and (2) the N4 imine nitrogen atom. Hence it is reasonable to assume

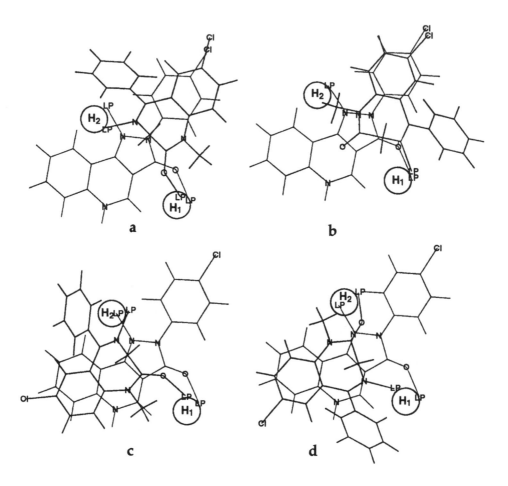

Figures 6a–d Four alternative alignments of diazepam (**72**) vs. 3H-4'-chloro-2-phenylpyrazolo[3,4-c]quinolin-3-one (CGS-9896, **50**). In alignments **6a,c** (left) receptor site H_1 interacts with the carbonyl group C2=O of diazepam while receptor site H_2 interacts with the imine nitrogen atom N4. In alignments **6b,d** (right) there is no direct interaction with H_2 and the interaction with H_1 is to the imine nitrogen atom N4 rather than to the carbonyl group C2=O. In alignment **6a,b** (top), ring-A of diazepam (**72**) is aligned with ring-D of pyrazoloquinoline **50**, while in alignments **6c,d** (bottom), ring-A of diazepam is overlaid with rings-A/B of the pyrazoloquinoline. Alignment **6a** (top-left) appears to be the most plausible of the four alternatives (see text).

that these two ligand sites interact with the receptor hydrogen bond donor sites H_1 and H_2. However, there are four possible orientations by which this interaction might occur and these are depicted in Figures 6a–d. In the first pair of alignments (Figures 6a,c) receptor site H_1 interacts with the carbonyl group C2=O of diazepam, while receptor site H_2 interacts with the imine nitrogen atom N4. In the second pair of alignments (Figures 6b,d) there is no direct interaction with H_2 and the interaction with H_1 is to the imine nitrogen atom N4 rather than to the carbonyl group C2=O.

Since interaction with receptor site H_1 is more critical than binding to site H_2, (as evidenced by the low affinity displayed by 5-des-aza pyridodiindole **44**), and since the carbonyl group C2=O is more critical to high affinity binding, we believe that the first pair of alignments (Figures 6a,c) is more plausible than the second (Figures 6b,d). This is supported by the observation that derivatives such as Ro 15-2201 (**97**) and Ro 15-0791 (**98**) which lack both a N5 nitrogen atom and a 2'-halogen substituent still retain high affinity, while elimination of the C2=O carbonyl oxygen atom as in medazepam (**83**) causes a marked decrease in affinity.

4.2.2 Alignment of the 1,4-benzodiazepine A-ring

Since the 1,4-benzodiazepines display pseudosymmetry with respect to their hydrogen bond accepting sites (a C2-axis passes through the ring-A centroid and bisects the C2=O carbonyl oxygen atom and the N4 imine nitrogen atom), it is conceivable that certain 1,4-benzodiazepines adopt a binding alignment which is opposite to that of other 1,4-benzodiazepines (Figures 6b,d). However, this seems unlikely. In order to achieve this alternative alignment, a conformational isomer of the 1,4-benzodiazepine ring system is required, yet it appears that only one conformational enantiomer of this series is readily accepted by the BzR.[81,120,121] This stereochemical requirement would suggest that all 1,4-benzodiazepine derivatives bind in the same relative orientation.

After aligning the benzodiazepines with sites H_1 and H_2, there remain two possible orientations for ring-A of the benzodiazepines with respect to ring-D (*i.e.*, receptor region L_1) or rings-A/B of the pyrazoloquinolines (compare Figures 6a,b vs. 6c,d). Alignments in which ring-A of the 1,4-benzodiazepines is superimposed with ring-D of the pyrazoloquinolines (Figures 6a,b) also place the substituent at position-7 of the 1,4-benzodiazepines in the vicinity of the *para*-substituents in ring D of the pyrazoloquinolines. The first pair of alignments (Figures 6a,b) is consistent with the observation that both of these substituents have a marked influence on the functional effects of these two classes of ligands. Furthermore, while electron withdrawing substituents at position-7 of 1,4-benzodiazepines appear necessary for high affinity and efficacy (see Tables 6–8), QSAR analysis indicates that the interaction between substituents at position-7 and the BzR probably is hydrophobic in nature.[47,49,122] This is further evidence that substituents at position-7 interact with the lipophilic receptor region L_2 while the ring-A of benzodiazepines interacts with L_1. The preference for electron withdrawing substituents on the A-ring may be an indirect one, namely strengthening a charge transfer interaction between the benzodiazepine ring-A and an electron rich aromatic ring in the BzR binding cleft.[37,122]

4.2.3 Overall alignment and definition of receptor region L_3

For the reasons outlined above, the pair of alignments depicted in Figures 6a,c are preferred over Figures 6b,d and the pair of alignments depicted in Figures 6a,b over Figures 6c,d. It is, therefore, believed that the alignment shown in Figure 6a is the most plausible of the four alternatives. Based on the alignment shown in Figure 6a, the C-ring of the 1,4-benzodiazepines occupies a region of space not occupied by the pyrazoloquinolines or β-carbolines and we tentatively assign this region as lipophilic area L_3, a receptor region that is compatible with and may enhance agonist activity.[50,51] A summary of the interactions of the major classes of BzR ligands with the inclusive pharmacophore model is listed in Table 13.

4.3   Comparison with Previously Published Models

Other research groups have employed somewhat different ligand functional groups and/or receptor descriptors in the derivation of their pharmacophore models[36] which makes direct comparisons with the model proposed above difficult. Nonetheless it is still possible to make rough, qualitative comparisons (see Table 13). Of the models that have been published previously, the Borea-Gilli model[42] would seem to correspond most closely to our preferred model illustrated in Figure 6a. In this model receptor site H_1 corresponds to the Borea-Gilli B1/B3 site, while H_2 corresponds to B2, L_2 to AG1, and L_3 to AG2. In the model illustrated in Figure 6a, a three centered hydrogen bonding interaction is proposed in which a single receptor site H_1 interacts with both the N2 pyridine nitrogen atom and the carbonyl oxygen atom of the 3-ester group of β-carbolines. In the Borea-Gilli model, however, two separate receptor sites B1 and B3 were proposed wherein the B1 site interacted with the oxygen atom C2=O of 1,4-benzodiazepines and B3 interacted with the N2 pyridine nitrogen atom of the β-carbolines. Furthermore the Borea-Gilli model lacks a site equivalent to A_2. Finally, the two models differ markedly in their explanation of functional effects caused by occupation of various regions of the receptor.

In contrast, most other workers have proposed models which match ring-A of 1,4-benzodiazepines with ring-A/B of the pyrazoloquinolines and β-carbolines (Figures 6c and 6d). For example, the models published by Fryer,[41] Wermuth,[43] and Gardner[44] most closely match that depicted in Figure 6c, while the models of Loew[37] and Codding[40] employ alignments which are roughly equivalent to that illustrated in Figure 6d. The two models reported by Crippen[38,39] differ from each other and do not use sites equivalent to H_1 and H_2, so direct comparison with our inclusive model is not possible.

4.4   Alignment of Other Structural Classes of BzR Ligands

Based on the alignment depicted in Figure 6a, the hydrogen bonding site H_1, and the lipophilic region L_1 have been established as points of interaction common to the agonist, antagonist and inverse agonist ligands in the inclusive pharmacophore/receptor model. However, the model requires a more precise definition of the relative positions of the hydrogen bonding sites H_1, A_2, and H_2 and lipophilic areas L_1, L_2, and L_3, as well as possible regions of steric repulsion. To accomplish these goals, the database of BzR ligands was expanded to ten different structural families (Tables 1–12). The pharmacophoric alignment has already been established for four of these families:

Table 6   1,4-Benzodiazepines[81]

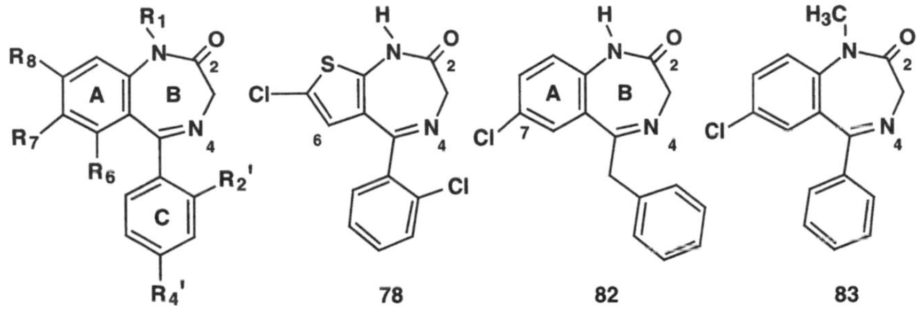

		R_1	R_6	R_7	R_8	R_2'	R_4'	IC_{50} (nM)	Profile
diazepam	**72**	Me	H	Cl	H	H	H	8.1	agonist
Ro 05-4435	**73**	H	H	NO_2	H	F	H	1.5	agonist
flunitrazepam	**74**	Me	H	NO_2	H	F	H	3.8	agonist
nitrazepam	**75**	H	H	NO_2	H	H	H	10	agonist
Ro 20-2533	**76**	Et	H	H	H	H	H	36	agonist
Ro 20-1815	**77**	Me	H	NH_2	H	F	H	65	agonist
Ro 08-9212	**78**							3.9	agonist
Ro 13-0699	**79**	Me	Cl	H	H	F	H	150	antagonist
Ro 13-0882	**80**	Me	Cl	H	Cl	F	H	300	antagonist
Ro 05-4864	**81**	Me	H	Cl	H	H	Cl	>5000	negative control
Ro 05-3580	**82**							>1000	negative control
medazepam	**83**							870	negative control

(1) β-carbolines (Table 1),[45–51]
(2) phenylpyrazoloquinolines (Table 3),[84]
(3) thienylpyrazoloquinolines (Tables 4 and 5)[85,87]
(4) 1,4-benzodiazepines (Table 6).[11,81]

4.4.1   1,2-annelated benzodiazepines

The ligands of the family of 1,2-annelated-benzodiazepines (Table 7) normally exhibit agonist activity.[11,81] The alignment and interaction of this family in the inclusive pharmacophore model is shown in Figure 7. Analogous to the procedure employed in the development of the agonist pharmacophore,[50] three points were employed in the least squares fitting of these structures to the other agonist ligands. These points were the centroid of ring-C, the lone pair of electrons on the nitrogen atoms of ring-A (N2 or N3), and the lone pair of electrons on the ring-B nitrogen atom (N5).[50] The length of the lone pair of electrons attached to the nitrogen atoms N2, N3, and N5 was extended to 1.8 Å to mimic an ideal hydrogen bond.[74–79] The nature of the interaction of the ligand with the receptor hydrogen bond donor site

Table 7   1,2-Annelated-benzodiazepines[81]

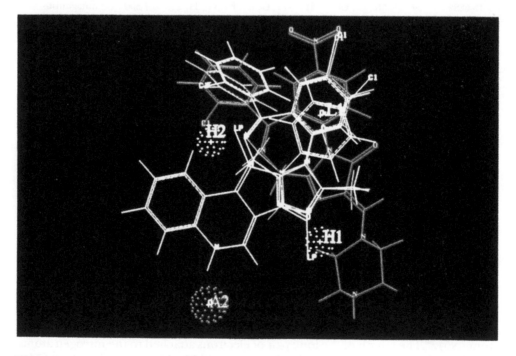

		Structure	R_8	R_1	R_2'	IC_{50} (nM)	Profile
brotizolam	**84**	a	Br	Me	Cl	1.2	agonist
Ro 17-4582	**85**	a	H	Me	Cl	3.5	antagonist
etizolam	**86**	a	Et	Me	Cl	3.1	agonist
midazolam	**87**					4.8	agonist
loprazolam	**88**					6.3	agonist
triazolam	**89**	b	Cl	H	H	4.0	agonist
U-35005	**90**	b	H	H	Cl	4.3	agonist

Figure 7   Alignment of CGS-9896 (**50**; green) with the agonists midazolam (**87**; violet), loprazolam (**88**; red), and triazolam (**89**; orange).

H_1 was determined by the type of ring-A annelation (triazole or imidazole). If the annelated A-ring was a triazole ring system as in brotizolam **84**, Ro 17-4582 **85**, or triazolam **89**, the imine nitrogen atom N2 interacts with the common receptor site descriptor H_1 (Figure 7). When ring-A was replaced with an imidazole heterocycle as in midazolam **87**, the interaction with H_1 occurred through the imine nitrogen atom N2. Loprazolam contains a completely different annelated ring system and interacts with the H_1 receptor site through the imine N3 nitrogen atom (Figure 7). The large piperazine side chain of loprazolam occupies a unique region which is not in the same plane as the phenyl ring of this molecule.

Clearly all these ligands interact with the H_2 donor group *via* the imine nitrogen atom N5. In addition to interactions with hydrogen bonding sites H_1 and H_2, the 1,2-annelated-benzodiazepines require lipophilic interactions with L_2 and L_3 to elicit potent agonist activity. In agreement with the proposed model, ligand Ro 17-4582 (**85**) has no substituents with which to fill the lipophilic pocket L_2, consequently this derivative can not exhibit a full agonist spectrum of activity; based on the lack of activity of **85** in the pentylenetetrazol protocol[81] it behaves as either an antagonist or weak agonist at BzR.

In a manner similar to the classical 1,4-benzodiazepines, the families of the 1- and 2-benzazepines which display agonist or antagonist activity[11] (see Table 8) interact with the H_1 receptor descriptor in the inclusive model through the imine nitrogen atom N3, as illustrated in Figure 8. In the case of the 1-benzazepines such as Ro

Table 8 1- and 2-Benzazepines[81]

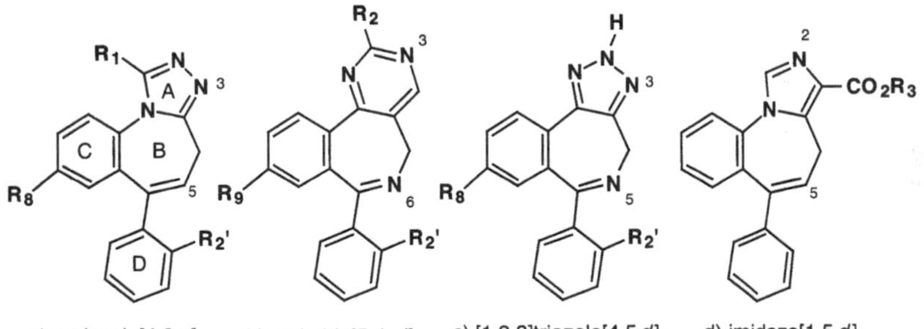

a) s-triazolo[4,3-*a*]-[1]-benzazepine   b) pyrimido[5,4-*d*]-[2]-benzazepine   c) [1,2,3]triazolo[4,5-*d*]-[2]-benzazepine   d) imidazo[1,5-*d*]-[1]-benzazepine

		Structure	R_1	R_2	R_3	R_8	R_9	R_2'	IC_{50} (nM)	Profile
Ro 14-0304	91	a	Me			Cl		F	6.5	agonist
Ro 14-7187	92	a	Me			H		H	410	antagonist
Ro 22-3245	93	b		H			Cl	Cl	2.8	agonist
Ro 22-1366	94	b		Me			Cl	F	4.0	antagonist
Ro 22-2466	95	c				Cl		F	1.9	agonist
Ro 22-3147	96	c				H		Cl	5.2	antagonist
Ro 15-2201	97	d			Me				1.5	antagonist
Ro 15-0709	98	d			Et				2.5	antagonist

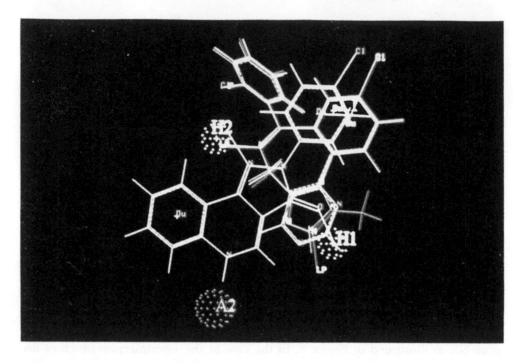

Figure 8   Alignment of CGS-9896 (**50**; green) with the agonists Ro 14-0304 (**91**; red), Ro 22-3245 (**93**; orange), and Ro 22-2466 (**95**; magenta).

14-0304 (**91**), the imine N5 nitrogen atom of the 1,2-annelated-benzodiazepines has been replaced by a methine moiety which can not interact with H_2. However, ligand **91** does exhibit agonist activity.[81] In this series of compounds, a halogen atom at position-2' on the phenyl ring-(D) appears to be required for potent activity (*e.g.*, **91** and **92**). These halogen atoms apparently act as surrogates for the imine nitrogen atom N5 (designated N4 in diazepam) and interact with H_2 (Figure 8). Moreover, a ligand such as Ro 13-9868 [81] lacks both an N(5) nitrogen atom and 2'-fluoro substituent resulting in decreased affinity at BzR with only a weak agonist response. In agreement with the model of the agonist pharmacophore, ligands Ro 14-7187 (**92**) and Ro 22-3147 (**96**) lack lipophilic substituents which can interact with the receptor site at L_2 and hence these analogs display antagonist activity. The pyrimidobenzazepine Ro 22-1366 (**94**) fulfills all the conditions required for agonist activity, indeed this derivative displays agonist activity.

4.4.2   *Triazolophthalazines*

The compounds of the 6-(alkylamino)-3-aryl-1,2,4-triazolo[3,4-*a*]phthalazine series (Table 9) have been reported to bind to the BzR with high affinity and elicit an agonist profile of activity.[88] These ligands (*e.g.*, **99–103**), which differ structurally from the compounds previously discussed, align and interact with the inclusive pharmacophore/receptor model, as shown in Figure 9. Phthalazine ligands were

Table 9 6-(Alkylamino)-3-aryl-1,2,4-triazolo[3,4-a]phthalazines[88]

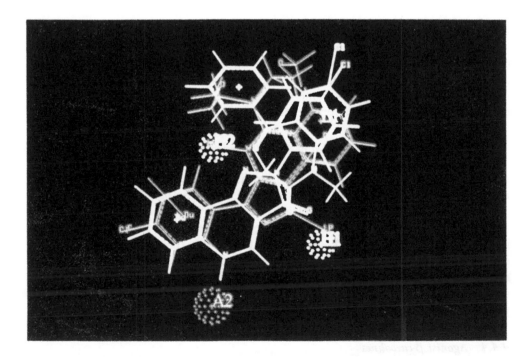

		R'_4	R'_3	K_i (nM)	Profile
ATP4'MeO	99	OMe	H	2.6	partial agonist
ATP4'Cl	100	Cl	H	17.0	partial agonist
ATP4'Me	101	Me	H	44.0	partial agonist
ATP4'F	102	F	H	20.0	partial agonist
ATP	103	H	H	66.0	partial agonist
ATP4'OH	104	OH	H	4.1	antagonist
ATP3'Me4'MeO	105	OMe	Me	1820	negative control
ATPNMA	106			5.4	antagonist

Figure 9 Superposition of the agonist CGS-9896 (**50**; green) with the 6-(alkylamino)-3-aryl-1,2,4-triazolo[3,4-a]phthalazines: **99** (orange), **100** (magenta), **102** (violet).

Table 10   (Imidazo[1,2-a]pyrimidin-2-yl)phenylmethanones[89]

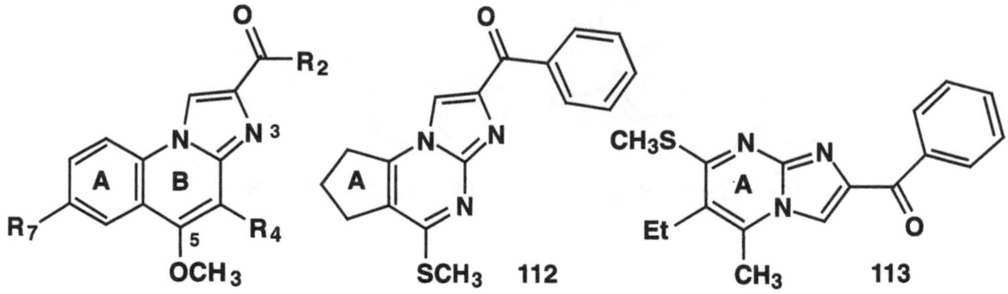

		R_2	R_4	R_7	K_i (nM)	Profile
IPP7Et	**107**	Ph	H	Et	14	partial agonist
IPP4Me	**108**	Ph	CH₃	Et	7	partial agonist
IPP	**109**	Ph	H	F	20	partial agonist
IPPy	**110**	2-Py	H	Et	9	partial agonist
IPM8Me	**111**	CH₃	H	Et	1300	negative control
IPPCP	**112**				35	partial agonist
IPPIBZ	**113**				12	partial agonist

found to best fit the model when hydrogen bonded with H_1 and H_2 through the imine N2 and N5 nitrogen atoms, respectively. This alignment permits the overlay of the di(methoxyethyl)-amine moiety with lipophilic area L_3 consistent with an agonist profile of activity.[88] Ligand **106** possesses a smaller side chain which cannot occupy both L_2 and L_3 to the same extent as the dialkoxyoyl analogs and this is consistent with its antagonist profile of activity.[88]

4.4.3   (Imidazo[1,2-a]pyrimidin-2-yl)phenylmethanones

Another series of agonist ligands are the (imidazo[1,2-a]pyrimidin-2-yl)phenyl-methanones and related compounds, depicted in Table 10.[89] These families align and interact with the inclusive model, as illustrated in Figure 10. Ligands from these two related series (*e.g.*, **107–113**) are fitted most easily to the model through interaction at the hydrogen bond donor site H_1 with the carbonyl group and a similar interaction between the hydrogen bond donor site H_2 with the imine nitrogen atom at N4. The series of phenylmethanones contains lipophilic substituents which are able to occupy the lipophilic receptor region of L_2, and hence display partial agonist activity.[50,51,83]

4.4.4   Agonist β-carbolines

The alignment of agonist β-carboline ligands [(*e.g.*, ZK-93423 (**1**), ZK-91296 (**2**)); Table 1] to agonists of other series has until recently been difficult. β-Carbolines

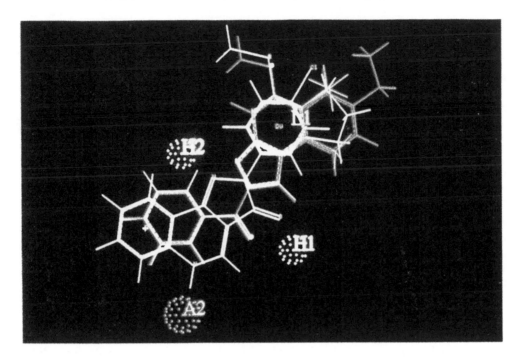

Figure 10 Superposition of CGS-9896 **50** (green) and (imidazo[1,2-*a*]pyrimidin-2-yl) phenylmethanones **107** (red), **108** (orange), and **109** (magenta).

have become important tools in the characterization of benzodiazepine receptors for these agents have demonstrated a continuum of activity from inverse agonist to antagonist to agonist. The 6-(benzyloxy)-4-(methoxymethyl)-β-carboline-3-carboxylic acid ethyl ester **1** (ZK-93423) initially was the only β-carboline reported to date that exhibited full agonist activity.[49] The synthesis of this 3,4,6-trisubstituted β-carboline is shown in Scheme III and described in Section 2.2.2. Replacement of the 6-benzyloxy group in ZK-93423 (**1**) by a methoxy function resulted in a high-affinity ligand **10** with no anticonvulsant/anxiolytic activity. However, this β-carboline did potently antagonize ($ED_{50} = 1.25$ mg/kg) the anticonvulsant effects of diazepam. Moreover, removal of the functionality at position-6 of **1** and substitution with a hydrogen atom also provided an antagonist **11**, as illustrated in Table 1.

Replacement of the 4-methoxymethyl group of the full agonist **1** with an ethyl group (see **13**) furnished a high affinity ligand **13** for BzR *in vitro*, yet this β-carboline appeared inactive in the tests employed to evaluate an agonist/inverse agonist profile (*anti*-PTZ) of activity. The corresponding β-carboline **12** which was devoid of a substituent at C4 was also inactive. These results when taken together support the requirement of an electronegative substituent at C4 of the β-carboline nucleus in order to elicit agonist activity. The oxygen atom of the 4-methoxymethyl group is of critical importance for agonists in the β-carboline series and directs the ligand into the active site (region) of the agonist pharmacophore through the formation of a hydrogen bond between the ether oxygen atom and H_2 of the receptor protein.

The agent must form hydrogen bonds with H_1 and H_2, as well as interact in lipophilic pockets L_2 and L_3 for a full agonist response.[50] The large phenyl group of the benzyloxy substituent of **1** (ZK-93423) is believed to lie in the same lipophilic pocket (L_3) as the 5-phenyl (ring-C) group of diazepam **72**.[50]

The SAR from the analogs depicted in Table 1 has shed light on the orientation of β-carbolines in the binding cleft[49,50] and it now appears that agonist and inverse agonist β-carbolines align with alternative orientations, as illustrated in Figure 11. Agonist β-carbolines also appear to interact at the H_1 donor group of the receptor protein through the imine N2 nitrogen atom at ring-C. Examination of the agonist model indicates that agonist β-carbolines must form an additional hydrogen bond with the receptor site at H_2 as well as fill the hydrophobic receptor pockets L_2 and L_3. For compounds **1** through **4** the interaction with H_2 takes place with the oxygen atom of the methoxymethyl group at position-4. The hydrogen bonding interactions with H_1 and H_2 direct the ligands in such a manner that ring-A of the β-carboline points toward the lipophilic areas L_1/L_2. In this orientation, the benzyloxy group at position-6 fills L_3 (Figure 11a). This alignment is similar to one proposed by Fryer.[41]

In contrast, the alignment of β-carboline inverse agonists is very different. This series is unique in that the ligands lack substituents with which to interact with the H_2 donor site; the A-ring of β-carbolines is no longer directed toward areas L_2 and L_3. As a result, a new orientation is obtained in which the N-H group of the indole ring N9 interacts with the hydrogen bond acceptor A_2. This orientation directs the β-carboline ring system parallel to H_1 and A_2, as illustrated in the picture of the inverse agonist derivative **16** (Figure 11b). However, neither of the two possible orientations of β-carbolines **15–18**[45–48] to the inclusive pharmacophore will allow for interaction with lipophilic site L_2 or L_3, and hence these compounds display inverse agonist activity.

The antagonist β-carbolines **5–9** interact with H_1 and A_2 in an orientation similar to that of inverse agonists. These analogs do not have substituents at position-4 capable of forming hydrogen bonds with H_2. However, the presence of bulkier ester groups (*e.g.*, *i*-propyl, *n*-butyl, *t*-butyl) increases the interaction at L_1 and L_2 which accounts for their antagonist activity. The antagonist β-carbolines **10** and **11**, however, possess substituents at position-4 capable of interacting with H_2 and therefore may adopt an orientation analogous to that of agonist β-carbolines at the BzR. The antagonist activity of **10** and **11** is a result of insufficient interaction with proposed lipophilic pockets designated L_2 and L_3 in the model.

On the basis of the SAR of the β-carbolines **4, 10–12, 31**, and **33** depicted in Table 1 and the model of the agonist pharmacophore, the synthesis of 6-(propyloxy)-4-(methoxymethyl)-β-carboline-3-carboxylic acid ethyl ester **4**, a new selective anxiolytic/anticonvulsant agent, was designed.[51] This β-carboline (6PBC) **4** inhibited PTZ-induced seizures in a dose-dependent fashion with an ED_{50} of 1.6 mg/kg and exhibited anxiolytic activity when evaluated in the elevated plus-maze paradigm. More importantly, **4** was devoid of muscle-relaxant effects in the inverted wire screen test and completely antagonized the myorelaxant actions of diazepam in this experimental model.[51]

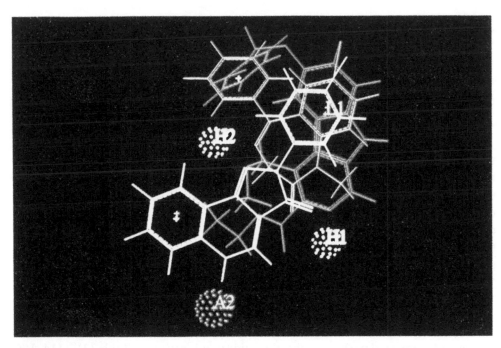

Figure 11a Alignment of agonist β-carbolines **1** (red) with diazepam (magenta), and CGS-9896 (**50**, green) in the inclusive pharmacophore model.

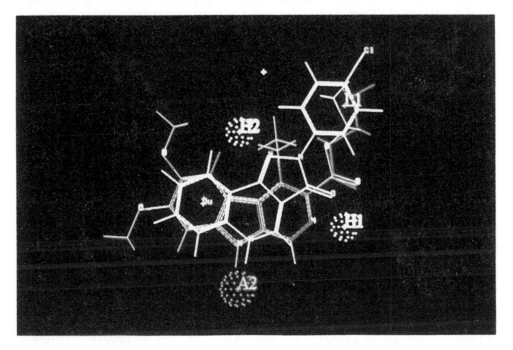

Figure 11b Alignment of inverse agonist β-carbolines **15** (magenta), **16** (orange), and **18** (red) with CGS-9896 (**50**, green) in the inclusive pharmacophore model.

Table 11 Imidazoquinoxalines[90]

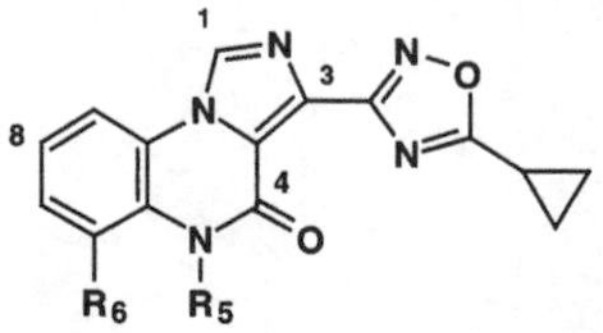

		R_5	R_6	K_i (nM)	Profile
U-78263	**114**	Me	H	7.5	antagonist
U-78280	**115**	Et	H	5.6	antagonist
U-78875	**116**	i-Pr	H	2.8	antagonist
U-82249	**117**	t-Bu	H	4.5	partial agonist
U-79098	**118**	i-Pr	Cl	1.4	partial agonist
U-81139	**119**	Me	CF_3	0.7	antagonist
U-89320	**120**	i-Pr	OCH_3	7.4	partial agonist

4.4.5 Imidazoquinoxalines

The recently reported partial agonists the imidazoquinoxalines (Table 11)[90] **114–120** have also been placed in the inclusive pharmacophore. These new ligands interact with H_1 and H_2 *via* the imine nitrogen atom N2 and the carbonyl oxygen function located at C4. The phenyl ring interacts with L_1 as well as a small interaction in L_2 and the N5-alkyl group interacts with lipophilic pocket L_3 (Figure 12). Analogs which possess large groups (isopropyl and t-butyl) at N5 occupy L_3, and as a consequence, exhibit agonist activity. The N-alkyl group apparently occupies less space in L_3 than the phenyl ring of diazepam, consequently these analogs exhibit a partial spectrum of agonist activity. Ligands which contain smaller N-alkyl groups (methyl and ethyl) do not interact with L_3, and as a result, they exhibit an antagonist profile of activity.

4.5 Benzodiazepine Molecular Yardsticks

4.5.1 Motivation

Recent evidence suggests that full occupation of the lipophilic region L_3 by the phenyl ring attached to position-6 of ZK-93423 (**1**) resulted in a full agonist spectrum of activity (anxiolytic, anticonvulsant, muscle-relaxant, sedative-hypnotic),[49–51] while partial occupation of this same region with a propyl group (C6) resulted in an anxiolytic/anticonvulsant (6-propyloxy-4-methoxymethyl-β-carboline 3-carboxylic acid ethyl ester **4** response devoid of muscle-relaxant activity, a so-called partial agonist profile.[50] Clearly much work must be carried out to confirm this hypothesis. In order to determine the size of lipophilic region L_2 and the effect of occupation of L_2 on agonist activity, a series of benzofused benzodiazepines have been designed (see Figure 13).

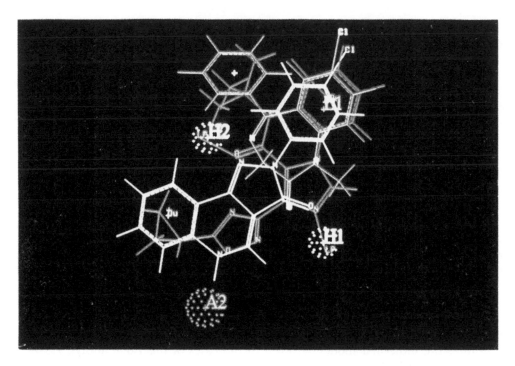

Figure 12 Superposition of CGS-9896 (**50**, green) and midazolam (**87**, magenta) with U-78875 (**116**, red) and **117** (orange).

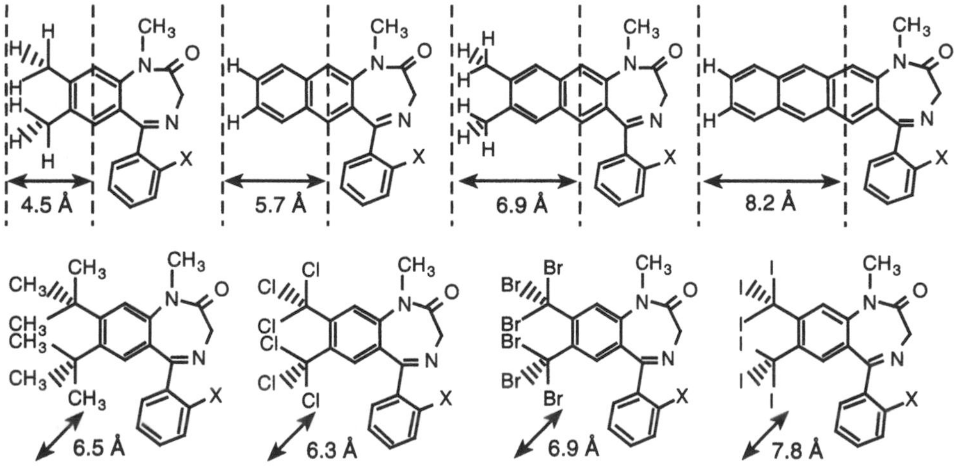

Figure 13 Molecular Yardsticks. Distances cited in the series at the top are between the centroid of the benzene ring and the edge of the van der Waals field of protons. Three dimensional molecular yardsticks pictured at the bottom illustrate distances orthogonal to the plane of the benzene A-ring. The values represent the van der Waals field of the *t*-butyl or trihalomethyl groups.

It is believed that for full agonist activity, BzR ligands such as diazepam (**72**) and ZK-93423 (**1**) must fully occupy the lipophilic region L_3 of the pharmacophore/ receptor model in order to elicit the full spectrum of agonist activity. Since the 6-methoxy analog of **1** (BC6OM, **10**) elicited only antagonist activity,[49] ligand 6PBC (**4**) was designed to permit only partial occupation of L_3. The result was the synthesis of an anxiolytic/anticonvulsant devoid of the muscle relaxant effects of diazepam.[51] Selectivity *in vitro* and *in vivo* at the BzR therefore appears highly dependent both on the size and the effect of occupation of the lipophilic pockets which comprise the binding cleft(s).[82,83,91] For this reason the synthesis of rigid analogs based on the structures of diindoles and benzofused benzodiazepines, termed "molecular yardsticks",[83,91] has become an important tool in the study and the realization of selectivity at the BzR.

The use of "molecular yardsticks" in the diindole series to define the boundary of the repulsive region S_1 was recently reported.[83] Potential 7,8-disubstituted molecular yardsticks in the 1,4-benzodiazepine series are depicted in Figure 13. The complementary probes in the 6,7- and 8,9-benzofused systems are illustrated here only for the benzofused benzodiazepines **131** and **127** (Figure 14) and their synthesis is described in Section 2.2.4. Molecular yardsticks (benzofused rings) employed here represent rigid probes to define the dimensions of the lipophilic pockets of the BzR in the absence of complications introduced by rotational freedom; this simplifies the molecular modeling. Even in the cases of the 7,8-dimethyl, 7,8-di-*t*-butyl or 7,8-bis-trihalo analogs rotational freedom of the symmetrically substituted carbon atoms does not complicate the molecular modeling.[83]

4.5.2 *Pharmacology*

The three analogs **123**, **127**, and **131** were screened for *in vitro* affinity to benzodiazepine receptor sites on rat cortical membranes, the data from which is depicted in Table 12. As can be seen from the data in Table 12, none of these analogs exhibited potent affinity to BzR. This is not surprising for examination of the ligand-receptor fit (Figure 14) for **123**, **127**, and **131** indicates that there is little room to spare in the binding site at S_1, L_2 and S_2, respectively. It is known, however, that substitution of a fluorine atom for a hydrogen nucleus at the 2'-position of the 1,4-benzodiazepines greatly enhances affinity and efficacy at the BzR.[81] This is important with reference to ligands **123**, **127**, and **131** for the 2'-fluorine substituent could interact at H_2 in place of the interaction at N4 permitting some ligand flexibility (plasticity) in the binding cleft. Attention then turned toward the preparation of the 2'-fluorobenzofused ligands **124**, **128**, and **132**.

The affinities of benzofused 1,4-benzodiazepines at BzR were evaluated by previously reported methods.[51,83] Illustrated in Table 12 are the IC_{50} values of these new "molecular yardsticks". The 8,9-benzofused ligands **125** and **127** exhibited low affinity for the BzR and this can be attributed to a negative steric interaction with the receptor protein at S_1 which is consistent with the results obtained in the pyrido-diindole and pyrroloimidazole series.[91] The lack of activity for these 8,9-benzofused ligands confirms the location of S_1 in the receptor binding cleft previously reported.[83]

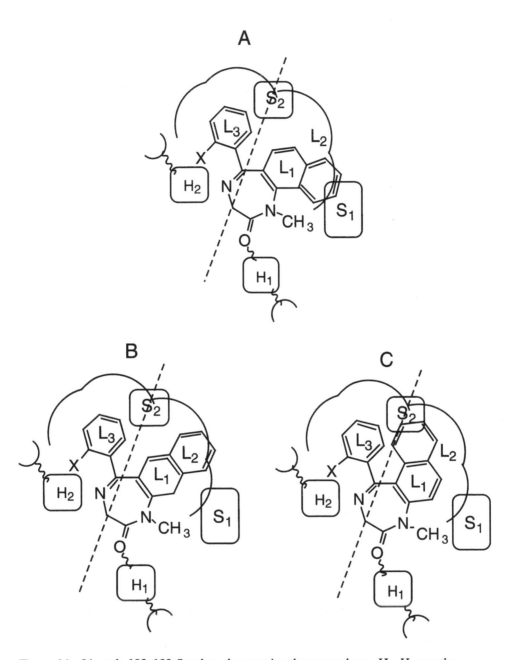

Figure 14 Ligands **123–132** fitted to the agonist pharmacophore. H_1, H_2 are the hydrogen bond donor sites on the receptor protein. L_1, L_2, and L_3, are the lipophilic pockets in the protein at the binding site. S_1 and S_2 are the areas of negative interaction between the ligand and the protein at the binding site.

Table 12 Benzofused benzodiazepines[91]

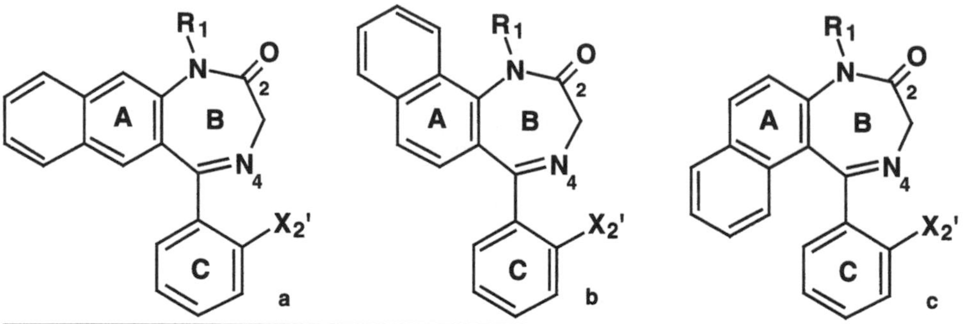

		Structure	R_1	X_2'	IC_{50} (nM)	Profile
7,8BzBDZ	**121**	a	H	H	400	ND
	122	a	H	F	>1000	ND
7,8BzBZD1Me	**123**	a	Me	H	>1000	ND
	124	a	Me	F	54	full agonist
8,9BzBDZ	**125**	b	H	H	>1000	ND
	126	b	H	F	152	ND
8,9BzBZD1Me	**127**	b	Me	H	>1000	ND
	128	b	Me	F	300	ND
6,7BzBZD	**129**	c	H	H	>1000	ND
	130	c	H	F	>1000	ND
6,7BzBZD1Me	**131**	c	Me	H	>1000	ND
	132	c	Me	F	>1000	ND

ND = not determined.

The 6,7-benzofused ligands **129** and **131** have IC_{50} values >1000 nM supporting the location of S_2 (a negative area of steric interaction between the boundaries of L_2 and L_3). These findings are in agreement with the recent results obtained with the pyridoimidazoles by Martin *et al.*[83,91] The low IC_{50} values of the linear 7,8-benzofused ligands **121** and **123** indicate only a very weak interaction at the BzR. Again, the rigid phenyl ring must experience a repulsive steric interaction with the receptor protein (Figure 14), presumably with the boundary of lipophilic pocket L_2. Because the fit of the 7,8-benzofused rings of **121** and **123** in lipophilic pocket L_2 was anticipated as a near acceptable interaction (see Figure 14), an additional modification of these rigid probes was required. It has been well documented, as mentioned above, that substitution of a fluorine atom for hydrogen at the 2'-position of the 1,4-benzodiazepines enhanced the affinity and efficacy of ligands at the BzR, consequently the 2'-fluoro analogs **124**, **128**, and **132** were screened. The effect of a fluorine substituent (2') on the activity of the 8,9-benzofused compounds (see **126** and **128**, Table 12) resulted in an increase in potency, while the effect on the 6,7-benzofused compounds (see **130** and **132**) was minimal. Substitution of the 2'-hydrogen atom with fluorine in the linear N(1)-methyl-7,8-benzofused system (see **132**, >1000 nM); however, resulted in a significant enhancement in affinity with the 2′-fluoro ligand **124** exhibiting an IC_{50} value of 54 nM. The increase in binding potency is significant in the context of the agonist pharmacophore/receptor model. The 2′-fluorine substituent is in close proximity to the required hydrogen bonding site H_2 in the proposed ligand binding cleft. It is possible that the required agonist hydrogen bonding interaction at H_2 with **124** now occurs *via* the 2′-fluorine atom rather than the imine nitrogen atom at position-4. This permits ligand **124** to move in the binding site just enough to fit into L_2 without loss of the important interactions at H_1, H_2, L_1, and L_3. This plasticity (flexibility) in the binding cleft is permitted because of the interaction of H_2 with the 2′-fluorine atom.

Because of the high affinity of ligand **124** to the BzR, as compared to the activity of the desfluoro analogs, the former was chosen for *in vivo* evaluation. Pharmacological studies in mice indicate that N(1)-methyl-5-(2′-fluorophenyl)-1,3-dihydronaphtho[2,3-*e*][1,4]diazepin-2-one **124** exhibits a full agonist spectrum of activity, analogous to that previously reported for diazepam.[51] The ED_{50} for the anticonvulsant effect of **124** was ≈ 15 mg/kg, and the ED_{50} for the myorelaxant effect was also ≈ 15 mg/kg.[91] In comparison to diazepam, ligand **124** exhibited a less potent anticonvulsant effect and a comparable ataxic effect. Apparently, full occupation of L_2 has potentiated the muscle relaxant effect of **124** evidently at the expense of the potency of the anticonvulsant/anxiolytic effect.[91] This is somewhat similar to the spectrum of activity elicited by the full agonist ZK-93423 [6-benzyloxy-4-methoxymethyl-β-carboline-3-carboxylic acid ethyl ester (**1**)] *vs.* the partial agonist 6PBC (**4**).[51] In the latter case, partial occupation of L_3 furnished an agent with anxiolytic and anticonvulsant activity devoid of muscle relaxant activity.[50,51] Although speculative, it is reasonable to hypothesize that ligand selectivity and intrinsic activity (efficacy) at different $GABA_A$ receptor isoforms will stem from different interactions in the lipophilic pockets of the BzR(s) receptor binding cleft(s) effected by selective BzR ligands. While it is not known whether transfected cells expressing various $GABA_A$ subunits assemble themselves with the same stoichiometry as native receptors, future studies with recombinant receptors should provide additional insights into the validity of this hypothesis.

4.6 Fluorescence Quenching Ligand BD 623

The fluorescence quenching ligand 8-fluoro-5,6-dihydro-5-methyl-6-oxo-4*H*-imidazo [1,5-*a*][1,4]benzodiazepine-3-carboxylic acid, 3-[(7-nitro-4-benzofurazanyl)-amino]propyl ester (BD 623) has been recently synthesized and shown to exhibit high affinity to BzR (K_i = 5.7 nM).[123] This ligand occupies the BzR and has been employed for fluorescence quenching experiments which may be superior to and eventually replace the radioligand binding assay in certain situations.[123] The alignment of BD 623 to the inclusive pharmacophore results in the occupation of lipophilic region L_1 by the A-ring of the imidazobenzodiazepine and interaction between the hydrogen bond donor group H_1 of the receptor protein and the 2-imine nitrogen atom of the ligand. The oxygen atom of the carboxyl group at C-3 probably interacts with receptor site H_1 to form a three centered hydrogen bond reminiscent of

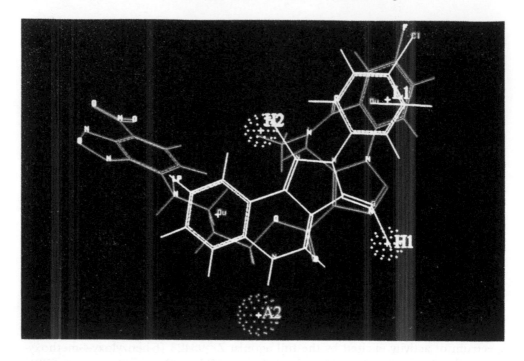

Figure 15 Alignment of fluorescence-quenching ligand BD 623 (red) with CGS 9896 (**50**, green) in the BzR pharmacophore.

β-carboline inverse agonists. The larger flexible side chain can be overlaid with CGS 9896 as shown in Figure 15. In this alignment, the bulky fluorescence quenching ligand BD 623 fits well into the inclusive pharmacophore.

5.   PHARMACOPHORE OF 'DIAZEPAM INSENSITIVE' SITE AT BZR

5.1   Background

The "diazepam-insensitive" (DI) subtype of benzodiazepine receptors (BzR) was initially described in rodent cerebellar membranes and cerebellar granule cell cultures.[124,125] The DI BzR represent 25–30% of cerebellar BzR, but are only a very minor constituent (0–5%) in other regions of the mammalian central nervous system.[126] DI are characterized by the low affinity (>1 μM) of prototypical 1,4-benzodiazepines [*e.g.*, diazepam (**72**), flunitrazepam (**74**)], some β-carbolines (*e.g.*, BCCM, **17**), triazolobenzodiazepines (*e.g.*, triazolam, **89**), and triazolopyridazines (*e.g.*, CL 218872) that are high affinity ligands at other "diazepam-sensitive" (DS) BzR isoforms.[126–129] These findings support the hypothesis that DI BzR represent a distinct subtype with more restrictive ligand binding requirements than other BzR isoforms. A pharmacological profile similar to that of native DI BzR can be reconstituted in cell cultures transfected with cDNA's encoding an $\alpha 6 \beta 2 \gamma 2$

Table 13 Summary of the principle interactions between the major structural classes of BzR ligands with the inclusive pharmacophore model (this work) and comparisons with other published models

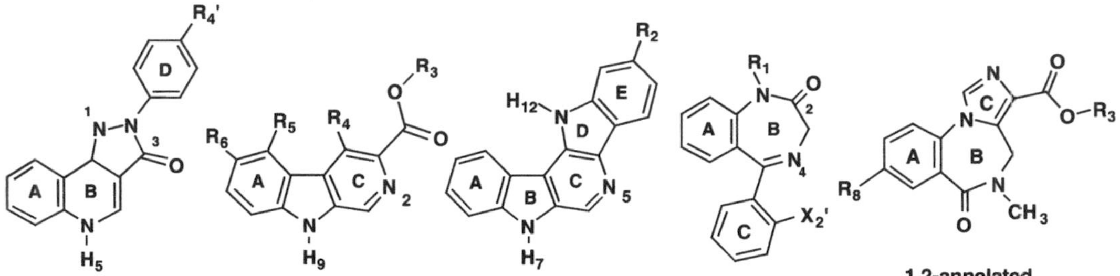

pyrazoloquinolines β-carbolines pyridodiindoles 1,4-benzodiazepines 1,2-annelated benzodiazepines

Model	Alignment[a]	Molecule	Ligand Site/(Receptor Site)						
			(H_1)	(A_2)	(H_2)	(L_1)	(L_2)	(L_3)	("A")[c]
this work	ref[b]	pyrazoloquinolines	C3=O	N5H	N1	Ring-D	R'_4	—	Ring-A
"	A	pyridodiindoles	N5	N7H	N12H	Ring-E	R'_2	—	Ring-A
"	A	inv. agonist β-carbolines	N2/C3=O	N9H	—	R_3	—	R_5/R_6	Ring-A
"	N/A[d]	agonist β-carbolines	N2	—	OR_4	Ring-A	—	C-R	—
"	A	1,4-benzodiazepines	C2=O	—	$N4/X'_2$	Ring-A	R_7	—	—
Borea[42]	ref[b]	pyrazoloquinolines	C3=O (B1)	—	N1 (B2)	Ring-D	R'_4 (AG1)	—	Ring-A
"	A	inv. agonist β-carbolines	C3=O (B1) N2 (B3)	—	—	—	—	—	Ring-A (IAG)
"	A	1,4-benzodiazepines	C2=O (B1)	—	N4 (B2)	Ring-A	R_7 (AG1)	Ring-C (AG2)	—
Fryer[41]	ref[b]	pyrazoloquinolines	C3=O (π_1)	—	—	—	—	—	Ring-A (A)
"	N/A[d]	agonist β-carbolines	N2 (π_1)	—	—	—	—	R_5/R_6 (π_2)	Ring-A (A)
"	C	1,4-benzodiazepines	C2=O (π_1)	—	—	—	—	Ring-C (π_2)	Ring-A (A)
Wermuth[43]	ref[b]	pyrazoloquinolines	C3=O (δ_1)	—	N1 (δ_2)	Ring-D (FRA)	—	—	Ring-A (PAR)
"	C	1,4-benzodiazepines	C2=O (δ_1)	—	N4 (δ_2)	—	—	Ring-C (OPR)	Ring-A (PAR)

Table 13 (*Continued*)

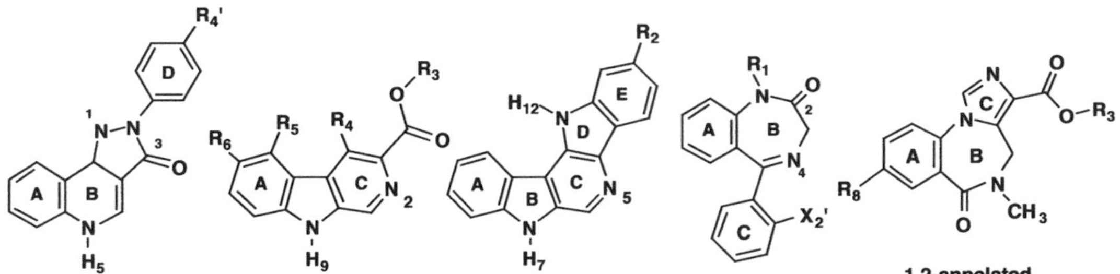

pyrazoloquinolines · β-carbolines · pyridodiindoles · 1,4-benzodiazepines · 1,2-annelated benzodiazepines

Model	Alignment[a]	Molecule	Ligand Site/(Receptor Site)						
			(H_1)	(A_2)	(H_2)	(L_1)	(L_2)	(L_3)	("A")[c]
Gardner[44]	ref[b]	pyrazoloquinolines	C3=O		N1	Ring-D	R_4'		Ring-A
"	C	agonist β-carbolines	N2	N9H				R_5/R_6	Ring-A
"	C	1,4-benzodiazepines	C2=O	N4				Ring-C	Ring-A
Codding[40]	ref[b]	pyrazoloquinolines	C3=O (2)	N5H (4)	—	Ring-D (3)	—	—	Ring-A (1)
"	D	β-carbolines	C3=O (2)	N9H (4)	—	R_3 (3)	—	—	Ring-A (1)
"	D	1,4-benzodiazepines	C2=O (2)	N1H (4)	—	—	—	Ring-C (*)	Ring-A (1)
Loew[37]	ref[b]	pyrazoloquinolines	C3=O	—	N1	—	—	—	Ring-A
"	D	1,4-benzodiazepines	N4	—	C2=O	—	—	—	Ring-A
"	D	1,2-annelated diazepines	N3	—	C4=O	—	—	—	Ring-A

[a] Alignments A, B, C, and D refer to Figures 6a–d respectively.

[b] The ref ("reference") alignment of all ligands in this table is to the pyrazoloquinoline series so that the alignment of the pyrazoloquinolines is by definition alignment-A.

[c] Receptor site "(A)" is not an explicit part of the inclusive pharmacophore model, however "A" is included in this table to allow for comparison with other published pharmacophore models.

[d] N/A = not applicable.

subunit configuration; moreover, cells transfected with an $\alpha 4$ subunit also have been defined as DI sites.[28] DS BzR can be reconstituted by combination of any one of the α-subunits $\alpha 1$–3, or 5 together with $\beta 2$ and $\gamma 2$ subunits.[130]

5.2 Alcohol Antagonism and the DI BzR

The DI BzR is of particular interest since the imidazodiazepines **167, 171,** and **190** exhibit high affinities ($K_i \leq 20$ nM) for this isoform[126] and these three compounds were previously shown to antagonize some of the behavioral and biochemical effects of ethanol.[131–133] These findings, coupled with a recent report which demonstrated that alcohol non-tolerant (ANT) rats lack DI receptors, suggest that the DI BzR may play a role in the ability of these ligands to reverse some of the neurochemical and behavioral effects of ethanol.[125] A better understanding of the requirements for ligand binding to DI and to DS sites should aid in the design and synthesis of more selective and higher affinity DI ligands. These agents may in turn be employed to further elucidate the physiological and pharmacological functions of this unique GABA$_A$ receptor isoform.

5.3 Arguments for Including the DI BzR in the Inclusive Model

Seeburg and coworkers have shown *via* site directed mutagenesis that a histidine to arginine amino acid substitution at position-101 (H101R) of the α-subunit of the complex $\alpha 2 \beta 2 \gamma 2$ can alter the pharmacology of this BzR isoform from DS to DI.[57] Conversely, a single amino acid substitution at the equivalent position in $\alpha 6 \beta 2 \gamma 2$ (R100H) can change the apparent phenotype from DI to DS. In addition, the $\alpha 1$- and $\alpha 6$-subunits display a high degree of overall homology (70%).[57] Finally, both the DS and DI BzR display identical stereochemical preferences for 1,4-diazepine ligands.[121] Taken together, this is strong evidence that the DS and DI binding sites are in fact located in equivalent positions in their respective receptors and therefore it should be possible to incorporate the DI binding site into an inclusive BzR pharmacophore model related to that of the DS site(s).

5.4 DI BzR SAR

The structure-affinity relationships of several 1,4-diazepines (*e.g.,* **192** and **196**) which contain one or more annealed ring systems was recently determined at DI BzR.[126,129] These studies were extended by evaluation of the affinities of additional 1,4-diazepines coupled with excluded volume[72,73] and CoMFA (Comparative Molecular Field Analysis)[134] studies to ascertain the determinants for selective, high affinity binding to DI BzR (Tables 14 and 15).[52,135]

5.4.1 Substitution at position-7 or -8

The affinities of a series of imidazo-1,4-diazepines for the DI and DS BzR are presented in Tables 14 and 15. Substitution at the 7- or 8-positions appears necessary

Table 14 Affinity of 5,6-dihydro-5-methyl-6-oxo-4H-imidazo[1,5a][1,4]benzodiazepine-3-carboxylic acid derivatives for the DS and DI benzodiazepine receptors[52,135]

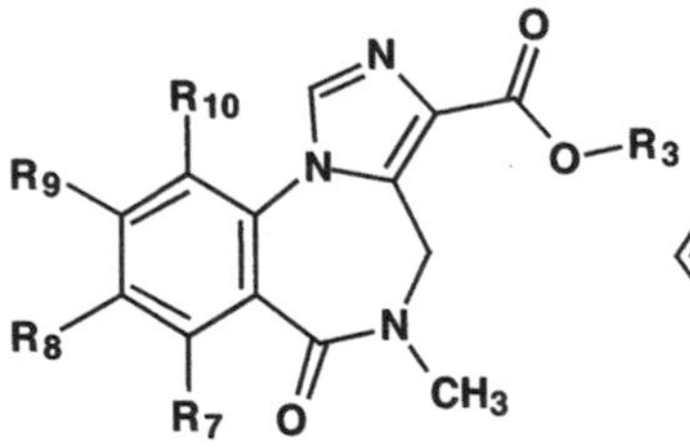

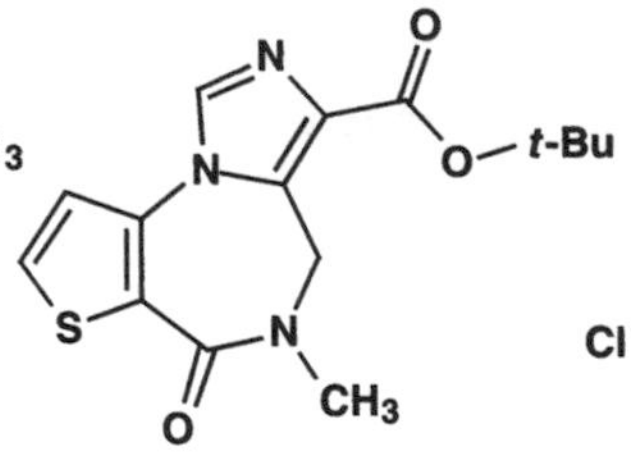

190

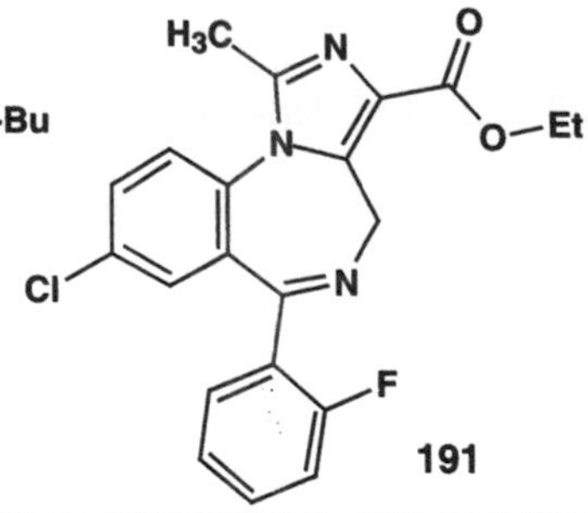

191

Compound		R_3	R_7	R_8	R_9	R_{10}	DS (nM)	DI (nM)	DI/DS	Ref
IBCM	165	CH$_3$					6.3	1409	224	52
Ro 14-7437	166	CH$_2$CH$_3$					1.3	214	165	52
Ro 15-3505	167	CH$_2$CH$_3$	Cl				0.2	20	100	52
Ro 15-5623	168	CH$_2$CH$_3$	F				0.8	102	128	52
IBCM8F	169	CH$_3$		F			3.1	239	77	52
IBCM8Cl	170	CH$_3$		Cl			29.8	123.5	4.1	52
Ro 15-4513	171	CH$_2$CH$_3$		N$_3$			5.3	3.1	0.6	52
Ro 15-1788	172	CH$_2$CH$_3$		F			0.8	58	73	52
Ro 15-1310	173	CH$_2$CH$_3$		Cl			5.4	16.9	3.0	52
IBCP8CL	174	CH$_2$CH$_2$CH$_3$		Cl			24.8	33.3	1.3	52
IBCIP8Cl	175	CH(CH$_3$)$_2$		Cl			10.5	8.8	0.8	52
IBCPE8Cl	176	CH(CH$_2$CH$_3$)$_2$		Cl			26.9	122.6	4.6	52
IBCCP8Cl	177	cyclopropylmethyl		Cl			9.7	39.8	4.1	52
IBCTB	178	C(CH$_3$)$_3$					1.1	21.2	19.2	52
IBCTB8Cl	179	C(CH$_3$)$_3$		Cl			4.0	1.7	0.4	135
IBCTB8Br	180	C(CH$_3$)$_3$		Br			16.0	2.8	0.17	135
IBCTB8I	181	C(CH$_3$)$_3$		I			7.0	1.5	0.21	135
IBCTB8NO	182	C(CH$_3$)$_3$		NO$_2$			14.0	10.8	0.77	135
IBCTB8NCS	183	C(CH$_3$)$_3$		NCS			2.7	3.7	0.73	135
IBCTB8N3	184	C(CH$_3$)$_3$		N$_3$			2.1	0.4	0.2	135
IBCTB8N3M	185	C(CH$_3$)$_3$		N(CH$_3$)$_2$			>3000	>1000	ND	135
IBCEM8Cl	186	CH$_2$C(CH$_3$)$_3$		Cl			499.3	300	0.67	135
IBCPM8Cl	187	CH$_2$CH$_2$C(CH$_3$)$_3$		Cl			184.0	>1000	ND	135
Ro 15-1746	188	CH$_2$CH$_3$			Cl		>1000	>10000	ND	52
Ro 15-3237	189	CH$_2$CH$_3$				Cl	>1000	>10000	ND	52
ITCTB	190						0.2	2.6	13.0	135
Ro 21-7254	191						2.2	2810	1277	52
diazepam	72						6.6	>1000	ND	135
triazolam	89						0.8	2086	2608	52

Table 15 Affinity of 12,12a-dihydro-9-oxo-9H,11H-azeto[2,1-c]imidazo[1,5-a][1,4] benzodiazepine-1-carboxylic acid and 11,12,13,13a-tetrahydro-9-oxo-9H-imidazo[1,5-a] pyrrolo[2,1-c][1,4]benzodiazepine-1-carboxylic acid derivatives for the DS and DI benzodiazepine receptors[135]

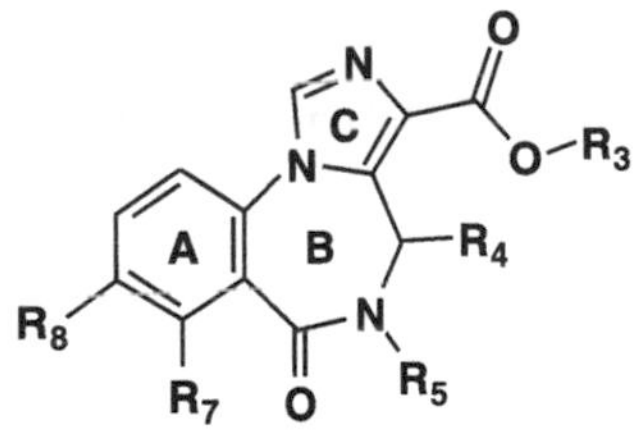

		R_3	R_4–R_5	R_7	R_8	DS (nM)	DI (nM)	DI/DS
Ro 17-1812	**192**	cyclopropylmethyl	$-(CH_2)_2-$	Cl		0.1	55.0	554
Ro 16-0858	**193**	CH_2CH_3	$-(CH_2)_2-$	Cl		0.2	34.0	170
Ro 16-0153	**194**	$C(CH_3)_3$	$-(CH_2)_2-$	Cl		1.1	15.0	14.0
Ro 17-2620	**195**	$C(CH_3)_3$	$-(CH_2)_2-$		F	1.8	16.9	9.0
Ro 16-6028	**196**	$C(CH_3)_3$	$-(CH_2)_3-$	Br		0.5	10.0	20.0
Ro 16-3607	**197**	$C(CH_3)_3$	$-(CH_2)_3-$	OCH_3		5.1	260	51.0
Ro 16-6624	**198**	$C(CH_3)_3$	$-(CH_2)_3-$	CH_2CH_3		1.0	164	164
Ro 15-4941	**199**	CH_2CH_3	$-(CH_2)_3-$	Cl		0.2	68.8	344
Ro 16-6127	**200**	$C(CH_3)_3$	$-(CH_2)_3-$	Cl	F	1.0	23.6	24.0
Ro 14-5975	**201**	CH_2CH_3	$-(CH_2)_3-$		Cl	72.6	53.8	0.7
Ro 16-3058	**202**	$C(CH_3)_3$	$-(CH_2)_3-$		Cl	65.2	26.8	0.4
Ro 17-3128	**203**	cyclohexyl	$-(CH_2)_3-$		Cl	391	165	0.42
Ro 17-3129	**204**	cyclopropylmethyl	$-(CH_2)_3-$		Cl	378	81.0	0.22
Ro 14-5974	**205**	CH_2CH_3	$[S]-(CH_2)_3-$			0.9	45.1	50.0
Ro 14-7527	**206**	CH_2CH_3	$[R]-(CH_2)_3-$			>100	>1,000	—

for high affinity binding to the DI BzR. Thus, analogs lacking substituents at either of these positions display relatively low affinity for the DI site [e.g., **165** ($K_i = 1409$ nM) and **166** ($K_i = 214$ nM)]. In contrast, substitution at the 9- and 10-positions are not well tolerated as evidenced by the low affinities ($K_i > 10,000$ nM) of the 9- and 10-chloro derivatives **188** and **189**, respectively. While substitutions at positions-7 and -8 enhance ligand affinity at the DI site, substitution at position-8 results in somewhat higher affinity for the DI site than does position-7 [e.g., 7-chloro-1,4-benzodiazepine **167** ($K_i = 20.9$ nM) vs. the 8-chloro analog **173** ($K_i = 16.9$ nM)]. In contrast, substitution at the 7-position increases ligand affinity at DS more than substitution at position-8 [compare **167** ($K_i = 0.2$ nM) vs. **173** ($K_i = 5.4$ nM)]. Hence, the net effect of substitution at the 8-position is to enhance selectivity for the DI site.

5.4.2 Position-3

As the size of the ester group at position-3 of the imidazo[1,5a][1,4]benzodiazepine ring system increases, ligand affinity at DI also increases. This is evident from the following series: methyl (**165**, $K_i = 1409$ nM), ethyl (**166**, $K_i = 214$ nM), and t-butyl ester (**178**, $K_i = 21.2$ nM). However, as the size of the alkyl group becomes larger than t-butyl, affinity is diminished: $CH(CH_2CH_3)_2$ (**176**, $K_i = 122.6$ nM), CH_2-t-Bu (**186**, $K_i = 299.6$ nM), and CH_2CH_2-t-Bu (**187**, $K_i > 1,000$ nM).

5.4.3 Ring-A

Replacement of the benzene A-ring of **178** by thiophene results in an increase in ligand affinity at DS [compare **178** ($K_i = 1.1$ nM) vs. **190** ($K_i = 0.2$ nM)]. However, an even greater increase in ligand affinity for the DI site is seen with the thiophene substitution **178** ($K_i = 21.2$ nM) vs. **190** ($K_i = 2.6$ nM)]. Hence, the net result of replacement of the A-ring with a thiophene unit is enhanced selectivity for the DI site.

5.4.4 Excluded and included volume analysis of the DI BzR

An excluded volume analysis[52] was performed to map out the surface of the DI receptor in the vicinity of bound ligands. The excluded volume map shows forbidden regions about the 1-, 6-, 7-, 9-, and 10-positions of the imidazo[1,5a][1,4]benzodiazepine ring system which correlates with the reduction in affinity for DI when large substituents are placed at any of these positions.

The receptor essential volume corresponds to receptor pocket(s) which, if occupied by the ligand, enhance affinity. The receptor essential volume map shows favorable regions in the vicinity of position-3 of the imidazo[1,5a][1,4]benzodiazepine ring system which correlates with the high affinity displayed by analogs which possess ester substituents of intermediate size in this position. Finally, the receptor essential volume displays a favorable region about the 8-position of the imidazo[1,5a][1,4]benzodiazepine ring system, consistent with the high affinity for DI that the 8-substituted analogs exhibit.

5.4.5 Active conformation and 3D-QSAR of DI ligands

In order to obtain more insight into the steric and electrostatic requirements for selective, high affinity binding to DI, a 3D-QSAR study was performed[52] using the CoMFA technique of Cramer *et al.*[134] As with all 3D-QSAR problems, the alignment rule and active conformations must first be determined. Since every analog examined in this study shares a common imidazo-1,4-diazepine substructure (see Tables 14 and 15), the alignment rule is trivial. The determination of the active conformation is more problematic. In particular, most of the analogs examined possessed a flexible ester function at the 3-position which can adopt either a low energy *syn* or *anti* conformation (N2—C3—C=O torsional angle = ~0 or ~180° respectively; see Figure 16). In addition, Ro 15-4513 (**171**) has a flexible azido

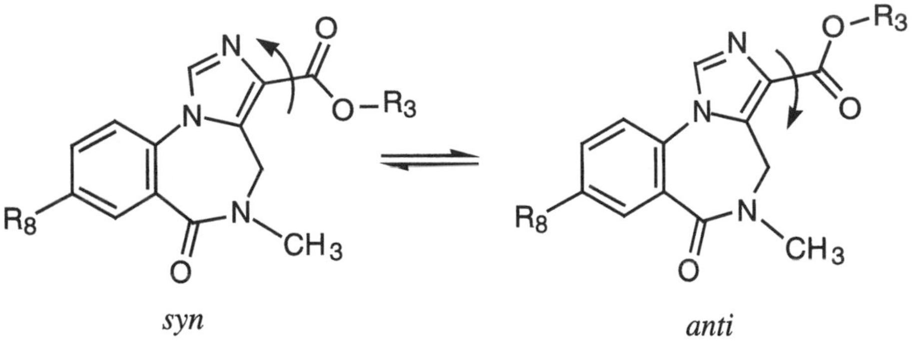

Figure 16 *Syn* and *anti* conformations of 3-ester substituent of imidazobenzodiazepines.

substituent at position-8 which again can adopt either a *syn* or *anti* conformation (C7–C8–N–N torsional angle = ~0 or ~180° respectively; see Figure 17). Since Ro 15-4513 (**171**) displays one of the highest affinities and selectivities for DI of any compound measured to date, it is crucial that the active conformation of this functional group be elucidated. Examination of the SAR data clearly shows that the DS site prefers substituents at position-7, therefore the *syn* conformation of the azido functionality of Ro 15-4513 (**171**) was used for the DS 3D-QSAR analysis. It was not clear, however, which conformation to employ for analysis at DI. Hence, both conformations of the azido group of **171** (Ro 15-4513) were examined in the 3D-QSAR DI study, and both the *syn* and *anti* conformations of the ester functions at the 3-position were examined. Only one set of conformations yielded a satisfactory cross validated correlation ($R^2 > 0.70$) which strongly implicates this set of conformations as being the "active" one. All other sets of conformations gave cross validated R^2's < 0.5. The active conformation of the ester groups appears to be the same for both the DI and DS sites, and that conformation is *anti*. In the case of Ro 15-4513 (**171**), the DI active conformation of the 8-azido group appears to be *anti*, in contrast to the active DS conformation which is almost certainly *syn*.

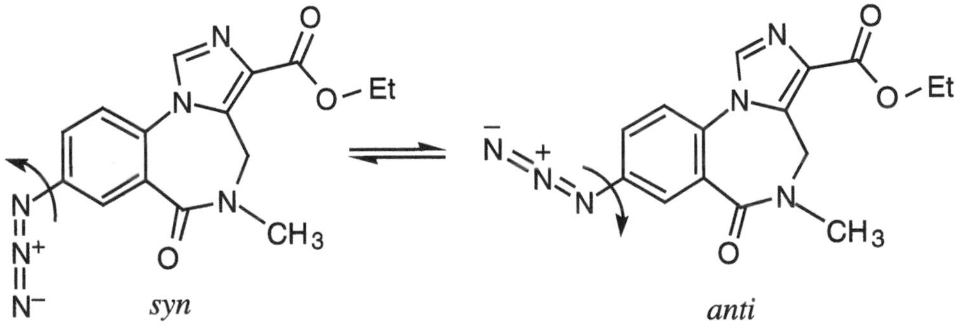

Figure 17 *Syn* and *anti* conformations of 8-azido substituent of imidazobenzodiazepine Ro 15-4513.

Excellent 3D-QSAR correlations were obtained for the negative logarithms of affinities at the diazepam-insensitive and -sensitive BzR sites [pK_i(DI) and pK_i(DS) respectively], and for the differences in their negative logarithms [pK_i(DI) — pK_i(DS)], the latter quantity being a measure of receptor selectivity.

We have previously demonstrated that CoMFA is capable of making quantitatively useful predictions of binding affinity for a series of 3-substituted β-carbolines at the DS site.[82] Efforts are currently underway to use CoMFA regression equations and maps derived in DI 3D-QSAR work[52] in the design of more potent and selective ligands for the DI site. It seems clear that lipophilic region L_3 in the DI site is smaller than that for the DS pharmacophore. Additional studies will be required to determine the actual difference in size between these two sites.

6.   INCLUSIVE PHARMACOPHORE/RECEPTOR MODEL

6.1   Alignment

The inverse agonist/antagonist and agonist pharmacophore/receptor models illustrated in Figures 1 and 2 govern inverse agonist versus agonist activity *in vivo* at BzR. These pharmacophoric definitions are important in the design of ligands which elicit one activity in preference to another. It is felt that inverse agonists (A_2, H_1, and L_1), antagonists, and agonists (H_1, H_2, L_1, L_2, and/or L_3) bind to the same domain of the BzR.[55,136] Although H_1 and L_1 are points of receptor interaction common to both inverse agonists and agonists, the remaining descriptors for the two activities are clearly different from each other.

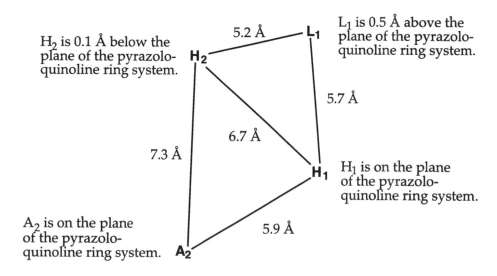

Figure 18   The schematic representation of the descriptors for the inclusive BzR pharmacophore.

The alignment of the ten different structural classes of benzodiazepine receptor ligands is based on the least squares fitting of three points. The first position (L_1) corresponds to the D-ring centroid of the CGS series and the fused benzene ring of the benzodiazepines. The second and third points (H_1 and H_2) correspond to lone pairs of electrons which extend from the imine nitrogen and carbonyl oxygen atoms, respectively. These lone pair vectors were elongated to 1.8 Å, the length of an ideal hydrogen bond. Hence, the end of these vectors represents the expected location of receptor hydrogen bond donating atoms on the receptor protein. The centroid of the end of these vectors was replaced with a hydrogen atom which represents the receptor hydrogen bond donor site H_1 or H_2. The position of A_2 corresponds to the receptor hydrogen bond acceptor site. The $N-H\cdots A_2$ hydrogen bond angle was set to 180°. Outlined in Figure 18 is a schematic two-dimensional representation of the inclusive pharmacophore.

6.2 Included Volumes

The alignment of 30 agonist ligands (Tables 1–12) which represent ten different structural families in the inclusive pharmacophore/receptor model is illustrated in Figure 19. This alignment was used to calculate the union of the active agonist volumes illustrated in Figure 20. For inverse agonist ligands, the pharmacophoric descriptors and alignment have previously been reported in detail.[47,48] The alignment of twelve inverse agonist ligands which represent five different families is illustrated

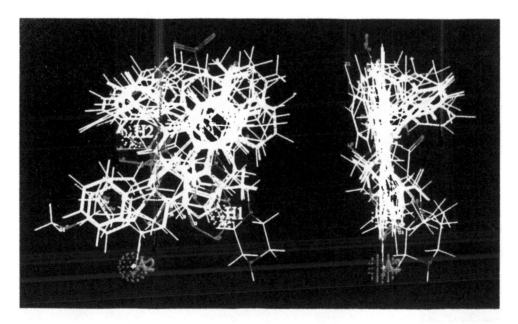

Figure 19 Stereoview of the superposition of 30 agonist ligands (Tables 1–12) (**1–4, 50, 51, 57, 58, 61, 66, 72, 74, 77, 78, 84, 87–89, 91, 93, 95, 99, 100, 102, 103, 107, 112, 113, 117, 124**).

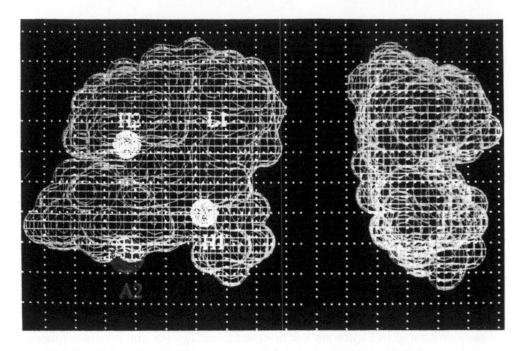

Figure 20 Stereoview of the union of the volumes of 30 agonist ligands (grid resolution is 2 Å).

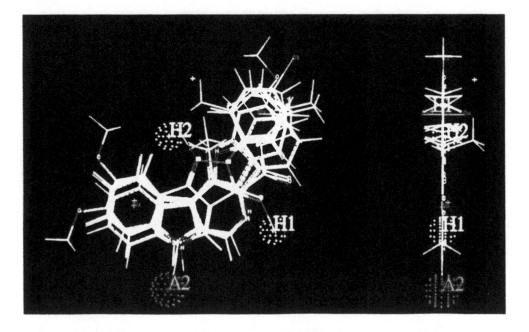

Figure 21 Stereoview of the superposition of 12 inverse agonist ligands (Tables 1–5) (**15, 16, 18, 20, 39, 40, 52, 63, 64, 69, 70, 71**).

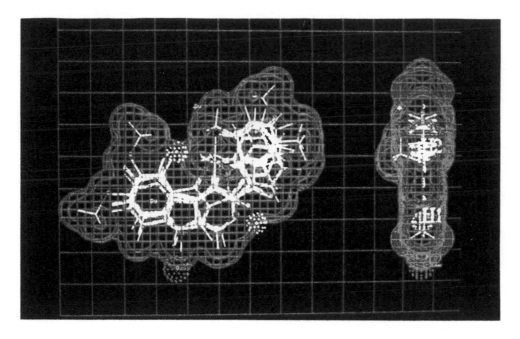

Figure 22 Stereoview of the union of volumes of 12 inverse agonist ligands (grid resolution is 2 Å).

in Figure 21 (Tables 1–5). The alignment was also used to calculate the union of active inverse agonist volumes (Figure 22). The alignment of fourteen DI ligands is illustrated in Figure 23. The alignment was used to calculate the union of DI volumes, as shown in Figure 24. These calculated volumes show the dimensions and borders of the agonist and inverse agonist pharmacophores from which the lipophilic areas L_1, L_2, and L_3 were determined (see Figures 20 and 22). Receptor essential volume analysis of lipophilic pockets L_2 and L_3 for agonist profiles have been illustrated previously.[50] The defined lipophilic areas correlate favorably with nineteen antagonist ligands described above. The superposition of the volumes of agonists and inverse agonists is shown in Figure 25.

6.3 DS vs. DI

Since *in vitro* binding data employed for this work was determined on rat cortical membranes, the potencies likely represent the weighted average of several $GABA_A$ receptor isoforms. The union of the agonist and inverse agonist pharmacophores illustrated in Figure 25 represents an inclusive pharmacophore of 'diazepam sensitive' (DS) sites which is distinctly different (with regard to size) from that of the 'diazepam insensitive' (DI) receptor/model (the major isoform of DI receptors contains an $\alpha6$ subunit, see Figures 24, 25). Clearly the DI site occupies a smaller receptor included volume than the DS site and appears to be a subset of the DS site (see Figure 26). This result is consistent with the Venn diagram reported by Doble and

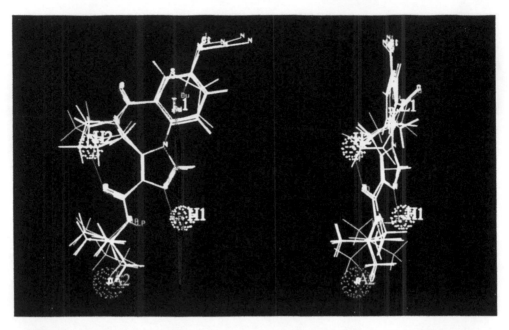

Figure 23   Stereoview of the superposition of 14 DI ligands (Tables 14–15) (**167, 171, 173, 175, 178–184, 190, 194, 196**).

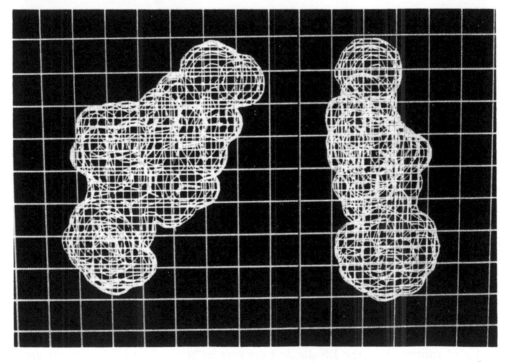

Figure 24   Stereoview of the union of volumes of 14 DI ligands (grid resolution is 2 Å).

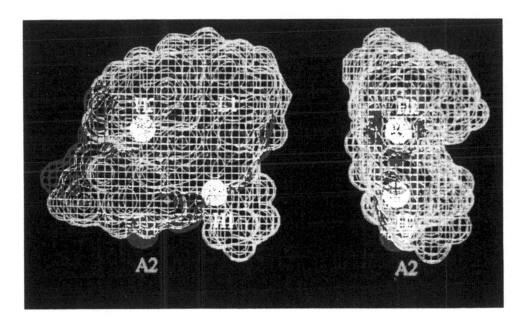

Figure 25 Superposition of the union of volumes of inverse agonists (Figure 22) and agonists (Figure 20).

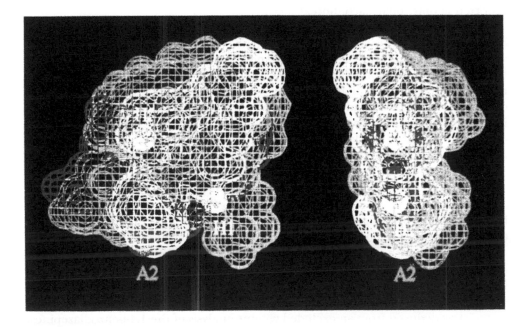

Figure 26 Superposition of DS included volume (Figures 20 plus 22) and DI included volume (Figure 24).

Martin wherein sapmazenil (Ro 15-4513, **171**) was shown to bind to all six recombinant receptors $[\alpha(1\text{-}6)\text{-}\beta 2\gamma 2]^{10}$ with high affinity.

Although speculative, as mentioned previously it is reasonable to predict that ligand selectivity and intrinsic activity (efficacy) at GABA$_A$ receptor isoforms will stem from different interactions in the lipophilic pockets of the BzR receptor binding cleft(s) effected by selective BzR ligands. The subtle topological differences (*i.e.*, L$_2$, L$_3$, S$_1$, S$_2$, etc.) between DS GABA$_A$ receptor subtypes can be determined when enough ligands are developed with high selectivity for a specific isoform. Within this context, both the stoichiometry and composition of native receptors must also be established. At this juncture it will be possible to subtract (*via* modeling[72,73]) the volume of ligands of different isoforms from the inclusive model in order to discover the topological differences between the various GABA$_A$ receptor isoforms. Ligands with rigid rings related to **124**, as well as molecular yardsticks similar to those in Figure 13 may be important in determining the differences between native BzR subtypes. Such studies are underway and will be reported in due course.

6.4 Negative Controls

Ligands with low affinity at the BzR (see Tables 1, 9 and 10) were used as negative controls to define areas of steric repulsion and to evaluate the significance of the hydrogen bond acceptor A$_2$ and donor H$_1$ descriptors. The evaluation of the hydrogen bond donor site H$_2$ has been established above and in the recently reported agonist pharmacophore.[50]

Ligands **31** and **32** were used to probe the hydrogen bond acceptor site A$_2$. These ligands do not bind with H$_2$ but rather align in a manner similar to that of the inverse agonist β-carbolines. In these ligands, a methyl group replaces the indole hydrogen atom changing the attractive interaction between the ligand and the receptor site A$_2$ to a strongly repulsive one. This accounts for the poor affinity displayed by ligands **31** and **32** (IC$_{50}$ > 5000 nM). The pyridodiindole derivative **41** is different in structure than the β-carbolines **31** and **32**. Nevertheless, ligand **41** also contains a N-methyl indole substituent which aligns closely with the indole of the inverse agonist β-carbolines (see Figure 27). Again the methyl group interferes with A$_2$ resulting in low affinity (IC$_{50}$ = 1163 nM) at the BzR.

The hydrogen bond donor site H$_1$ was evaluated with the pyridodiindole analog **44**. This analog has a ring system similar to that of the parent (pyridodiindole) **39** but the imine N5 nitrogen atom has been replaced by a methine moiety. In the absence of this heteroatom, the receptor site H$_1$ cannot engage in hydrogen bonding to the ligand, and hence this ligand displays low affinity (IC$_{50}$ = 1970 nM). Other examples which illustrate the importance of acceptor A$_2$ and donor H$_1$ have also been reported in publications on the inverse agonist pharmacophore.[45-48]

The location of sites S$_1$ and S$_2$ of steric repulsion between the receptor and the ligands was deduced from an excluded volume analysis of the 1,4-benzodiazepines; **81**, **82**, and **121-132** and the homologs of pyridodiindoles, **42** to **49**. Ligands **42** to **49** and **121** to **132** were designed and synthesized to probe the dimensions of the

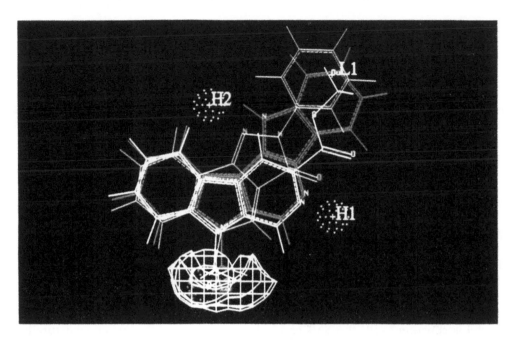

Figure 27 Superposition of the negative controls **32** (yellow) and **41** (red) with CGS-8216 (**52**, green).

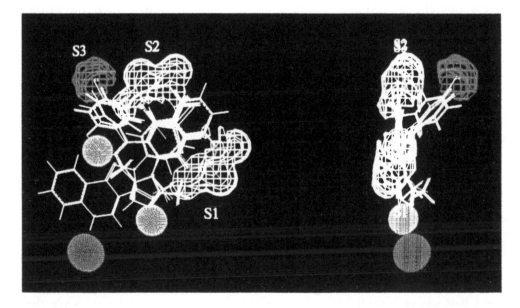

Figure 28 Superposition of the agonists (**72, 78, 84, 86, 87, 50, 124**) with negative controls **81** (white), **128** (magenta), and **132** (green). Regions of steric repulsion S_1, S_2 and S_3 were derived from the subtraction of the included receptor volume from the union of volumes of the negative controls.

lipophilic pocket (L_2) of the BzR.[48,137] Although these ligands were designed with the correct pharmacophoric descriptors to interact with H_1, A_2, and H_2, these compounds displayed poor affinities at BzR. The specific point(s) used in the binding interaction depend on the structural characteristics of the ligands. The sites of steric repulsion (S_1, S_2, and S_3) are illustrated in Figure 28. The region S_1 is located in the same plane of the heterocyclic core of the β-carbolines and pyridodiindoles. This region was also reported in the agonist pharmacophore.[50] Repulsive region S_2 is located between lipophilic regions L_2 and L_3 and S_3 is located in the direction of area L_3 and above the plane of lipophilic area L_1 (Figure 28).[83,91]

7.   CONCLUSIONS

7.1   Summary of the Inclusive Model

The unified pharmacophore/receptor model proposed in this work defines in detail the structural requirements necessary for high affinity binding to the DS and the DI BzR isoforms and the requirements for DS selectivity. The model also defines the requirements for agonist, antagonist, or inverse agonist activity. This unified model is consistent with the SAR of 136 ligands which represent ten structurally diverse families of compounds. Furthermore, this inclusive model is in partial agreement with previously proposed models.[37,38,40-44,86] as well as consistent with our previously proposed model,[45-52] but is more comprehensive (see Figure 29).

The model proposed in this work is unique in the alignment used and in the definitions of the positions of the receptor hydrogen bonding sites, lipophilic pockets, and regions of steric repulsion. The receptor sites H_1, A_2, and H_2 help direct and anchor the ligands at the receptor site and likely contribute to efficacy. Interactions between the ligand and the hydrophobic receptor regions L_1, L_2, and L_3 determine the efficacy and functional activity of the bound ligand. In addition, these hydrophobic interactions probably constitute the major driving force for binding potency. Finally, regions S_1 and S_2 of steric repulsion place limits on the size that BzR ligands may possess and still display potent receptor affinity.

7.2   Future Directions

As mentioned previously, BzR pharmacology is complicated by the existence of multiple receptor isoforms. The inclusive pharmacophore model for DS BzR is based on the weighted average of these receptor subtypes. Ligands such as zolpidem and alpidem as well as benzodiazepines related to and including Ro15-1788 fit well in the inclusive pharmacophore. Receptor isoforms whose pharmacology resembles that of previously reported BzR-I and BzR-II receptors have been expressed: the type-I BzR was constructed from an $\alpha 1\beta 2\gamma 2$ combination of subunits while the type-II zolpidem-insensitive[14] BzR was composed of $\alpha 2\beta 2\gamma 2$, $\alpha 3\beta 2\gamma 2$, and $\alpha 5\beta 2\gamma 2$ receptor isoforms.[138,139] As SAR is accumulated for each of the cloned receptor isoforms, separate pharmacophore models will need to be developed for each.

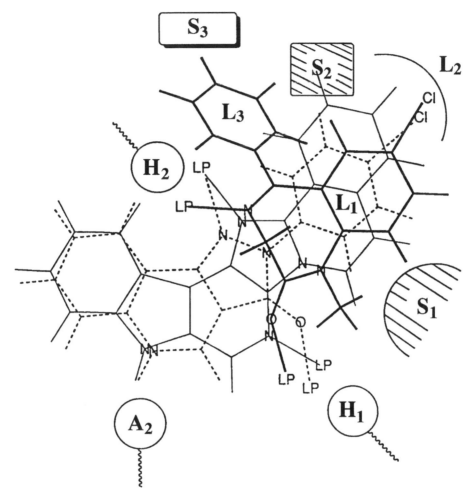

Figure 29 The pyrazolo[3,4-*c*]quinolin-3-one ligand CGS-9896 (**50**, dotted line),
diazepam (**72**, thick line), and diindole (**49**, thin line) fitted to a schematic
representation of the inclusive pharmacophore model for the BzR. The sites H_1 and H_2
designate hydrogen bond donor sites on the receptor protein while A_2 represents a
hydrogen bond acceptor site. Interaction with the lipophilic pocket L_1, as well as with
H_1, and A_2 is required for potent inverse agonist activity. Agonist activity requires
interaction with H_1, H_2, L_1, L_2, and/or L_3. Receptor descriptors S_1, S_2, and S_3 are
regions of negative steric repulsion.

While these individual models may differ in detail from the inclusive model presented
here, these separate models will nevertheless share many features in common with
the inclusive model. In this vein recent binding data from our laboratory[140] suggests
that lipophilic pocket designated L2 (in the inclusive pharmacophore) is smaller
in cloned $\alpha 1\beta 2\gamma 2$ sites than present in the related $\alpha 5\beta 2\gamma 2$ sites. Consequently, the
attractive lipophilic-lipophilic interactions between the methyl group on zolpidem

and lipophilic pocket L2 at $\alpha1\beta2\gamma2$ are responsible at least in part for the high affinity of zolpidem at the so called Bz1 (ω1) site. However, L2 is much larger in size in the cloned $\alpha5\beta2\gamma2$ series and this attractive interaction is absent resulting in little or no affinity of zolpidem at the cloned $\alpha5\beta2\gamma2$ channels. The receptor subtype SAR and pharmacophore models which result will permit the design of more selective ligands for each of these subtypes and may allow one to decouple the broad spectrum of effects exhibited by the current class of non-selective ligands. An outgrowth of this approach will be new therapeutic opportunities.

A direct result of the BzR pharmacophore/receptor model has been the design and synthesis of a new partial agonist,[51] as well as partial inverse agonists.[47] The data obtained from these new ligands will be useful in refining the model and may ultimately lead to better drugs for treatment of anxiety disorders and of other maladies associated with neurotransmission in the CNS.

8.   ACKNOWLEDGEMENTS

This work was supported by a grant from NIMH (MH 46851). We wish to acknowledge the efforts of many coworkers as well as collaborators including Dr. Phil Skolnick without whose enthusiasm and dedication this chapter would not have been possible.

9.   REFERENCES

1.   Squires, R.F. and Braestrup, C. (1977) Benzodiazepine receptors in rat brain. *Nature,* **266**, 732–734.
2.   Mohler, H. and Okada, T. (1977) Benzodiazepine receptors: demonstration in the central nervous system. *Science,* **198**, 849–851.
3.   Gavish, M. and Snyder, S.H. (1980) Soluble benzodiazepine receptors: GABAergic regulation. *Life Sci.,* **26**, 579–82.
4.   Gavish, M. and Snyder, S.H. (1981) Gamma-aminobutyric acid and benzodiazepine receptors: copurification and characterization. *Proc. Natl. Acad. Sci. USA,* **78**, 1939–42.
5.   Squires, R. (1988) *GABA and Benzodiazepine Receptors.* CRC Press: Boca Raton, Florida, Vols. 1 and 2.
6.   Witkin, J.M., Barrett, J.E., Cook, J.M. and Larscheid, P. (1986) Differential antagonism of diazepam-induced loss of the righting response. *Pharmacol. Biochem. Behav.,* **24**, 963–5.
7.   Ninan, P.T., Insel, T.M., Cohen, R.M., Cook, J.M., Skolnick, P. and Paul, S.M. (1982) Benzodiazepine receptor-mediated experimental "anxiety" in primates. *Science,* **218**, 1332–4.
8.   Mendelson, W.B., Cain, M., Cook, J.M., Paul, S.M. and Skolnick, P. (1983) A benzodiazepine receptor antagonist decreases sleep and reverses the hypnotic actions of flurazepam. *Science,* **219**, 414–6.
9.   Venault, P., Chapouthier, G., de-Carvalho, L.P., Simiand, J., Morre, M., Dodd, R.H. and Rossier, J. (1986) Benzodiazepine impairs and β-carboline enhances performance in learning and memory tasks. *Nature,* **321**, 864–6.
10. Doble, A. and Martin, I.L. (1992) Multiple benzodiazepine receptors: no reason for anxiety. *Trends Pharmacol. Sci.,* **13**, 76–81.
11. Haefely, W., Piere, L., Polc, P. and Schaffner, R. (1981) *Handbook of Experimental Pharmacology, Part II.* Springer-Verlag: Berlin.
12. Greenblatt, D.J. and Shader, R.I. (1974) *Benzodiazepines in Clinical Practice.* John Wiley and Sons: New York.

13. Corda, M.G., Blaker, W.D., Mendelson, W.B., Guidotti, A. and Costa, E. (1983) β-Carbolines enhance shock-induced suppression of drinking in rats. *Proc. Natl. Acad. Sci. USA*, **80**, 2072–6.

14. Schofield, P.R., Darlison, M.G., Fujita, N., Burt, D.R., Stephenson, F.A., Rodriguez, H., Rhee, L.M., Ramachandran, J., Reale, V., Glencorse, T.A., Seeburg, P.H. and Barnhard, E.A. (1987) Sequence and functional expression of the GABA$_A$ receptor shows a ligand-gated receptor super-family. *Nature*, **328**, 221–7.

15. Olsen, R.W. and Tobin, A.J. (1990) Molecular biology of GABA$_A$ receptors. *FASEB J.*, **4**, 1469–80.

16. Levitan, E.S., Schofield, P.R., Burt, D.R., Rhee, L.M., Wisden, W., Kohler, M., Fujita, N., Rodriguez, H.F., Stephenson, A., Darlison, M.G., Barnhard, E.A. and Seeburg, P.H. (1988) Structural and functional basis for GABA$_A$ receptor heterogeneity. *Nature*, **335**, 76–9.

17. Wisden, W., Laurie, D.J., Monyer, H. and Seeburg, P.H. (1992) The distribution of 13 GABA$_A$ receptor subunit mRNAs in the rat brain. I. Telencephalon, diencephalon, mesencephalon. *J. Neurosci.*, **12**, 1040–1062.

18. Burt, D. and Kamatchi, G. (1991) GABA$_A$ receptor subtypes: from pharmacology to molecular biology. *FASEB J.*, **5**, 2916–2923.

19. Pritchett, D.B., Sontheimer, H., Shivers, B.D., Ymer, S., Kettenmann, H., Schofield, P.R. and Seeburg, P.H. (1989) Importance of a novel GABA$_A$ receptor subunit for benzodiazepine pharmacology. *Nature*, **338**, 582–5.

20. Wong, G., Sei, Y. and Skolnick, P. (1992) Stable expression of type I γ-aminobutyric acid$_A$/benzodiazepine receptors in a transfected cell line. *Mol. Pharmacol.*, **42**, 996–1003.

21. Skolnick, P. and Wong, G. (1993) Drug-receptor interactions in anxiety in imidazopyridines. In *Anxiety disorders: a novel experimental and therapeutic approach*, Bartholini, G., Garreau, M., Morselli, P. and Zivkovic, B., Ed.; Raven Press: New York; pp. 23–32.

22. Matsushita, A., Kawasaki, K., Matsubara, K., Eigyo, M., Shindo, H. and Takada, S. (1988) Activation of brain function by S-135, a benzodiazepine receptor inverse agonist. *Prog. Neuro-Psychpharmacol. Biol. Psychiatry*, **12**, 951–966.

23. Holley, L.A., Miller, J.A., Chmielewski, P., Dudchenko, P. and Sarter, M. (1993) Interactions between the effects of basal forebrain lesions and chronic treatment with MDL 26,479 on learning and markers of cholinergic transmission. *Brain Res.*, **610**, 181–183.

24. Sarter, M., Moore, H., Dudchenko, P., Holley, L.A. and Bruno, J.P. (1992) Cognition enhancement based on GABA-cholineric interactions. In *Neurotransmitter interactions and cognition function*, Levin, E.D., Deeker, M. and Butcher, L.L., Ed.; Birkhaüser: Boston; pp. 329–354.

25. Wafford, K.A., Whiting, P.J. and Kemp, J.A. (1993) Differences in affinity and efficacy of benzodiazepine receptor lignds at recombinant γ-aminobutyric acid$_A$ receptor subtypes. *Mol. Pharmacol.*, **43**, 240–4.

26. Wafford, K.A., Bain, C.J., Whiting, P.J. and Kemp, J.A. (1993) Functional comparison of the role of γ-subunits in recombinant human γ-aminobutyric acid$_A$/benzodiazepine receptors. *Mol. Pharmacol.*, **44**, 437–442.

27. Hadingham, K.L., Wingrove, P., Le-Bourdelles, B., Palmer, K.J., Ragan, C.I. and Whiting, P.J. (1993) Cloning of cDNA sequences encoding human α2 and α3 γ-aminobutyric acid$_A$ receptor subunits and characterization of the benzodiazepine pharmacology of recombinant α1-, α2-, α3-, and α5-containing human γ-aminobutyric acid$_A$ receptors. *Mol. Pharmacol.*, **43**, 970–5.

28. Luddens, H., Pritchett, D.B., Kohler, M., Killisch, I., Keinanen, K., Monyer, H., Sprengel, R. and Seeburg, P.H. (1990) Cerebellar GABA$_A$ receptor selective for a behavioural alcohol antagonist. *Nature*, **346**, 648–51.

29. Pritchett, D.B., Luddens, H. and Seeburg, P.H. (1989) Type I and type II GABA$_A$-benzodiazepine receptors produced in transfected cells. *Science*, **245**, 1389–92.

30. Klepner, C.A., Lippa, A.S., Benson, D.I., Sano, M.C. and Beer, B. (1979) Resolution of two biochemically and pharmacologically distinct benzodiazepine receptors. *Pharmacol. Biochem. Behav.*, **11**, 457–62.

31. Depoortere, H., Zivkovic, B., Lloyd, K.G., Sanger, D.J., Perrault, G., Langer, S.Z. and Bartholini, G. (1986) Zolpidem, a novel nonbenzodiazepine hypnotic. I. Neuropharmacological and behavioral effects. *J. Pharmacol. Exp. Ther.*, **237**, 649–58.

32. Pritchett, D.B. and Seeburg, P.H. (1990) Gamma-aminobutyric acid$_A$ receptor alpha 5-subunit creates novel type II benzodiazepine receptor pharmacology. *J. Neurochem.*, **54**, 1802–4.

33. Gardner, C.R. (1988) Pharmacological profiles *in vivo* of benzodiazepine receptor ligands. *Drug Dev. Res.*, **12**, 1–28.

34. Sieghart, W., Eichinger, A., Richards, J.G. and Mohler, H. (1987) Photoaffinity labeling of benzodiazepine receptor proteins with the partial inverse agonist [^{3}H]Ro 15-4513: a biochemical and autoradiographic study. *J. Neurochem.*, **48**, 46–52.

35. Wisden, W., Herb, A., Wieland, H., Keinanen, K., Luddens, H. and Seeburg, P.H. (1991) Cloning pharmacological characteristics and expression patterns of the GABA$_A$ receptor α4-subunit. *FEBS Lett.*, **289**, 227–30.

36. Villar, H.O., Davies, M.F., Loew, G.H. and Maguire, P.A. (1991) Molecular models for recognition and activation at the benzodiazepine receptor: a review. *Life Sci.*, **48**, 593–602.

37. Villar, H.O., Uyeno, E.T., Toll, L., Polgar, W., Davies, M.F. and Loew, G.H. (1989) Molecular determinants of benzodiazepine receptor affinities and anticonvulsant activities. *Mol. Pharmacol.*, **36**, 589–600.

38. Ghose, A.K. and Crippen, G.M. (1990) Modeling the benzodiazepine receptor binding site by the general three-dimensional structure-directed quantitative structure-activity relationship method REMOTEDISC. *Mol. Pharmacol.*, **37**, 725–34.

39. Crippen, G.M. (1981) Distance geometry analysis of the benzodiazepine binding site. *Molecular Pharm.*, **22**, 11–19.

40. Codding, P.W. and Muir, A.K.S. (1985) Molecular structure of Ro 15-1788 and a model for the binding of benzodiazepine receptor ligands. *Mol. Pharm.*, **28**, 178–184.

41. Fryer, R.I., Cook, C., Gilman, N.W. and Walser, A. (1986) Conformational shifts at the benzodiazepine receptor related to the binding of agonists, antagonists, and inverse agonists. *Life Sci.*, **39**, 1947–57.

42. Borea, P.A., Gilli, G., Bertolasi, V. and Ferretti, V. (1987) Stereochemical features controlling binding and intrinsic activity properties of benzodiazepine-receptor ligands. *Mol. Pharmacol.*, **31**, 334–344.

43. Tebib, S., Bourguignon, J.J. and Wermuth, C.G. (1987) The active analog approach applied to the pharmacophore identification of benzodiazepine receptor ligands. *J. Comput.-Aided Mol. Des.*, **1**, 153–70.

44. Gardner, C.R. (1992) A review of recently-developed ligands for neuronal benzodiazepine receptors and their pharmacological activities. *Prog. Neuro-Psychopharmacol. & Biol. Psychiat.*, **16**, 755–781.

45. Allen, M.S., Hagen, T.J., Trudell, M.L., Codding, P.W., Skolnick, P. and Cook, J.M. (1988) Synthesis of novel 3-substituted β-carbolines as benzodiazepine receptor ligands: probing the benzodiazepine receptor pharmacophore. *J. Med. Chem.*, **31**, 1854–61.

46. Hagen, T.J., Guzman, F., Schultz, C. and Cook, J.M. (1986) Synthesis of 3,6-disubstituted β-carbolines which possess either benzodiazepine antagonist or agonist activity. *Heterocycles*, **24**, 2845–2855.

47. Allen, M.S., Tan, Y.C., Trudell, M.L., Narayanan, K., Schindler, L.R., Martin, M.J., Schultz, C., Hagen, T.J., Koehler, K.F., Codding, P.W., Skolnick, P. and Cook, J.M. (1990) Synthetic and computer-assisted analyses of the pharmacophore for the benzodiazepine receptor inverse agonist site. *J. Med. Chem.*, **33**, 2343–57.

48. Zhang, W. and Cook, J.M. (1993) Novel organic reactions in the search for anxioselective anxiolytics at the benzodiazepine receptor In *Drug Design for Neuroscience*, Kozikowski, A., Ed.; Raven Press: New York; pp. 87–117.

49. Hollinshead, S.P., Trudell, M.L., Skolnick, P. and Cook, J.M. (1990) Structural requirements for agonist actions at the benzodiazepine receptor: studies with analogues of 6-(benzyloxy)-4-(methoxy-methyl)-β-carboline-3-carboxylic acid ethyl ester. *J. Med. Chem.*, **33**, 1062–9.

50. Diaz-Arauzo, H., Koehler, K.F., Hagen, T.J. and Cook, J.M. (1991) Synthetic and computer assisted analysis of the pharmacophore for agonists at benzodiazepine receptors. *Life Sci.*, **49**, 207–16.

51. Diaz-Arauzo, H., Evoniuk, G.E., Skolnick, P. and Cook, J.M. (1991) The agonist pharmacophore of the benzodiazepine receptor. Synthesis of a selective anticonvulsant/anxiolytic. *J. Med. Chem.*, **34**, 1754–1756.

52. Wong, G., Koehler, K.F., Skolnick, P., Gu, Z.Q., Ananthan, S., Schönholzer, P., Hunkeler, W., Zhang, W. and Cook, J.M. (1993) Synthetic and computer-assisted analysis of the structural requirements for selective, high affinity ligand binding to 'diazepam-insensitive' benzodiazepine receptors. *J. Med. Chem.*, **36**, 1820–30.

53. Gardner, C.R. (1988) Functional *in vivo* correlates of the benzodiazepine agonist-inverse agonist continuum. *Prog. Neurobiology*, **31**, 425–476.

54. Schweri, M., Cain, M., Cook, J., Paul, S. and Skolnick, P. (1982) Blockade of 3-carbomethoxy-β-carboline induced seizures by diazepam and the benzodiazepine antagonists, Ro 15-1788 and CGS 8216. *Pharmacol. Biochem. Behav.*, **17**, 457–60.

55. Skolnick, P., Schweri, M., Kutter, E., Williams, E. and Paul, S. (1982) Inhibition of [^{3}H]diazepam and [^{3}H]3-carboethoxy-β-carboline binding by irazepine: evidence for multiple "domains" of the benzodiazepine receptor. *J. Neurochem.*, **39**, 1142–6.

56. Pritchett, D.B. and Seeburg, P.H. (1991) γ-Aminobutyric acid type A receptor point mutation increases the affinity of compounds for the benzodiazepine site. *Proc. Natl. Acad. Sci. USA*, **88**, 1421–5.

57. Wieland, H.A., Luddens, H. and Seeburg, P.H. (1992) A single histidine in GABA$_A$ receptors is essential for benzodiazepine agonist binding. *J. Biol. Chem.*, **267**, 1426–9.

58. Haefely, W., Martin, J.R. and Schoch, P. (1990) Novel anxiolytics that act as partial agonists at benzodiazepine receptors. *Trends Pharmacol. Sci.*, **11**, 452–6.

59. Bertolasi, V., Feretti, V., Gilli, G. and Borea, P.A. (1984) Structure of methyl-β-carboline-3-carboxylate (βCCM), $C_{13}H_{10}N_2O_2$. *Acta Crystallogr., Sect. C: Cryst. Struct. Commun.*, **40**, 1981–1983.

60. Camerman, A. and Camerman, N. (1972) Stereochemical basis of anticonvulsant drug action. II. Molecular structure of diazepam. *J. Am. Chem. Soc.*, **94**, 268–272.

61. Ferretti, V., Bertolasi, V., Gilli, G. and Borea, P.A. (1985) Stereochemistry of benzodiazepine-receptor ligands. II. Structures of two 2-arylpyrazolo[4,3-c]quinolin-3-ones: CGS-8216, $C_{16}H_{11}N_3O$, and CGS-9896, $C_{16}H_{10}ClN_3O$. *Acta Crystallogr., Sect. C: Cryst. Struct. Commun.*, **C41**, 107–110.

62. Hempel, A., Camerman, N. and Camerman, A. (1987) Benzodiazepine stereochemistry: crystal structures of the diazepam antagonist Ro 15-1788 and the anomalous benzodiazepine Ro 5-4864. *Can. J. Chem.*, **65**, 1608–1612.

63. Muir, A.K.S. and Codding, P.W. (1985) Structure-activity studies of β-carbolines. 3. Crystal and molecular structures of methyl β-carboline-3-carboxylate. *Can. J. Chem.*, **63**, 2752–2756.

64. Nakai, H. (1990) Structure of 2-(5-methyl-3-thienyl)-2H,5H-pyrazolo[4,3-c]quinolin-3-one. *Acta Crystallogr., Sect. C: Cryst. Struct. Commun.*, **46**, 1951–3.

65. Nakai, H. (1990) Structure of 2-(4-methylthien-2-yl)-2,5-dihydro-3H-pyrazolo[4,3-c]quinolin-3-one. *Acta Crystallogr., Sect. C: Cryst. Struct. Commun.*, **C46**, 1143–5.

66. Neidle, S., Webster, G.D., Jones, G.B. and Thurston, D.E. (1991) Structures of two DNA minor-groove binders, based on pyrrolo[2,1-c][1,4]benzodiazepines. *Acta Cryst.*, **C47**, 2678–2680.

67. Shiro, M. (1990) Structure of 2-(5-methylthien-2-yl)-2,5-dihydro-3H-pyrazolo[4,3-c]quinolin-3-one. *Acta Crystallogr., Sect. C: Cryst. Struct. Commun.*, **C46**, 1152–3.

68. Pearlman, R.S., Balducci, R., McGarity, C., Rusinko III, A. and Skell, J. (1990) CONCORD. University of Texas, Austin, TX.

69. Frisch, M.J., Head-Gordon, M., Trucks, G.W., Foresman, J.B., Schlegel, H.B., Raghavachari, K., Robb, M.A., Binkley, J.S., Gonzalez, C., Defrees, D.J., Fox, D.J., Whiteside, R.A., Seeger, R., Melius, C.F., Baker, J., Martin, L.R., Kahn, L.R., Stewart, J.J.P., Topiol, S. and Pople, J.A. (1990) Gaussian 90. Gaussian, Inc., Pittsburgh, PA.

70. Frisch, M.J., Trucks, G.W., Head-Gordon, M., Gill, P.M.W., Wong, M.W., Foresman, J.B., Johnson, B.G., Schlegel, H.B., Robb, M.A., Replogle, E.S., Gomperts, R., Anderes, J.L., Raghavachari, K., Binkley, J.S., Gonzalez, C., Martin, L.R., Fox, D.J., Defrees, D.J., Baker, J., Stewart, J.J.P. and Pople, J.A. (1992) Gaussian 92, Revision A. Gaussian, Inc., Pittsburgh, PA.

71. Mohamadi, F., Richards, N.G.J., Guida, W.C., Liskamp, R., Caufield, C., Chang, G., Hendrickson, T. and Still, W.C. (1990) MacroModel — An integrated software system for modeling organic and bioorganic molecules using molecular mechanics. *J. Comput. Chem.*, **11**, 440–467.

72. Marshall, G.R., Barry, C.D., Bosshard, H.E., Dammkoehler, R.A. and Dunn, D.A. (1979) The conformational parameter in drug design: the active analog approach. In *Computer-Assisted Drug*

Design; ACS Symposium Series 112; Olson, E.C. and Christoffersen, R.E., Ed.; American Chemical Society: Washington D.C.; Vol. 112; pp. 205–226.

73. Sufrin, J.R., Dunn, D.A. and Marshall, G.R. (1981) Steric mapping of the L-methionine binding site of ATP:L-methionine S-adenosyltransferase. *Mol. Pharmacol.*, **19**, 307–13.

74. Boobbyer, D.N., Goodford, P.J., McWhinnie, P.M. and Wade, R.C. (1989) New hydrogen-bond potentials for use in determining energetically favorable binding sites on molecules of known structure. *J. Med. Chem.*, **32**, 1083–94.

75. Ippolito, J.A., Alexander, R.S. and Christianson, D.W. (1990) Hydrogen bond stereochemistry in protein structure and function. *J. Mol. Biol.*, **215**, 457–71.

76. Murray-Rust, P. and Glusker, J.P. (1984) Directional hydrogen bonding to sp2 and sp3-hybridized oxygen atoms and its relevance to ligand-macromolecular interactions. *J. Am. Chem. Soc.*, **106**, 1018–1025.

77. Thanki, N., Thornton, J.M. and Goodfellow, J.M. (1988) Distributions of water around amino acid residues in proteins. *J. Mol. Biol.*, **202**, 637–57.

78. Tintelnot, M. and Andrews, P. (1989) Geometries of functional group interactions in enzyme-ligand complexes: guides for receptor modelling. *J. Comput.-Aided Molec. Des.*, **3**, 67–84.

79. Vedani, A. and Dunitz, J.D. (1985) Lone-pair directionality in H-bond potential functions for molecular mechanics calculations. *J. Am. Chem. Soc.*, **107**, 7653–7658.

80. Lippke, K.P., Schunack, W.G., Wenning, W. and Müller, W.E. (1983) β-Carbolines as benzodiazepine receptor ligands. 1. Synthesis and benzodiazepine receptor interaction of esters of β-carboline-3-carboxylic acid. *J. Med. Chem.*, **26**, 499–503.

81. Haefely, W., Kyburz, E., Gerecke, M. and Mohler, H. (1985) Recent advances in the molecular pharmacology of benzodiazepine receptors and in the structure-activity relationships of their agonists and antagonists. In *Adv. Drug Res.*; Testa, B., Ed.; Academic Press: New York; Vol. 14; pp. 165–322.

82. Allen, M.S., LaLoggia, A.J., Dorn, L.J., Martin, M.J., Costantino, G., Hagen, T.J., Koehler, K.F., Skolnick, P. and Cook, J.M. (1992) Predictive binding of β-carboline inverse agonists and antagonists via the CoMFA/GOLPE Approach. *J. Med. Chem.*, **35**, 4001–4010.

83. Martin, M.J., Trudell, M.L., Diaz-Arauzo, H., Allen, M.S., Deng, L., Schultz, C.A., Tan, Y.-C., Bi, Y., Narayanan, K., Dorn, L.J., Koehler, K.F., Skolnick, P. and Cook, J.M. (1992) Molecular yardsticks: rigid probes to define the spatial dimensions of the benzodiazepine receptor binding site. *J. Med. Chem.*, **35**, 4105–4117.

84. Yokoyama, N., Ritter, B. and Neubert, A.D. (1982) 2-Arylpyrazolo[4,3-c]quinolin-3-ones: novel agonist, partial agonist, and antagonist of benzodiazepines. *J. Med. Chem.*, **25**, 337–9.

85. Shindo, H., Takada, S., Murata, S., Eigyo, M. and Matsushita, A. (1989) Thienylpyrazoloquinolines with high affinity to benzodiazepine receptors: continuous shift from inverse agonist to agonist properties depending on the size of the alkyl substituent. *J. Med. Chem.*, **32**, 1213–7.

86. Fryer, R.I., Zhang, P., Rios, R., Gu, Z.Q., Basile, A.S. and Skolnick, P. (1993) Structure-activity relationship studies at benzodiazepine receptors (BzR): a comparison of the substituent effects of pyrazoloquinolinone analogs. *J. Med. Chem.*, **36**, 1669–73.

87. Takada, S., Shindo, H., Sasatani, T., Chomei, N., Matsushita, A., Eigyo, M., Kawasaki, K., Murata, S., Takahara, Y. and Shintaku, H. (1988) Thienylpyrazoloquinolines: potent agonists and inverse agonists to benzodiazepine receptors. *J. Med. Chem.*, **31**, 1738–45.

88. Tarzia, G., Occelli, E., Toja, E., Barone, D., Corsico, N., Gallico, L. and Luzzani, F. (1988) 6-(Alkylamino)-3-aryl-1,2,4-triazolo[3,4-a]phthalazines. A new class of benzodiazepine receptor ligands. *J. Med. Chem.*, **31**, 1115–23.

89. Clements-Jewery, S., Danswan, G., Gardner, C.R., Matharu, S.S., Murdoch, R., Tully, W.R. and Westwood, R. (1988) (Imidazo[1,2-a]pyrimidin-2-yl)phenylmethanones and related compounds as potential nonsedative anxiolytics. *J. Med. Chem.*, **31**, 1220–6.

90. Petke, J.D., Im, H.K., Im, W.B., Blakeman, D.P., Pregenzer, J.F., Jacobsen, E.J., Hamilton, B.J. and Carter, D.B. (1992) Characterization of functional interactions of imidazoquinoxaline derivatives with benzodiazepine-γ-aminobutyric acid$_A$ receptors. *Mol. Pharmacol.*, **42**, 294–301.

91. Zhang, W., Koehler, K.F., Harris, B., Skolnick, P. and Cook, J.M. (1994) The synthesis of benzofused benzodiazepines employed as probes of the agonist pharmacophore of benzodiazepine receptors. *J. Med. Chem.*, **37**, 745–57.

92. Cain, M., Weber, R.W., Guzman, F., Cook, J.M., Barker, S.A., Rice, K.C., Crawley, J.N., Paul, S.M. and Skolnick, P. (1982) β-Carbolines: synthesis and neurochemical and pharmacological actions on brain benzodiazepine receptors. *J. Med. Chem.*, **25**, 1081–1091.

93. Ungemach, F. and Cook, J.M. (1978) The spiroindolenine intermediate: A review. *Heterocycles*, **9**, 1089–1119.

94. Coutts, R.T., Micetich, R.G., Baker, G.B., Benderly, A., Bewhurst, T., Hall, T.W., Locock, A.R. and Pyrozko, J. (1984) Some 3-carboxamides of β-carbolines and tetrahydro-β-carbolines. *Heterocycles*, **22**, 131–142.

95. Nielsen, M., Schou, H. and Braestrup, C. (1981) [³H]Propyl β-carboline-3-carboxylate binds specifically to brain benzodiazepine receptors. *J. Neurochem.*, **36**, 276–285.

96. Hagen, T.J., Skolnick, P. and Cook, J.M. (1987) Synthesis of 6-substituted β-carbolines that behave as benzodiazepine receptor antagonists or inverse agonists. *J. Med. Chem.*, **30**, 750–753.

97. Neef, G., Eder, U., Huth, A., Rahtz, D., Schmiechen, R. and Seidelmann, D. (1983) Synthesis of 4-substituted β-carbolines. *Heterocycles*, **20**, 1296–1313.

98. Shannon, H.E., Guzman, F. and Cook, J.M. (1984) β-Carboline-3-carboxylate-*t*-butyl ester: A selective Bz₁ benzodiazepine receptor antagonist. *Life Sci.*, **35**, 2227–2236.

99. Sandrin, J., Soerens, D., Hutchins, L., Richfield, E., Ungemach, F. and Cook, J.M. (1976) Pictet-Spengler condensations in refluxing benzene. *Heterocycles*, **4**, 1101–1105.

100. Trudell, M.L., Basile, A.S., Shannon, H.E., Skolnick, P. and Cook, J.M. (1987) Synthesis of 7,12-dihydropyrido[3,4-*b*:5,4-*b*′]diindoles. A novel class of rigid, planar benzodiazepine receptor ligands. *J. Med. Chem.*, **30**, 456–458.

101. Trudell, M.L., Lifer, S.L., Tan, Y.C., Martin, M.J., Deng, L., Skolnick, P. and Cook, J.M. (1990) Synthesis of substituted 7,12-dihydropyrido[3,2-*b*:5,4-*b*′]diindoles: rigid planar benzodiazepine receptor ligands with inverse agonist/antagonist properties. *J. Med. Chem.*, **33**, 2412–2420.

102. Fryer, R.I. (1989) Ligand interactions at the benzodiazepine receptor. In *Comprehensive Med. Chem.*; Emmett, J., Ed.; Pergamon Press, Inc.: Oxford, U.K.; Vol. 3; pp. 539–566.

103. Fryer, R.I., Gu, Z.Q. and Wang, C.G. (1991) Synthesis of novel substituted 4*H*-imidazo[1,5-*a*][1,4]-benzodiazepines. *J. Heterocycl. Chem.*, **28**, 1661–1669.

104. Zhang, W., Liu, R. and Cook, J.M. (1993) The regiospecific synthesis of ortho aminonaphthophenones via the addition of carbanions to napthoxazin-4-ones. *Heterocycles*, **36**, 2229–2236.

105. Tomioka, Y., Ohkubo, K. and Yamazaki, M. (1985) Studies on aromatic nitro compounds. V. A simple one-pot preparation of *o*-aminoarylnitriles from some aromatic nitro compounds. *Chem. Pharm. Bull.*, **33**, 1360–1366.

106. Ohshima, T., Tomioka, Y. and Yamazaki, M. (1981) Studies on aromatic nitro compounds. II. Reaction of 2-nitronaphthalene with malononitrile in the presence of base. *Chem. Pharm. Bull.*, **29**, 1292–1298.

107. Sternbach, L.H. (1971) 1,4-Benzodiazepines. Chemistry and some aspects of the structure activity relation. *Angew. Chem. Int. Ed.*, **10**, 34–43.

108. Clemence, F., Le Martret, O. and Collard, J. (1984) New route to N-aryl and N-heteroaryl derivatives of 4-hydroxy-3-quinoline carboxamides. *J. Heterocyclic Chem.*, **21**, 1345–1353.

109. Hino, K., Kawashima, K., Oka, M., Nagai, Y., Uno, H. and Matsumoto, J. (1989) A novel class of antiulcer agents. 4-Phenyl-2-(1-piperazinyl)-quinolines. *Chem. Pharm. Bull.*, **37**, 110–115.

110. Braestrup, C., Nielsen, M.C. and Olsen, C.E. (1980) Urinary and brain β-carboline-3-carboxylates as potent inhibitors of brain benzodiazepine receptors. *Proc. Natl. Acad. Sci. USA*, **77**, 2288–2292.

111. Nielsen, M., Gredal, O. and Braestrup, C. (1979) Some properties of ³H-diazepam displacing activity from human urine. *Life Sci.*, **25**, 679–686.

112. Squires, R.F. (1981) GABA receptors regulate the affinities of anions required for brain specific benzodiazepine binding. *Adv. Biochem. Psychopharmacol.*, **26**, 129–138.

113. Basile, A.S. (1993) The role of benzodiazepine receptor ligands in the pathogenesis of hepatic encephalopathy. In *Naturally Occuring Benzodiazepines: Structure, Distribution, and Function*. Izquierdo, I. and Medina, J., Ed.; Ellis Horwood: New York; pp. 89–114.

114. Jones, E.A., Yuraydin, C. and Basile, A.S. (1993) Benzodiazepine antagonists and the management of hepatic encephalopathy. In *Advances in Hepatic Encephalopathy and Metabolic Nitrogen Exchange*, Capoccia, L., Ed.; in press.

115. Havoundjian, H., Reed, G.F., Paul, S.M. and Skolnick, P. (1987) Protection against the lethal effects of pentobarbital in mice by a benzodiazepine receptor inverse agonist, 6,7-dimethoxy-4-ethyl-3-carbomethoxy-β-carboline. *J. Clin. Invest.*, **79**, 473–477.

116. LaLoggia, A.J., Skolnick, P. and Cook, J.M. unpublished results.

117. Ariens, E.J. (1987) Stereochemistry in the analysis of drug action. Part II. *Med. Res. Rev.*, **7**, 367–87.

118. Ariens, E.J. (1988) Stereochemical implications of hybrid and pseudo-hybrid drugs. Part III. *Med. Res. Rev.*, **8**, 309–20.

119. Trullas, R., Ginter, H., Jackson, B., Skolnick, P., Allen, M.S., Hagen, T.J. and Cook, J.M. (1988) 3-Ethoxy-β-carboline: a high affinity benzodiazepine receptor ligand with partial inverse agonist properties. *Life Sci.*, **43**, 1193–1197.

120. Gilman, N.W., Rosen, P., Earley, J.V., Cook, C.M., Blount, J.F. and Todaro, L.J. (1993) Atropisomers of 1,4-benzodiazepines. 2. Synthesis and resolution of imidazo[1,5-a][1,4]benzodiazepines. *J. Org. Chem.*, **58**, 3285–3298.

121. Fryer, R.I., Zhang, P., Lin, K.-Y., Upasani, R.B., Wong, G. and Skolnick, P. (1993) Conformational similarity of diazepam-sensitive and -insensitive benzodiazepine receptors determined by chiral pyrroloimidazobenzodiazepines. *Med. Chem. Res.*, **3**, 183–191.

122. Greco, G., Novellino, E., Silipo, C. and Vittoria, A. (1992) Study of benzodiazepine receptor sites using a combined QSAR-CoMFA approach. *Quant. Struct.-Relat.*, **11**, 461–477.

123. Havunjian, R.H., DeCosta, B.R., Rice, K.C. and Skolnick, P. (1990) Characterization of benzodiazepine receptors with a fluorescence-quenching ligand. *J. Biol. Chem.*, **265**, 22181–22186.

124. Malminiemi, O. and Korpi, E.R. (1989) Diazepam-insensitive [^{3}H]Ro 15-4513 binding in intact cultured cerebellar granule cells. *Eur. J. Pharmacol.*, **169**, 53–60.

125. Uusi-Oukari, M. and Korpi, E.R. (1990) Diazepam sensitivity of the binding of an imidazobenzodiazepine, [^{3}H]Ro 15-4513, in cerebellar membranes from two rat lines developed for high and low alcohol sensitivity. *J. Neurochem.*, **54**, 1980–7.

126. Wong, G. and Skolnick, P. (1992) High affinity ligands for 'diazepam-insensitive' benzodiazepine receptors. *Eur. J. Pharmacol. Mol. Pharm. Sec.*, **225**, 63–8.

127. Turner, D.M., Sapp, D.W. and Olsen, R.W. (1991) The benzodiazepine/alcohol antagonist Ro 15-4513: binding to a GABA$_A$ receptor subtype that is insensitive to diazepam. *J. Pharmacol. Exp. Ther.*, **257**, 1236–42.

128. Wong, G., Skolnick, P., Katz, J. and Witkin, J. (1993) Transduction of a discriminative stimulus through a diazepam-insensitive GABA$_A$ receptor. *J. Pharmacol. Exp. Ther.*, **266**, 570–6.

129. Korpi, E.R., Uusi-Oukari, M. and Wegelius, K. (1992) Substrate specificity of diazepam-insensitive cerebellar [^{3}H]Ro 15-4513 binding sites. *Eur. J. Pharmacol.*, **213**, 323–9.

130. Luddens, H. and Wisden, W. (1991) Function and pharmacology of multiple GABA$_A$ receptor subunits. *Trends Pharmacol. Sci.*, **12**, 49–51.

131. Lister, R.G. (1988) Antagonism of the behavioral effects of ethanol, sodium pentobarbital, and Ro 15-4513 by the imidazodiazepine Ro 15-3505. *Neurosci. Res. Commun.*, **2**, 85–92.

132. Lister, R.G. and Durcan, M.J. (1989) Antagonism of the intoxicating effects of ethanol by the potent benzodiazepine receptor ligand Ro 19-4603. *Brain Res.*, **482**, 141–4.

133. Harris, C.M. and Lal, H. (1988) Central nervous system effects of the imidazobenzodiazepine Ro 15-4513. *Drug Dev. Res.*, **13**, 187–203.

134. Cramer, R.D., Patterson, D.E. and Bunce, J.D. (1988) Comparative Molecular Field Analysis (CoMFA). 1. Effect of Shape on Binding of Steroids to Carrier Proteins. *J. Am. Chem. Soc.*, **110**, 5959–5967.

135. Gu, Z.Q., Wong, G., Dominguez, C., de-Costa, B.R., Rice, K.C. and Skolnick, P. (1993) Synthesis and evaluation of imidazo[1,5-a][1,4]benzodiazepine esters with high affinities and selectivities at diazepam insensitive (DI) benzodiazdepine receptors. *J. Med. Chem.*, **36**, 1001–6.

136. Skolnick, P. and Paul, S. (1983) New concepts in the neurobiology of anxiety. *J. Clin. Psychiatry*, **44**, 12–20.

137. Naryanan, K. and Cook, J.M. (1990) Molecular yardsticks: synthesis of higher homologs of 7,12-dihyropyrido[3,4-b: 5,4-b']diindole. Probing the dimensions of the benzodiazepine receptor inverse agonist site. *Heterocycles*, **31**, 203–208.

138. Benavides, J., Peny, B., Ruano, D., Vitorica, J. and Scatton, B. (1993) Comparative autoradiographic distribution on central ω (benzodiazepine) modulatory site subtypes with high, intermediate, and low affinity for zolpidem and alpidem. *Brain Res.*, **604**, 240–250.

139. Mertens, S., Benke, D. and Mohler, H. (1993) GABA$_A$ receptor populations with novel subunit combinations and drug binding profiles identified in brain by $\alpha 5$ and δ-subunit-specific immuno-purification. *J. Biol. Chem.*, **268**, 5965–5973.

140. Liu, R., Zhang, P., McKernan, R., Wafford, K. and Cook, J.M. (1996) Synthesis of novel imidazobenzo-diazepines selective for the $\alpha 5 \beta 2 \gamma 2$ (Bz5) GABA$_A$/benzodiazepine receptor subtype. *Med. Chem. Res.*, in press.

INDEX